Complete Confined Spaces Handbook

Complete Confined Spaces Handbook

John F. Rekus, MS, CIH, CSP
Safety and Health Consultant
Riderwood, Maryland

Cover photograph courtesy of Don Allen and Roco Rescue.

Library of Congress Cataloging-in-Publication Data

Rekus, John.
Complete confined spaces handbook / John Rekus.
p. cm.
Includes bibliographical references and index.
ISBN 0-87371-487-3
1. Industrial safety--Handbooks, manuals, etc. 2. Industrial toxicology--Handbooks, manuals, etc. 3. Indoor air pollution--Handbooks, manuals, etc. I. Title.
T55.R45 1994
604.7--dc20

94-14496
CIP

Direct all inquiries to CRC Press, Inc., 2000 Corporate Blvd., N.W., Boca Raton, Florida 33431.

No claim to original U.S. Government works
International Standard Book Number 0-87371-487-3
Library of Congress Card Number 94-14496
Printed in the United States of America 3 4 5 6 7 8 9 0
Printed on acid-free paper

To Mary Jo who probably knows more about
occupational safety and health than any other
romance novelist in the world, and to
the Jesuits who encouraged me to write.

A special thanks to Paul Trattner, MA, who produced many of the elegant line drawings, and to the following technical reviewers whose kind assistance, continual encouragement, and insightful comments contributed greatly to the quality of this book.

Michael J. Blotzer, MS, CIH, CSP
Industrial Hygienist and Author
Cleveland, Ohio

Paul K. Buckmaster, MS, CIH
Deputy Chief, Occupational Safety and Health
US Department of Defense
Ft. Mead, Maryland

K. Paul Clifford, PhD
Director of Research and Development
Mosaic Industries
Newark, California

M. Jay Rupp, CIH
Manager, Safety and Industrial Hygiene
Martin-Marietta Corporation
Essex, Maryland

Robert Spielvogel, MS, CIH, CSP
Director of Occupational Health and Safety
Clean Harbors
Braintree, Massachusetts

FOREWORD

You just received the worst possible news you could get as a safety professional. An employee of your company has just been killed in a confined space accident. It's difficult to express the feelings experienced upon hearing the words "he's dead." It's even more chilling if you helped develop the confined space entry program. I know, because I've been there.

Inevitably you identify the cause of the accident and see in hindsight how simple it would have been to prevent: better metering techniques, a different means of extraction, better communication—the list goes on. Simple yet complex, these seemingly isolated and unlikely interactions that converged to cause the accident.

Then comes the self-doubt, questioning why you weren't able to identify the cause, or interrupt the sequence of events that led to the accident. After all, in retrospect the cause is so obvious. Unfortunately, the obvious isn't always apparent before it manifests.

What tools has the safety professional had to help understand the complexities of confined spaces? An ANSI standard, an OSHA regulation, some trade and professional journal articles. These tools were often informative, but seldom instructive. They assisted us in identifying the obvious, but not the subtle aspects of entering confined spaces.

John Rekus fills this void by dissecting the essential elements needed to understand the full range of confined space hazards. Both atmospheric and physical hazards are thoroughly discussed. By examining the factors that lead to hazards in confined spaces, John lays the foundation for formulating a comprehensive understanding of confined space risks. This knowledge is vital for anyone charged with the task of developing an entry program.

Each of the program elements needed to identify and prevent confined space accidents is discussed in individual chapters. Lockout/tagout, ventilation, personal protective equipment, respiratory protection, emergency planning, and employee training—John covers them all. But he not only presents the basic information needed to understand each subject, he also identifies many of the often overlooked nuances of each topic. This definitive text is destined to link John Rekus' name to confined spaces just as Frank Patty's name is linked to industrial toxicology.

John draws on numerous case studies to illustrate both the subtle and the obvious. These case studies will no doubt serve as feed stock for those developing or conducting confined space training programs.

Two chapters that are particularly invaluable regardless of the reader's role in developing a confined space program are "Establishing a Confined Space Program" and "Employee Training and Education." The information in these chapters can be applied to any type of safety and health training or program.

John distills his experience, knowledge and education into the key points needed to master each discipline. In essence, he uses the same entertaining, thought-provoking, stimulating energy typical of his noteworthy presentations to describe and explain key elements of confined spaces.

His style enables the reader to learn, comprehend, and develop a comprehensive understanding of the subject, while at the same time being entertained and stimulated. This is an enviable accomplishment for anyone, but when you stop to consider that the subject is confined spaces, *it's even more amazing.*

Robert Spielvogel, MS, CIH, CSP
Director of Occupational Health & Safety
Clean Harbors Inc., Braintree, MA

I first became interested in confined spaces almost twenty years ago when, over a three-week period in July 1977, six Maryland workers died in three separate confined space accidents. The following year, the Commissioner of Labor, Harvey Epstein, issued a confined space regulation.

Although the state's regulation wasn't perfect, it was better than nothing. And nothing is exactly what OSHA had until January 1993. That's when, after almost twenty years of effort, the federal safety and health agency issued its confined space rule.

The new confined space rule, like many of the standards OSHA has issued over the last decade, is performance-oriented. That means that it sets broad goals which tell you what you have do, but it doesn't tell you how to do it.

Performance standards are a mixed blessing. While they afford lots of flexibility, they offer very little guidance. I believe that my book fills this void by providing you with the guidance you'll need to develop and implement an effective confined space program.

I wanted to reach out beyond the safety and health community and make my book accessible to other people such as plant engineers, line supervisors, and rank-and-file workers. As a result, I opted for a more personal and casual writing style than that found in most textbooks.

I also wanted this book to be as self-contained as possible. I've tried to include virtually all the material you need inside one cover. I tried to be detailed enough to be useful, yet not so detailed as to be cost prohibitive.

Since everybody likes a good story, I've liberally sprinkled the text with case histories. I hope some of you are able to benefit from the mistakes of others. After all, those who don't learn from history are doomed to repeat it.

Finally, I'd like to know what you think. Even though the ink is barely dry, I already have lots of ideas for the second edition. Please let me know what you liked and didn't like about my book. I'm always looking for more case histories, amusing anecdotes or just interesting observations you may have made.

You can write to me at PO Box 158, Riderwood, MD, 21139-0158, call me at (410) 583-7954, FAX me at (410) 593-7955—or for you those of you in cyberspace—EMAIL me at Compuserve 73547,3377.

John F. Rekus, MS, CIH, CSP

ABOUT THE AUTHOR

John Rekus is a self-styled Renaissance man, author, lecturer, world traveler and bon vivant with over 20 years of experience in occupational safety and health spanning industry, government, education and consulting.

He holds a BS in chemistry from Loyola College in Baltimore, and an MS in industrial safety from Indiana University of Pennsylvania. He is Board Certified in Comprehensive practice by both the American Board of Industrial Hygiene and the American Board of Certified Safety Professionals. He is also certified as a Fire Inspector II by the National Fire Protection Association.

His professional memberships include: the American Industrial Hygiene Association, the American Academy of Industrial Hygiene, the American Society of Safety Engineers, The National Fire Protection Association and the American Chemical Society.

Mr. Rekus has authored more than fifty professional papers and magazine articles on occupational safety and health. In 1991, the American Society of Safety Engineers presented him with the Scrivener writing award for his article "Confined Space Hazards in the Coatings and Linings Industry," which appeared in the *Journal of Protective Coatings and Linings*.

He presently resides in Riderwood, Maryland, where he is an independent consultant specializing in OSHA compliance, training for workers and managers, and of course, confined spaces.

Table of Contents

CHAPTER 1

Introduction and Overview

THE CONFINED SPACE PROBLEM

Confined space accidents don't happen often, but when they do, they're usually fatal. Even more alarming is that many confined space incidents involve *multiple fatalities!* How is it that these accidents which occur so seldom, claim so many lives in a single event?

As evidenced by reports issued by the Occupational Safety and Health Administration (OSHA) and the National Institute for Occupational Safety and Health (NIOSH), the factor that most often turns a single confined space incident into a multiple death catastrophe is an unsuccessful rescue attempt. These ill-fated rescue efforts are usually made by well intentioned but untrained employees, who being unaware of invisible hazards posed by confined spaces, respond to an emergency *emotionally* rather than *rationally.*

A typical scenario involves a worker who enters a space such as a chemical storage tank for cleaning or inspection. Suddenly, he is overcome by either an oxygen-deficient atmosphere, or a toxic air contaminant. A second worker, seeing a fellow employee unconscious at the bottom of the tank, enters to rescue him. He too is overcome. A third, fourth, or even a fifth worker may make further fatally unsuccessful rescue attempts.

The domino effect of worker after worker entering and being overcome has been well documented, and as many as six people have been killed in a single incident. In one particularly tragic case, the lives of three generations of family farmers were snuffed out one-by-one as each tried to rescue others who had been previously overcome in a manure pit. According to newspaper accounts, a prolonged summer heat wave caused manure in the pit to decompose much faster than normal. Carbon dioxide, ammonia and methane formed by the decomposing manure apparently displaced the air in the pit creating an oxygen deficient atmosphere.

Eye witnesses explained that the 28-year old son of a farmer entered the pit to replace a shear pin on an agitator. This task had been performed many times in the past without incident. When the farmer's son attempted to climb out, he was overcome and fell to the bottom. The farmer, his grandson, a nephew and another son entered one at a time, each attempting to rescue those who had gone before. None of them survived.

My hope in writing this book is to prevent accidents like this from happening by arming you with the information you'll need to effectively manage confined space hazards. To do this, I'll use a three pronged approach.

First, I'll help you to develop the knowledge and skills needed to recognize a wide variety of confined space hazards. Second, I'll explain how you can use this knowledge and skill to develop an effective confined space entry program. Third, I'll show you how to implement this program through effective training of workers and supervisors.

Since valuable lessons can be learned from past confined space catastrophes, I have included a variety of case studies to illustrate and reinforce many key points. You might want to incorporate some of these case studies into your training programs. Since people can easily remember stories, case studies provide a great tool for getting workers to remember why it's in their best interest to follow confined space safety procedures.

CONFINED SPACES DEFINED

Confined space definitions have been published by national standards setting organizations, government agencies and industrial trade associations (Table 1-1). In time, the OSHA general industry definition will probably supersede all others, but OSHA's definition is relatively new compared to definitions offered by the American National Standards Institute (ANSI), the American Petroleum Institute (API), NIOSH and some state OSHA program. In fact, years before the federal confined space regulation even existed, 5 of OSHA's 25 state-plan states were enforcing their own confined space regulations.

It's also interesting to note that OSHA defines confined spaces differently in maritime operations and construction than in general industry. Both the maritime and construction standards date to the early 1970s and were among the first regulations adopted by OSHA. In its formative years, OSHA was allowed to adopt any existing federal or national consensus standards without the elaborate rule-making process

Table 1-1. Confined space definitions

American National Standards Institute. The definition in the 1989 edition of the American National Standards Institute's Safety Requirements for Confined Spaces (ANSI Z-117.1-1989) states that a confined space is "...an enclosed area that has the following characteristics:

- Its primary function is something other than human occupancy, and
- Has restricted entry and exit, and
- May contain potential or known hazards."

American Petroleum Institute. In its Guidelines to Confined Space Work in the Petroleum Industry, the American Petroleum Institute explains that "...confined spaces are normally considered enclosures with known or potential hazards and restricted means of entrance or exit. These enclosures are not normally occupied by people or well ventilated."

National Institute for Occupational Safety and Health. NIOSH defines a confined space as "...a space which by design has limited openings for entry and exit; unfavorable natural ventilation which could contain or produce dangerous air contaminants, and which is not intended for continuous human occupancy." NIOSH also acknowledges that varying degrees of hazard may exist under different conditions by describing three separate classes of spaces.

Class A spaces are those that present situations which are immediately dangerous to life or health. These include spaces that are either deficient in oxygen or contain explosive, flammable, or toxic atmospheres.

Class B spaces do not present an immediate threat to life or health; however, they have the potential for causing injury or illness if protective measures are not used.

Class C spaces are those where any hazards posed are so insignificant that no special work practices or procedures are required.

California. In California a confined space is "...a space defined by the concurrent existence of the following conditions: (A) existing ventilation is insufficient to remove dangerous air contaminants and/or oxygen deficiency which may exist or develop (B) ready access or egress for the removal of a suddenly disabled employee is difficult due to the location and/or size of the opening(s)."

Kentucky. Kentucky defines a confined spaces as "...a space having the following characteristics: (1) limited means of exit or entry (2) ventilation of the space is lacking or inadequate allowing for the potential accumulation of toxic air contaminants, flammable or explosive agents, and/or depletion of oxygen."

Maryland. Maryland's definition says that "a confined space is a space having limited means of entry or egress and so enclosed that adequate dilution ventilation is not obtained by natural air movement, or mechanically induced movement. In order to be a confined space for the purpose of this regulation a space must be subject to the accumulation of toxic or combustible agents or to a deficiency of oxygen."

Michigan. In Michigan "confined space means a space which is not intended for continuous human occupancy and because of its physical construction or use may have one or more of the following: (a) unacceptable air quality (b) the risk of engulfment by loose particles or bulk materials (liquid or solid) present."

Oregon. While regulations promulgated by the Oregon Department of Insurance and Finance list and describe precaution and work practices that should be followed when working in tanks and vats, they do not specifically define the term "confined space."

Virginia. Virginia defines a confined space as "...any space not intended for continuous employee occupancy, having limited means of egress and which is also subject to either the accumulation of an actual or potentially hazardous atmosphere as defined in this subsection or a potential for engulfment as defined in this subsection." Engulfment is defined as "the surrounding and effective capture of a person by finely divided particulate matter or liquid." Hazardous atmospheres are defined as those where:

- Gases and vapors are in excess 10% of the lower explosive limit.
- Oxygen levels are less than 19.5% or greater than 23% by volume.
- Contaminant concentrations exceed OSHA permissible exposure limits listed in Subpart Z of 29 CFR 1910.
- Conditions may result in acute or immediately severe health effects.

Washington. Washington's definition is the most concise and says that a confined space is any space "...having a limited means of egress which is subject to the accumulation of toxic or flammable contaminant or an oxygen deficient atmospheres."

required today. Since the maritime and construction regulations were respectively part of the federally mandated *Longshore and Harbor Workers' Compensation Act* and *The Construction Safety Act*, they were easily adopted by OSHA.

In contrast, the more recently issued general industry standard was subjected to a rigorous rule-making process that included notices in the Federal Register, public debate and lengthy administrative hearings.

OSHA General Industry Definition

OSHA incorporated common elements found in existing confined definitions when it published its definition in 29 CFR 1910.146, which states that a confined space is "...a space that:

1. Is large enough and so configured that an employee can bodily enter and perform assigned work; and
2. Has limited or restricted means for entry or exit (for example, tanks, vessels, silos, storage bins, hoppers vaults, and pits are spaces that might have limited means of entry); and
3. Is not designed for continuous human occupancy."

Note that for a space to be considered a *confined space*, it must meet *all three* of these criteria. As we will see later though, confined spaces may be classified as either permit-required spaces or non-permit spaces depending on the nature of the hazards they pose.

OSHA Maritime Definition

OSHA maritime standard differentiates between confined spaces and enclosed spaces. A confined space is defined in 29 CFR 1915.4 (p) as "... a compartment of small size and limited access such as a double bottom tank, coffer dam, or other space which by its size and confined nature can readily create or aggravate a hazardous exposure." An enclosed space, on the other hand, as defined in 29 CFR 1915.4 (q) is "... any space other than a confined space which is enclosed by bulkheads and overhead. It includes cargo holds, tanks, quarters and machinery and boiler spaces."

OSHA Construction Definition

Paragraph 29 CFR 1926.21(b)(6)(i)(10) of OSHA construction standards defines a confined space as "...any space having limited means of egress, which is subject to accumulation of toxic or flammable contaminants or has an oxygen deficient atmospheres."

It further stipulates that confined spaces include, but are not limited to, storage tanks, process vessels, bins, boilers, ventilation and exhaust duct, sewers, underground utility vaults, tunnels, pipelines and open top spaces such as pits and tubs.

CONFINED SPACE VARIABILITY

Confined spaces vary widely both in their physical characteristics and in the reasons for which they are entered. Some typical reasons for entering these spaces are listed in Table 1-2. Since the type and magnitude of the hazards posed to entrants varies from space to space, it is essential that individual differences be thoroughly evaluated to assure that the unique hazard posed by each space is adequately controlled.

Variations in Size

Some confined spaces, such as valve pits, sewer lines and utility manholes, are only slightly larger than their occupants. While these spaces may be large enough for one or two people to work comfortably, there is usually very little room for them to move around. While this limitation might only present a minor inconvenience during normal activities, it could easily become a major hindrance in an emergency.

Other spaces, like barges, ships' holds and chemical process vessels, may be dozens of feet across and many stories high. Although these spacious environments afford tremendous freedom of movement, they may also contain hazards which could not exist in smaller spaces. For example, scaffolding installed inside very large petroleum storage tanks introduces the potential for falls from elevations. Similarly, new and unfamiliar hazards may be introduced when different crafts work inside the same space. For example, pipe fitters working in one end of a large horizontal tank may be exposed to mists and vapors generated by painters working at the other end.

Variations in Shape

Water towers, chemical reactors and petroleum storage tanks are often built in regular geometric shapes like boxes, spheres and cylinders. These

Table 1-2. Typical reasons for confined space entry

- Cleaning to remove sludge and other waste materials
- Inspection of physical integrity and process equipment
- Maintenance such as abrasive blasting and application of surface coatings
- Tapping, coating, wrapping and testing of underground sewage, petroleum, steam and water piping systems
- Installing, inspecting, repairing and replacing valves, piping, pumps, motors, etc. in below ground pits and vaults
- Repair, including welding and adjustments to mechanical equipment
- Adjusting and aligning mechanical devices and components
- Checking and reading meters, gauges, dials, charts and other indicators
- Installing, splicing, and repairing, and inspecting electric, telephone, and fiber optic cables
- Rescue of workers who are injured or overcome inside the space.

spaces are usually open and free from internal barriers and obstruction. On the other hand, chemical tank trucks, process furnaces, industrial boilers and ships' holds may be irregularly shaped, and divided into smaller sections by walls, tubes, bulkheads and baffle plates (Figure 1-1). The task of evaluating these convoluted spaces for atmospheric hazards is much more difficult since toxic gases and vapors may become trapped in hard-to-reach pockets. Life-lines may also be cut on projections or tangled up on interior obstacles like pipes, cable hangers and mechanical fittings.

Variations in Function

Confined spaces are found in so many different industrial operations that it is virtually impossible to identify them all. However, a variety of example spaces covering a broad cross-section of industries is presented in Table 1-3. An estimate of the number of confined spaces and number of confined space entrants in various industries is presented in Table 1-4.

Open surface tanks like those used for electroplating, caustic cleaning or acid pickling are frequently overlooked as confined spaces under the misguided notion that they pose no hazard since they are open to the atmosphere. This erroneous assumption resulted in the deaths of three workers who were killed while cleaning an open-topped electroplating tank. Apparently the acid they used to clean a tank reacted with dried cyanide reside on the walls to produce deadly hydrogen cyanide gas.

In another case, a maintenance worker was almost overcome when he climbed into a large open-topped shipping container to inspect it. Investigation revealed that the container's cargo had been packed with dry-ice (solid carbon dioxide). When the ice sublimed the resulting carbon dioxide gas displaced the air in the bin.

CONFINED SPACE HAZARDS

Confined space hazards may be broadly divided into two categories: atmospheric hazards and physical hazards (Table 1-5). Atmospheric hazards include oxygen-deficiency, oxygen-enrichment, explosive gases and vapors, and "toxic" air contaminants. Physical hazards, on the other hand, include such things as moving mechanical equipment, energized electrical conductors, ionizing and non-ionizing radiation, heat, cold, in-flowing fluids, and finely divided solids like grain or saw dust that can engulf and trap a victim. Even gravity can be a hazard if tools and equipment fall through an elevated opening onto the heads of workers below.

Extent of Hazards

Some spaces, like underground environmentally controlled telecommunications equipment chambers, present few, if any, hazards. Conversely, chemical reactors may be plagued with a litany of hazards, including mechanical agitators, steam jackets, nitrogen inerting systems and a network of pipes which could flood unsuspecting workers with solvents, acids, caustics or other hazardous materials. Between these two extremes are innumerable tanks, pits, containers and vessels, each of which presents its own unique set of hazards.

Even innocent looking street manholes may contain explosive concentrations of naturally occurring methane gas formed by the fermentation of organic debris like weeds, leaves and grass. As shown in Figure 1-2, street manholes could also contain hazardous concentrations of toxic or combustible vapors and gases which inadvertently migrate into the space from sources such as pipe lines, leaking underground storage tanks or spills resulting from hazardous materials transportation accidents.

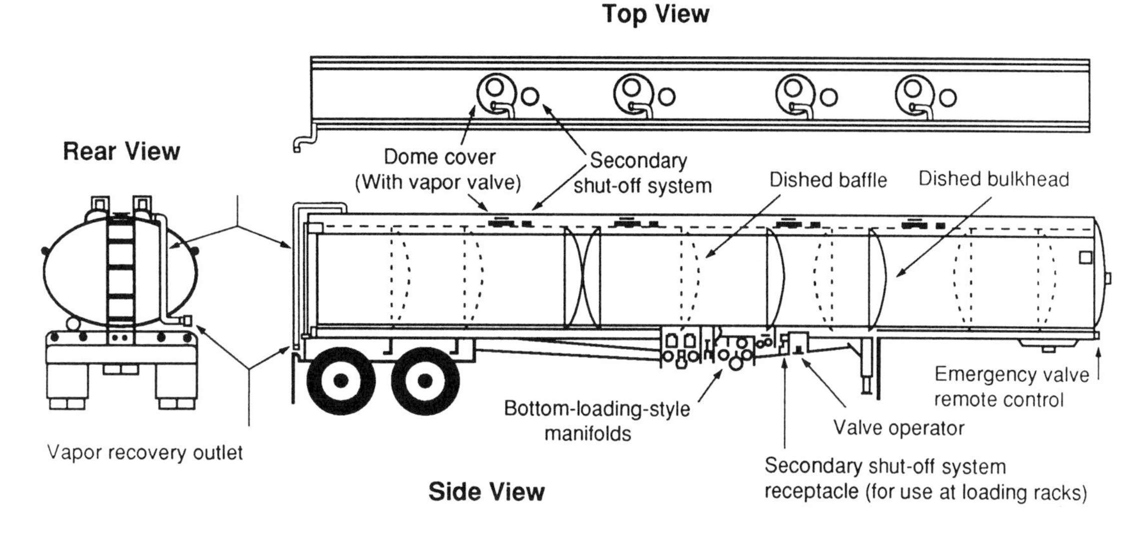

Figure 1-1. A tank truck like this provides an example of a space with interior baffles which can restrict an entrant's movements. (Paul Trattner)

Storm water systems and sanitary sewers present a particularly insidious hazard since anyone can dump anything into them at any time. As a result, they may contain any of the tens of thousands of chemicals of commerce. While it may be argued that the potential for contamination is greater in industrial settings simply because larger quantities of chemicals are present, the hazard posed by residential neighborhoods should not be discounted.

Many household products such as paint, bleach, ammonia, battery acid and lacquer thinner are hazardous chemicals. If a homeowner decides to dispose of these materials by pouring them down the drain, he might be creating a serious hazard for the county sewer inspection team working outside his house. Similarly, a maintenance crew working in a storm drain could be affected by flooding during a sudden rain storm, or by flammable vapors resulting from gasoline that leaks into the drain from a tank truck accident a half a mile away.

Hazards to Rescue Personnel

In addition to all of the hazards posed to entrants, rescue personnel must cope with a myriad of physical and psychological stresses as they race against time to save the lives of incapacitated victims. Furthermore, when emergency personnel go into spaces to rescue workers who have been overcome, they are entering environments that pose a clear and present danger!

Even under the best of circumstances, the task of removing unconscious victims can be grueling. But in many situations rescuers find their job made even more difficult by the cumbersome protective equipment they must wear. A typical self-contained breathing apparatus, for example, weighs about thirty pounds. As shown in Figure 1-3, their size and shape also makes them difficult to maneuver through small entry portals. Clearly rescue operations expose additional personnel to risks which could have been prevented by a comprehensive entry program.

THE ACCIDENT RECORD

The exact number of workers killed and injured each year in confined space accidents is unknown. However there are a number of published reports that provide some insight to the scope and magnitude of the confined space problem. While the details of these accidents vary from case to case, many of them have certain things in common. Some of the similarities include:

- Mechanical equipment such as mixers being accidentally switched on
- Hazardous liquids and gases entering spaces through valves that were inadvertently opened
- Hazardous atmospheres inside the space

Table 1-3. Selected examples of confined spaces by industry

Agriculture
Manure pits
Hoppers and silos
Fertilizer storage tanks
Conveyor enclosures
Spray tanks

Construction Industry
Caissons
Box beams
Sewers
Pits
Trenches
Excavations
Crawl spaces

Food and Kindred Products
Retorts
Tubs and kettles
Basins
Cold rooms
Ovens
Flour bins
Air scrubbers
Batch cookers
Caustic soda tanks
Clay hoppers
Conditioners
Continuous cookers
Extractors
Heated lard tanks
Heated sugar bins
Holding bins
Hydrogenators
Meal bins
Meal dryers
Mixers
Tallow tanks

Tobacco Processing
Cooling towers
Rotating drums
Dryers
Sanding drums

Textiles
Bleaching ranges
J-boxes
Kiers
Die kettles
Bale presses
Dye becks
Sizing tanks
Steam boilers

Paper and Pulp
Chip bins
Barking drums
Rag cookers
Acid towers
Digesters
Beaters
Hydropulpers
Stock chests
Adhesive tanks
Bleach tanks
Chip silos
Furnaces
Machine chests
Mix tanks
Resin tanks
Clay mix tanks

Printing and Publishing
Ink tanks
Solvent tanks

Petroleum and Chemicals
Reactors
Storage tanks
Distillation columns
Cooling towers
Dike areas
Fire water tanks
Precipitators
Scrubbers
Crystallizers
Spray dryers

Rubber Products
Solvent tanks
Shredders
Furnaces
Ovens
Mixers

Leather Products
Dye vats
Tanning tanks
Sludge pits

Stone, Clay, Glass and Concrete Products
Kilns
Aggregate bins
Cement silos
Crushers
Dryers
Hoppers
Mills
Sand bins

Primary Metals
Blast furnaces
Cupolas
Coal bins
Coke bunkers
Annealing furnaces
Slag pits
Water treatment tanks
Submarine cars
Gas holders
Soaking pits
Acid pickling tanks
Plating tanks

Fabricated Metals
Paint dip tanks
Degreasers
Caustic cleaning tanks
Drying ovens
Shot blasting enclosures
Enclosed assemblies
Sludge tanks

Machinery
Boilers
Conveyors
Dust collectors
Tunnels

Electronic Industry
Degreasers
Gas cabinets
Plating/rinse tanks

Transportation
Aircraft wing tanks
Test chambers
Cargo tank trucks
Rail tank cars

Electric, Gas and Sanitary Services
Cable vaults
Manholes
Meter vaults
Transformer vaults
Bar screen enclosures
Chemical pits
Incinerators
Pump stations
Regulators
Sludge pits
Wet wells
Valve pits
Digesters
Grease traps
Lift stations
Sewage ejectors
Storm drains

Maritime Operations
Barges
Ships holds
Boilers
Fuel tanks
Fresh water tanks
Bilges
Compartments

Note that this is only a partial list of typical spaces in a cross-section of industries. It is not all inclusive and other types of spaces may be present even though they are not noted above.

Table 1-4. Profile of affected establishments with confined spaces

Standard Industrial Code	Industry	Establishments with Permit Spaces	Number of Permit Spaces	Number of Employees	Number of Permit Entrants
07	Agriculture services	10,864	79,821	62,990	25,748
13	Oil and gas extraction	10,000	12,477	155,660	11,239
20	Food and kindred products	10,236	142,727	805,247	99,420
21	Tobacco products	69	776	37,845	2,007
22	Textile mill products	1,491	17,062	186,752	27,831
24	Wood products (except furniture)	10,290	39,409	146,042	31,035
25	Furniture and fixtures	5,254	26,012	224,589	35,424
26	Paper and paper products	4,397	95,533	475,171	46,208
27	Printing and publishing	47	206	2,196	94
28	Chemicals and allied products	8,098	170,982	593,738	71,962
29	Petroleum refining	1,644	93,700	104,704	15,560
30	Rubber products	6,282	143,818	319,262	143,522
31	Leather and leather products	151	514	6,395	1,055
32	Stone, clay glass and concrete	12,290	116,708	366,454	110,568
33	Primary metals industry	2,788	35,521	463,942	56,669
34	Fabricated metal products	8,441	88,507	346,800	33,959
35	Machinery except electrical	4,330	34,670	437,200	116,987
36	Electrical/electronic equipment	6,610	176,895	892,336	111,087
37	Transportation equipment	3,302	1,085,966	1,043,403	31,706
38	Instruments and related products	64	901	7,296	514
39	Miscellaneous manufacturing	885	31,267	18,926	5,744
42	Motor freight transportation	14,585	201,680	201,679	40,336
40	Electric, gas sanitary service	28,444	1,575,170	410,290	263,217
50	Wholesale trade (durable)	2,753	3,965	36,485	3,359
51	Wholesale trade (nondurable)	36,913	411,095	358,647	194,454
54	Food stores	10,073	10,073	318,010	10,073
59	Miscellaneous retail	7,149	28,201	57,923	10,694
65	Real estate (commercial)	13,583	45,190	391,923	12,442
70	Hotels and other lodging	5,099	77,672	163,323	80,442
72	Personal services	3,577	24,604	198,447	7,154
76	Miscellaneous repair services	752	802	3,718	652
78	Motion picture	11	33	16,500	66
80	Health services	8,252	71,709	3,357,391	27,308
84	Museum, botanical gardens, zoos	130	1,183	7,338	781
	Total	238,853	4,844,849	12,218,622	1,629,201

Source: Federal Register, Vol. 48 No. 9.

Table 1-5. Confined space hazards

Atmospheric Hazards
• Oxygen deficiency atmospheres
• Oxygen enriched atmospheres
• "Toxic" or irritating atmospheres
Physical Hazards
• Fixed and portable mechanical equipment
• Electrically energized conductors
• Fluids: liquids, powders and gases
• Thermal condition: hot or cold
• Engulfment by finely divided material
• Ionizing and non-ionizing radiation
• Contact with corrosive substances

- Lack of attendants outside the space
- Inadequate rescue planning and emergency response

OSHA Reports

One OSHA report which summarized an in-house review of 122 confined space accident investigations conducted between 1974 and 1982 found that asphyxiating or toxic atmospheres were responsible for 173 deaths. In another report summarizing accident investigations conducted between 1974 and 1979 the agency found that 50 incidents involving fires or explosions were responsible for 78 fatalities. With the promulgation of its permit-required confined space standard, OSHA expects to prevent 54 deaths and more than 10,700 injuries per year (Table 1-6).

NIOSH Reports

A NIOSH study which reviewed more that 20,000 accident reports filed over a 3-year period found that 234 deaths and 193 injuries were linked to 276 confined space incidents (Table 1-7). But employees who work in confined spaces are not the only ones at risk. NIOSH investigations show time and time again that the one thing which most frequently turns a single confined space accident into a multiple death catastrophe is unsuccessful rescue attempts. In fact, in one study NIOSH revealed that almost 60% of those killed in confined spaces were rescuers. This same study indicated that supervisors who, according to conventional wisdom, are supposed to be more knowledgeable about job safety, weren't! This is evidenced by the fact that almost half of the accidents NIOSH investigated involved the death of at least one supervisor. In three of the incidents two supervisors died, and in one incident *three supervisors* died!

In the cases they investigated, NIOSH researchers also found that the greatest risk was posed by spaces where the potential for hazardous atmospheres was either not known or not recognized. This observation was supported by subsequent interviews with victims' employers and coworkers, 84% of whom were either unaware of the hazards associated with confined spaces, or who did not know that they were entering confined spaces. A summary of accident statistics gleaned from the NIOSH reports listed in Table 1-8 are presented in Figures 1-4 through 1-5.

Union Reports

The United Automobile, Aerospace and Agricultural Implement Workers of America (UAW) conducted a study of confined space accidents that occurred at facilities represented by their union between 1974 and 1989. A survey of accident reports submitted to the UAW Health and Safety Department indicated that 26 confined space incidents had occurred over this period. These incidents resulted in 61 non-fatal injuries and 19 fatalities. Profiled highlights of the study are shown in Tables 1-9 and 1-10.

OSHA CONFINED SPACE STANDARDS

Now that we've laid a foundation consisting of a summary of confined space definitions, confined space hazards and confined space accident statistics, I'd like introduce some of the OSHA regulations that govern confined space entry.

As noted previously, OSHA defines confined spaces differently for industrial, construction and maritime operations. In fact, each of these employment sectors is regulated by a different part of the *Code of Federal Regulations* (CFR) regulations. Part 1910, for example, covers general industry, while Parts 1915 and 1926 cover maritime and construction operations, respectively.

While the focus of this book is primarily directed at the permit-required confined space standard in Part 1910, it is important to understand that this is not the only OSHA standard governing confined space entry. Other general industry standards related to paper mills, telecommunications and gain handling—to mention a few—also include specific confined space entry provisions.

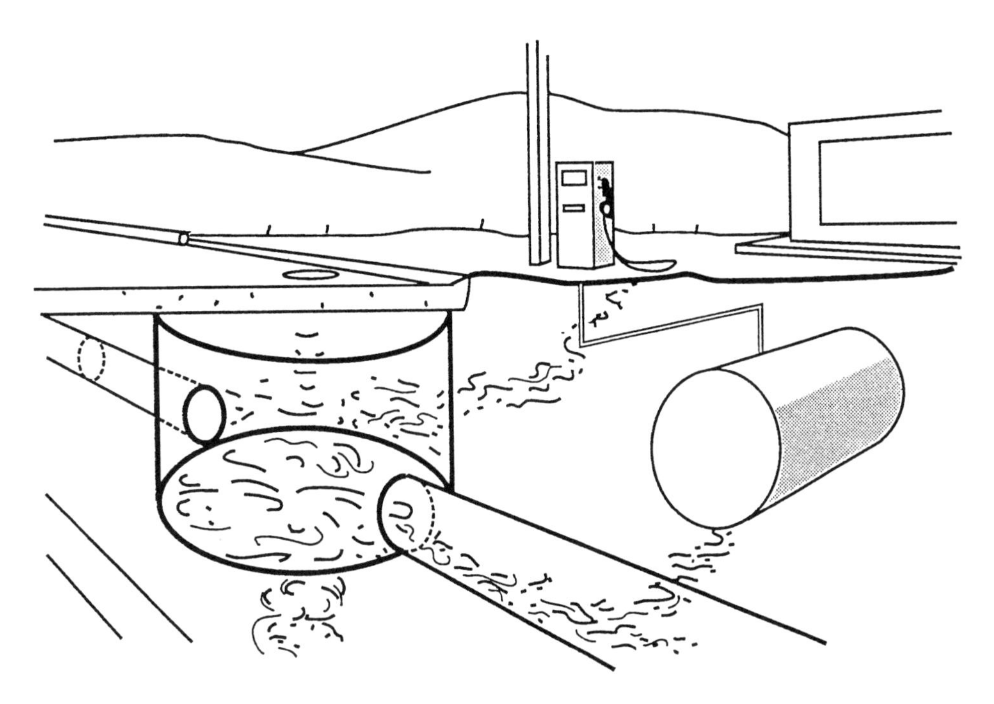

Figure 1-2. Hazardous gases and vapors leaking from underground storage tanks can migrate into street manholes. (Paul Trattner)

Figure 1-3. Rescuers are often encumbered by protective gear like the self-contained breathing apparatus shown here. (Greg Valcourt)

Although the construction standards do not specifically include confined space regulations, they do generally address confined space hazards. For example, both the trenching and underground construction regulations include provisions for atmospheric testing, stand-by personnel and emergency planning. These requirements are very similar to those stipulated for entry into confined spaces in general industry.

In a similar vein, Part 1915 prescribes detailed requirements for confined space work performed on ships, barges and other sea going vessels. Readers who have a particular interest in maritime confined spaces can find additional information on this topic in National Fire Protection Standard 306, *Control of Gas Hazards on Vessels.*

Since my experience suggests that many people are not aware of OSHA's other confined space standards, I'd like to briefly summarize their requirements for you. My goal is to provide you with a broad overview of the standards, not a reprint of the Code of Federal Regulations. In this light, it's important to understand that different standards impose different demands, some are very specific, others are fairly general. These differences are reflected in the level

Table 1-6. Estimation of fatalities and injuries prevented by OSHA permit-required confined space standard

Standard Industrial Code	Industry	Illness and Injuries Avoided		
		Fatalities	Lost-work Days	Non-Lost-Work Days
07	Agriculture services	0.3	18	17
13	Oil and gas extraction	6.8	163	120
20	Food and kindred products	2.7	404	513
21	Tobacco products	0.3	51	64
22	Textile mill products	0.3	51	64
24	Wood products (except furniture)	1.0	152	192
25	Furniture and fixtures	0.0	—	—
26	Paper and paper products	0.7	101	128
27	Printing and publishing	0.0	—	—
28	Chemicals and allied products	4.1	607	709
29	Petroleum refining	0.0	—	—
30	Rubber products	0.7	101	128
31	Leather and leather products	1.0	152	192
32	Stone, clay glass and concrete	0.7	101	128
33	Primary metals industry	2.0	303	384
34	Fabricated metal products	5.1	758	961
35	Machinery except electrical	1.0	152	192
36	Electrical/electronic equipment	0.0	—	—
37	Transportation equipment	2.0	303	384
38	Instruments and related products	0.3	51	64
39	Miscellaneous manufacturing	0.3	51	64
42	Motor freight transportation	5.1	223	163
40	Electric, gas sanitary service	6.8	297	217
50	Wholesale trade (durable)	0.3	42	48
51	Wholesale trade (nondurable)	1.0	125	145
54	Food stores	0.0	0	0
59	Miscellaneous retail	0.3	42	48
65	Real estate (commercial)	0.0	0	0
70	Hotels and other lodging	0.0	0	0
72	Personal services	0.0	0	0
76	Miscellaneous repair services	3.1	374	437
78	Motion picture	0.0	0	0
80	Health services	0.3	42	48
84	Museum, botanical gardens, zoos	0.0	0	0
	Total	53.7	5,041	5,908

Source: Federal Register, Vol. 48, No. 9.

of detail provided by the summaries below. A more elaborate summary is provided for those standards that contain detailed requirements than for those that offer only general guidelines.

OSHA GENERAL INDUSTRY STANDARDS

OSHA's most comprehensive confined space regulations are contained on 29 CFR 1910.146 *Permit-Required Confined Spaces for General Industry.* However, there are seven other general industry standards which also address confined spaces. Two of the standards, *Open Surface Tanks* and *Welding* are "horizontal standards." That means that they apply to all industries from auto repair to zoo keeping. The other five standards are "vertical standards" and apply only to paper mills, bakeries, textile mills, telecommunications facilities and grain handling operations.

For the most part, the requirements of these industry-specific standards supersede the more sweeping general industry rule. However, if the entry is related to a space not regulated by the industry-specific standard, then the general industry rule prevails.

For example, entry into bins and silos at grain handling facilities is regulated by the grain handling

Table 1-7. NIOSH study of confined space accidents

Type of Accident	Events	Injured	Killed
Atmospheric condition	80	72	78
Explosion or fire in space	15	49	15
Explosion or fire at point-of-entry	23	20	32
Trapped in unstable material	16	0	16
Electrocution or electric shock	11	2	9
Caught in or crushing of	10	3	10
Struck by falling objects	15	1	14
Falls while in confined space	27	26	1
Ingress or egress	33	30	3
Insufficient maneuverability	15	15	0
Eye injury	10	10	0
Contact with temperature extremes	7	4	3
Noise	1	1	0
Vibration	1	1	0
Stress from excess exertion	12	0	12
Totals	276	234	193

Adapted from Criteria for a Recommended Standard Working in Confined Spaces, NIOSH, Pub. 80–106.

standard, 29 CFR 1910.272. However, if employees of a grain handling facility enter spaces *other than those used for handling grain,* then those entries would be covered by the general industry rule. For example, workers who enter LP gas tanks and used to fuel grain dryers would be covered by 1910.146 not 1910.272.

Similarly, most telecommunications confined space entries involve street manholes. Since these spaces are specifically addressed by the telecommunications standard, 29 CFR 1910.268, its requirements preempt the general industry rule. However, if telecommunications workers enter spaces other than manholes—for example, underground gasoline storage tanks—the general industry rule would apply. The general industry rule also applies to telecommunications manholes that contain hazards other than those associated with telecommunications, for example, moving mechanical equipment, radioactive sources or spilled hazardous chemicals.

Permit-Required Confined Space Standard: 29 CFR 1910.146

Recognizing that all confined spaces do not present the degree of hazard, the *OSHA Permit-Required General Industry Standard* differentiates between two types of confined spaces: permit-required spaces and non-permit spaces.

This decision was largely influenced by testimony given during the OSHA rule-making process which made it quite clear that different confined spaces presented different levels of risk. For example, entries into chemical reactors which contained a myriad of hazards such as toxic reside, agitators and slippery walking surfaces clearly pose greater potential for harm then do entries into manholes containing telephone lines.

In fact, OSHA's own telecommunications standard acknowledges the limited hazard posed by work in communications manholes by prescribing very elementary entry precautions. As simple as these precautions are, their adequacy is supported by the telecommunications industry's excellent safety record. Testimony provided during the rule-making process indicated that over a forty year period, only a handful of accidents had resulted from the millions of entries made into hundreds of thousands of telecommunications manholes located throughout the country. It was further pointed out that most of these accidents were traffic related and involved top-side workers being struck by vehicles.

Permit-Required Spaces

Permit-required spaces are confined spaces which are hazardous to enter unless special precautions are taken. The standard defines a permit-required confined space as "...a confined space that has one or more of the following characteristics:

1. Contains or has the potential to contain hazardous atmospheres;
2. Contains a material that has the potential for engulfing an entrant;

Table 1-8. NIOSH fatal accident circumstances and epidemiology reports related to confined space entry

Report Title	Report No.
Fire at a waste water treatment plant	84-11
Two confined space fatalities during construction of a sewer line	84-13
Two rescuers die in fracturing tank in West Virginia gas field	85-02
Confined space incident kills two workers, company employee and rescuing fireman	85-05
Worker dies in 20,000 gallon gasoline bulk tank while wearing closed circuit self-contained breathing apparatus in Vermont	85-09
One dead, one injured in elevator fall at construction site	85-13
Use of sulfuric acid in septic tank leaves one dead and one critical in Pennsylvania	85-23
27-year old dies inside 6 million gallon storage tank	85-26
Rescue effort results in fatality for a wine manufacturing plant worker in Illinois	85-27
Three sanitation workers and one policeman die in an underground sewage pumping station in Kentucky	85-31
Construction worker dies as a result of applying coating material in a confined space	85-33
City water worker dies as a result of being overcome by natural gas vapors while reading water meter in confined space in Ohio	85-40
Two sanitation employees die in confined space in Kentucky	85-44
Worker killed in cave-in at Ohio excavation site	85-45
Three fire fighters killed fighting silo fire in Ohio	85-49
Worker dies in fermentation tank in Montana	86-13
Steel worker dies in industrial waste pit in Pennsylvania	86-15
Truck driver suffocates in sawdust bin in Pennsylvania	86-19
Foundry worker dies in Indiana	86-23
Three dead in confined space incident in New York	86-34
Two workers die in underground valve pit in Oklahoma	86-37
Three dead, one critical in industrial septic tank in Georgia	86-38
28-year old dies in rescue attempt in drainage pit in Illinois	86-48
Insufficient oxygen level in sewer claims the life of plumbing contractor in Georgia	86-54
Owner/foreman of construction company dies in 15-foot deep manhole in California	87-05
Two dead, five injured, in confined space incident in Oregon	87-06
Worker dies while cleaning freon 113 degreasing tank in Virginia	87-17
Two workers die in digester unit in New Mexico	87-23
Worker dies while repairing vacuum evaporator in Virginia	87-25
Worker dies after lifting access cover on acid reclaim storage tank in Virginia	87-26
Truck driver dies while cleaning out inside of tanker in South Carolina	87-27
Digester explosion kills two workers at waste water plant in Pennsylvania	87-33
Farm worker asphyxiated in grain silo in Indiana	87-39
One dead, one near miss in sewer in Kentucky	87-45
Confined space fatality at waste water treatment plant in Indiana	87-46

Table 1-8. Continued

Report Title	Report No.
Worker dies inside filtration tank in Michigan	87-47
Farmer dies in Indiana	87-49
Tractor-trailer repairman dies while welding interior wall of a tanker in Indiana	87-50
Parks and recreation director dies in oxygen deficient atmosphere in West Virginia	87-57
73-year old self-employed pump service contractor dies in Maryland	87-59
Mechanic asphyxiated within steam service passageway	87-64
Two construction workers die inside sewer manhole in Indiana	87-67
Two supervisors die in manhole in South Carolina	88-01
Three construction supervisors die from asphyxiation in manhole	88-36
Construction sub-contractor asphyxiated in manhole	88-44
Two maintenance workers die after inhaling hydrogen sulfide in manhole	89-28

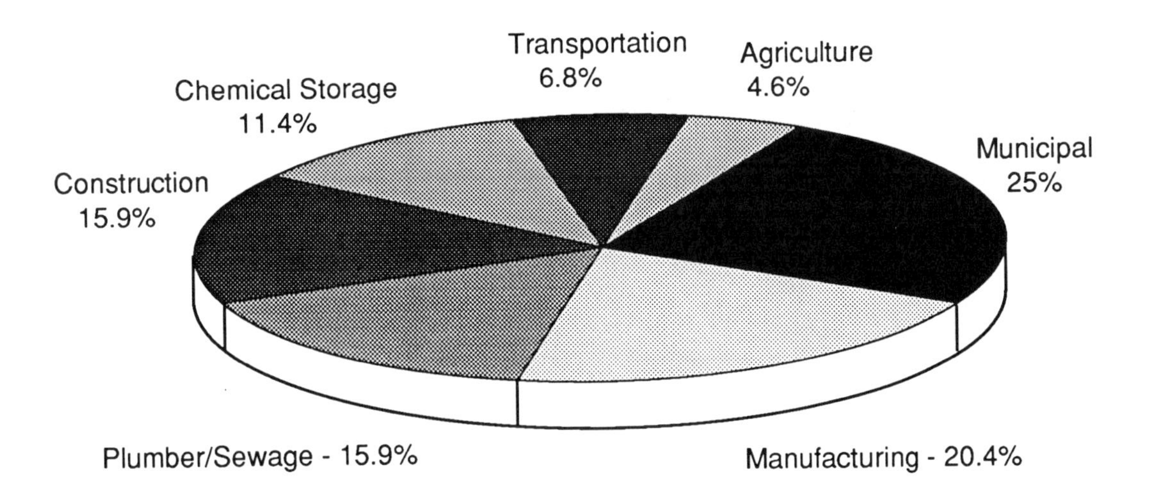

Figure 1-4. Industry distribution of confined space fatalities investigated by NIOSH between 1984 and 1988.

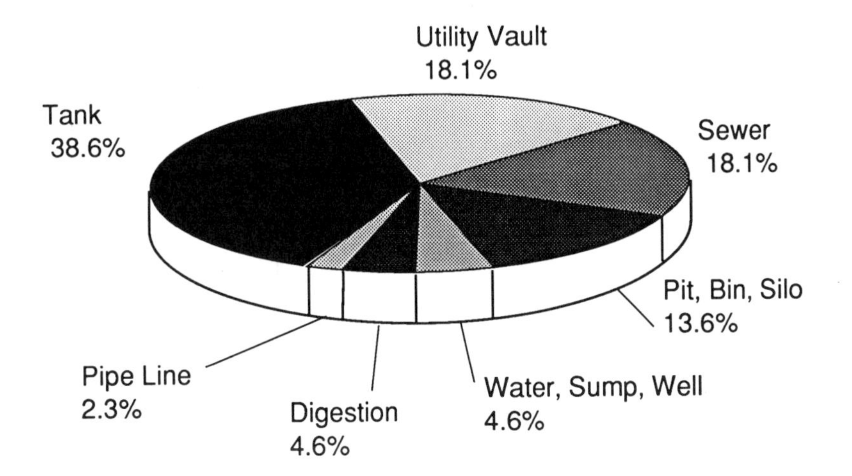

Figure 1-5. Distribution of confined space by type for fatality investigations conducted by NIOSH 1984–1988.

Table 1-9. UAW study of confined space accidents 1974–1989 distribution plant type operations

Plant Type	Incidents	Non-fatal	Fatal
Foundry	7	6	19
Stamping	1	2	20
Assembly	9	6	12
Steel	2	2	0
Warehousing	1	0	3
Machining	3	1	5
Chemical	2	1	2
Aerospace	1	1	0

Table 1-10. UAW study of confined space accidents 1974–1989 distribution by type of space and hazard

Type of Space	Hazard	Fatal	Non-fatal
Furnaces/ Cupolas	Gases	2	22
Pits	Vapors	5	24
	Gases	1	0
	Engulfment	1	0
	Mechanical	1	0
Sand Mullers	Mechanical	2	0
Vehicles	Vapors	1	12
Tanks	Vapors	3	1
Hoppers/ Bins	Engulfment	2	0
Dust Collectors	Dusts	0	2
	Gases	1	0

3. Has an internal configuration such that an entrant could be trapped or asphyxiated by inwardly converging walls or by a floor which slopes downward and tapers to a smaller cross-sectional area; or
4. Contains any other recognized serious safety or health hazards."

Note that a confined space has to contain only one of these elements to be considered a permit-required confined space. As indicated in Table 1-11, entry into permit-spaces is generally governed by a comprehensive program that includes provisions for detailed hazard evaluation, written entry permits, employee training, and emergency planning. However, abbreviated procedures may be substituted for spaces that present only atmospheric hazards that can be eliminated by ventilation.

These procedures are explained in detail in Chapter 8, but they generally include:

- Establishing methods to assure that manhole covers can be removed safely
- Protection of man hole openings with, railings or barricades
- Continuously ventilating the space during entry to assure an acceptable atmosphere
- Initial and periodic atmospheric testing
- Preparation of an entry certificate

Non-Permit Spaces

Non-permit confined spaces are confined spaces that do not actually or potentially contain hazards that could cause death or serious physical harm. Examples of non-permit spaces include: drop ceilings, mechanical equipment closets and motor control cabinets.

If all of the hazards posed by a permit space can be eliminated, the space may be reclassified as non-permit spaces. The practical advantage of reclassifying permit-spaces to non-permit spaces is that they can be entered without the need for permits, attendants, and other permit program elements listed in Table 1-11. Possible candidates for reclassification include mixers, material bins, industrial boilers and chemical storage tanks.

For example, the mechanical hazards posed by rotating mixing drums such as cement trucks can be eliminated by locking-out the drive mechanism before the drums are entered. Similarly, the engulfment hazard posed by material bins can be eliminated by emptying the bins before entry. Tanks and boilers can be isolated, drained, flushed and ventilated to remove all traces of atmospheric hazards. If the hazards can be eliminated without having to enter the vessel, it can be readily reclassified as a non-permit space.

If, on the other hand, the vessel is configured in a way that requires it to be entered, either to eliminate or evaluate hazards such as testing the atmosphere in

Table 1-11. Elements of permit-required confined space program specified by 29 CFR 1910.146

- Procedures for identifying hazards associated with entry
- Procedures, method and practices used to control confined space hazards
- A written permit system
- Specialized equipment, such as air sampling instruments, ventilation equipment, rescue gear, personal protective equipment, etc. required for entry
- Designation of employees who have an active role in the entry, e.g., authorized entrants, attendants, entry supervisor
- Provision for testing and evaluating the space to ensure that conditions are suitable for entry
- Coordination of contractor activities
- Emergency response procedures, including provision for rescue equipment and an attendant
- Employee training and information
- Annual program review to assure continued effectiveness

areas that cannot be reached from outside, the initial entry would be considered a permit-entry, and as such, would be governed by a comprehensive permit-entry program. However, once the hazards are eliminated and air monitoring confirms the absence of atmospheric hazards, the space may be reclassified as a non-permit space. Of course, the reclassification remains valid only as long as all hazards are eliminated.

When spaces are reclassified, the basis for determining that all hazards have been eliminated must be documented. The documentation must contain the identity of the space, the date and the signature of the certifying individual. It must also be made available for review by workers before they enter the space.

Telecommunications: 29 CFR 1910.268

OSHA standard 29 CFR 1910.268 (o) addresses telecommunications work performed on underground lines in manholes and unvented vaults. It includes provisions for guarding manhole openings, ventilation and atmospheric testing. These requirements are virtually identical to alternative entry provisions discussed above and are described in detail in paragraph (c)(5)(ii) of the permit-required confined space standard.

Pulp, Paper, and Paperboard Mill: 29 CFR 1910.261

OSHA's standard for pulp, paper, and paperboard mills includes some general provisions that cover all mill-related confined space entries and some very detailed requirements for entry into specific process equipment such as continuous barking drums, rag cookers, acid tanks and digesters.

General Entry Work Practices

Lifelines and safety harnesses must be used by employees who enter closed vessels, tanks, chip bins and similar areas. A person must also be stationed outside the vessel in a position to handle the lifeline and to summon assistance in case of an emergency. The air in the vessel must be tested for oxygen and for flammable and toxic gases before entry. Self-contained breathing apparatus must be present, and equipment and valves that could injure the entrant if operated unexpectedly must be locked in the "safe" position.

Specific Entry Work Practices

Continuous Barking Drums. The driving mechanisms on continuous barking drums be locked-out at the disconnect whenever employees enter the drum.

Ragcokers. All steam, water and other control valves must be blanked, or locked and tagged in the off position.

Acid tanks. Intake lines and acid lines must be blanked. Tanks must be washed and free from acid before an employee enters. Fresh air must be blown into the acid tanks, and entrants must be provided with air-supplied respirators, harnesses and lifelines.

Chip and Sawdust Bins. Unless workers are equipped with safety harnesses and lifelines, arches that form in bins and chip lofts must be broken down with steam or compressed air before employees enter.

Digesters. Valves leading to digesters must be locked out and tagged. Fresh air must be constantly blown into the digester while employees are working inside. Supplied-air respirators must also be available in the event that the fresh air supplies fail.

Stock Chests. All control devices must be locked or tagged out when workers enter stock chests. Low voltage extension lights must be provided when cleaning, inspecting or performing other work in stock chests.

Lead Burning. The standard requires that fresh air be directed into the space so that air reaches the face of the workers whenever lead burning is performed. However, local exhaust ventilation that captures lead fumes at their source may be a more effective method of controlling exposure.

Textile Operations: 29 CFR 1910.262

Paragraphs (p) and (q) of 29 CFR 1910.216 require that valves controlling steam and other injurious gases or liquids feeding kiers and continuous bleaching range J-boxes be equipped with devices that allow them to be locked out or isolated before employees enter.

Bakery Equipment: 29 CFR 1910.263

OSHA standard 29 CFR 1910.263(d)(6) requires that covers on flour handing equipment and storage bins be provided with hasps and locks so employees may lock covers in the open position when they enter the bins. The inside and outside of storage bins five or more feet deep must also be equipped with a stationary safety ladder. The outside ladder must reach from floor level to the top of the bin, and the inside ladder must reach from the top of the bin to the bottom.

Grain Handling Facilities: 29 CFR 1910.272

A permit is required for entry into grain silos, bins or tanks, unless the person who would normally issue the permit is present during the entire entry operation. All mechanical, electrical and hydraulic equipment which presents a danger to entrants must be disconnected, blocked, locked-, or tagged-out out of service. The atmosphere in the space must be tested for combustible and toxic agents if they are suspected of being present. Oxygen tests must also be made unless there is continuous natural air movement, or unless continuous forced air ventilation is provided before and during the entry.

Employees who enter silos and bins from the top must use either a full body harness with a life line, or a boatswain chair. An observer who is in communication with the entrants must be stationed outside the space and equipped to provide assistance in an emergency.

Open Surface Tanks: 29 CFR 1910.94

The open-surface tank standard 29 CFR 1910.94 requires that tanks be drained and ventilated before employees enter them for inspection or repair. The atmosphere in the tank must be tested for oxygen and air contaminants before entry. An attendant equipped with a lifeline that permits entrants to be lifted out of the tank must also be provided.

Welding: 29 CFR 1910.252

OSHA welding standards require that a minimum general ventilation rate of 2,000 cubic feet per minute per welder be maintained in confined spaces. If local exhaust is used, the capture velocity at the point of work must be at least 100 feet per minute.

Welding machines and compressed gas cylinders may not be taken into confined spaces. When work stops for a prolonged period of time (such as lunch) electrodes must be removed from their holders and gas flow to torches must be shut off from outside the space. When practical, torches, cables and hoses must be removed from the space.

A preplanned rescue procedure must be in place, and an attendant equipped with the means to quickly remove welders in an emergency must be stationed outside the space.

OSHA CONSTRUCTION STANDARDS

While the current OSHA construction standards do not specifically address confined space entry, there are three regulations that speak to confined space related hazards. These standards are:

- 29 CFR 1926.651 *Excavations, Trenching and Shoring*
- 29 CFR 1926.800 *Underground Construction*
- 29 CFR 1926.956 *Underground Lines*

Excavations, Trenching and Shoring: 29 CFR 1926.651

Special precautions must be taken when employees enter trenches deeper than four feet that are dug in locations where atmospheric hazards may be present. These locations include places such as landfills, hazardous waste sites, chemical plants, refineries and underground storage tank removal projects.

Specific precautions that must be taken include atmospheric testing, ventilation and respiratory protection. An emergency plan must be in effect and rescue equipment must be available on site. Workers who enter bell-bottom pier holes or other deep confined footing excavations must wear lifelines and full body harnesses. The lifeline must be separate from any material handling lines and must be manned by an attendant whenever an entrant is in the excavation.

Underground Construction 29 CFR 1926.800

OSHA Standard 29 CFR 1926.800 applies to construction of underground tunnels, shafts, chambers and passageways. Underground construction, like trenching, presents a variety of hazards similar to those associated with "classic" confined spaces. To address these hazards, the underground construction standard includes very detailed provisions for employee safety.

Employee Training

Employees must be instructed in how to recognize and avoid underground construction hazards. When appropriate, the training program must include discussion of:

- Air monitoring procedures
- Ventilation methods
- Communications procedures
- Flood control
- Hazardous mechanical equipment
- Personal protective equipment
- Use of explosives
- Fire prevention
- Emergency procedures

Atmospheric Testing

A competent person must test the air for oxygen, combustibles and toxic containments as often as necessary to assure that conditions are suitable for work. When rapid excavation machines are used, a continuous flammable gas monitor must also be operated at the face of the excavation, with the sensor placed as high and as close as practical to the machine's cuter head.

If air monitoring indicates hydrogen sulfide (H_2S) concentrations greater than 5 ppm, a test for H_2S must be conducted at least at the beginning and in the middle of each shift until the H_2S concentration is shown to be less than 5 ppm for 3 consecutive days. Whenever H_2S levels exceed 10 ppm, workers must be informed and continuous monitoring must be implemented. When H_2S levels exceed 20 ppm, monitors must be installed to provide a visual and aural alarm that signals the need for additional protective measures such as respirators, increased ventilation or evacuation.

Ventilation

At least 200 cubic feet per minute of fresh air must be provided for each employee underground. The direction of air flow must be reversible, and flow rates of at least 30 feet per minute must be maintained in areas where blasting or rock drilling is performed and in other areas where harmful dusts, fumes, mists vapors or gases are produced. After blasting, ventilation must be provided for a sufficient period of time to exhaust smoke and contaminants from the tunnel or shaft.

Communications

A powered communications system must be provided whenever natural unassisted voice communication is ineffective between the work face, the bottom of the shaft and the surface. All shafts which are being developed, or which are used for personnel access or for hoisting, must be equipped with two effective means of communication, at least one of which must be a voice system.

Powered communications systems must be operated on an independent power supply and must be installed so that disruption of one phone or signal location does not affect the operation of the entire system. Communications systems must be tested at the beginning of each shift and as often as necessary during the shift to ensure that they work.

Above-Ground Attendant

A designated above-ground attendant must be provided. The attendant is responsible for controlling access to the job site, keeping an accurate count of employees underground through a check-in/check-out system and securing immediate aid in the event of an emergency.

Rescue Teams

At least one 5-person rescue team must be available within 30 minutes travel time to the tunnel portal. On job sites with more than 25 people working underground at any one time, a second 5-person team must available within 2 hours travel distance.

Underground Lines: 29 CFR 1926.956

Street manholes or unvented vaults must be protected with a barrier, temporary cover or other suitable guard. Manholes must be ventilated and tested for oxygen and combustible gases. A continuous supply of forced ventilation must be provided when employees are working in the manhole.

OSHA MARITIME STANDARDS

Confined space hazards are addressed in 3 subparts of OSHA maritime standards 29 CFR 1915. These subparts and the specific confined space related standards they include are

Subpart B: Explosive and Other Dangerous Atmospheres

- 29 CFR 1915.12 *Precautions Before Entering*
- 29 CFR 1915.13 *Cleaning and Other Cold Work*
- 29 CFR 1915.14 *Certification Before Hot Work Begins*
- 29 CFR 1915.15 *Maintaining Gas-free Conditions*

Subpart C: Surface Preparation and Preservation

- 29 CFR 1915.32 *Toxic Cleaning Solvents*
- 29 CFR 1915.33 *Chemical Paint and Preservative Removers*
- 29 CFR 1915.35 *Painting*
- 29 CFR 1915.36 *Flammable Liquids*

Subpart D: Welding, Cutting and Heating

- 29 CFR 1915.51 *Ventilation and Protection in Welding and Cutting*
- 29 CFR 1915.53 *Welding, Cutting and Heating in Way of Preservative Coatings*
- 29 CFR 1915.54 *Welding, Cutting and Heating of Hollow Metal Containers and Structures*
- 29 CFR 1915.55 *Gas Welding and Cutting*
- 29 CFR 1915.56 *Arc Welding and Cutting*

The specific requirements contained in each of hese subparts are summarized below.

Subpart B: Explosive and Dangerous Atmospheres

Regulations in Subpart B require that a competent person test shipboard confined spaces for oxygen and combustibles prior to entry. The spaces must also be cleaned of flammable or toxic materials as thoroughly as practical, and ventilation must be provided to keep vapor concentrations of any cleaning solvents to less than 10% of the lower explosive limit (LEL). Exhausted vapors must be kept from accumulating around vessels, docks or piers. "No Smoking" signs must be posted, and metal parts of air moving devices and all duct work must be bonded to the vessels deck. Employees are not allowed to perform hot work on pipe lines or potentially hazardous areas such as fuel tanks, heating coils, pumps, fittings and other appurtenances on cargo tanks that contain flammable liquids or gases until a certificate has been issued by a marine chemist.

Subpart C: Surface Preparation

Subpart C requires that either natural ventilation or mechanical exhaust be used to remove vapors. Vapor concentrations must also be kept within acceptable levels for the duration of the work.

Hardened preventative coatings may not be removed by flames unless employees are protected by air-supplied respirators. Employees working in confined spaces who are continuously exposed to paint sprays containing toxic vehicles or solvents must also be protected with air-line respirators.

Sufficient exhaust ventilation must be provided to keep the concentration of solvent vapors below 10% of the LEL, and a competent person must test frequently enough to insure concentrations remain within acceptable levels.

Exhaust ducts must discharge clear of all working areas and away from sources of possible ignition. Periodic tests must be made to ensure that the exhausted vapors are not accumulating in other areas within or around the vessel or dry dock.

If the ventilation system fails, or if the concentration of solvent vapors exceeds 10% of the LEL, painting must be stopped and the space evacuated until the concentration falls below 10% of the LEL. If the concentration does not decrease when painting stops, additional ventilation must be provided.

Ventilation must be continued after the completion of the painting until the space is gas-free. The final determination as to whether or not it is gas-free is made after the ventilation has been shut down for at least 10 minutes.

Subpart D: Welding, Cutting and Heating

When welding is performed in shipboard confined spaces, sufficient mechanical ventilation must be provided to keep fume concentrations within acceptable levels. Contaminated air must be exhausted from the work space and discharged clear of the source of intake air. When sufficient ventilation cannot be achieved without blocking the means of egress, employees in the space must be equipped with air-supplied respirators. In this case, an employee must also be stationed outside the space to maintain communication with those inside and to assist them in an emergency.

Local exhaust or air-supplied respirators must be provided if welding, cutting or heating is performed on lead-, cadmium- or mercury-coated base metals. Work on beryllium-containing metals must be performed using both local exhaust and air-supplied respirators.

SUMMARY AND CONCLUSIONS

While confined spaces may differ in their size, shape and function they share many common elements. In general, they are not usually intended for continuous human occupancy, have limited means of egress and entry, and may be subject to physical and atmospheric hazards. Confined spaces may be classified as permit-required spaces or non-permit spaces.

Non-permit confined spaces are confined spaces that do not actually or potentially contain hazards that could cause death or serious physical harm. Examples of non-permit spaces include: drop ceilings, mechanical equipment closets and motor control cabinets.

Permit-required confined spaces, on the other hand, are confined spaces which are hazardous to enter unless special precaution are taken. However, if all of the hazards posed by a permit space can be eliminated, the space may be reclassified as a non-permit space.

For a confined space to be considered a permit-required space it must possess one or more of the following characteristics:

- Actual or potentially hazardous atmospheres
- Material that can engulf an entrant
- An internal configuration that could trap or asphyxiate an entrant due to inwardly converging walls, or floors which slope downward and taper to a smaller cross-sectional area
- Recognized serious safety or health hazards

Workers who enter permit-required confined spaces may be exposed to a wide spectrum of potential hazards. Some of these hazards may not be readily apparent, others may not be present all of the time. Numerous case histories have demonstrated that failure to recognize and control these hazards often results in fatal accidents. The death toll is further increased by unsuccessful rescue attempts that claim the lives of those trying to render aid.

The most comprehensive OSHA standard related to confined space entry is 29 CFR 1910.146. Among other things, this standard requires a comprehensive written program for characterizing and controlling the hazards associated with entry.

It should be noted though that OSHA construction and maritime standards also address confined space entry. Exact regulatory requirements vary from industry to industry, but most confined space standards include provisions for atmospheric testing, ventilation, control of hazardous energy and emergency planning.

Although the construction standards do not contain specific regulations governing confined space entry, requirements imposed by standards for trenches, excavations, tunnels and manholes all include provisions that address confined space-related hazards.

OSHA maritime standards also recognize the hazards posed by confined space entry and, as such, contain very detailed requirements for confined work performed on board ships, barges and other seagoing vessels.

The next four chapters will discuss the nature of atmospheric and physical hazards that may exist in confined spaces. Building on this foundation, Chapter 6 will explain the elements of the comprehensive entry program required by 29 CFR 1910.146. Proceeding from there, Chapters 7 though 13 will elaborate on selected program requirements such as control of hazardous energy, ventilation, atmospheric testing, permits, personal protective equipment and emergency planning. Finally, Chapter 14 will explain how a confined space program can be implemented through effective employee training.

REFERENCES

Amos, D.J., *Labor's View on Confined Space,* paper presented at the National Safety Council Congress and Exposition, Las Vegas, NV, October 1990.

American National Standard for Safety Requirements for Confined Spaces, ANSI Z117.1-1989 (New York, NY: American National Standards Institute, 1989).

Associated Press, Despite Tragedy Farming Family Goes On, *The Herald Mail,* Sunday August, 13, 1989.

Arc Welding and Cutting, Code of Federal Regulations, Vol. 29, Part 1910.254.

Arc Welding and Cutting, Code of Federal Regulations, Vol. 29, Part 1915.56.

Certification Before Hot Work Begins, Code of Federal Regulations, Vol. 29, Part 1915.14.

Chemical Paint and Preservative Removers, Code of Federal Regulations, Vol. 29, Part 1915.33.

Cleaning Petroleum Storage Tanks, API Publication 2015, (Washington, DC: American Petroleum Institute, 1985).

Cleaning and Other Cold Work, Code of Federal Regulations, Vol. 29 Part 1915.13.

Cleaning Open-Top and Covered Floating Roof Tanks, API Publication 2015B, (Washington, DC: American Petroleum Institute, 1981).

Cloe, W.W., *Selected Occupational Fatalities and Asphyxiating Atmospheres in Confined Spaces as Found in Reports of OSHA Fatality/Catastrophe Investigations* (Washington, DC, U.S. Department of Labor, Occupational Safety and Health Administration, 1985).

Cloe, W.W., *Selected Occupational Fatalities Related to Welding and Cutting as Found in Reports of OSHA Fatality/Catastrophe Investigations* (Washington, DC, U.S. Department of Labor, Occupational Safety and Health Administration, 1988).

Cloe, W.W., *Selected Occupational Fatalities Related to Lockout/Tagout as Found in Reports of OSHA Fatality/Catastrophe Investigations* (Washington, DC, U.S. Department of Labor, Occupational Safety and Health Administration, 1988).

Cloe, W.W., *Selected Occupational Fatalities Related to Fixed Machinery Found in Reports of OSHA Fatality/Catastrophe Investigations* (Washington, DC, U.S. Department of Labor, Occupational Safety and Health Administration, 1978).

Cloe, W.W., *Selected Occupational Fatalities Related to Ship Building and Repairing as Found in Reports of OSHA Fatality/Catastrophe Investigations* (Washington, DC, U.S. Department of Labor, Occupational Safety and Health Administration, 1990).

Confined Spaces, *California General Industry Safety Orders,* Title 8, Article 108 Section 5156.

Confined Spaces, *Virginia Regulation,* 425-02-12.

Confined Spaces, *Washington Administrative Code,* Chapter 296-62-145.

Confined Space Entry, *Kentucky Occupational Safety and Health Standards,* 803 KAR 2:200.

Confined Space Entry, *Michigan Occupational Safety and Health Standards for General Industry,* Part 90.

Comments of the National Institute For Occupational Safety and Health on the Occupational Safety and Health Administration's Notice of Proposed Rulemaking on Permit-required Confined Spaces, U.S. Department of Health and Human Services, Public Health Service, October 4, 1989.

Entering Tanks, Vats and Similar Confined Spaces, *Oregon Administrative Rules,* Chapter 437, Division 41, Section 175.

Gas Welding and Cutting, Code of Federal Regulations, Vol. 29, Part 1915.55.

Grain Handling Facilities, Code of Federal Regulations, Vol. 29, Part 1910.272.

Guidelines for Confined Space Work in the Petroleum Industry, API Publication 2217, (Washington, DC: American Petroleum Institute, 1984).

Guide for Controlling the Lead Hazard Associated with Tank Entry and Cleaning, API Publication 2015A, (Washington, DC: American Petroleum Institute, 1982).

Hot Work, Code of Federal Regulations, Vol. 29, Part 1926.14.

Maintaining Gas-Free Conditions, Code of Federal Regulations, Vol. 29, Part 1915.15.

Maryland Occupational Safety and Health Standard for Confined Spaces, Code of Maryland Regulations, Section 09.12.33.

NIOSH Alert Request for Assistance in Preventing Worker Deaths of Farm Workers in Manure Pits (Cincinnati, OH: National Institute for Occupational Safety and Health, DHHS/NIOSH Publication 90-103, 1990).

NIOSH Alert Request for Assistance in Preventing Entrapment and Suffocation Caused by the Unstable Surfaces of Stored Grain and Other Materials (Cincinnati, OH: National Institute for Occupational Safety and Health, DHHS/NIOSH Publication No. 88-102, 1987).

Oxygen-Fuel Gas Welding and Cutting, Code of Federal Regulations, Vol. 29, Part 1910.253.

Painting, Code of Federal Regulations, Vol. 29, Part 1915.36.

Permit-Required Confined Spaces, Code of Federal Regulations, Vol. 29, Part 1910.146.

Pettit, T.A., P.M. Gussey, and R.S. Simmons, *Criteria for a Recommended Standard: Working in Confined Spaces* (Cincinnati, OH: National Institute for Occupational Safety and Health, DHEW/NIOSH Publication 80-106, 1979).

Pettit, T.A. and H. Linn, *A Guide to Safety in Confined Spaces,* (Cincinnati, OH: National Institute for Occupational Safety and Health, DHHS/NIOSH Publication No. 87-113, 1987).

Precautions Before Entering Compartments and or Spaces, Code of Federal Regulations, Vol. 29, Part 1915.12.

Precautions Before Entering, Code of Federal Regulations, Vol. 29, Part 1915.12.

Pulp, Paper and Paperboard Mills, Code of Federal Regulations, Vol. 29, Part 1910.261.

Safety Training and Education, Code of Federal Regulations, Vol. 29, Part 1910 1926.21.

Telecommunications, Code of Federal Regulations, Vol. 29, Part 1910.268.

Textiles, Code of Federal Regulations, Vol. 29, Part 1910.262.

Toxic Cleaning Solvents, Code of Federal Regulations, Vol. 29, Part 1915.32.

Trenching, Code of Federal Regulations, Vol. 29, Part 1926.651.

Underground Construction, Code of Federal Regulations, Vol. 29, Part 1926.800.

Underground Lines, Code of Federal Regulations, Vol. 29, Part 1926.956.

Ventilation, Code of Federal Regulations, Vol. 29, Part 1910.94.

Ventilation and Protection in Welding, Cutting and Heating, Code of Federal Regulations, Vol. 29, Part 1915.51.

Welding, Cutting and Heating in Way of Preservative Coatings, Code of Federal Regulations, Vol. 29, Part 1926.354.

CHAPTER 2

Oxygen-Deficient and Oxygen-Enriched Atmospheres

INTRODUCTION

We can survive for weeks without food, days without water, but only minutes without air. But the air we breathe is not a single substance, it is a mixture of gases. While it consists mostly of nitrogen (78%) and oxygen (21%), other gases, including argon, neon, krypton, carbon dioxide and hydrogen, are also present in small quantities (Table 2-1). Of these gases, oxygen alone is essential for life.

Oxygen is consumed by a variety of cellular chemical processes which produce carbon dioxide as a waste. While all cells die when deprived of oxygen, some are more critically affected than others. Brain cells, for example, begin to die within 4–6 minutes of being deprived of oxygen. Since the brain cannot produce new cells, the damage is permanent. Skin and liver cells, on the other hand, are less critically affected by a diminished oxygen supply because of their ability to generate new cells to replace those that die.

Oxygen levels must be maintained within well-defined limits, and either too little or too much of it in a confined space can be disastrous. The effects of too little oxygen are largely health-related, with symptoms ranging from a mild headache to permanent brain damage depending on the degree of oxygen deficiency. An overabundance of oxygen in a space increases fire risks. Although oxygen is not combustible itself, oxygen-enriched atmospheres alter the burning characteristics of combustible materials making them easier to ignite and faster burning once ignited.

In this chapter we will explore some of the ways that oxygen-deficient and oxygen-enriched atmospheres may be created in confined spaces. Later, in Chapters 8 and 9, we will learn how oxygen-related hazards can be evaluated and controlled using instrumentation and ventilation.

OXYGEN-DEFICIENT ATMOSPHERES

The physiological effects associated with different oxygen concentrations are illustrated in Figure 2-1. This illustration, as well as the OSHA regulation and many confined-space guidance documents, indicates that 19.5% oxygen by volume is the lowest level acceptable for entry. While this value is widely touted as a handy rule-of-thumb, establishing an acceptable oxygen level *solely* on the basis of volume-percent is overly simplistic and, in some cases, perhaps even dangerous. In order to understand why this is so, we must first know how oxygen is transferred from the lungs to the bloodstream.

The Respiratory System

The respiratory system (Figure 2-2) consists of a single airway that branches into smaller and smaller passages similar to the roots of a tree. It is traditionally divided into two sections: the upper respiratory tract, which extends from the nose to the larynx (vocal cords); and the lower respiratory tract, which extends from the larynx to the alveoli.

The alveoli are small grape-like clusters at the end of the respiratory tract (Figure 2-3). It is at this level that the exchange of gases occurs. Each of the alveoli is separated from the blood-carrying capillaries by a barrier only two cells thick. Since this barrier is permeable to gases, oxygen and carbon dioxide can readily pass through it. The driving force for this gas exchange is the pressure difference that exists on opposite sides of the alveolar walls. This pressure differential allows oxygen to flow from the alveolar spaces into the blood, while carbon dioxide flows from the blood into the alveolar spaces.

Partial Pressure Influences

Normal atmospheric air at sea level has a pressure of 760 millimeters of mercury (mm Hg). Since air contains approximately 21% oxygen, oxygen's contribution to the total pressure, in other words its *partial pressure,* is 21% of 760 mm Hg or about 159 mm Hg.

As fresh air enters the upper respiratory tract, it is humidified. The humidified air now contains water vapor which lowers the partial pressure of oxygen. Since the vapor pressure of water at body temperature is about 47 mm Hg, the partial pressure of oxygen in the upper respiratory tract can be calculated as shown below:

Atmospheric pressure mm Hg	-	Pressure of water vapor mm Hg	×	Decimal fraction of oxygen in atmospheric air	=	Partial pressure of oxygen mm Hg
(760 mm Hg - 47 mm Hg)			×	0.21	=	150 mm Hg

Table 2-1. Composition of air

Constituent	% by Volume
Nitrogen	78.03
Oxygen	20.99
Argon	0.94
Carbon dioxide	0.03
Hydrogen	0.01

Adapted from *Handbook of Compressed Gases,* Van Nostrand Reinhold (1966).

Once in the alveolar spaces, the oxygen partial pressure is further affected by carbon dioxide which has passed into the alveoli from the bloodstream. In the average person, the pressure of carbon dioxide in the alveoli is about 40 mm Hg. This means that oxygen partial pressure is lowered by another 40 mm Hg. In other words, it goes from 150 to 110 mm Hg as indicated in Figure 2-4.

Once oxygen gets into the blood it attaches to hemoglobin molecules which carry it to the cells. At an alveolar partial pressure of 110 mm Hg, hemoglobin is saturated with oxygen, which means that it's carrying about all the oxygen it can. However, the saturation level is affected by the alveolar partial pressure and, as indicated in Figure 2-5, a decrease in the oxygen partial pressure produces a corresponding decrease in hemoglobin saturation. It is important to note at this point that physiologists generally agree that the effects of oxygen deficiency begin to manifest at partial pressures of about 60 mm Hg.

Relevance to Confined Space Entry

So what's all this got to do with confined spaces you might ask? Well, the oxygen partial pressure inside a confined space may be lower than the 159 mm Hg found in ambient air. This lowering could result from two things.

First, inert gases like argon and nitrogen may enter a space and displace some of the atmospheric air. When this happens, the amount of oxygen, and thus its partial pressure, goes down. For example, assume that nitrogen leaks into the space lowering the oxygen level to 19.5%. The oxygen partial pressure is now 19.5% of 760 mm Hg or 148 mm Hg. When we subtract out the partial pressure contributions of water vapor and carbon dioxide, the oxygen partial pressure in the alveolar spaces is now down to about 100 mm Hg.

Since the hemoglobin saturation point is 110 mm Hg, the blood is not quite carrying the optimum quantity of oxygen. However, a partial pressure of 100 mm Hg is still 40 mm greater than the 60 mm physiological danger point.

The second, and perhaps more subtle, way that oxygen partial pressure may be lowered is through increases in altitude (Table 2-2). High altitude air contains the same percentage of oxygen as air at sea level, 21%, but the barometric pressure at altitude is less than it is at sea level. The barometric pressure at 5,000 feet, for example, is 632 mm Hg vs. 760 at sea level. That means that the oxygen partial pressure at 5,000 feet is about 133 mm Hg vs. 160 mm Hg at sea level (21% of 632 mm Hg = 133 mm Hg). If we again subtract out the contribution for water vapor and carbon dioxide, we will find that alveolar oxygen partial pressure at this altitude is about 83 mm Hg, vs. 110 mm Hg at sea level. However, at an oxygen level of 19.5%—the level widely touted as "safe for entry"—the oxygen partial pressure in the alveoli drops to about 74 mm Hg. Recalling that the effects of oxygen deficiency will generally manifest at 60 mm Hg, it is clear that the margin of safety under these conditions has narrowed considerably.

While this discussion may appear to be "academic," the effects of decreased oxygen partial pressure becomes an important consideration in a few special circumstances. For example, consider a coastal area tank-cleaning crew that lands a contract to clean tanks in Denver or Salt Lake City. The work crew, unlike the residents of these areas, is not acclimatized to the "thinner air."

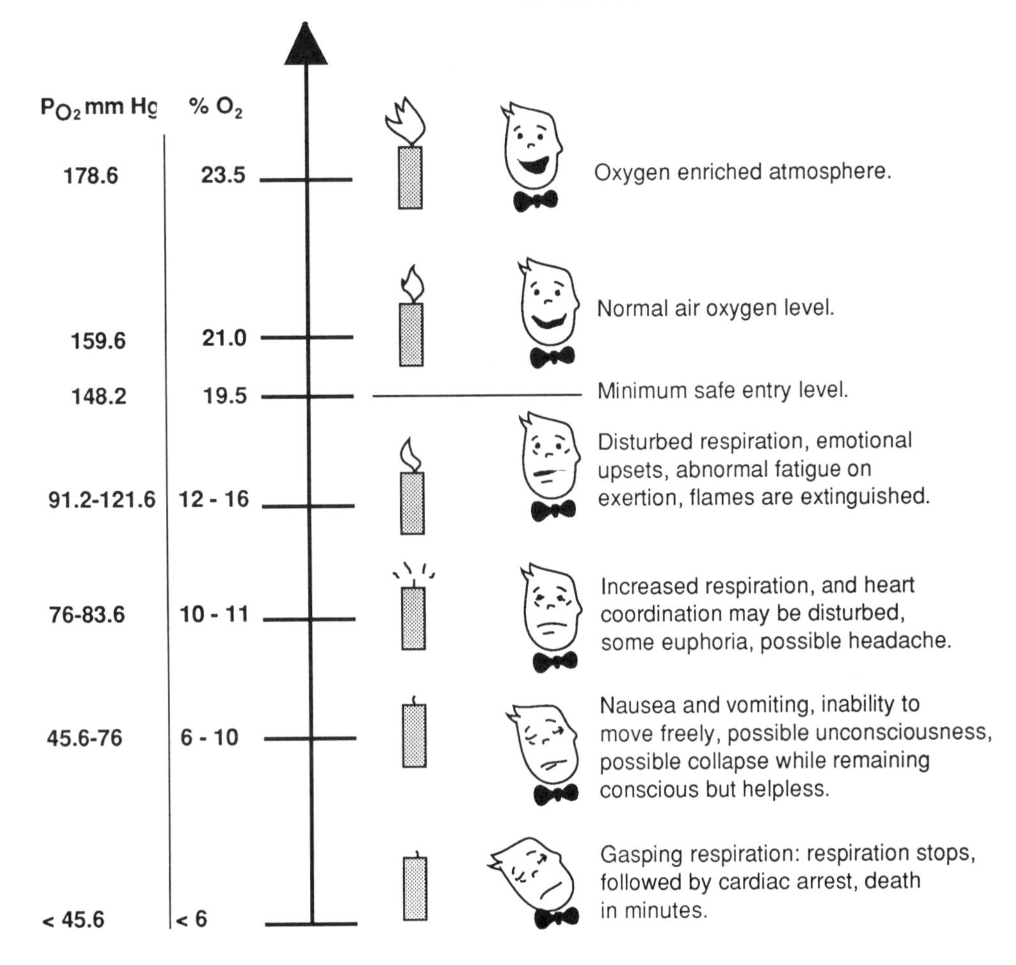

Figure 2-1. Physiological effects of oxygen at different levels. (Paul Trattner)

Consequently, after only mild exertion, they may suffer a variety of adverse effects, including reduced peripheral vision, abnormal fatigue and shortness of breath. While these impairments are inconsequential in ordinary environments, they could impede escape or contribute to accidents in a confined space.

CAUSES OF OXYGEN DEFICIENCY

Oxygen-deficient atmospheres may occur in confined spaces as a result of ambient oxygen being

- *Consumed* by chemical reactions like oxidation (rusting and naturally occurring fermentation)
- *Displaced* by inert gases like argon, carbon dioxide and nitrogen
- *Adsorbed* by porous surfaces like activated charcoal

Oxygen Deficiency Through Consumption

Oxygen may be consumed by industrial processes like welding, torch cutting and brazing that utilize open flames. Oxygen may also be consumed by fuel-fired space heaters that are used for comfort heating or to warm the air to the temperature required to cure some protective coatings.

Space heaters require a certain amount of make-up air to replace that which is consumed by combustion. If adequate make-up air is not provided, the combustion process will be less efficient, resulting in increased carbon monoxide production. As we will see in Chapter 4, carbon monoxide interferes with the body's ability to transport oxygen since it combines with hemoglobin about 200 times more easily than oxygen. Entrants are then faced with the double-barreled problem of both oxygen deficiency and carbon monoxide.

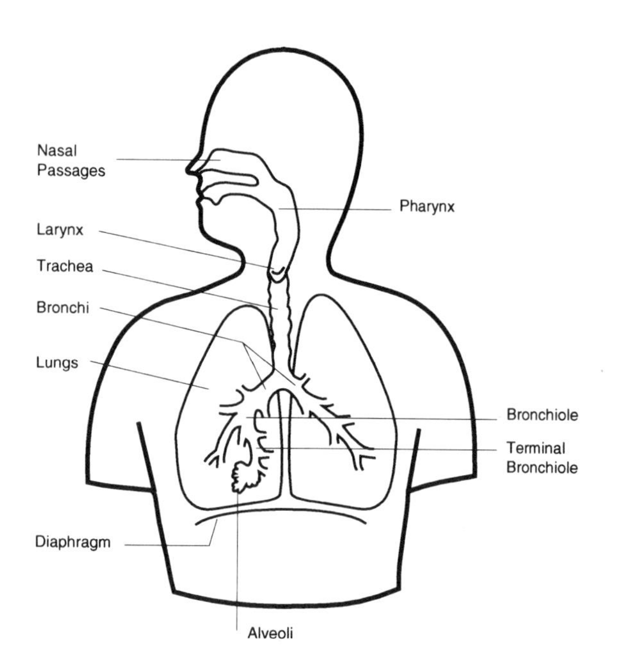

Figure 2-2. The respiratory system. (Paul Trattner)

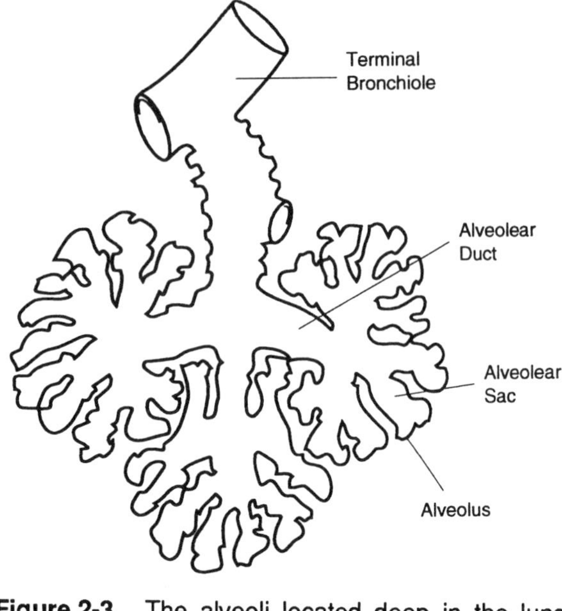

Figure 2-3. The alveoli located deep in the lungs play a critical role in the exchange of oxygen and carbon dioxide. (Paul Trattner)

Oxygen Consumption by Rusting

Oxygen may also be consumed by naturally occurring chemical reactions. Rusting and organic fermentation provide two classic examples of oxygen consumption by this mechanism.

Iron oxide, commonly called rust, is formed when moisture and oxygen in air combined with iron or steel

$$2Fe + 3O_2 \xrightarrow{\text{Moisture}} 2Fe_2O_3$$

While rust is frequently considered an unwelcome nuisance, its formation in a confined space may lead to fatal consequences. One case suggestive of oxygen consumption by this mechanism is presented below.

Case 2-1: Oxygen Deficiency Due to Rusting

A worker was assigned the task of chipping some concrete away from a below-ground water valve so that a control key could be inserted into it from street level. A witness later reported seeing the worker lift the manhole cover and climb down the rebar ladder into the 4 foot square by 11 foot deep valve vault.

The worker immediately climbed back out, paused and then climbed down again. When he reentered, he was overcome and fell to the bottom of the vault. Even though rescue services arrived promptly, the worker was pronounced dead shortly after arrival at the hospital.

Subsequent investigation revealed that the space was dry, free of debris and well-illuminated by natural light. However, heavy accumulations of rust were noted on both the steps of the rebar ladder and on the 24-inch cast iron valve. Measurements taken 2 hours after the incident showed oxygen levels between 16 and 17%. Although this level is below that required for "safe" entry, it should not have proved fatal (see Table 2-2).

However, if the oxygen level was 16–17% *two hours after the incident,* it seems reasonable to believe that it was *even lower* when the worker entered since ambient air would have diffused into the vault over the intervening two hours. While the actual oxygen concentration will never be known, it is conceivable that it could have been as low as 6–10%, explaining the employee's sudden loss of consciousness.

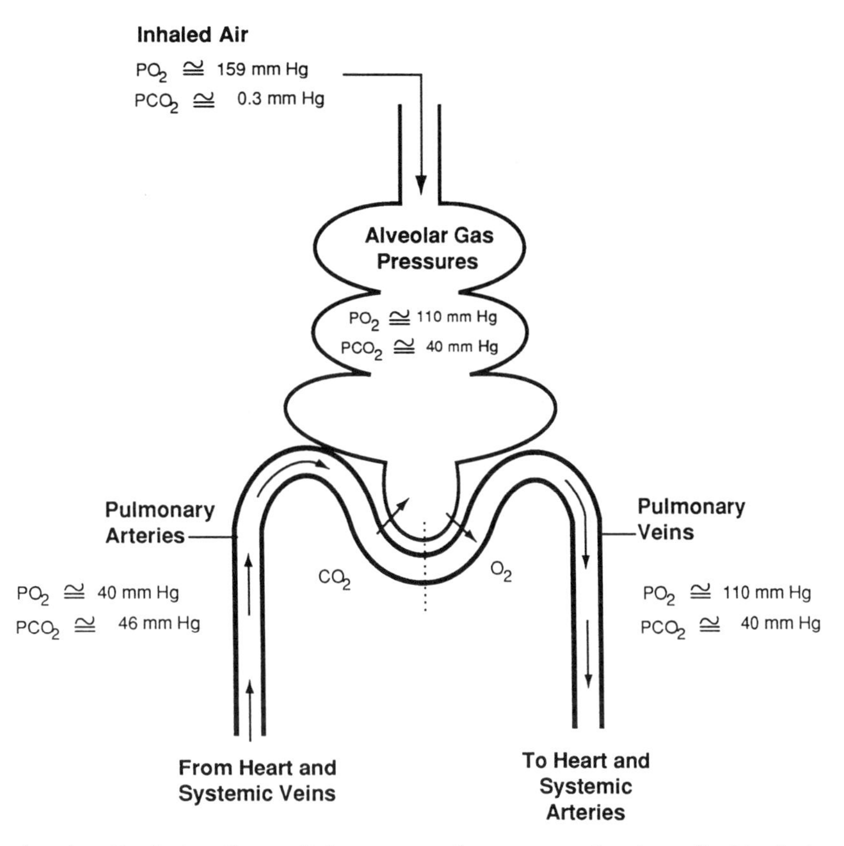

Figure 2-4. This drawing illustrates the partial pressure of oxygen and carbon dioxide during gas exchange in the lungs. (Paul Trattner)

The medical examiner's report shed no further light on this hypothesis but indicated instead that there was no evidence to suggest either a heart attack or carbon monoxide poisoning. The report ultimately listed the cause of death as "undetermined."

Having no other clues upon which to base their opinion, investigators concluded that the most probable cause was oxygen deficiency created by the heavy rust formation on the cast iron pipe and rebar ladder.

Case 2-2: Oxygen Depletion Through Corrosion

Two chemical plant maintenance workers were assigned the task of repairing a leaking steam line in a magnesium chloride ($MgCl_2$) brine evaporator (Figure 2-6). This was a routine job which had been performed without incident many times during the previous 35 years.

In preparation for entry, the evaporator, which measured 12 feet in diameter and 54 feet high, was washed, drained and vented. It was then closed until the day it was scheduled for repair. At that time, only the top manhole was removed, even though the plant's entry procedures specified that the bottom manhole also be opened to help cool and dehumidify the space.

The first worker entered and collapsed shortly thereafter. The second worker entered in an attempt to rescue the first and was also overcome. The second entrant died but the first recovered completely.

Measurements made about an hour and a half after the incident showed oxygen levels of about 12%. Since inspection of the piping system and analysis of gas samples ruled out the possibility of displacement by an inert gas, investigators theorized that the oxygen deficiency may have been produced by rapid corrosion of the steel. They tested this theory by placing pieces of the steam tubes in a glass vessel that simulated the evaporator. Analysis of the air two

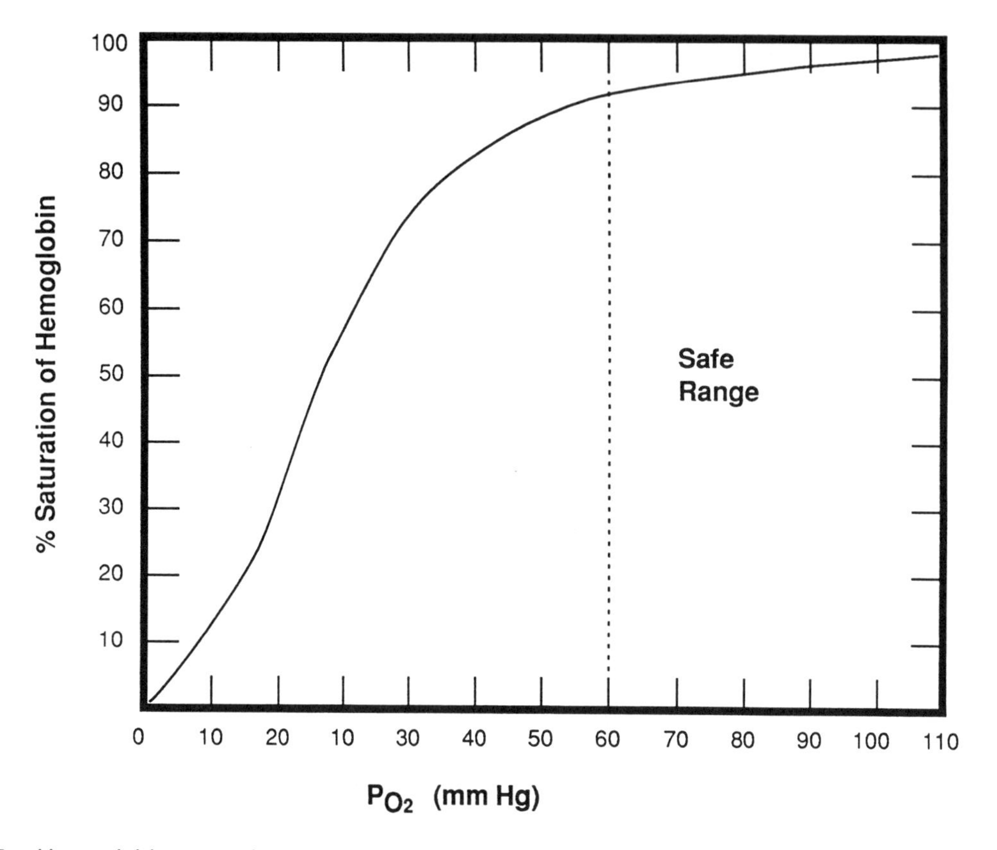

Figure 2-5. Hemoglobin saturation curve. (Paul Trattner)

days later showed that the oxygen level had dropped to 15%.

Subsequent full-scale tests showed that oxygen levels dropped to 10% after only 8 hours, and to less than 2% after 16 hours. The mechanism investigators proposed to explain the reaction is

$$Fe_3O_4 \text{ and/or } Fe \xrightarrow{Cl^-,\ O_2} FeOCl$$

$$FeOCl \xrightarrow[OH^-]{MgCl_2} 4Mg(OH)_2 \cdot FeOCl \cdot XH_2O$$

Additional experiments found that high corrosion rates resulting in an oxygen deficiency could also be produced with other salt solutions including KCl, NaCl, $CaBr_2$ and $MgSO_4$.

Oxygen Consumption by Fermentation

Fermentation is an enzymatically controlled chemical reaction in which microorganisms digest organic matter. In some cases, like grain fermentation in an oxygen-limiting silo, ambient oxygen is consumed through a reaction that produces nitrogen dioxide as a by-product. In other cases, oxygen-consuming microorganisms which are ubiquitous in the environment produce carbon dioxide as a metabolic waste

$$\text{Organic Matter} + O_2 \xrightarrow[\text{enzymes}]{\text{microrganisms}} CO_2$$

Since nitrogen dioxide and carbon dioxide are both about one and a half times heavier than air, they sink to low points in a space, further reducing the oxygen content by displacing the ambient air.

Two representative cases of oxygen deficiencies created by fermentation are summarized below.

Case 2-3: Fermentation of Organic Debris

A laborer entered a 20-foot deep storm drain to remove grass clippings, leaves and other debris that

Table 2-2. Variation oxygen partial pressure with altitude

Altitude (feet)	Barometric Pressure (mm Hg)	Oxygen Partial Pressure (mm Hg)
Sea level	760.0	159.0
1,000	732.9	153.5
2,000	706.7	148.0
3,000	681.1	142.7
4,000	656.4	137.5
5,000	632.4	132.5
6,000	609.1	127.6
7,000	586.9	122.9
8,000	564.6	118.3
9,000	543.7	113.9
10,000	522.7	109.5

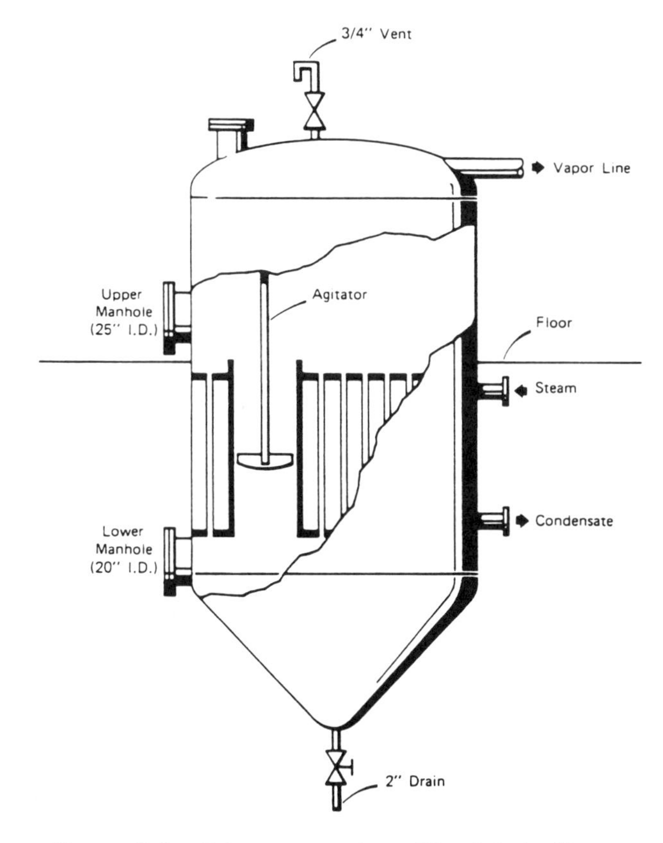

Figure 2-6. Brine evaporator. (Reprinted with permission, *American Industrial Hygiene Association Journal* Vol. 48, No. 6, July 1987, pg. 608.)

were obstructing the flow of high-pressure flushing water. A short time later, a second worker observed the laborer unconscious at the bottom of the manhole. He too entered and was overcome.

A third employee, sensing trouble when he saw the second man enter, ran over and found both workers collapsed at the bottom. He too entered, but must have realized that he made a mistake, because he was later seen trying to crawl out the top.

Unfortunately, just as another onlooker ran to help him over the edge of the manhole, he too fell to the bottom. The fire department eventually arrived and recovered the three bodies. However, the rescue was not without further incident. The fire captain involved in recovering the bodies was almost overcome when he removed the facepiece of his self-contained breathing apparatus to more clearly bark orders to top-side firefighters.

Atmospheric tests conducted about an hour after the accident showed 1–2% oxygen and 120,000 parts per million (ppm) carbon dioxide. Investigators attributed these results to fermentation of the grass and leaves which had accumulated in the space.

Case 2-4: Molasses Fermentation

A feed mill employee set out to repair a defective motor in an underground molasses pit. The pit was 21 feet long, 7 feet deep and 10 wide, and contained about 1 to 2 feet of molasses. The worker entered through a single 21-inch square manway and immediately called out for help. A manager and another worker entered in response to the worker's cry. All three were fatally overcome.

Measurements made 10 days later showed that the atmosphere in the pit contained 1% oxygen, 3% carbon dioxide and 5,000 ppm ethanol, strongly suggesting that the molasses had fermented.

Oxygen Deficiency Through Displacement

Inert gases like argon, carbon dioxide and nitrogen are sometimes used to displace air from the head spaces of storage vessels in order to prevent stored products from being damaged by oxidation. Inert gases may also be used to purge pipelines, tanks and vessels of flammable atmospheres prior to "hot-work" like welding, torch cutting and brazing. Finally, as previously discussed, oxygen may be displaced by heavier-than-air gases such as carbon dioxide formed by naturally occurring chemical reactions.

The following two case studies graphically demonstrate how displacement by inert gases may lead to fatal consequences.

Case 2-5: Oxygen Displacement by Nitrogen During Tank Coating

A mechanical contractor's work crew, consisting of 2 pipe fitters and a laborer, was assigned the task of cleaning and coating the inside of a 14-foot high, 39-inch diameter tank. One worker entered the tank to take measurements and was quickly overcome. A second employee entered in an attempt to rescue the first, but he quickly became short of breath and made a hasty retreat. A third worker was then lowered into the tank with an opened cylinder of welding oxygen. He attached a harness to the unconscious worker, who was pulled out by those outside. The harness was then lowered down to the rescuer who was similarly removed.

The rescued employee could not be revived and eventually died of asphyxiation. The two pipe fitters and two company employees who assisted with the rescue were hospitalized, treated and released.

Atmospheric measurements made 3 hours after the incident showed an oxygen level of 9.6%. Investigators also determined that the company had purged the tank with nitrogen four days before the incident, but had not informed the contractor of its condition. No one had tested the tank's atmosphere prior to entry, no lifelines, safety harnesses, or breathing apparatus were provided and an attendant had not been stationed outside the space.

Case 2-6: Oxygen Displacement by Argon in Nuclear Power Plant

After finishing work on a 31-inch diameter vertical pipe connecting a steam generator and a pump, welders in a nuclear power plant purged a 20-foot section of the line with argon. Two pipe fitters were later assigned the job of exhausting the argon. This task consisted of removing foam plugs from both ends of the pipe and using either compressed air or a blower to force the argon out of the pipe.

One of the pipe fitters climbed down a ladder suspended above the pump end of the pipe and pushed the plug with his foot to dislodge it. He removed the plug and handed it up to his coworker. When the coworker looked down again he did not see the other pipe fitter. Being concerned, he climbed down the ladder to investigate. He found his associate unconscious, and because he too was feeling numb and dizzy, he climbed back up the ladder and called for help.

Air was blown into the pump area and a foreman climbed down with a rope. He too was overcome, apparently because insufficient time was allowed for ventilation prior to his entry. Another foreman who entered a short while later was not adversely affected. He tied the rope around the legs of the pipe fitter and the other foreman and they were both pulled out.

All four workers were taken to the hospital where the two foreman and second pipe fitter were treated and released. The first pipe fitter who entered the space died a few days later. The cause of death was determined to be asphyxiation resulting from the air in the pump area being displaced by the argon used to purge the pipe.

Oxygen Deficiency Through Adsorption

Under some conditions the oxygen level in a confined space may be reduced through adsorption by porous materials like activated carbon. One such case is summarized below.

Case 2-7: Oxygen Adsorption at Water Treatment Plant

A newly constructed water filtration tank 17-feet tall and 12½ feet in diameter was half-filled with a slurry of activated carbon and water. The water was then drained off through a bottom outlet and the tank was closed to protect it from the weather. The following morning two workers entered the tank to smooth out the carbon bed and adjust an interior sprinkler mechanism. When they did not appear at lunch time, coworkers went looking for them. Their bodies were found on top of the carbon bed. A rescuer who entered the tank without respiratory protection suffered no ill effects. Subsequent tests indicated that the tank atmosphere contained 21% oxygen and no hydrogen sulfide.

The tank was closed and not reopened again until the next day. Atmospheric tests at that time again revealed no hydrogen sulfide, but, surprisingly, oxygen levels had dropped to 12%. Other tanks in the area were checked. Some which had been closed for several days showed oxygen levels of only 2%.

Investigators discovered that the dry, activated charcoal had no effect on the oxygen level. However, the damp carbon, which had previously been considered to be a non-hazardous material, had apparently selectively adsorbed ambient oxygen, creating a hazardous atmosphere.

OXYGEN-ENRICHED ATMOSPHERES

Oxygen Enrichment Defined

The OSHA confined space standard defines oxygen-enriched atmospheres as those with an oxygen content greater than 23.5% by volume. While not flammable itself, oxygen alters the burning characteristics of many materials, making them both easier to ignite and faster burning once ignited (Table 2-3). Atmospheres enriched with oxygen also permit flammable gases and vapors to ignite over a much wider range of concentrations than is possible in ordinary air.

Causes of Oxygen Enrichment

Oxygen-enriched atmospheres may be inadvertently created in a confined space by poorly designed or malfunctioning oxygen storage and dispensing equipment. They may also result from leakage around couplings, fittings and hoses of oxy-fuel gas welding equipment. Occasionally, they have been created when workers used oxygen to ventilate a space or to power pneumatic hand tools under the mistaken belief that oxygen and air were the same. Fires which have subsequently occurred in these spaces have burned with great speed and intensity.

One oxygen-enriched fire that some readers may recall occurred on January 27, 1967 as the Apollo 13 command module sat on the launch pad at Kennedy Space Center. The command module was pressurized with 100% oxygen and when a fire broke out, the spacecraft's interior was almost instantly consumed by the blaze. All three astronauts on board were killed.

Figures 2-7 and 2-8 illustrate the relative ignition energy levels and flame spread rates for combustible materials in oxygen-enriched atmospheres. Ignition energy provides an indication of how easily materials will ignite, and the flame spread rate indicates how fast they burn once ignited. Both of these characteristics are functions of oxygen concentration and atmospheric pressure, and materials will ignite more easily and burn more quickly as either the oxygen concentration or atmospheric pressure increases.

The following three case studies provide striking examples of the spectacular fires produced by oxygen-enriched atmospheres in confined spaces.

Case 2-8: Oxygen Used to Ventilate Tank During Repair

A bulge formed on a newly constructed tank after the manhole flange was installed by welding. The repair plan called for two men to enter the tank with a jack while a third was to heat the flange from outside. After the two workers entered, the manhole cover was bolted closed. In order to improve the air quality inside, welding oxygen—which the welders called "air"—was blown in through an opening at the top of the tank. A passerby noticed through an opening in the tank wall that the hair of one of the welders was on fire. The manhole cover was opened and one worker scrambled out with his hair and clothing blazing rapidly. The second worker had collapsed and was unconscious. Rescuers had to eventually invert the tank to remove him.

Both welders suffered serious burns. One died a short time later, the other was hospitalized for several months. One rescuer also sustained severe burns to his hands.

Case 2-9: Compressed Oxygen Used to Dislodge Welding Slag

A fire occurred suddenly in a shipboard fuel tank, killing one worker and seriously injuring a second. The tank was new and unused. The employees had entered it to mop up some water and remove welding slag with pneumatic needle guns. A ventilation fan was available, but the workers chose not to use it because of the cold weather.

Investigators found an oxy-propane torch inside the space at the scene of the fire. Since it was not needed for the job, they suspected workers used it to ward off the early morning chill. When they no longer needed it for heat, they apparently used its high-pressure oxygen jet to blow scale off the wall. Oxygen accumulated and when one of the men lit a cigarette, he was immediately engulfed in flames. The second man tried to help him, but he was also engulfed in flames.

Case 2-10: Leaking Welding Oxygen Lines in Barge Wing Tank

Two pipe fitters were instructed to finish installing a 10 foot pipe in a barge wing tank. After stopping by their shop to pick up tools and supplies, they climbed down a ladder to the tank.

Table 2-3. Effects of oxygen on ignition temperature and ignition energy for selected materials

Material	Minimum Ignition Temperature				Minimum Ignition Energy	
	Air		Oxygen		Air (mJ)	Oxygen (mJ)
	°F	°C	°F	°C		
Acetylene	581	305	565	296	0.017	0.0002
Acetone	869	465	—	—	1.15	0.0024
Butane	550	288	532	278	0.25	0.009
Cyclopropane	932	500	849	454	0.18	0.001
Hexane	437	225	424	218	0.288	0.006
Hydrogen	968	520	752	400	0.017	0.0012
Gasoline	824	440	600	316	—	—
Kerosene	440	227	420	216	—	—
Octane	428	218	406	208	—	—
Propyl alcohol	824	440	622	328	—	—

Adapted from NFPA-53M, Manual on Fire Hazards in Oxygen-Enriched Atmospheres, National Fire Protection Association, Quincy, MA (1990).

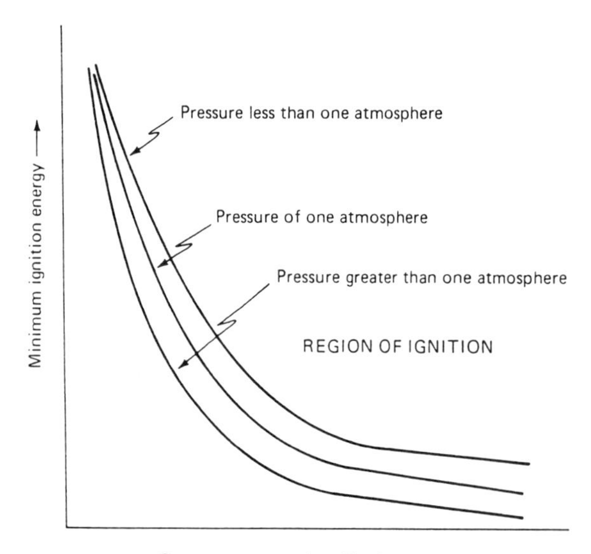

Figure 2-7. Effects of oxygen pressure and concentration on ignition energy. (Reprinted with permission from NFPA 53M, *Fire Hazards in Oxygen-Enriched Atmospheres.* 1990 National Fire Protection Association, Quincy, MA 02269.)

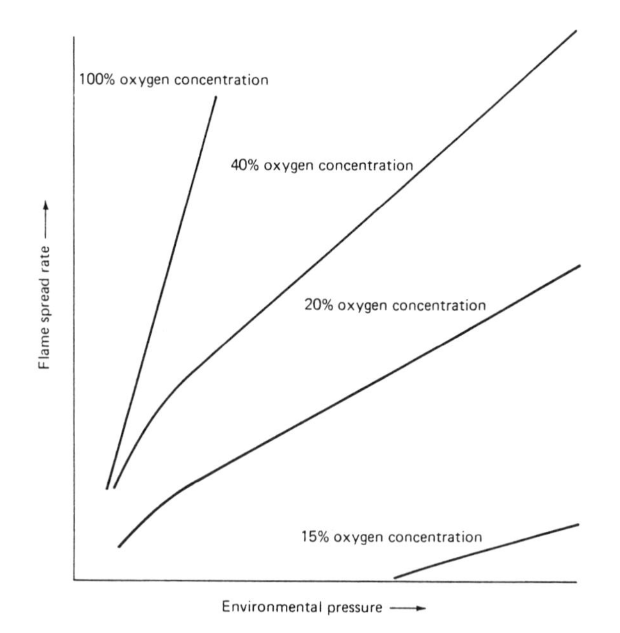

Figure 2-8. Effects of oxygen pressure and concentration on flame spread. (Reprinted with permission from NFPA 53M, *Fire Hazards in Oxygen-Enriched Atmospheres,* 1990, National Fire Protection Association, Quincy, MA 02269.)

A short time later, employees working on another deck saw smoke billowing out of the tank. Three men who were nearby climbed down the ladder to investigate and discovered that the pipefitters' clothes were on fire. When the clothes of the first investigator also started to burn, all three men scurried back up the ladder. When the two other workers attempted to extinguish their companion's burning clothes, he panicked, ran down the deck and was completely enveloped by flames. Both he and the two pipe fitters died later of severe burns.

Investigators determined that oxygen leaked into the tank from the oxy-acetylene hoses left behind by the previous crew of burners. The pipe fitters' clothes apparently became saturated with oxygen and were ignited by either smoking, a spark from their tools or drops of molten metal from welding and flame cutting being done on the deck above.

SUMMARY AND CONCLUSIONS

This chapter focused on two atmospheric hazards that may be encountered in confined spaces: oxygen deficiency and oxygen enrichment. Two other atmospheric hazards, flammable and toxic atmospheres, will be discussed in the following two chapters.

Spaces containing less than 19.5% oxygen (partial pressure 148 mm Hg) are considered to be deficient and those containing more than 23.5% (178 mm Hg) are considered to be enriched. Oxygen deficiencies may be produced by chemical reactions, displacement by inert gases or adsorption by certain porous materials. Oxygen enrichment, on the other hand, results largely from the improper use of welding oxygen in confined spaces.

Many of the accidents described in this chapter's case histories would probably have been prevented by a comprehensive entry program that included provisions for employee training, ventilation and atmospheric testing. Since each of these topics requires an in-depth discussion, they will be treated in separate chapters.

REFERENCES

American National Standard Practices for Respiratory Protection, Z.2-1980 (New York, NY: American National Standards Institute, 1980).

An oxygen-deficient atmosphere created by rapid rusting of an empty vessel, Chementor, *Chem. Eng.* 91(6): 28, March 1984.

Cloe, W.W., Selected Occupational Fatalities Related to Welding and Cutting as Found in Reports of OSHA Fatality/Catastrophe Investigations (Washington, DC, U.S. Department of Labor, Occupational Safety and Health Administration, 1990).

Cloe, W.W., Selected Occupational Fatalities Related to Ship Building and Repairing as Found in Reports of OSHA Fatality/Catastrophe Investigations (Washington, DC, U.S. Department of Labor, Occupational Safety and Health Administration, 1990).

Cloe, W.W., Selected Occupational Fatalities Related to Toxic and Asphyxiating Atmospheres in Confined Work Spaces as Found in Reports of OSHA Fatality/Catastrophe Investigations (Washington, DC, U.S. Department of Labor, Occupational Safety and Health Administration, 1990).

Cloe, W.W., Selected Occupational Fatalities Related to Fire and or Explosions in Confined Work Spaces as Found in Reports of OSHA Fatality/Catastrophe Investigations (Washington, DC, U.S. Department of Labor, Occupational Safety and Health Administration, 1990).

Fire Hazards in Oxygen Enriched Atmospheres, NFPA 53M (Quincy, MA: National Fire Protection Association, 1990).

Lavoisier, A.L., La traite Elementaire de Chemie, 1789, translated by R. Kerr, Elements of Chemistry, facsimile reprint, Dover Publications, Inc. New York, 1965.

Melcher, R.G., C.E. Crowder, J.C. Tou, and D.I. Townsend, Oxygen depletion in corroded steel vessels, *Am. Ind. Assoc. J.* 48(7):608–612 (1987).

Pettit, T. and H. Linn, A Guide to Safety in Confined Spaces, (Cincinnati, OH: U.S. Department of Health and Human Services, Centers for Disease Control, National Institute for Occupational Safety and Health NIOSH/DHHS Publication 87-113, 1987).

Pettit, T.A., P.M. Gussey and R.S. Simmons, Criteria for a Recommended Standard: Working in Confined Spaces (Cincinnati, OH: National Institute for Occupational Safety and Health, DHEW/NIOSH Publication 80-106, 1979).

Shukitt, B.L., R.L. Bruse and D.R. Banderet, Cognitive Performance, Mood States, and Altitude Symptomology in 13–21% Oxygen Environments, U.S. Army Research Institute of Environmental Medicine, Natick, MA, 1988.

CHAPTER 3

Flammable Atmospheres

INTRODUCTION

Fire may be described as the self-sustaining oxidation of a fuel that produces light and heat. It was once believed that only three things had to be present for fire to occur: fuel, oxygen and heat. The simultaneous combination of these three elements was referred to as the fire triangle (Figure 3-1). While the fire triangle served as a working model for decades, recent research now indicates that a fourth element, a self-sustaining chemical reaction, must also be present.

The new model is depicted as a tetrahedron (Figure 3-2) with the phenomenon of fire resulting from the interaction of all four elements. Since fire is the result of an oxidation reaction, it will not manifest unless a sufficient quantity of oxygen is present. But the presence of oxygen alone, however, will not cause ignition. For ignition to occur, the fuel must be exposed to a source of heat. This heat provides the energy needed to initiate the oxidation reaction. Once the reaction starts, the rapidly oxidizing fuel provides additional heat, causing the reaction to sustain itself either until all of the fuel is consumed, or until the oxygen level is reduced to a point where combustion is no longer possible.

In this chapter we will examine some basic principles of fire dynamics and explain how they can be applied to control fire and explosion hazards in confined spaces.

CASES OF EXPLOSIONS IN CONFINED SPACES

Regrettably, the four elements of the fire tetrahedron come together all too often in confined spaces. A few cases where they combined with explosive consequences are discussed below.

Case 3-1: Washing Rail Tankcar

A tank car which previously contained LP gas (propane and butane) was steam cleaned in preparation for welding. Combustible gas measurements indicated a "safe" environment and the exterior of the tank was welded without incident. Several days later, a worker, who was not assigned the task, took the initiative of removing some residue from the car's interior. He reentered the car, used a trouble lamp to inspect the inside walls, and then climbed out to get a water hose.

He inserted the hose and reentered the car. Seconds later there was an explosion. He was killed instantly and two other employees working nearby were hospitalized with severe burns. Since the water had not been turned on at the nozzle, static electricity was ruled out as an ignition source. The cause was ultimately attributed to the non-explosion proof trouble light igniting flammable vapors that evaporated from the sediment at the bottom of the car.

Case 3-2: Explosion After Painting Tank

After a painter finished spraying the interior of a 21-foot long by 11-foot diameter horizontal tank, he passed a ladder and portable electric light through the manhole to a second man standing outside. This employee had just laid the ladder down when he heard a muffled explosion. Turning toward the tank, he saw flames inside and noticed that the painter's clothes were on fire. After being pulled out of the tank, the painter explained that he inadvertently bumped his spray gun on a second lamp inside. When the bulb broke, it ignited the paint vapors. The painter died in the hospital three days later.

Case 3-3: Unapproved Space Heater

The interior of a water tank was sand blasted and sprayed with solvent-based preservative coating. Since the ambient temperature was lower than that required for curing the coating, warm air had to be forced into the space. This was usually done by blowing air into the space through a piece of flexible duct connected to an electric heater. However, this day, the crew discovered that it did not have the ductwork. Over the objections of the workers, the supervisor used a piece of rope to lower the heater into the tank. When he turned on the power switch, the vapors exploded.

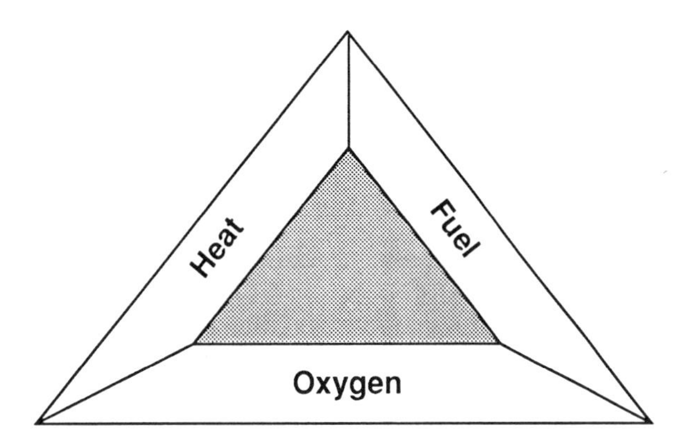

Figure 3-1. The fire triangle. (Paul Trattner)

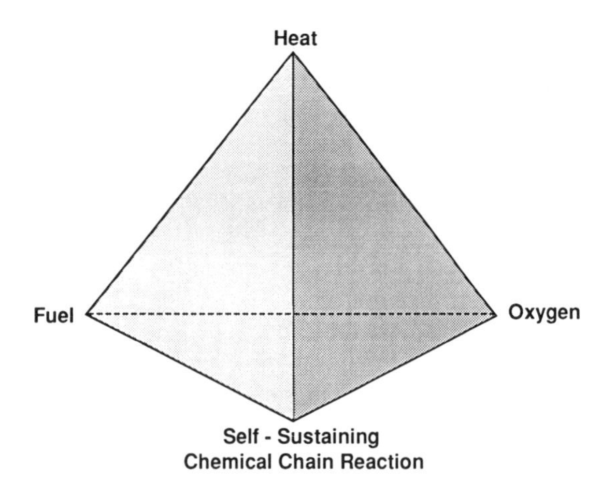

Figure 3-2. The fire tetrahedron. (Paul Trattner)

DETONATIONS AND DEFLAGRATIONS

In addition to producing light and heat, the combustion process also generates gaseous by-products that include oxides of nitrogen, carbon monoxide and carbon dioxide. If a fire propagates quickly enough, the pressure exerted by these rapidly evolving gases can rise to a point where a vessel will burst apart with explosive force. The resulting explosion is classified as a deflagration or detonation, depending on how fast the flame front was moving.

Explosions where the expanding flame front travels slower than the speed of sound (1100 feet per minute) are called deflagrations. Those where the flame fronts velocity equals or exceeds the speed of sound are called detonations. An important practical distinction between these two types of explosions is the fact that the after-effects of detonation tend to be more severe because of the shattering effect they have on most construction materials.

Both detonation and deflagration hazards may exist in confined spaces due to the presence of ignitable quantities of dusts, mists, gases and vapors. However, dusts—except those in long, narrow pipelines—have a greater tendency for deflagration than detonation.

FLAMMABLE GASES AND VAPORS

One of the most frequently encountered sources of explosive atmospheres is the residue left in supposedly "empty" tanks. As shown in Table 3-1, even tanks that are 99% empty may still contain substantial quantities of residue which could evaporate to form an explosive vapor-air mixture.

Many processes like spray painting, adhesive application, solvent cleaning and stress-crack dye penetrant testing employ products that may produce flammable vapors. Flammable liquids and gases may also leak into a space through pipes, valves and fittings attached to the space. Materials that migrate into utility vaults, storm drains and sewers from highway spills and leaking underground storage tanks or pipelines present another source of fuel. It has also been known for centuries that the natural decomposition of organic matter can generate explosive methane gas.

Flammable gases are formed when some metals react with acidic or alkali solutions. For example, both bare steel and galvanized materials (zinc coated) will react with sulfuric acid to produce hydrogen. Hydrogen is also produced when aluminum comes in contact with caustic soda (sodium hydroxide) solutions which are frequently used as cleaning and degreasing agents. Without knowledge of these hazards,

Table 3-1. Residue remaining in tanks that are 99% empty.

Gallons	
Tank Size	**Residue**
50,000	500
40,000	400
30,000	300
20,000	200
10,000	100
5,000	50
2,500	25
1,000	10

employees could unwittingly create an explosive atmosphere when they place galvanized or aluminum ladders in tanks containing acidic or alkaline residues.

Many serious fires and explosions in confined spaces may have been averted if those responsible for managing confined space entries had had a working knowledge of fire chemistry and dynamics. Clearly then, anyone responsible for developing a confined space entry program should have an understanding of basic fire concepts, especially those related to flammable liquids.

Flash Point, Fire Point and Ignition Point

The first thing to understand about flammable liquid fires is that it is not the liquid itself that burns, but rather the liquid's vapors. Furthermore, a flammable liquid will not give off a sufficient quantity of vapor to support combustion until it is heated to a certain temperature called the flash point.

Flash point may be defined more exactly as the lowest temperature at which a liquid produces sufficient vapor to cause a momentary flame in a specific test apparatus. If the liquid's temperature is below the flash point it simply will not produce enough vapor to ignite.

For example, methyl alcohol has a flash point of 52°F. If a match were put to an ice-cold glass of methyl alcohol, the match would go out. However, if the glass were put onto a hot plate and warmed to 52°F, the vapors formed at the mouth would ignite momentarily when the match was brought close to the edge.

Fire point is a temperature slightly above the flash point. At the fire point, the liquid produces vapors at a rate sufficient to sustain steady burning.

Ignition temperature, sometimes called the "auto-ignition temperature" or "auto-ignition point," is the temperature at which a flammable liquid will ignite under its own heat energy. In the case of methyl alcohol, this temperature is 725°F.

As indicated by Figure 3-3, the ignition temperature for straight chain hydrocarbons decreases with increasing carbon chain length. In other words, the auto-ignition temperature of methane > ethane > propane > butane > pentane > hexane > heptane, etc.

Flash Point Test Methods

Flash point determinations are important since they provide a relative index of the degree of hazard

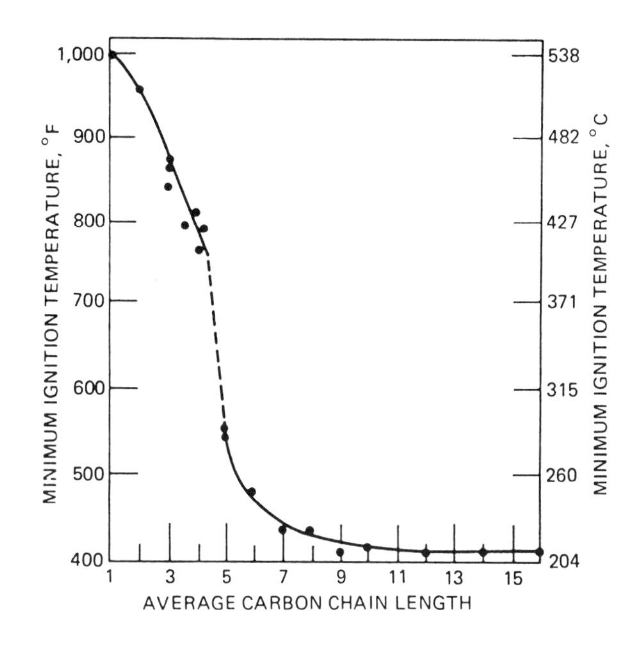

Figure 3-3. Minimum ignition temperatures of hydrocarbons of various carbon chain lengths. (Reprinted with permission from *Fire Protection Handbook,* 17th edition, (1991) National Fire Protection Association, Quincy, MA 02269.)

posed by a flammable liquid. Liquids with low flash points tend to be more hazardous since they also have low boiling points and are consequently more volatile. Since they evaporate more quickly, they form flammable fuel-air mixtures more easily than high flash-point liquids. The four most commonly used flash-point testers are shown in Figure 3-4.

The specific details of conducting flash point determinations for each testing method are provided in the applicable standard of American Society for Testing and Materials, but, in general, the material under test is slowly heated while a flame is applied to the vapor space at periodic intervals. The flash point is the temperature at which a fire flashes briefly when the flame is applied.

It should be noted that the viscosity or "thickness" of the test liquid is a factor that must be considered in selecting the appropriate test method. Examples of high viscosity liquids include paints, adhesives and mastics while gasoline, kerosene and rubbing alcohol are examples of low viscosity liquids. Viscosity is measured in units called "Saybolt Universal Seconds" (SUS). This term stems from the Saybolt Apparatus and stop watch which are used in making viscosity measurements.

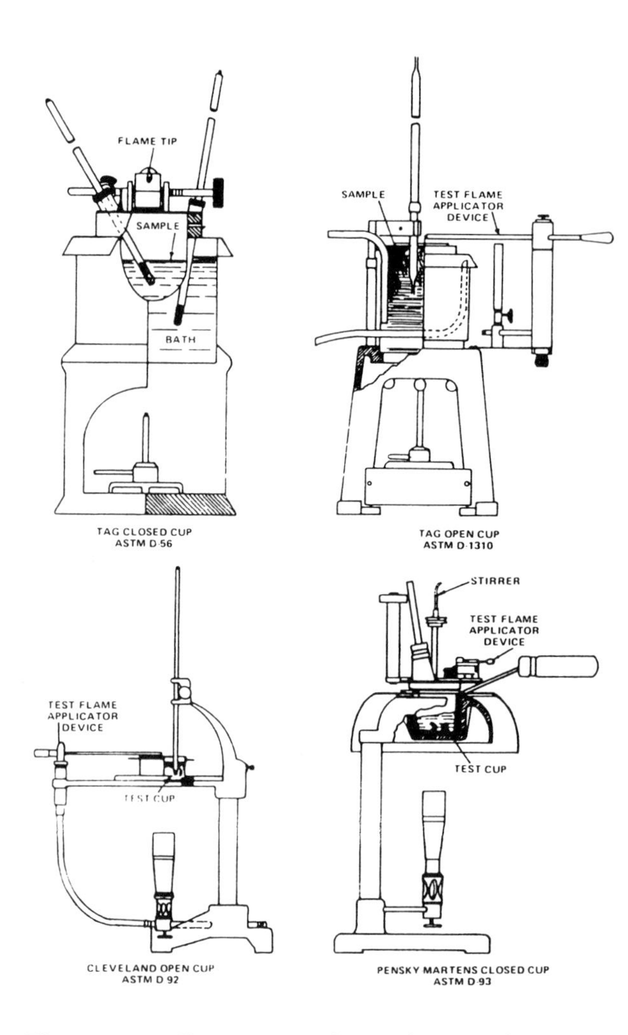

Figure 3-4. Four commonly used testers for determining flash points of flammable or combustible liquids. (Reprinted with permission from *Fire Protection Handbook,* 17th edition, (1991) National Fire Protection Association, Quincy, MA 02269.)

The *Tag Closed Cup Tester* is used to determine flash points of liquids that have a viscosity less than SUS at 100°F. The method is accurate up to 200°F, but the liquid cannot contain any suspended solids nor can it have a tendency to form a surface film under test conditions.

The Pensky Martens Closed Cup Tester is used to determine flash points of liquids with viscosities greater than or equal to 45 SUS at 100°F. It is accurate at temperatures over 200°F and can be used for liquids that contain suspended solids or have a tendency to form a surface film under test conditions.

The Setaflash Closed Cup Tester may be used for petroleum products, paints, enamels, varnishes, lacquers and related materials having flash points between 32 and 230°F.

While the *Tag Open Cup Tester* and *Cleveland Open Cup Tester* are largely historical artifacts, they are still used occasionally in special applications like grading flammable liquids for transportation. Flash points obtained using open cup methods are generally higher than the closed cup determinations for the same substance. The reason for this difference lies in the fact that vapors are not confined in the open cup.

It is important to note that flash point measurements are not "physical constants." Instead, they vary depending on the atmospheric pressure, the ambient oxygen concentration, and the purity of the sample specimen.

Flammable (Explosive) Limits

At the flash point temperature, the vapor concentration of a flammable liquid is just high enough to support combustion. For combustion to be sustained, the vapor concentration must be between two specific levels called the upper and lower flammable limits (also referred to as explosive limits). These limits are usually expressed in terms of per cent by volume. The typical relationship that exists between percent by volume of a flammable liquid and the liquid's flash point and vapor pressure is represented graphically in Figure 3-5.

The lower flammable limit (LFL) may be defined as the lowest concentration of a gas or vapor that must be present to support combustion in the presence of a source of ignition.

The upper flammable limit (UFL), on the other hand, is the highest concentration of a gas or vapor that will support combustion. The region between these two extremes is referred to as the flammable range (Figure 3-6).

Figure 3-7 shows that the lower and upper and flammable limits for acetone are 2.6 and 12.8%, respectively. Concentrations of acetone vapor below 2.6% are too "lean" to burn, while those above 12.8% are too "rich." However, any point between these two limits is ignitable.

Note from Figure 3-7 that not only do different materials have different upper and lower limits, but that the range of limits also varies from substance to substance. Some materials like carbon disulfide have very wide ranges while others like benzene are much more narrow. Substances with a wide flammable

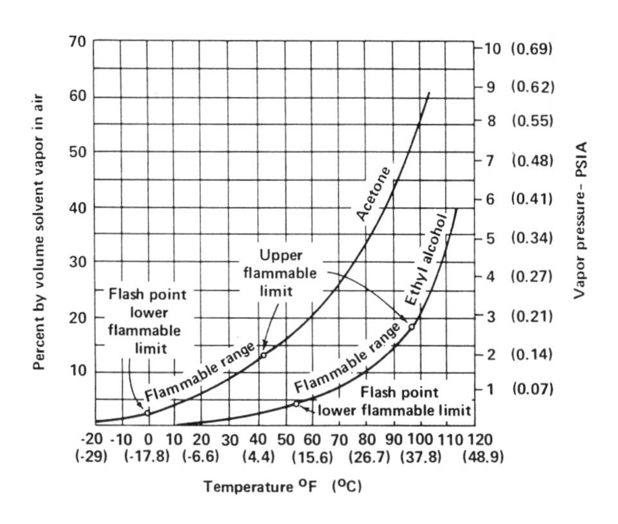

Figure 3-5. Relationship of temperature to vapor pressure and volume percent of acetone and ethyl alcohol. (Reprinted with permission from *Principles of Fire Protection Chemistry,* 1st edition, (1976) National Fire Protection Association, Quincy, MA 02269.)

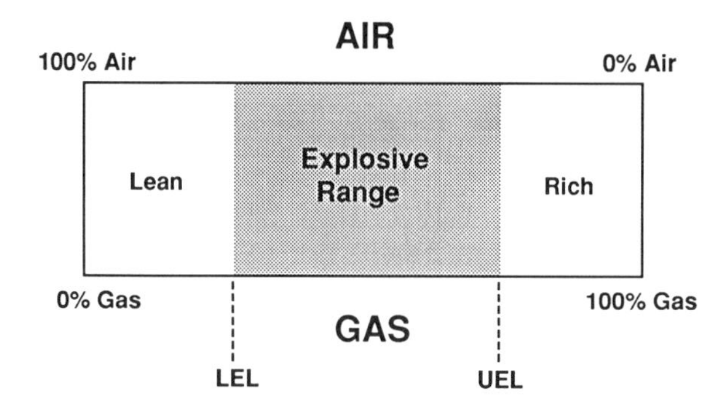

Figure 3-6. Upper and lower flammable limits. Concentration of gases and vapors less than the LFL are too "lean" to burn. Those whose concentrations are above the UFL are too "rich" to burn. (Paul Trattner)

range are considered to be more hazardous since it is possible for them to be ignited over a wide range of circumstances.

The explosive limits for hundreds of chemicals of commerce can be found in NFPA-325M, *Fire Hazard Properties of Flammable, Liquids, Gases and Volatile Solvents,* available from the National Fire Protection Association in Quincy, MA.

Variations with Temperature and Pressure

While explosive limits are commonly treated as fixed values, they actually vary depending on the magnitude of ignition energy and the ambient temperature and pressure.

The explosive range of gases and vapors expands and contracts with increasing and decreasing temperature. Increasing the temperature reduces the lower explosive limit and raises the upper explosive limit. Decreasing the temperature has the opposite effect, increasing the lower explosive limit and reducing the upper explosive limit.

The lower explosive limit is generally insensitive to changes in pressure except when they are well below atmospheric. Conversely, the upper limit is unaffected by sub-atmospheric pressures, but broadens as the pressure increases.

Figures 3-8 and 3-9 illustrate these effects under equilibrium conditions. It should be noted, however, that equilibrium can only exist in closed systems, since open systems are continually releasing vapors to the surrounding air. Consequently, temperature and pressure effects are really only relevant in situations involving closed tanks, pipes and similarly sealed process equipment where vaporization takes place at a given temperature and pressure.

Ignition Energy Variability

Figure 3-10 illustrates the relationship that exists between fuel concentration and the amount of energy required for ignition. Note that while the explosive range for propane extends between 2 and 10%, less energy is need for ignition of a mixture for mid-range concentrations than for those at either end. Similarly, the amount of energy needed to ignite hydrogen decreases by about 2 orders of magnitude as concentration increases from the lower explosive limit (4%) to a level of about 28%. Above 28% the amount of energy needed for ignition begins to increase, reaching a maximum at the upper explosive level.

Table 3-2 provides additional insight to the variability of energy needed for ignition by illustrating the effect that temperature has on the energy required for ignition. Note that vapors at elevated temperatures require far less ignition energy than those at ambient temperatures. This is not surprising since some energy is imparted to the material by heating. Consequently, the quantity of energy that must be provided by an external source is reduced.

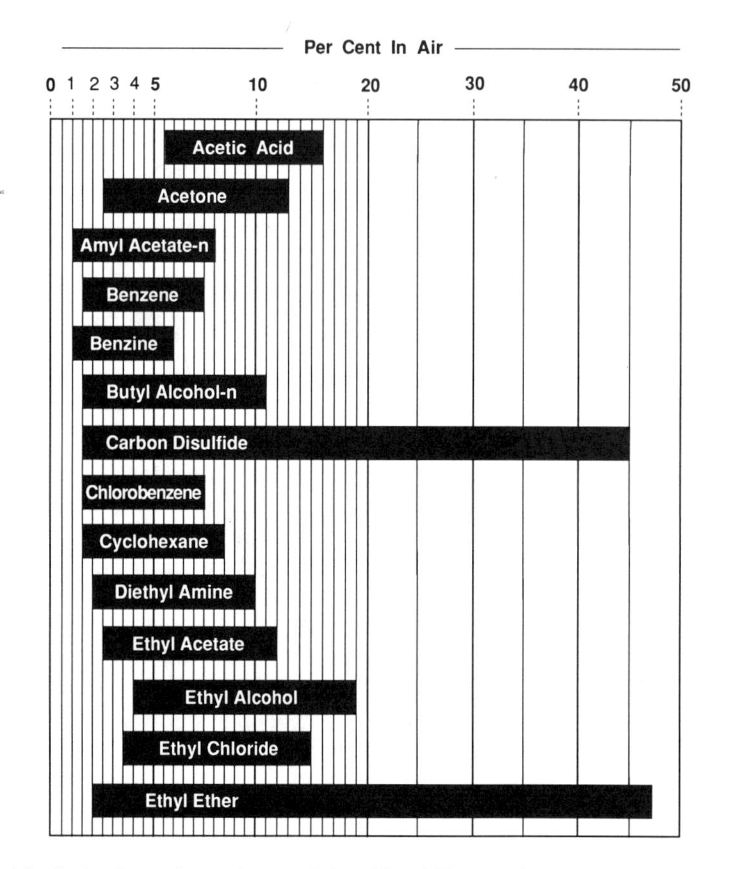

Figure 3-7. Flammable limits for selected materials. (Paul Trattner)

Vapor Density

Vapor density, which is another property of flammable liquids, may be defined as the weight ratio of a volume of flammable vapor compared to an equal volume of air. Since most flammable vapors are heavier than air, they tend to hug the ground and flow like water to low spots. If they happen upon an open flame, a hot surface or other source of ignition, they ignite and flash back to their point of origin like a fuse on a stick of dynamite.

The vapor density of a flammable liquid can be calculated by dividing its molecular weight by that of air. Since air consists of approximately 21% oxygen and 79% nitrogen, its molecular weight is determined to be 29.

21% oxygen	$0.21 \times 32 =$	6.72
79% nitrogen	$0.79 \times 28 =$	21.12
		$28.84 \cong 29$

Thus the vapor density for any vapor can be calculated by dividing its molecular weight by 29

$$\text{Vapor Density} = \frac{\text{Molecular Weight of Substance}}{29}$$

COMBUSTIBLE DUSTS

Combustible dusts may be present in confined spaces such as coal bunkers, chemical storage bins, process hoppers and grain silos. They include agricultural products, chemicals, plastics, drugs, pharmaceuticals and metals.

The relative hazard posed by a combustible dust depends on its ease of ignition and the severity of the resulting explosion. During the 1960s, the U.S. Bureau of Mines incorporated these factors in an explosibility index.

The ease of ignition, in other words the *ignition sensitivity,* depends on the ignition temperature and the minimum ignition energy

Ignition Sensitivity

$$= \frac{(T_I \times E_m \times C_m) \text{ for Pittsburgh coal dust}}{(T_I \times E_m\ C_m) \text{ for sample dust}}$$

where

T_I = Ignition temperature
E_m = Minimum energy
C_m = Minimum concentration

The *explosion severity,* on the other hand, depends on the maximum rate of pressure rise and the maximum explosion pressure that is achieved

Explosion severity

$$= \frac{(E_M \times P_M) \text{ for sample dust}}{(E_M \times P_M) \text{ Pittsburgh coal dust}}$$

where

E_M = Maximum explosive pressure
P_M = Maximum rate of pressure rise

Pittsburgh coal dust in a concentration of 0.50 oz. per ft.^{3} (500 g/M^3) is used as a reference to allow comparison between different materials.

The explosibility index of any material can be calculated by multiplying the ignition sensitivity by the explosion severity. As shown in Table 3-3, explosions may then be classified as weak, moderate, strong or severe.

As evidenced by Table 3-4 the quantity of dust needed to produce an explosive atmosphere is orders of magnitude larger than those allowed by OSHA health standards. Furthermore, the concentration of dust required to achieve an explosive atmosphere would so severely obscure vision that entry into the space would be highly unlikely. But suspended dust is not the only concern. Another less obvious hazard is the dust that accumulates on ledges, pipes, I-beams and other surfaces in the space. If this settled dust is disturbed, for instance when a surface is struck with a hammer or hit with a blast of compressed air, an ignitable dust cloud may be produced. If a source of ignition such as a hot surface or sparks from welding or burning is also present, the dust cloud could ignite. The initial ignition could then produce a chain reaction in which the expanding hot gases continue to disturb and ignite additional settled material with explosive consequences.

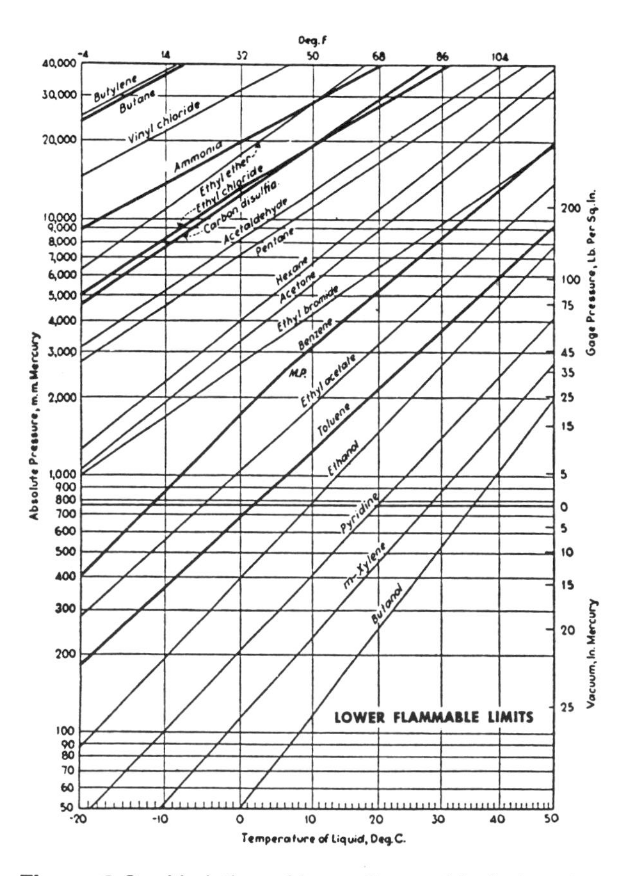

Figure 3-8. Variation of lower flammable limits with temperature and pressure. This chart is applicable only to flammable liquids or gases in equilibrium in a closed container. Mixtures of vapor and air will be *too lean* to burn at *temperatures below* and at *pressures above* the values shown by the line on the chart for any substance. Conditions represented by points to the *left* of and *above* the respective lines are nonflammable. Points where the diagonal lines cross the zero gauge pressure line (760 mm Hg absolute pressure) indicate flash point temperatures at normal atmospheric pressure. (Reprinted with permission from *Fire Protection Handbook,* 17th edition, 1991, National Fire Protection Association, Quincy, MA 02269.)

Ignition Susceptibility

The ease with which dust will ignite depends on a number of factors including:

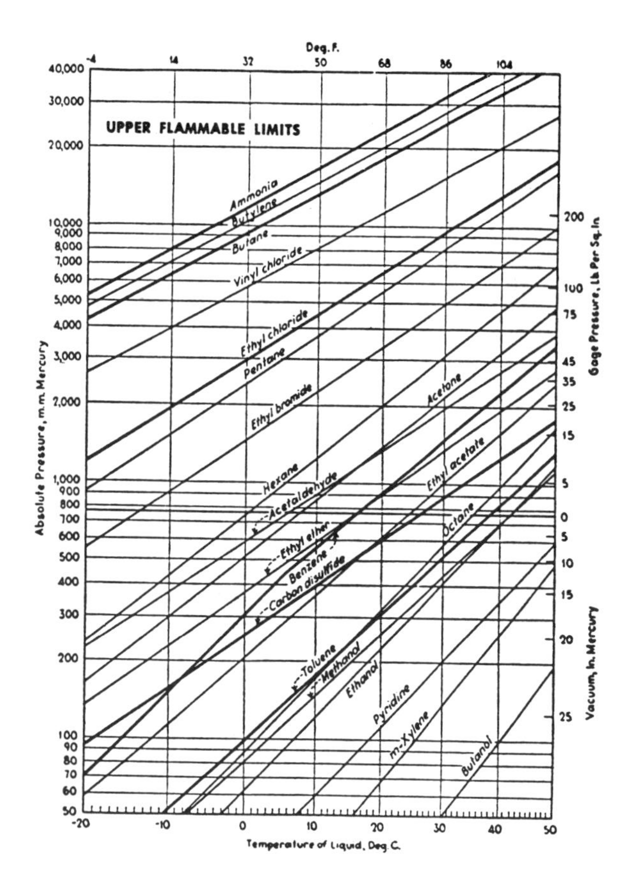

Figure 3-9. Variations of upper flammable limits with temperature and pressure. This chart is applicable only to flammable liquids or gases in equilibrium in a closed container. Mixtures of vapor and air will be too rich to be flammable at *temperatures above* and at *pressures below* the values shown by the line on the chart for any substance. Conditions represented by points to the *right* of and *below* the respective lines are accordingly nonflammable. (Reprinted with permission from *Fire Protection Handbook,* 17th edition, 1991, National Fire Protection Association, Quincy, MA 02269.)

- The dust's chemical composition
- The size, shape and surface structure of individual particles
- The degree of uniformity among suspended dust particles
- The initial temperature and pressure
- The amount of energy needed for ignition

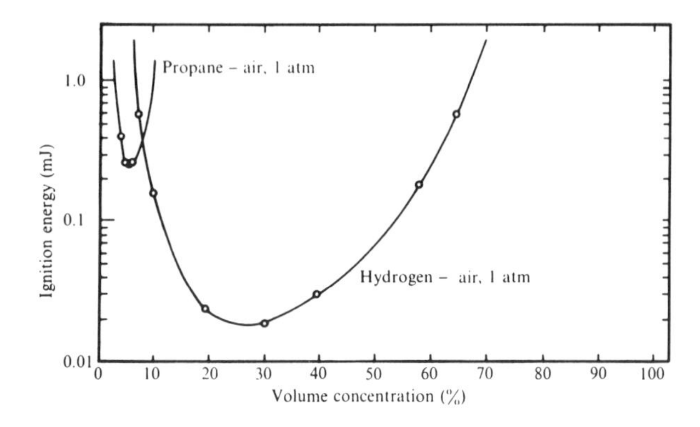

Figure 3-10. Relationship between fuel concentration and ignition energy. (Reprinted with permission, copyright Instrument Society of America 1978. From *Electrical Instruments in Hazardous Locations.*)

Small dust particles are easier to ignite than large dust particles, and the rate of pressure rise is faster for small particles than for large ones. The lower explosive limit, the ignition temperature and the ignition energy also decrease as particle size decreases.

Moisture tends to increase the ignition temperature of a dust cloud because some of the heat is lost in vaporizing the water. However, moisture in the air has little to no effect on the course of a deflagration once ignition has occurred. As a practical matter, moisture is not an effective tool for preventing dust explosions since most ignition sources have more than enough heat to vaporize the moisture and ignite the dust. For moisture to be effective, the dust would have to be damp enough to prevent formation of a dust cloud.

The presence of inert solid powder reduces the combustibility of dust because it absorbs heat. Inert material also reduces the rate of pressure rise and increases the minimum dust concentration required for an explosion. However, the amount of inert material necessary to prevent an explosion is usually considerably higher than the amount of foreign material that would normally be tolerated from a product quality perspective.

Table 3-2. Effect of temperature on ignition energy

Fuel	Temperature (°C)	Ignition Energy (mJ)
Carbon disulfide	25	0.067
	100	0.05
n-Heptane	25	1.45
	100	0.67
	171	0.32
Iso-octane	25	2.7
	100	1.1
	171	0.32
n-Pentane	-30	4.5
	-20	1.45
	25	0.78
	100	0.42
	171	0.23
	175	0.25
Propane	-40	1.17
	-30	0.97
	-20	0.84
	25	0.84
	57	0.42
	82	0.36
	100	0.35
	204	0.14
Propylene oxide	25	0.24
	100	0.15
	182	0.09

Table 3-3. Classification of explosions

Type of Explosion	Ignition Sensitivity	Explosion Severity	Explosibility Index
Weak	<0.2	<0.5	<0.1
Moderate	0.2 – 1.0	0.5 – 1.0	0.1 – 1.0
Strong	1.0 – 5.0	1.0 – 2.0	1.0 – 10.0
Severe	>5.0	>2.0	>10.0

Explosive Concentration

As with gases and vapors, there is a lower explosive level, or minimum dust concentration dust, that must be present in the air before an explosion will occur. The explosion pressure and rate of pressure rise are at a minimum at the lower explosive level and rise to a maximum value at some optimum concentration. As dust concentrations increase beyond the optimum point, the explosion pressure and rate of pressure rise then tend to decrease. It is also interesting to note that the maximum rate of rise and maximum pressure do not necessarily occur at the same concentration.

While it is customary to express the lower explosive limit of combustible dusts in units of weight per unit volume, the numbers are meaningless without information on particle size since minimum explosive concentrations vary with particle diameter. Other factors that can affect the lower explosive limit of a dust cloud are the oxygen level, the degree of sample purity, turbulence of the dust cloud, uniformity of the dispersion and the strength of ignition source.

There is some question as to whether or not there is a clear-cut upper explosive limit for dusts because it is difficult to establish upper limits experimentally. It should be noted that the results listed in Table 3-4 were obtained from dust small enough to pass a No. 200 mesh screen (74 microns or smaller).

SOURCES OF IGNITION

Possible sources of ignition that could be found in confined spaces include: open flames, arcs from electrical equipment, hot surfaces, static electricity and frictional sparks. Because combustible dusts generally require about 20 to 50 times more energy than flammable gases or vapors they are more difficult to ignite. As a frame of reference, ignition energies for dust clouds typically range from 10 and 40 mJ vs. about 0.2 to 10 mJ required for gases and vapors.

Open Flames

Open flames are perhaps the most obvious source of ignition since they have an unfailing ability to heat gases and vapors to their ignition point. In a confined space, open flames may be present in the form of welding torches, space heaters, or as smoking materials like matches and cigarette lighters.

Electrical Arcing

Electric arcs are formed when current-carrying elements are interrupted. This interruption may occur accidently when loose wires and terminal connections short out, or by design such as when relay contacts and switches are opened and closed. Arcing may be especially severe when inductive circuits like motors are involved, and arc temperatures can easily approach 2000°F as evidenced by arc-related melting of wiring and other copper components which have a melting point of 1083°F.

Table 3-4. Explosion characteristics for selected dusts

	Explosibility Index	Ignition Sensitivity	Explosion Severity	Maximum Explosion Pressure (psig)	Max. Rate of Pressure Rise (psi/s)	Ignition Temperature		Min. Cloud Ignition Energy (J)	Minimum Explosion Concentration	
						Cloud (°C)	Layer (°C)		oz./ft^3	mg/M^3
Cellulose	2.8	1.0	2.8	130	4,500	480	270	0.080	0.055	550,000
Charcoal, hardwood	1.3	1.4	0.9	83	1,300	530	180	0.20	0.140	1,400,000
Coal, Pittsburgh ref.	1.0	1.0	1.0	90	2,300	610	170	0.060	0.55	550,000
Coal, bituminous	4.1	2.2	1.9	101	4,000	610	180	0.30	0.050	500,000
Coffee	<0.1	0.2	0.1	38	150	720	270	0.15	0.085	850,000
Corn	6.9	2.3	3.0	113	6,000	400	250	0.04	0.055	550,000
Cornstarch	9.5	2.8	3.4	106	7,500	400	—	0.04	0.045	450,000
Grain dust	9.2	2.8	3.3	131	7,000	430	230	0.03	0.055	550,000
Nylon polymer	>10	6.7	1.8	95	4,000	500	430	0.020	0.030	300,000
Polycarbonate	8.6	4.5	1.9	96	4,700	710	—	0.025	0.025	250,000
Rice	0.3	0.5	0.5	47	700	510	450	0.10	0.085	850,000
Soy flour	0.7	0.6	1.1	94	800	550	340	0.10	0.06	60,000
Sugar, powdered	9.6	4.0	2.4	109	5,000	370	400	0.03	0.045	450,000
Sulfur	>10	20.2	1.2	78	4,700	190	220	0.015	0.035	350,000
Wheat flour	4.1	1.5	2.7	97	2,800	440	440	0.06	0.05	500,000
Wood flour	9.9	3.1	3.2	113	5,500	470	260	0.040	0.035	350,000

Source: U.S. Bureau of Mines.

Table 3-5. Color temperature of incandescent surfaces

Color	Approximate Temperature	
	°F	°C
Dull red	930 – 1100	500 – 600
Dark red	1100 – 1470	600 – 800
Bright red	1470 – 1830	800 – 1000
Yellow-red	1830 – 2190	1000 – 1200
Bright yellow	2190 – 2550	1200 – 1400
White	2550 – 2910	1400 – 1600

Adapted from Turner, C.F. and McCreery, J.W., *The Chemistry of Fire and Hazardous Materials,* Allyn & Bacon, Boston (1981).

Frictional Sparks

Frictional sparks may be produced when a metal object strikes a hard surface. Familiar examples include steel tools hitting or scraping other steel objects, or steel objects hitting concrete. Frictional sparks result when the heat of impact causes freshly exposed surfaces of metal particles to rapidly oxidize. This oxidation produces additional heat which warms the particle to incandescence. While the temperatures of incandescence for steel may approach 2500°F, the potential for ignition depends on the particle's total heat content. This may be quite small since many mechanical sparks have a limited surface area which cools quickly.

Hot Surfaces

If exposed surfaces are large enough and hot enough they too can become sources of ignition. As shown in Table 3-5, the color of an incandescent material can be used to approximate the surface temperature. Hot surfaces that may be reasonably encountered in confined spaces include steam lines, resistance heaters and exposed light bulbs. As demonstrated by Figures 3-11 and 3-12, the surface temperatures of electric light bulbs can approach 300°C, the ignition temperature of some gases and vapors.

Static Electricity

Static electricity is created by the flow of electrons that results from the contact and separation of

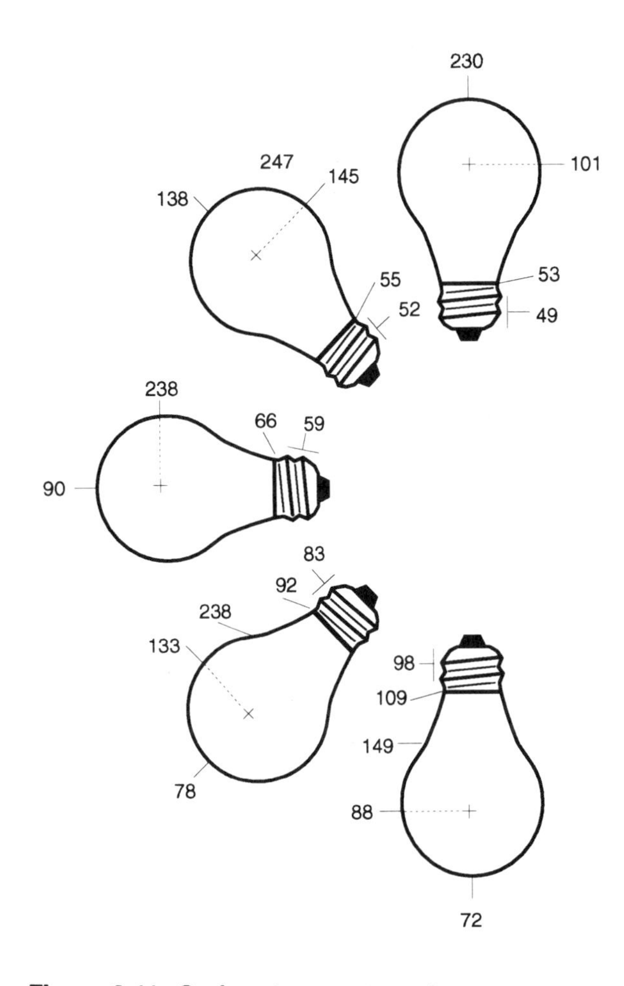

Figure 3-11. Surface temperature of 100-watt CC-B or A-19 incandescent filament lamp in still air at 25°C ambient. All temperature in degrees Celsius. (From IES *Lighting Handbook,* Illuminating Engineering Society of America, used with permission.)

two materials. The flow of electrons is usually not the same between the materials and if they are insulated from each other it is possible for an excess of electrons to accumulate on one surface when they are separated. When this highly charged material comes in contact with a grounded surface, the difference in charge between the two bodies is equalized with the resulting production of an static electric spark.

Processes generally recognized as being potential sources of static electricity include fluid flow through pipes, contact and separation between belts and pulleys and pneumatic transfer of finely divided solid

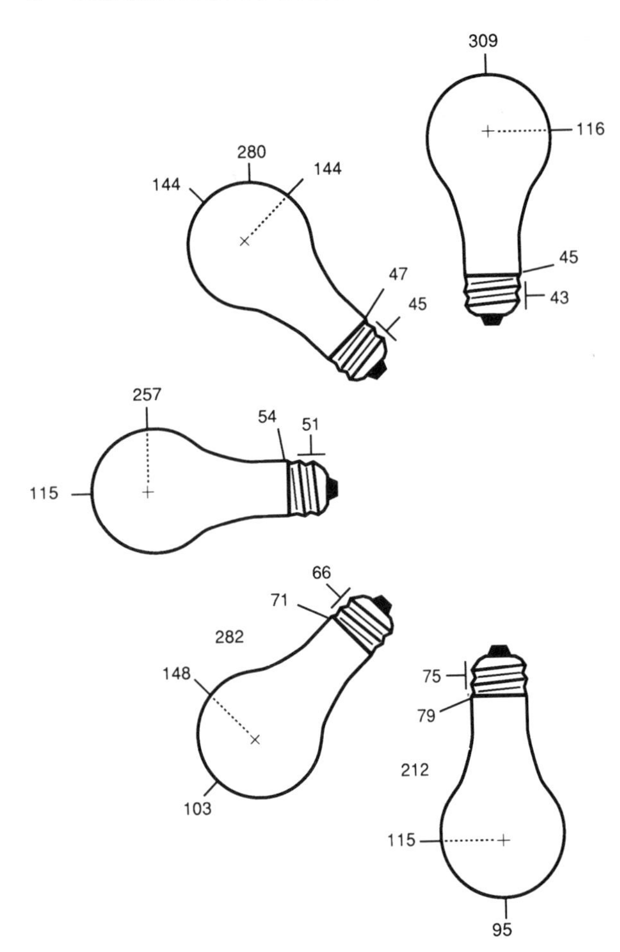

Figure 3-12. Surface temperature of 500-watt CC-8 or PS-35 incandescent filament lamp in still air at 25°C ambient. All temperature in degrees Celsius. (From IES *Lighting Handbook,* Illuminating Engineering Society of America, used with permission.)

materials. Static currents can generally be controlled by electrically bonding and grounding components that are likely to accumulate a charge.

Chemical Reactions

Some exothermic chemical reactions produce enough heat to ignite combustible materials. For example, the heat released when glycerine is mixed with finely divided potassium permanganate is great enough to produce a flame. While heat releasing reactions are rarely encountered in properly prepared confined spaces, one that should be noted is the formation of deposits of finely divided iron polysulfide that may occur in steel storage tanks used to store sulfur-containing "sour" stocks or aromatic tars.

Iron polysulfide is a pyrophoric material which means that it can ignite spontaneously on contact with air. Deposits of iron polysulfide which accumulate on tank walls will react exothermically when exposed to air. Unless the heat is dissipated quickly, spot temperatures may rise high enough to ignite flammable residue. This hazard may be controlled by wetting the inside surface of the tank with water applied by a fog nozzle positioned at an open man way. Since water flowing through the nozzle is a potential source of static generation, the nozzle should be bonded to the shell of the tank.

CONTROL OF IGNITION HAZARDS

Ignition hazards in confined spaces can be controlled through the use of non-sparking hand tools, approved electrical equipment and special work practices during welding and flame cutting.

Non-Sparking Tools

Anyone who has ever used a grinding wheel knows that hot particles and sparks are emitted when steel tools impact hard surfaces. The thermal energy in some of these particles may be sufficient to ignite flammable vapors. This hazard can be controlled through the use of special tools made of brass or copper-beryllium alloys. While these tools are commonly called "non-sparking," the term is misleading since even "non-sparking" tools can emit sparks on impact. However, the spark energy is so low that it is incapable of igniting all but a few flammable vapors, such as carbon disulfide, which requires very little energy for ignition.

Approved Electrical Equipment

"Hazardous locations" as defined by the National Electrical Code®, (NEC®) are areas subject to accumulations of flammable gases and vapors, or combustible dusts, fibers or flyings. Hazardous locations are also referred to as "classified locations" since they are divided into three classes on the basis of the nature of hazard. Class I locations are those containing gases and vapors, Class II contain combustible dusts and Class III contain combustible fibers and flyings. These three broad classes are further characterized by "groups" of materials that pose similar fire and explosion hazards, and "divisions" which indicate whether the hazard exists under normal operating conditions (Division 1) or only under abnormal conditions (Division 2).

Gaseous materials are grouped into four categories (Groups A, B, C and D) on the basis of similarities in their maximum experimental safe gap, explosion pressure and auto-ignition temperature. Dusts, on the other hand, are divided into three groups (E, F and G) depending on their layer ignition temperature and conductivity. Grouping will be explained in more detail in Chapter 9.

Electrical equipment such as ventilation blowers, portable lights and power tools may all be sources of ignition unless they have been designed for use in a hazardous location. Equipment that has been designed for use in hazardous locations can be identified by a label or nameplate that lists the types of atmospheres that it can be used in by class, division and group. As a practical matter, atmospheres most likely to be encountered during confined space entries will normally be limited to Class I. To be approved for use in a Class I atmosphere, electrical equipment must be intrinsically safe, explosion-proof or purged and pressurized.

Purged and Pressurized Equipment

Purged and pressurized is typically used for instrument and process control. Flammable atmospheres are excluded from equipment cabinets housing spark-producing devices by pressurizing the cabinet's interior with either compressed air or an inert gas such as nitrogen. The cabinet is also fitted so that a pressurization failure shuts off the spark-producing devices and sounds an alarm. Specific design requirements for purged and pressurized equipment may be found in NFPA-396.

Intrinsically Safe Equipment

Equipment that is intrinsically safe is designed in such a way that sparks or heat produced either in normal use or under fault conditions will not ignite the most easily ignitable concentration of specified gases and vapors. Many industrial hygiene instruments used to evaluate workers' exposure to air contaminants fall into this category of equipment.

Explosion-Proof Equipment

Explosion-proof design recognizes that flammable vapors can enter enclosures as a result of normal breathing that takes place as metal enclosures expand and contract in response to changes in ambient temperature. Flammable gases and vapors can enter an enclosure either through threaded connections or through small gaps between mated surfaces of the enclosure. If a flammable mixture as to be ignited by a spark-producing device inside the enclosure, the flames propagated through the enclosure openings could result in secondary ignition of the surrounding atmosphere.

Explosion-proof equipment addresses this hazard in two ways. First, the equipment is constructed to withstand the explosion forces resulting from internal ignition (Figure 3-13). Second, explosion-proof devices are designed so that hot combustion gases are vented through threaded connections (Figure 3-14) or finely machined air gaps in enclosure covers (Figure 3-15). When combustion gases leave the enclosure, they have been cooled below the ignition temperature of the surrounding atmosphere and do not possess sufficient energy to cause secondary ignition.

Special Hot-Work Precautions

Gas welding and cutting presents the potential for particularly severe fire and explosion hazards in confined spaces. Oxygen and fuel gas can accumulate in a space due to leaking hoses, valves and fittings. If levels of fuel gas reach the lower flammable limit, they can ignite with explosive consequences. Oxygen, while not combustible itself, greatly alters the burning characteristics of many materials, making them easier to ignite and more rapid-burning once ignited. Clothing that has become saturated with oxygen can burst into flames when exposed to a hot surface, an open flame or even a lit cigarette.

An individual knowledgeable in hot work hazards should inspect the confined space before permitting welding or oxy-fuel gas cutting. This person should also list, either on the entry permit or on a separate hot work permit, any special precautions that should be followed to assure that the work is performed safely.

Vessel Inerting

Fire hazards posed by exterior welding or cutting on vessels that previously contained flammable gases or liquids can be controlled by limiting the oxygen in the space with nitrogen or carbon dioxide (Table 3-6).

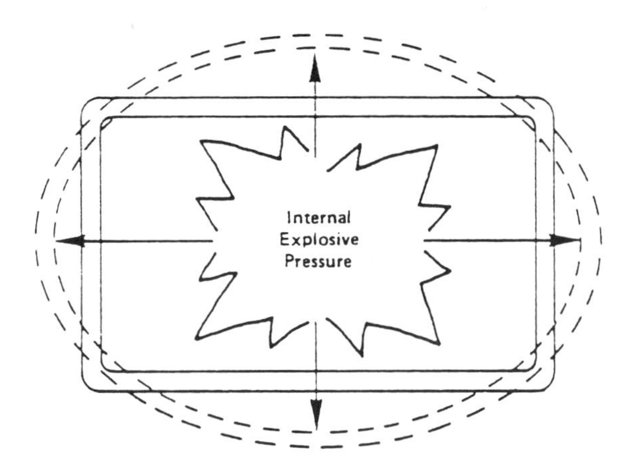

Figure 3-13. Explosion-proof equipment is designed to contain the explosion force generated inside the enclosure when gases or vapors ignite. (OSHA)

Torches and Control Valves

Torches should not be taken into a confined space until they are ready to be used and they should be removed whenever they are not actually in use, or when they will not be used for a substantial period of time, such as during lunch, coffee breaks or overnight. If it is not feasible to remove torches, such as when they are used regularly, yet intermittently, over an extended period of time, the torch control valves should be closed and the gas supply positively shut off from outside the space. This will eliminate the possibility of gas escaping through leaking hoses or improperly closed valves.

Hoses and Regulators

Parallel lengths of oxygen and fuel-gas hose may be taped together for convenience provided that no more than 4 inches per foot are covered. Hoses showing evidence of leaks, burns, wear or other defects which render them unfit for service must be repaired or replaced.

Fuel gas and oxygen hoses must be distinguishable from one another and may not be interchanged. Contrast may be achieved by using different colors or by using hoses with different surface characteristics that are readily distinguishable by touch. Hose that has been subject to flashback or show evidence of severe wear or damage should be tested at twice the normal pressure to which it is subjected, but not less than 300 psi. Hose couplings should be of a type that cannot be unlocked or disconnected by a straight pull without a rotary motion.

Before connecting a regulator to a cylinder valve, the valve should be cracked (opened briefly, then closed) to clear it of any dirt or debris that might enter and damage the regulator. Unless cylinders are secured on a special cart, regulators should be removed and valve-protection caps replaced before cylinders are moved. The cylinder valve should be closed and the gas released from the regulator before it is removed.

Compressed Gas Cylinders

Compressed gas cylinders should be legibly marked to identify their contents. They should be kept away from heat sources and never be placed where they might become part of an electric circuit. Cylinder valves should be closed at the completion of the work, when cylinders are empty, and before cylinders are moved. Valve-protection caps should be in place and hand-tight except when cylinders are connected for use.

Cylinders must not be dropped or allowed to strike each other violently. Valve caps may not be used for lifting cylinders. Acetylene cylinder valves must not be opened more than 1-1/2 turns, and 3/4 of a turn is preferable. Cylinders that are not equipped with fixed hand wheels must have special keys, handles or nonadjustable wrenches on their valve stems while they are in service so the gas flow can be shut off in an emergency. If multiple cylinders are connected together then at least one key or handle is required for each manifold.

Fire Prevention and Protection

Cutting or welding should be permitted only in areas determined to be fire-safe by someone knowledgeable on the special hazards posed by hot work. This person should have the authority to establish a fire watch and require that specific precautions be taken prior to the work.

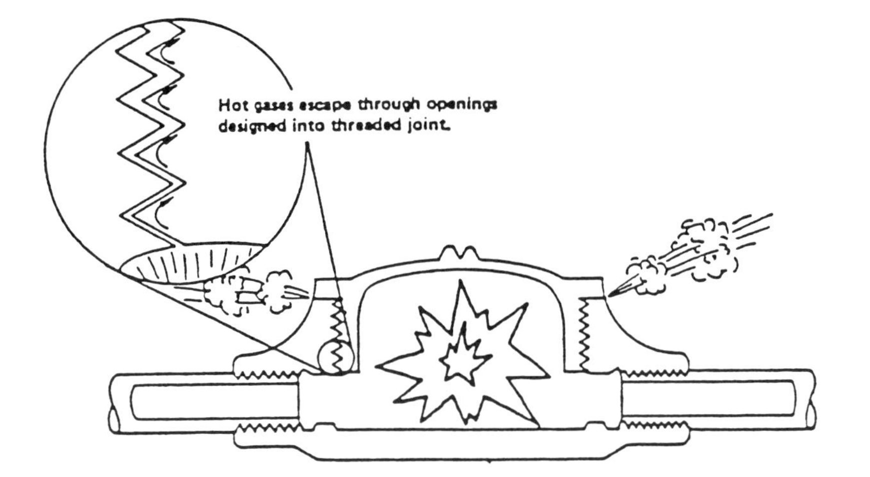

Figure 3-14. Explosion-proof equipment venting hot gases through threads. (OSHA)

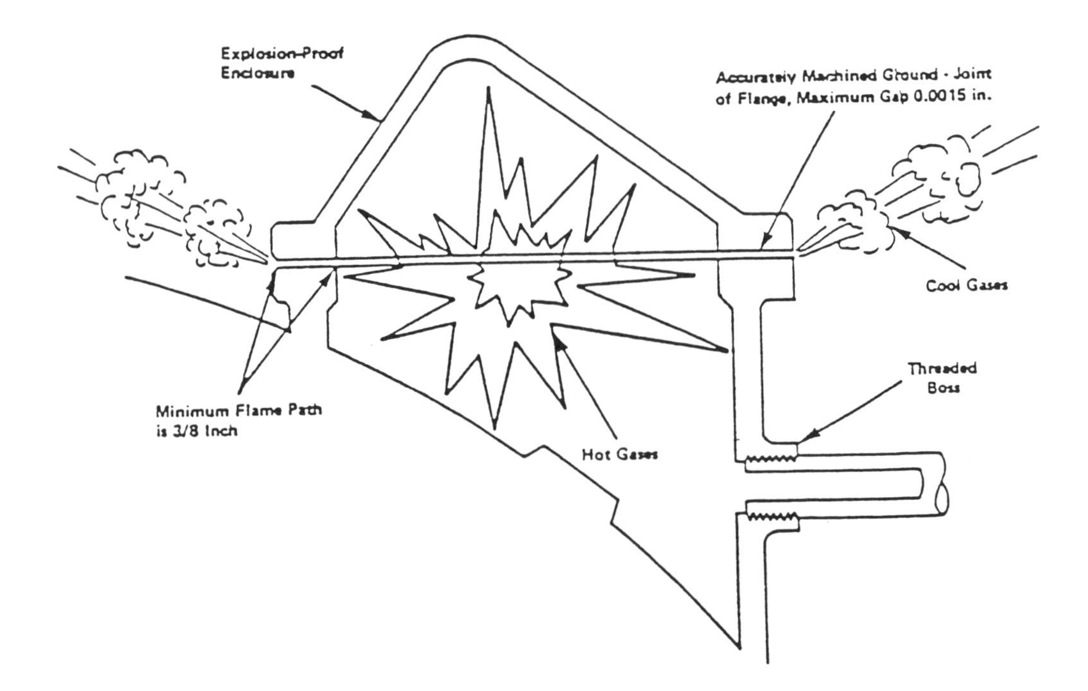

Figure 3-15. Explosion-proof equipment venting air gases through an air gap. (OSHA)

Fire Safety Precautions. When practical, all combustible equipment and materials should be removed from the space before hot work begins. Items that cannot be removed should be protected from heat, sparks and slag by fire-resistant coverings or shielded with sheet metal. Precautions should be taken to prevent the escape of sparks and hot slag through holes and openings in the space that cannot be closed off. Ducts, conveyors and other systems that might carry sparks to distant combustibles should also be protected or shut down. If hot work is to be done on vessel walls, pipes or other metal surfaces that may transfer heat, precautions must be taken to prevent ignition of combustibles that may be affected by thermal conduction or radiation.

Establishing A Fire Watch. Fire watchers should be present whenever welding or cutting is done in areas where a major fire could develop, or where heat

Table 3-6. Maximum allowable oxygen level required to prevent ignition

	N_2-Air		CO_2-Air	
	O_2 % Above Which Ignition Can Take Place	Maximum Recommended O_2 %	O_2 % Above Which Ignition Can Take Place	Maximum Recommended O_2 %
Acetone	13.5	11	15.5	12.5
Benzene	11	9	14	11
Butane	12	9.5	14.5	10.5
Carbon disulfide	5	4	8	6.5
Cyclopropane	11.5	9	14	11
Ethyl alcohol	10.5	8.5	13	10.5
Gasoline				
73–100 Octane	12	9.5	15	12
100–130 Octane	12	9.5	15	12
115–145 Octane	12	9.5	14.5	11.5
Hexane	12	9.5	14.5	11.5
Hydrogen	5	4	6	5
JP–1 Fuel	10.5	8.5	14	11
JP–3 Fuel	12	9.5	14	11
JP–4 Fuel	11.5	9	14	11
Kerosene	11	9	14	11
Methane	12	9.5	14.5	11.5
Methyl alcohol	10	8	13.5	11
Propane	11.5	9	14	11

Adapted from NFPA-53M.

transferred by radiation or conduction could ignite combustible materials adjacent to the space.

Hoselines or portable extinguishers should be readily available. Fire watchers should be trained in their use and know the procedures for sounding an alarm. They should also understand that while they are responsible for *watching* for fires in all exposed areas, they should only attempt to *extinguish* those that are within the capacity of the available equipment. Otherwise they should sound the alarm.

Since smoldering can occur long before the flaming stage of a fire, a fire watch should be maintained for at least a half hour after completion of welding or cutting operations to detect and extinguish any slowly developing fires.

SUMMARY AND CONCLUSIONS

Flammable atmospheres may be produced by the evaporation of flammable residue, by flammable materials used in the space or by naturally occurring reactions such as the formation of methane gas. They may also exist due to suspended combustible dusts in spaces such as coal bunkers, storage bins and grain silos. Settled dust which is disturbed can also present an explosion hazard if it is ignited by sources such as open flames, welding torches or hot sparks.

Three conditions must exist for a fire to occur. First, there must be sufficient oxygen. Second, there must be a source of ignition. Third, there must be sufficient fuel present as a gas, vapor or dust.

Fires and explosions in confined spaces may be prevented by a variety of methods, including work practices that limit the use of flammable materials, atmospheric testing to evaluate potentially hazardous atmospheres and ventilation that eliminates the hazardous condition.

REFERENCES

Best, R., Two grain elevator explosions kill five in Missouri, *Fire Journal,* 72:50–54 (1978).

Cloe, W.W., Selected Occupational Fatalities and Asphyxiating Atmospheres in Confined Spaces as Found in Reports of OSHA Fatality/Catastrophe Investigations, U.S. Department of Labor, Occupational Safety and Health Administration, Washington, DC (1985).

Classification of Class Gases, Vapors and Dust for Electrical Equipment in Hazardous Locations NFPA-497A Quincy, MA: National Fire Protection Association (1986).

Cloe, W.W., Selected Occupational Fatalities Related to Welding and Cutting as Found in Reports of OSHA Fatality/Catastrophe Investigations, U.S. Department of Labor, Occupational Safety and Health Administration, Washington, DC (1988).

Cote, A.E., (ed.), Fire Protection Handbook, 16th Ed., Quincy, MA: National Fire Protection Association (1986).

Cutting and Welding Processes, NFPA-51B, Quincy, MA: National Fire Protection Association (1989).

Dorset, H.G., Jr. and Nagy, J., Dust Explosibility of Chemicals, Drugs, Dyes, and Pesticides, Report of Investigation 7132, U.S. Bureau of Mines (1968).

Fire Hazards in Oxygen-Enriched Atmospheres, NFPA-53M, Quincy, MA: National Fire Protection Association (1990).

Fire Hazard Properties of Flammable Liquids, Gases and Volatile Solids, NFPA-325M, Quincy, MA: National Fire Protection Association (1991).

Fires and Dust Explosions in Grain Elevators and Facilities Handling Bulk Raw Agricultural Commodities, NFPA-61B, Quincy, MA: National Fire Protection Association (1989).

Flammable and Combustible Liquids Code, NFPA-30, Quincy, MA: National Fire Protection Association (1990).

Flammable and Combustible Liquids in Manholes and Sewers NFPA-328 Quincy, MA: National Fire Protection Association (1987).

Hartman, I., et. al., Recent Studies of the Explosibility of Corn Starch, RI 4725 USDI, Bureau of Mines, Pittsburgh, PA (1950).

Hazardous Chemical Reactions, NFPA-491M Quincy, MA: National Fire Protection Association (1991).

Henry, M., Flammable and Combustible Liquids, in *Fire Protection Handbook* 16th Ed., Quincy, MA: National Fire Protection Association (1986).

Jacobson, M., J. Nagy, A.R. Cooper, and F.J. Ball, Explosibility of Agricultural Dusts, Report of Investigation 5753, U.S. Bureau of Mines (1961).

Jacobson, M., A.R. Cooper and J. Nagy, Explosibility of Metal Powders, Report of Investigation 6516, U.S. Bureau of Mines (1964).

Jacobson, M., J. Nagy and A.R. Cooper, Explosibility of Dusts Used in the Plastics Industry, Report of Investigation 5971, U.S. Bureau of Mines (1962).

Lathrop, J.K., 54 Killed in two grain elevator explosions, *Fire Journal* 72:29–35 (1978).

Magison, E.C., Electrical Instruments in Hazardous Locations, Research Triangle Park, NC: Instrument Society of America (1978).

Mertzberg, et. al., The Flammability of Coal Dust-Air Mixtures Lean Limits, Flame Temperature, Ignition Energies, and Particle Size Effects, RI 8360, U.S. Bureau of Mines (1979).

Nagy, J., et. al., Pressure Development in Laboratory Dust Explosions, RI 6561 U.S. Bureau of Mines (1964).

National Electrical Code, NFPA-70, Quincy, MA: National Fire Protection Association (1993).

Palmer, K.N., Dust Explosions and Fires, London: Chapman & Hall Ltd. (1973).

Purged and Pressurized Enclosures for Electrical Equipment Used in Hazardous (Classified) Locations, NFPA-496, Quincy, MA: National Fire Protection Association (1990).

Prevention of Fire and Dust Explosions in The Chemical, Dye, Pharmaceutical and Plastics Industries, NFPA-654, Quincy, MA: National Fire Protection Association (1988).

Static Electricity, NFPA-77, Quincy, MA: National Fire Protection Association (1988).

Schwab, R.F., Dusts in Fire Protection Handbook, 16th Ed., Quincy, MA: National Fire Protection Association (1986).

Underground Leakage of Flammable and Combustible Liquids, NFPA-329, Quincy, MA: National Fire Protection Association (1987).

CHAPTER 4

Toxic Atmospheres

Any of tens of thousands of hazardous materials may be found inside of a confined space. Some of these materials are formed by planned chemical reactions. Others are produced by naturally occurring chemical reactions such as fermentation. Still others may occur as a result of reactions that occur accidentally. A toxic atmosphere may result if these materials become airborne. But chemical reactions are not the only means by which a space's atmosphere may be contaminated. Operations such a welding, spray painting and abrasive blasting may also produce atmospheric hazards. Hazardous substances may also get into a space through inadvertent leaks and spills.

The effects that hazardous materials have on the body varies from substance to substance. Some materials irritate the eyes and respiratory tract, others may cause subtle damage to the brain, liver or kidneys. Some materials may trigger adverse effects after brief exposures at relatively low levels. Others produce adverse effects only after repeated exposures at moderate levels.

The huge number of chemicals and diversity of their applications may at first seem overwhelming. However, it is possible to gain a clearer perspective of the hazards posed by chemicals by identifying a few basic toxicological concepts.

These concepts can be applied in the field to identify the nature and type of contaminants that are most likely to be found in a particular space. This chapter employs a 2-step approach that will help you develop this skill. First, it explains a strategy that you can use to determine what materials may be present in a space. Second, it explains some basic toxicological principles that you can apply when trying to evaluate airborne chemical hazards.

ASSESSING TOXIC ATMOSPHERES

While thousands of chemicals may be encountered in confined spaces, those most likely to be present in a specific space can be determined by asking five basic questions.

1. What did the space previously contain?
2. What reactions could have occurred in the space?
3. What operations will be performed in the space?
4. What materials will be brought into the space?
5. What materials may have inadvertently entered the space?

Previous Contents

Knowledge of what the space last contained is important since some material could remain as a residue after the space is drained. Even small traces of volatile residues can evaporate and form hazardous levels of vapor. Tars and other sticky substances pose a special problem since they tend to adhere to interior surfaces and may off-gas for long periods of time.

Consequently, one of the first considerations in assessing potentially toxic atmospheres in confined spaces is to determine what the space last contained. Once that has been established, the hazards posed by the former contents can be fully evaluated.

Possible Chemical Reactions

Chemical reactions occur when two or more materials interact with each other to form a new substance. For example, when sodium hydroxide (NaOH) is mixed with hydrochloric acid (HCl) they react to form sodium chloride (NaCl) or table salt and water (H_2O).

A generalized expression for a simple chemical reaction is

$$A + B \rightarrow C$$

The expression shows that the materials A and B have reacted to form a new material, C.

But not all reactions proceed as directly as the one above. Some involve a very complex set of side-reactions that may occur between the individual constituents of the starting materials.

A + B → C (desired product)
↳ D (undesired side-reaction product)

As suggested in the above equation, these side-reactions may produce small amounts of materials other than the desired product.

In commercial chemical production facilities, reactions are carried out under tightly controlled conditions. The chemical plant operator is aware of the nature of the reactants and knows the status of the reaction from readings obtained from an array of process sensors and instruments. In addition, chemical engineers who are responsible for the process will be aware of any undesired side-reaction products.

But not all chemical reactions take place under such close scrutiny. Some reactions occur naturally, others occur accidentally. For instance, recall the two case histories reported in Chapter 2 where an oxygen deficiency was created by steel surfaces rusting and corroding. Other case histories in Chapter 2 discussed situations where assorted materials fermented naturally, producing carbon dioxide which displaced ambient oxygen. But carbon dioxide is not the only product produced by naturally occurring chemical reactions. Hydrogen sulfide, a gas which interferes with the body's ability to transfer oxygen on a cellular level, is a contaminant formed naturally in sewers and wastewater treatment plants.

Tools and equipment that entrants take into a space can also react with reside and sludge to form toxic air contaminants. In one case, workers were overcome by arsine gas when the aluminum ladder they used to enter an evaporating tank reacted with a solution of sodium arsenite ($NaAsO_2$) and sodium hydroxide ($NaOH$) that remained on the tank floor. Apparently, the sodium hydroxide contacted the ladder and created hydrogen which subsequently combined with the arsinite to form highly toxic arsine gas.

Because the reactants and products of commercial chemical reactions are known, their hazards are easier to evaluate than those associated with reactions that occur naturally or by accident. For purposes of this discussion, commercial chemical reactions can be divided into two types: batch and continuous.

In batch processing, the starting reactants, catalysts and other feedstocks are first metered into a vessel. The mixture may then be heated, stirred or subjected to changes in temperature or pressure that help the starting materials to react with one another. After a prescribed period of time, the reaction proceeds to completion and the product is removed from the reactor.

In continuous reaction processes, a steady stream of starting materials is fed into the reaction vessel. The feed materials then intermingle and react as they flow through the vessel. When the mix of materials exits, any unreacted feedstocks, catalysts and reaction by-products are separated from the product and are removed or recycled back into the reactor.

The importance of having a basic awareness of these chemical processing methods becomes clear when one considers that residue left in any reactor could be composed of both the product and traces of unreacted feed stocks. In other cases, the residue could also contain unwanted by-products that may have formed as a result of side-reactions. In this case, the hazard assessment process must identify each of the by-products.

Operations Performed

The operations performed in the space also influence the nature of the airborne hazards posed to entrants. For example, solvent cleaning, degreasing and spray painting may all produce chemical vapors. Welding and abrasive blasting may result in exposure to toxic fumes and dusts. Even a simple task like squeegeeing waste sludge into a suction line may create localized pockets of hazardous organic vapors when the sludge is agitated.

Inadvertent Entry of Materials

Some thought must also be given to the possibility that hazardous materials may have inadvertently entered the space. For example, a utility manhole situated in the vicinity of an automotive service station may contain gasoline vapors that migrated from a leaking underground storage tank. While the concentration of these vapors may be below the lower explosive limit, they could still pose a potential inhalation hazard.

Carbon monoxide from vehicle exhaust is another hazard that might inadvertently find its way into street manholes, especially on busy city streets with a lot of stop-and-go traffic. Unwanted material may also drip into a space through unprotected openings or leak in through attached pipes.

It can be seen, then, that a thorough physical survey of the space's surrounding environment will be necessary to identify potential routes by which contaminants can enter the space from outside.

TOXICOLOGICAL PRINCIPLES

Tens of thousands of chemicals are used in the industrial environment and their toxicity may vary from relatively harmless to highly toxic. Some, like hydrogen cyanide and phenol, can be life-threatening even at short, low-level exposures. Other materials, like aluminum fume, present a problem only after repeated exposures at relatively high levels. Factors such as the contaminant's concentration, its route of entry into the body, the duration of exposure and an individual's health status also play a role in establishing the level of hazard.

Given the diversity of chemicals in the industrial environment, it is essential that anyone supervising a confined space entry program have a basic knowledge of toxicological principles. This knowledge is essential for two reasons. First, it is necessary in order to clearly and convincingly describe the nature of chemical hazards to entrants. Second, it is virtually impossible to evaluate potentially hazardous atmospheres without a knowledge of exposure standards that are established on the basis of toxic effects.

Toxicity vs. Hazard

Toxicity may be defined as the inherent ability that all substances have to harm the body. In other words, everything is toxic. Think about it, even things like table salt, baking soda and alcoholic beverages can kill you if they are consumed in large enough quantities. Water, which is essential for life, can also be toxic as evidenced by the fact that hundreds of people drown in it every year. Hazard, on the other hand, is the *probability* that a material will cause harm.

The subtle distinction between toxicity and hazard can be clarified by using potassium cyanide as an example.

Potassium cyanide is a highly toxic material. As little as a taste of it could cause death in a very short time. But a bottle of potassium cyanide that is sitting on a shelf in a chemical laboratory poses very little hazard. In order for it to present a hazard, someone must come in contact with its contents. In the case just described, we have a situation where a *highly toxic material poses very little hazard* because no one is exposed to it.

Conversely, acetone, which is a constituent in many consumer products including adhesives, paints and nail polish, is generally considered to have a relatively low order of toxicity. Yet exposure to acetone in a confined space can be exceptionally hazardous because it can adversely affect the central nervous system, producing dizziness and narcosis. An entrant who didn't evacuate promptly after the onset of any of these symptoms may later become incapacitated and unable to self-rescue.

While the range of toxicity for different materials extends from the relatively harmless to the extremely toxic, the relative toxicity of a specific substance remains constant. On the other hand, the hazard posed by a toxic material varies with the exposure situation. As discussed above, there are some cases where highly toxic materials present very little hazard, and others where relatively harmless substances can pose serious hazards.

Dose-Response Relationship

Loosely described, "dose" is the amount of a chemical someone is exposed to, and "response" is the effect that the chemical has on the body. In general, the greater the dose the subject receives, the greater the effect it will have on the body.

The dose-response relationship (Figure 4-1) begins with the assumption that there is some small dose, below which, no adverse effects will occur. This point is known as the threshold of response. At levels slightly above this threshold the body may experience some adverse effects, but it recovers without permanent injury. At higher dose levels the body may incur permanent damage. And at still higher levels, death will result. Medicinal use of aspirin provides a good example of the dose-response relationship. Two aspirin taken orally will relieve headache pain in about fifteen minutes. A spoonful, on the other hand, may produce severe stomach distress, and consumption of a few bottles of aspirin could be fatal.

As explained previously, all materials are not equally toxic. Instead, different materials exhibit different degrees of toxicity. For the most part, these

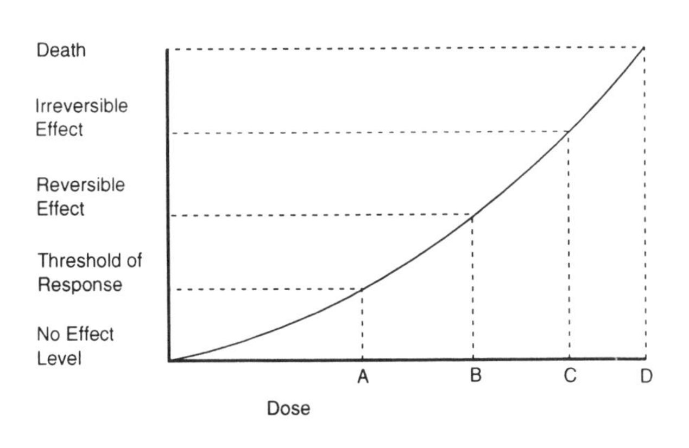

Figure 4-1. Dose-response relationship. (Paul Trattner.)

degrees of toxicity are established on the basis of animal studies. The results of these studies are then extrapolated to predict probable lethal doses for humans.

One of the most frequently conducted toxicological tests is the determination of amount of a material that will kill half of a test population of animals. This value is called the LD_{50} since it is the lethal dose for 50% of the test subjects. LD_{50}s are expressed in units of milligrams of material per kilograms of animal body weight (mg/kg) and provide a relative index of the degree of toxicity. In other words, materials with low LD_{50}s are generally more toxic than those with high LD_{50}s. The LD_{50} for thousands of other chemicals can be found in the *Registry of Effects of Chemical Substances* published by the National Institute for Occupational Safety and Health.

Local vs. Systemic Effects

Toxic materials that have an adverse effect at the site where the material contacts the body are said to have a local effect. Some examples of local effects include skin and eye irritation resulting from a caustic splash, and nose, throat and lung irritation brought on by exposure to gases like ammonia, ozone or phosgene.

In contrast, those effects that manifest at a location other than the site of contact are called systemic effects. Exposure to carbon monoxide provides a good illustration of systemic effects. Although carbon monoxide enters the body through inhalation, it has virtually no effect on either the lungs or respiratory tract. Instead, once it gets into the blood, it interferes with hemoglobin's ability to transport oxygen throughout the body. As a consequence, the victim is chemically asphyxiated even though the ambient air contains enough oxygen to support life.

It is important to note that the local and systemic effects a particular chemical exert on the body are usually quite different. For example, cadmium fumes formed during welding can be irritating to the lungs and respiratory tract (local effect) but it can also produce irreparable damage to the kidneys (systemic effect). The local and systemic effects of a few selected materials are presented in Table 4-1.

Acute vs. Chronic

The terms acute and chronic may be applied to both the duration of exposures and the time it takes the effects to appear. Acute exposures are those that take place over a short period of time, usually minutes or hours or days, while chronic exposures may occur over months, years or even decades.

Similarly, acute effects are those which appear shortly after exposure—usually within one to three days—chronic effects, on the other hand, may not show up for months or years.

In the context of contaminant concentration, acute effects are often associated with a single, brief exposure to a relatively high contaminant concentration, while chronic effects are associated with repeated exposures to relatively low concentrations. Like the differences observed between a particular substance's local and systemic effects, the acute and chronic effects of exposure are usually quite different. For example, an exposure to a high concentration of benzene vapor for a short period of time (acute exposure) will produce narcotic effects. On the other hand, low-level, long-term exposures (chronic exposure) to benzene have been associated with the development of leukemia. Similarly, acute exposure to many chlorinated solvents adversely affects the brain and central nervous system, while chronic exposure may damage the liver and kidneys.

EXPOSURE STANDARDS

The history of occupational exposure standards can be traced to 1938 when a group of governmental industrial hygienists met to discuss their work. This informal group continued to meet annually, and in 1945, the now fledgling American Conference of Governmental Industrial Hygienists (ACGIH) adopted a list of recommended limits for 150 materials developed by one of its members, Warren A. Cook. Cook's initial list has been updated and expanded over the years and the present list contains recommended limits for almost 650 materials.

In the early stages of development, these limits were known as Maximum Allowable Concentrations, Maximum Permissible Concentrations and Maximum Acceptable Concentrations. The feeling then was that these were levels which were never to be exceeded. However, as time went on the philosophy changed. Eventually it was agreed that short-term exposure to slightly higher concentrations could be allowed without significant risk of harm—provided that the total exposure during the day was sufficiently below the maximum level. This approach eventually led ACGIH to develop the concept of the Threshold Limit Value.

Table 4-1. Comparison of local and systemic effects of selected materials

Material	Local	Systemic
Amyl alcohol	Mild irritant to skin, eyes, and upper respiratory tract	Low levels produce vertigo and muscular weakness, narcotic at higher levels
Carbon tetrachloride	Repeated skin contact may lead to dry, scaly fissured dermatitis	Central nervous system depression and gastrointestinal symptoms, liver and kidney damage
Carbon disulfide	Irritating to skin, eyes and mucous membranes; skin contact may cause blistering with second and third degree burns	May cause neurological and cardiovascular disorders, headache, dizziness, diminished mental and motor ability with staggering gait and less of coordination
Cadmium	Respiratory tract irritant	Kidney damage, mild anemia, increased incidence of prostrate cancer
Cobalt	Mildly irritating to the eyes, slightly irritating to the skin but may product allergic skin response	Asthma-like disease with cough which may progress to interstitial pneumonia with marked fibrosis
Hydrogen cyanide	Mild irritant to eye, nose and upper respiratory tract	Inactivates certain enzymes essentials for oxygen transport on a cellular level
n-Heptane	Dermatitis and irritation of the mucous membrane; aspiration of liquid may produce chemical pneumonitis, pulmonary edema and hemorrhage	Narcosis resulting in vertigo, incoordination, intoxication characterized by hilarity, slight nausea, loss of appetite, and persistent gasoline taste in the mouth
Phenol	Has marked corrosive effects on body tissues; prolonged contact with skin can cause severe burns	Weakness, sweating, headache, frothing of the nose and mouth, dark-colored urine, damage to the liver and kidneys
Vinyl chloride	Skin irritant, evaporating liquid may cause frostbite	Depresses central nervous system resulting in symptoms resembling alcohol intoxication; liver damage, human carcinogen

Source: Occupational Diseases, A Guide to their Recognition, National Institute for Occupational Safety and Health.

Threshold Limit Values

The term Threshold Limit Value® (TLV) is the registered trademark of the ACGIH and refers to the airborne concentration of a contaminant to which it is believed that almost all workers may be repeatedly exposed, day-after-day, over a working lifetime without developing adverse effects. It is important to note, however, that TLVs are not sharp lines of distinction between "safe" and "unsafe" conditions. Because of individual differences in physiology, metabolism and biochemistry, some workers may experience adverse effects at levels below the TLV, while others will not be adversely affected even at levels well above the TLV.

TLVs are based on industrial experience as well as experimental studies conducted on both humans and animals. As a consequence, the basis for establishing a TLV at a particular level varies depending on the substance. For example, the TLVs for ammonia, chlorine and sulfur dioxide were established on the basis of eye and respiratory tract irritation. On the other hand, the TLVs for ethyl butyl ketone, ethyl ether and toluene were established on the basis of anesthetic or narcotic effects; still others like toluene 2,4-diisocyanate and benzene were chosen to prevent sensitization or systemic effects such as changes in the blood or damage to the liver, kidney or central nervous system.

Since the amount and nature of toxicological information varies from substance to substance, TLVs must not be used as a relative index of toxicity. For example, ammonia, with a TLV of 25 parts per million (ppm), should not be considered 4 times more toxic than toluene which has TLV of 100 ppm. The 25 ppm TLV for ammonia was established to prevent eye and respiratory tract irritation while the 100 ppm TLV for toluene was selected largely to prevent narcotic effects.

Each year the ACGIH publishes a pocket-sized booklet which includes both the *Threshold Limit Values for Chemical Substances and Physical Agents* and *Biological Exposure Indices.* A portion of a typical page from the booklet is shown in Figure 4-2. Note that TLVs for gases and vapors are given in units of ppm and milligrams per cubic meter (mg/m^3), but TLVs for particulates are listed only in units of mg/m^3.

The ACGIH also publishes a companion to the TLV booklet called *The Documentation of the Threshold Limit Values.* Although *The Documentation* is an essential text for rational application of the TLVs, its existence is virtually unknown outside the circle of professional industrial hygienists.

The Documentation is essentially a reference text which identifies and summarizes the information the ACGIH TLV Committee used to establish each TLV. Most importantly, it explains the rational for each TLV by explaining how and why specific values were chosen. Examples of the documentation for two materials are provided in Figures 4-3 and 4-4. Given the wide range of reasons for establishing TLVs, it is essential that the latest edition of *The Documentation* always be consulted to determine the basis for a specific TLV.

Copies of the latest TLV booklet and *The Documentation* may be obtained from the American Conference of Governmental Industrial Hygienists, 1300 Kemper Meadow Drive, Cincinnati, OH 45240.

	ADOPTED VALUES			
	TWA		STEL	
Substance [CAS #]	ppm[a)]	mg/m^3[b)]	ppm[a)]	mg/m^3[b)]
tert-Butyl acetate [540-88-5] (1987)	200	950	—	—
n-Butyl acrylate [141-32-2] (1978)	10	52	—	—
n-Butyl alcohol [71-36-3] — Skin (1977)	C 50	C 152	—	—
sec-Butyl alcohol [78-92-2] (1990)	100	303	—	—
‡tert-Butyl alcohol [75-65-0] (1976)	100	303	(150)	(455)
n-Butylamine [109-73-9] — Skin (1976)	C 5	C 15	—	—
••tert-Butyl chromate, as CrO_3 [1189-85-1] — Skin (1977)	—	C 0.1	—	—
•n-Butyl glycidyl ether (BGE) [2426-08-6] (1981)	25	133	—	—
n-Butyl lactate [138-22-7] (1977)	5	30	—	—
Butyl mercaptan [109-79-5] (1977)	0.5	1.8	—	—
o-sec-Butylphenol [89-72-5] — Skin (1980)	5	31	—	—

Figure 4-2. TLVs for selected materials. (From *1992-1993 Threshold Limit Values for Chemical Substances and Physical Agents.* American Conference of Governmental Industrial Hygienists, used with permission.)

Time Weighted Averages (TWAs)

Most TLVs are expressed as 8-hour time-weighted averages. The time-weighting aspect allows excursions above the established limit, provided that those excursions are offset by exposure below the TLV. Since the contaminant level varies with time, the calculation used for averaging the exposure must incorporate both concentration and time components.

TWAs for Single Substance Exposures. In the simplest situation, a worker exposed all day to a constant 50 ppm carbon monoxide level would have an 8-hour time-weighted average exposure of 50 ppm. If conditions changed so that he was exposed to 100 ppm for the first half of the day, but received no exposure during the second half, his 8-hour time-weighted average exposure would still be 50 ppm.

Rather than calculating a simple average by dividing 8 hours into 100 ppm, consideration must be given to the length of time the worker was exposed to each concentration. In the example above, the overall exposure would be calculated to reflect the exposure conditions of 4 hours at 100 ppm and 4 hours at 0 ppm as shown below.

$$\frac{(100 \text{ ppm} \times 4 \text{ hours}) + (0 \text{ ppm} \times 4 \text{ hours})}{8 \text{ hours}}$$

$$= \frac{400 \text{ ppm} - \text{hours}}{8 \text{ hours}} = 50 \text{ ppm}$$

Similarly, an exposure at 200 ppm for 2 hours and no exposure for the remaining 6 hours also results in an 8-hour time-weighted average of 50 ppm.

$$\frac{(200 \text{ ppm} \times 2 \text{ hours}) + (0 \text{ ppm} \times 6 \text{ hours})}{8 \text{ hours}}$$

$$= \frac{400 \text{ ppm - hours}}{8 \text{ hours}} = 50 \text{ ppm}$$

The 8-hour time-weighted average for any exposure can be calculated using the equation:

$$\frac{C_1T_1 + C_2T_2 + C_3T_3 \cdots C_nT_n}{T_1 + T_2 + T_3 \cdots T_n}$$

= Time weighted average

where C_1, C_2, etc. indicate contaminant concentration and T_1, T_2, etc. indicate the exposure time.

As shown in Figure 4-2, the TLV-TWA values appear in the first two columns of the TLV booklet. The CAS notation to the right of the TWA column is the Chemical Abstract Service number. The CAS number is a unique number assigned by the American Chemical Society to each chemical substance.

TWAs for Exposures to Mixtures. As noted previously, employees are usually not exposed to a single substance, but rather to a mixture of materials. Lacquer thinners, for example, may contain an assortment of materials such as acetone, methyl ethyl ketone and toluene. While the TLV for each individual material may not be exceeded, each contributes toward the total exposure. In order to determine the total exposure, the contribution of each material must be considered by comparing its exposure concentration to the TLV. As shown below, if the sum of these ratios exceeds unity, an overexposure to the mixture exists.

$$\frac{C_1}{TLV_1} + \frac{C_2}{TLV_2} + \frac{C_3}{TLV_3}$$

= 1 (over exposure exists)

where

C_1 = concentration of contaminant 1
C_2 = concentration of contaminant 2
C_3 = concentration of contaminant 3

and

TLV_1 = TLV of substance 1
TLV_2 = TLV of substance 2
TLV_3 = TLV of substance 3

Short-Term Exposure Limits

One of the limitations of 8-hour time-weighted averaging is that it does not take into account situations in which there is exposure to a high concentration of material, but for only a short period of time. This single, high level exposure, could in itself result in adverse effects, even though the 8-hour time-weighted average is below the Threshold Limit Value. For example, the 8-hour time-weighted average of a worker exposed to a 3200 ppm of toluene for 15 minutes would be approximately 100 ppm. Since the TLV for toluene is also 100 ppm, this exposure would not appear to be a problem. In reality, however, an exposure of 3200 ppm for 15 minutes would produce adverse health effects including narcosis, dizziness and eye and throat irritation.

This limitation posed by 8-hour time-weighted averaging may be addressed through the use of short-term exposure limits (STELs). STELs are concentrations above the 8-hour average to which workers may be exposed for short periods of time without suffering from irritation, irreversible tissue damage or narcosis that could increase the chance of accidental injury or impair self-rescue. For toluene, this level is 150 ppm. It is important to note that STELs are not independent exposure limits. Instead, they supplement the 8-hour time-weighted average values in situations where an acute adverse response may result from exposure to materials with a TLV established on the basis of chronic exposure.

STELS are defined as 15-minute time-weighted averages that should not be exceeded even if the 8-hour time-weighted average is below the TLV. Exposures at the STEL should not be permitted for longer that 15 minutes and should occur no more than 4 times during the day with at least 1 hour between 2 successive exposures.

As shown in Figure 4-2, STEL values appear in the third and fourth columns of the TLV booklet.

Ceiling Concentrations

The nature of the adverse effects of some materials is such that neither a time-weighted average nor a STEL provides adequate protection. In other words,

BENZYL CHLORIDE

CAS: 100-44-7

α-Chlorotoluene

C_7H_7Cl

$C_6H_5-CH_2Cl$

TLV–TWA, 1 ppm (5.2 mg/m^3)

1954: TLV–TWA, 1 ppm, proposed

1956–present: TLV–TWA, 1 ppm

1991: Documentation revised

Chemical and Physical Properties

Benzyl chloride is a colorless, refractive liquid with a pungent odor. The stabilized form of benzyl chloride contains a fixed amount of a sodium carbonate solution or propylene oxide. Chemical and physical properties include:

Molecular weight: 126.58
Specific gravity: 1.100 at 20°C
Boiling point: 179°C
Freezing point: –39°C
Vapor pressure: 1.0 torr at 22°C
Flash points: 67°C, closed cup; 74°C, open cup
Lower explosion limit: 1.1% by volume in air
Autoignition temperature: 525°C
Solubility: insoluble in water; miscible with most organic solvents
Reactivity: very reactive; unless stabilized, it undergoes a Friedel–Crafts-type condensation when exposed to certain metals, liberating hydrogen chloride

Major Uses or Sources of Occupational Exposure

Benzyl chloride is a chemical intermediate in the manufacture of dyes, plasticizers, lubricants, gasoline additives, pharmaceuticals, tanning agents, and quaternary ammonium compounds.

Animal Studies

Acute

Two-hour LC_{50} values of 80 ppm and 150 ppm benzyl chloride are cited for the mouse and rat, respectively.[1] Back et al.[2] reported that all mice and rats survived a 1-hour exposure at 400 ppm. The difference in the results of these two studies cannot be explained. Rabbits and cats exposed 8 hours/day for 6 days at 95 ppm showed eye and respiratory tract irritation, while a dog died following 8 hours at 380 ppm.[3] Skin sensitization in guinea pigs has been reported.[4]

Chronic/Carcinogenicity

Weekly, subcutaneous, high dose (80 mg/kg) administration of benzyl chloride for 51 weeks resulted in injection site sarcomas, with lung metastases, in rats; the mean induction time was 500 days. At half this dosage, there were some local sarcomas but no metastases.[3,5] The National Institute for Occupational Safety and Health (NIOSH) concluded that the presently available data are insufficient upon which to base a firm conclusion as to the carcinogenic potential of benzyl chloride.[3]

In a study by Lijinsky,[6] benzyl chloride was administered by gavage in corn oil at a dose of 50 or 100 mg/kg body weight (mice) and 15 or 30 mg/kg (rats) 3 times/week for 2 years. A statistically significant increased incidence of papillomas and carcinomas of the forestomach was observed in mice of each sex. The only statistically significant increased incidence of neoplasms in the rats (female only) was for thyroid C-cell tumors. A few neoplasms of the forestomach were observed in male rats. Based on the subcutaneous and gavage studies, benzyl chloride was evaluated by the International Agency for Research on Cancer (IARC)[7] to have limited evidence for carcinogenicity in animals.

Genotoxicity Studies

The IARC review of benzyl chloride[7] reported that the substance did not induce micronuclei in mice treated *in vivo*. It induced DNA strand breaks, but not unscheduled DNA synthesis or chromosomal aberrations in cultured human cells. Conflicting results were obtained for the induction of sister-chromatid exchanges in human cells. In cultured rodent cells, benzyl chloride induced sister-chromatid exchanges, chromosomal aberrations, mutation and DNA strand breaks. It induced somatic and sex-linked recessive lethal mutations in Drosophila; mitotic recombination, gene conversion, mutation and DNA damage in fungi; and mutation and DNA damage in bacteria.[7]

Human Studies

According to Smyth,[8] "This [benzyl chloride] is a potent lacrimator irritating to the eye, nose, and throat and capable of causing lung edema. . . . It may be inferred that the liquid causes corneal injury. . . . The 1 ppm threshold limit can be derived from older human sensory data. It is undoubtedly low enough to prevent lung injury."

From references cited in the NIOSH criteria document for benzyl chloride,[3] exposure at 1.5 ppm for 5 minutes can result in slight conjunctivitis, and 8 ppm is the eye irritation threshold for a 10-second exposure. A single breath of air containing 35 ppm of benzyl chloride will reportedly cause nasal irritation. Flury and Zernik[9] reported that a 1-minute exposure at 16 ppm was intol-

Figure 4-3. Documentation for benzyl chloride. (From *Documentation of the Threshold Limit Values,* American Conference of Governmental Industrial Hygienists, used with permission.)

erable to man.

Studies in two plants making benzyl chloride in the United States indicated vapor concentrations generally were between 0.02 and 0.1 ppm, with one sample showing 0.27 ppm.[3] Slightly higher levels, up to 0.4 ppm, were reported from a benzyl alcohol production facility in the Soviet Union.

TLV Recommendation

Based on the available data up to 1980, the TLV Committee recommends a TLV–TWA of 1 ppm (5.2 mg/m^3) for occupational exposure to benzyl chloride. This value should prevent lung injury and irritation of the eye, nose, and throat. At this time, no STEL is recommended until additional toxicological data and industrial hygiene experience become available to provide a better base for quantifying on a toxicological basis what the STEL should be. The reader is encouraged to review the section on *Excursion Limits* in the "Introduction to the Chemical Substances" of the current TLV/BEI Booklet for guidance and control of excursions above the TLV–TWA even when the 8-hour TWA is within the recommended limits. In view of the recent reports on the carcinogenicity[6] and genotoxicity[7] of benzyl chloride, this substance is under review by the TLV Committee.

Other Recommendations

OSHA PEL: The OSHA PEL–TWA for benzyl chloride is 1 ppm. Benzyl chloride was one of the 160 substances for which the PEL was unchanged and was not evaluated during the 1989 OSHA rulemaking on air contaminants — permissible exposure limits.[10] The OSHA PEL is consistent with the recommendation by ACGIH.

NIOSH REL/IDLH: NIOSH established a REL–TWA for benzyl chloride of 5 mg/m^3 (1 ppm) as a 15-minute ceiling. NIOSH [Ex 8-47, Table N7] did not concur with the OSHA PEL since NIOSH recommends a ceiling value for benzyl chloride.[11] NIOSH established 10 ppm as the IDLH value for benzyl chloride.

ACGIH Rationale for TLVs that Differ from the PEL or REL: The NIOSH REL, as a 1 ppm ceiling, is predicated on a single report of slight conjunctivitis in workers exposed at 6–8 mg/m^3 for 5 minutes. This study failed to report information relating to sampling and analytical methods, number or percentage of workers affected, or possible exposures to other chemicals. This study also reported severe neurologic symptoms and hematologic signs attributed by the authors as liver dysfunction in workers exposed at 10 mg/m^3 (2 ppm). The TLV of 1 ppm (5 mg/m^3) has been recommended by ACGIH to provide protection against the significant risk of irritation to the eyes and mucous membranes and systemic toxicity of benzyl chloride.

NTP Studies: NTP has not reported long-term toxicology and carcinogenesis effects studies on benzyl chloride. Benzyl chloride was weakly positive in the *Salmonella* assay, positive in the mouse lymphoma assay, and positive in cultured Chinese hamster cells for increased frequencies of both chromosomal aberrations and sister-chromatid exchanges.

Carcinogenic Classification

IARC: Group 2B, possibly carcinogenic in humans.
MAK: Group B, justifiably suspected of having carcinogenic potential.

Other Nations

Australia: 1 ppm (1990); Federal Republic of Germany: 1 ppm, short-term level 2 ppm, 5 minutes, 8 times per shift; Group B, justifiably suspected of having carcinogenic potential; Pregnancy Group D, a trend of data indicate potential for embryo/fetotoxicity (1990); Sweden: 1 ppm, short-term value 2 ppm, 15 minutes, carcinogenic (1984); United Kingdom: 1 ppm (1991).

References

1. Mikhailova, T.V.: Comparative Toxicity of Chloride Derivatives of Toluene — Benzyl Chloride, Benzal Chloride and Benzotrichloride. Fed. Proc. 24:T877–880 (1965).

2. Back, K.C.; Thomas, A.A.; MacEwen, J.D.: Reclassification of Materials Listed as Transportation Health Hazards. Report TSA-20-72-3. U.S. Department of Transportation, Office of Hazardous Materials, Office of the Assistant Secretary for Safety and Consumer Affairs, Washington, DC (1972).

3. National Institute for Occupational Safety and Health: Criteria for a Recommended Standard — Occupational Exposure to Benzyl Chloride. DHEW (NIOSH) Pub. No. 78-182; NTIS No. PB-81-226-698. National Technical Information Service, Springfield, VA (1978).

4. Landsteiner, K.; Jacobs, J.: Studies on the Sensitization of Animals with Simple Chemical Compounds — II. J. Exp. Med. 64:625–639 (1936).

5. Preussman, R.: Direct Alkylating Agents as Carcinogens. Food Cosmet. Toxicol. 6:576–577 (1968).

6. Lijinsky, W.: Chronic Bioassay of Benzyl Chloride in F344 Rats and (C57BL/6JxBALB/c) F1 Mice. J. Natl. Cancer Inst. 76:1231–1236 (1986).

7. International Agency for Research on Cancer: IARC Monographs on the Evaluation of the Carcinogenic Risk of Chemicals to Humans, Suppl. 7, Overall Evaluations of Carcinogenicity: An Updating of IARC Monographs Volumes 1 to 42, pp. 148–149. IARC, Lyon France (1987).

8. Smyth, Jr., H.F.: Improved Communication — Hygienic Standards for Daily Inhalation. Am. Ind. Hyg. Assoc. Q. 17(2):129–185 (1956).

9. Flury, G.; Zernik, F.: Schadliche Gase, p. 339. J. Springer, Berlin (1931).

10. U.S. Department of Labor, Occupational Safety and Health Administration: 29 CFR Part 1910, Air Contaminants; Final Rule. Fed. Reg. 54(12):2925 (January 19, 1989).

11. National Institute for Occupational Safety and Health: Testimony of NIOSH on the Occupational Safety and Health Administration's Proposed Rule on Air Contaminants: 29 CFR Part 1910, Docket No. H-020; Table N7 (Appendix A) (August 1, 1988).

Figure 4-3. Continued.

CYCLOHEXANOL

CAS: 108-93-0

Cyclohexyl alcohol; Hexahydrophenol; Hydralin; Hydroxycyclohexane

$C_6H_{12}O$

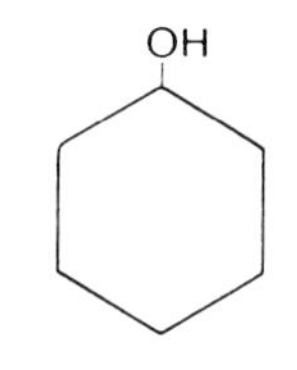

Skin

TLV–TWA, 50 ppm (206 mg/m^3)

1946–1947: MAC–TWA, 100 ppm

1948–1960: TLV–TWA, 100 ppm

1961–present: TLV–TWA, 50 ppm

1986–present: Skin notation

1991: Documentation revised

Chemical and Physical Properties

Cyclohexanol is a colorless, viscous liquid or a sticky solid with a camphor- or menthol-like odor. An odor threshold of 0.15 ppm has been reported.[1] Chemical and physical properties include:[2]

Molecular weight: 100.16
Specific gravity: 0.962 at 20°C
Melting point: 23°–25°C
Boiling point: 161°C
Vapor pressure: 1.0 torr at 20°C
Vapor density: 3.46 (air = 1.0)
Closed cup flash point: 68°C
Autoignition temperature: 300°C
Lower explosive limit: 2.4% by volume in air
Solubility: soluble in water (3.6 g/100 g at 20°C); miscible with aromatic hydrocarbons, ethanol, ethyl acetate, linseed oil, and petroleum solvents

Cyclohexanol is prepared by the catalytic air oxidation of cyclohexane or by the catalytic hydrogenation of phenol. Oxidation of cyclohexane to a mixture of cyclohexanol and cyclohexanone is the most common process in industry and yields a product known as K-A oil (a ketone, alcohol, cyclohexanone–cyclohexanol mixture).[3]

Major Uses or Sources of Occupational Exposure

Cyclohexanol is used as the source of adipic acid in the production of nylon. Other uses are in the production of lacquers, paints, varnishes, degreasers, plastics and plasticizers, soaps and detergents, textiles, and insecticides.[2–4]

Animal Studies

Acute

The single, oral LD_{50} of cyclohexanol for the rat is 2.06 g/kg,[5] and the minimumm lethal dose is between 2.2 and 2.6 g/kg.[6] Intravenously, the LD_{50} for a single administration in the mouse is 270 mg/kg.[7] When given intramuscularly, the LD_{50} is 1.0 g/kg.[7] Instilled into the eye of the rabbit, moderately severe irritation and corneal injury, which was reversible, occurred.[8] While only slightly irritating to the rabbit skin, cyclohexanol can be absorbed through the skin in toxic amounts which, at high concentrations and if applied to extensive skin areas, will result in tremors, narcosis, hypothermia, and death.[8]

Subchronic

Repeated or prolonged contact of cyclohexanol with the skin of rabbits elicited temporary erythema and superficial sloughing of the skin when cyclohexanol as a 15% ointment was applied for 1 hour/day over a period of 10 days. Death, preceded by tremors, narcosis, and hypothermia, occurred on the 11th day following initial skin application.[8]

Inhalation of "high concentrations" of cyclohexanol for 6 hours/day, 5 days/week for 5 to 11 weeks resulted in intoxication manifested by conjunctival irritation, lacrimation, salivation, incoordination, narcosis, mild convulsions, and death in 50% of rabbits exposed at concentrations of 1229 ppm and 997 ppm for 150 and 300 hours, respectively.[9] One monkey and eight rabbits survived concentrations below 700 ppm for 300 hours.[9] Toxic degenerative changes were found in the brain, heart, liver, and kidneys of the rabbits exposed at the 997 ppm and 1229 ppm concentrations.[9] Rabbits exposed at a concentration of 145 ppm suffered only slight degenerative changes in the liver and kidneys.[9]

Reproductive/Developmental

The toxicity of cyclohexanol was investigated in two strains of mice by Gondry.[10] Pregnant and nonpregnant mice were given 0.1%, 0.5%, or 1.0% cyclohexanol in the diet through gestation, lactation, and weaning for several generations. Growth and mortality of the animals were monitored. The 1.0% concentration produced a significant increase in mortality and growth of the offspring during the 21 days after birth. In TB strain mice, the percentage of mortality in first- and second-generation cyclohexanol-treated offspring was 14.1% and 53.5%, respectively, as compared with 11.9% for first- and second-generation controls. In first generation offspring of MNRI mice, the mortality of treated mice was 43.1%, while the mortality for the controls was 12.2%. The author[10] concluded that the toxicity of cyclohexanol is carried into successive generations but that normal growth resumes when administration of the cyclohexanol

Figure 4-4. Documentation for cyclohexanol. (From *Documentation of the Threshold Limit Values.* American Conference of Governmental Industrial Hygienists, used with permission.)

is interrupted.

Genotoxicity Studies

Collin,[11] in a review of the mutagenicity of cyclohexanol and related compounds, concluded that no members of this class of chemicals can be considered mutagenic. Cyclohexanol was not mutagenic when tested in several strains of *Salmonella typhimurium* in concentrations of 500 μg/plate either in the presence or absence of a rat liver homogenate activation system.[7]

Human Studies

Data on the toxic effects of cyclohexanol on humans are very limited. Nelson et al.[12] exposed ten human subjects to various concentrations of cyclohexanol for 3–5 minutes. The estimated acceptable concentration in air for 8 hours was reported by the volunteer subjects to be less than 100 ppm. No person was actually exposed to any of the test concentrations for 8 hours.

In a group of 174 women and 279 men who were exposed daily to less than the "permitted" concentration of cyclohexanol, 114 individuals showed nonspecific disturbances of the autonomic nervous system during a 2-year period, while only 8 out of 100 persons in a nonexposed control group had similar disturbances.[7]

In a 48-hour closed patch test using 4% cyclohexanol in petrolatum, there was evidence of erythema or edema in human subjects.[7]

TLV Recommendation

Using the estimated acceptable air concentration for human subjects[12] and the skin absorption data from animals,[8] a TLV–TWA of 50 ppm with a skin notation is recommended for cyclohexanol. At this time, no STEL is recommended until additional toxicological data and industrial hygiene experience become available to provide a better base for quantifying on a toxicological basis what the STEL should be. The reader is encouraged to review the section on *Excursion Limits* in the "Introduction to the Chemical Substances" of the current TLV/BEI Booklet for guidance and control of excursions above the TLV–TWA, even when the 8-hour TWA is within the recommended limits.

Other Recommendations

OSHA PEL: OSHA established a PEL–TWA for cyclohexanol of 50 ppm and added a skin designation.[13] The OSHA PEL is consistent with the recommended ACGIH TLV.

NIOSH REL/IDLH: NIOSH [Ex 8-47, Table N1] established a REL–TWA for cyclohexanol of 50 ppm, with a skin notation, by concurrence with the OSHA PEL.[14] NIOSH established an IDLH value of 3500 ppm for this substance.

NTP Studies: NTP has not conducted long-term toxicology and carcinogenesis effects studies on cyclohexanol. Cyclohexanol was negative in the *Salmonella* assay.

Other Nations

Australia: 50 ppm, skin (1990); Federal Republic of Germany: 50 ppm, short-term level 100 ppm, 30 minutes, 4 times per shift (1990); Sweden: 50 ppm, 15-minute short-term value 75 ppm (1984); United Kingdom: 50 ppm (1991).

References

1. Amoore, J.E.; Hautala, E.: Odor as an Aid to Chemical Safety: Odor Thresholds Compared with Threshold Limit Values and Volatilities for 214 Industrial Chemicals in Air and Water Dilution. J. Appl. Toxicol. 3:272–290 (1983).

2. The Merck Index, 10th ed., p. 391. M. Windholz, Ed. Merck & Co., Inc., Rahway, NJ (1983).

3. Fisher, W.B.; Van Peppen, J.F.: Cyclohexanol and Cyclohexanone. In: Kirk–Othmer Concise Encyclopedia of Chemical Technology, pp. 339–340. M. Grayson, Ed. John Wiley & Sons, New York (1985).

4. Sittig, M.: Cyclohexanol. In: Handbook of Toxic and Hazardous Chemicals and Carcinogens, 2nd ed., pp. 279–280. Noyes Publications, Park Ridge, NJ (1985).

5. Bar, F.; Griepentrog, F.: Die Situation in der Gesundheitlichen Beurteilung der Aromaliscerungsmittel fur Lebensmittel. Med. Ernahr. 8:244–251 (1967).

6. Treon, J.F.; Crutchfield, W.E.; Kitzmiller, K.V.: The Physiological Response of Rabbits to Cyclohexane, Methylcyclohexane, and Certain Derivatives of These Compounds. I. Oral Administration and Cutaneous Application. J. Ind. Hyg. Toxicol. 25:199–214 (1943).

7. Rowe, V.K.; McCollister, S.B.: Alcohols. In: Patty's Industrial Hygiene and Toxicology, 3rd Rev. ed., Vol. 2C, Toxicology, pp. 4643–4649. G.D. Clayton and F.E. Clayton, Eds. John Wiley & Sons, New York (1982).

8. Pohl, J.: The Toxicity of Several Aromatic Hydration Products (Tetralin, Hexalin, and Methylhexalin). Zentr. Gewerbehyg. Unfallverhuet 1:91 (1924).

9. Treon, J.F.; Crutchfield, W.E.; Kitzmiller, K.V.: The Physiological Response of Animals to Cyclohexane, Methylcyclohexane, and Certain Derivatives of These Compounds. II. Inhalation. J. Ind. Hyg. Toxicol. 25:323–347 (1943).

10. Gondry, E.: Studies on the Toxicity of Cyclohexylamine, Cyclohexanone, and Cyclohexanol, Cyclamate Metabolites. Eur. J. Toxicol. 5:227–238 (1972).

11. Collin, J.P.: Cytogenetic Effect of Cyclamate, Cyclohexanone, and Cyclohexanol. Le Diabete 19:215–221 (1971).

12. Nelson, K.W.; Ege, J.F.; Ross, M.; et al.: Sensory Response to Certain Industrial Vapors. J. Ind. Hyg. Toxicol. 25:282–285 (1943).

13. U.S. Department of Labor, Occupational Safety and Health Administration: 29 CFR Part 1910, Air Contaminants; Final Rule. Fed. Reg. 54(12):2718–2719, 2931 (January 19, 1989).

14. National Institute for Occupational Safety and Health: Testimony of NIOSH on the Occupational Safety and Health Administration's Proposed Rule on Air Contaminants: 29 CFR Part 1910, Docket No. H-020; Table N1 (Appendix A) (August 1, 1988).

Figure 4-4. Continued.

a singe exposure above the TLV may pose an acute or immediate irreversible health effect. For these materials, the ACGIH has established ceiling concentrations which are levels that may *never* be exceeded.

The TLVs for materials with ceiling concentrations are preceded with a capital letter "C" as shown for *n*-butyl alcohol, *n*-butylamine and cadmium oxide fume shown in Figure 4-2.

"Skin" Notation

The "skin" notation that appears after some chemical names such as *tert*-butyl chromate and *o*-sec-butyl phenol in Figure 4-2 means that the worker's overall exposure may be affected by absorption of the material through the skin including mucous membranes and the eye. The skin designation is intended to call attention to the fact that adequate precautions must also be taken to protect the rest of the body from exposure. This consideration is especially important in situations where there may be relatively high concentrations of materials that possess a low TLV.

OSHA Permissible Exposure Limits (PELs)

When Congress passed the Occupational Safety and Health Act it empowered OSHA with the authority to adopt by reference any existing national consensus standards like ANSI, NFPA or any other standards that had been previously adopted by other federal regulations. Since 1968, TLV had been adopted by the Walsh-Healy Public Contracts Act; OSHA adopted them and incorporated them into Part 29 of the Code of Federal Regulations as Section 1910.93. During the mid-1970s, they were recodified as Section 1910.1000.

The OSHA standards contained in 29 CFR 1910.1000 employ the same concepts of 8-hour time-weighted averaging, short-term exposure limits and ceiling concentrations. The primary difference between the ACGIH's TLVs and OSHA's PELs is that the *TLVs are recommendations* while the *PELs are legal mandates.*

While ACGIH annually reviewed and updated the TLVs, OSHA continued to enforce the 1968 version. That changed in 1989 when, in an unprecedented move, OSHA adopted a later edition of the TLVs. The adoption of the 1989 TLVs automatically added 164 new materials and reduced the legal standard for 212 other substances. A 4-year phase period was also provided to allow employers ample time to come into compliance.

While this dramatic change was hailed with great fanfare in some circles, not everyone was pleased with it. Industry groups, unions and individual companies all filed suit to stop the adoption process.

Some industry associations argued that OSHA had not followed proper rule-making procedures in adopting the TLVs. The AFL-CIO complained that the new PELs were not protective enough. A handful of individual companies and a few trade associations claimed that the PELs proposed for some substances were not supported by scientific evidence.

All of the petitions were consolidated into a single case heard before the 11th Circuit Court. The court ruled that OSHA's adoption of the TLVs, while well-intentioned, was flawed because the agency had not followed proper rule-making procedures. OSHA immediately asked for a rehearing, but the Court denied the request. OSHA's last avenue of hope was to appeal the case to the U.S. Supreme Court. However, when the administration changed, the appeal fell between the cracks and was never filed. So in March 1993, the PELS reverted back to their 1968 values.

ATMOSPHERES IMMEDIATELY DANGEROUS TO LIFE OR HEALTH

Atmospheric concentrations deemed to be immediately dangerous to life or health (IDLH) for selected materials are presented in Table 4-2. Although IDLH values are not actually exposure standards, it is relevant to mention them here since they provide another frame of reference for assessing hazardous atmospheres.

The term immediately dangerous to life or health was originally coined by the Bureau of Mines as part of its respirator certification program. An IDLH concentration was the maximum level of atmospheric contamination from which a worker could escape in the event of respirator failure without loss of life, or without immediate or delayed irreversible health effects. The maximum time allowed for escape, and thus for exposure, was considered to be thirty minutes.

The OSHA confined space standard, on the other hand, defines an IDLH atmosphere as "... any condition that poses an immediate or delayed threat to life or that would cause irreversible adverse health effects or that would interfere with an individual's ability to escape unaided from a permit space."

Table 4-2. IDLH values for selected materials

Acetaldehyde	10,000	Hydrazine	80
Acetic acid	1,000	Hydrogen chloride	100
Acetone	20,000	Hydrogen selenide	2
Acrolein	5	Hydrogen sulfide	300
Allyl alcohol	150	Isoamyl acetate	3,000
Ammonia	500	Isoamyl alcohol	8,000
n-Amyl acetate	4,000	Isobutyl alcohol	8,000
Aniline	500	Isopropyl alcohol	20,000
Arsine	6	Isopropyl ether	10,000
Benzene	2,000	Methyl acetate	10,000
Benzyl chloride	10	Methyl cellusolve	2,000
2-Butoxy ethanol	700	Methyl chloroform	1,000
Butyl alcohol	100	Methylcyclohexanol	10,000
Carbon dioxide	50,000	Methylene chloride	5,000
Carbon disulfide	500	Methyl ethyl ketone	3,000
Carbon tetrachloride	300	Methyl isobutyl ketone	3,000
Chlorine	25	Methyl isocyanate	20
Chlorine dioxide	10	Nitrobenzene	200
Chlorobenzene	2,400	Nitrogen dioxide	50
Chloroform	1,000	Octane	3,750
Cyclohexane	10,000	Ozone	10
Cyclohexanol	3,500	Pentane	5,000
Cyclohexanone	5,000	2-Pentanone	5,000
1,2 Dichloroethylene	4,000	Phenol	100
Dichloroethyl ether	250	Phosgene	2
Diethylamine	2,000	Phosphine	200
Dimethyl sulfate	10	Phosphorous trichloride	50
Ethanolamine	1,00	*n*-Propyl acetate	8,000
2-Ethoxyethanol	6,000	Propyl alcohol	4,000
Ethyl acetate	400	Propyleneimine	500
Ethyl acrylate	2,000	Propylene oxide	2,000
Ethyl benzene	2,000	Pyridine	3,600
Ethylene dibromide	400	Quinone	75
Ethylene oxide	800	Stibine	40
Ethyl ether	19,000	Styrene	5,000
Ethyl formate	8,000	Sulfur dioxide	100
Formaldehyde	100	Toluene	2,000
Formic acid	100	Toluene 2,4-diisocyanate	10
Furfural	250	1,1,2-Trichloroethane	500
Furfural alcohol	250	Trichloroethylene	1,000
Heptane	4,250	Triethylamine	1,000
Hexane	5,000	Vinyl toluene	5,000
Hexone	3,000	Xylene	10,00

Source: NIOSH Pocket Guide to Chemical Hazards (NIOSH 90-177)

It is important to note that some materials—like cadmium vapor and hydrogen fluoride—may produce immediate transit effects which, even if severe, pass without medical attention. However, they are followed by sudden, and perhaps even fatal, collapse 12 to 72 hours later. In other words, the victim may feel "normal" from recovery of the transient effects until collapse.

PHYSICAL STATES OF MATTER

Airborne contaminants may exist in any of five physical states: dust, fume, vapor, mist or gas. Since each state presents unique hazards, personnel responsible for evaluating confined spaces must be able to recognize them and describe their characteristics. This knowledge is especially critical when it comes to making atmospheric measurements and selecting appropriate ventilation techniques.

Airborne Dusts

Dusts are solid particles generated by processes like cutting, sawing, grinding, sanding or abrasive blasting. They may also be produced when granular materials are blended, transferred or conveyed. Dusts may be further characterized on the basis of their toxicity and particle size.

Some, such as starch, Portland cement and Plaster-of-Paris, are biologically inert and are largely nuisances which irritate the body through simple mechanical action. Others, like cyanide salts and pesticides, may produce adverse systemic effects.

Particles larger than 10 micrometers in diameter are referred to as non-respirable since they are usually removed by the nasal passages and upper respiratory tract before the airstream reaches the lungs. However, particles smaller than 10 micrometers can find their way down into the deep lung passages where they can either accumulate, or dissolve and pass into the bloodstream.

Fumes

Fumes are solid particles that are formed in much the same way as snow. As water evaporates, the vapor travels to the upper atmosphere where it cools, condenses, and eventually falls back to earth as a liquid—rain. However, if the weather is cold enough, the condensate freezes and falls to earth as a solid—snow.

Metals and plastics are the two most common fume-forming materials. Both normally exist as solids, but on heating they melt and form a liquid. This liquid gives rise to vapors which, when cooled, condense into solid particles called fumes. Fumes are almost entirely uniform in size and shape, being spherical and about 1 micrometer in diameter. Target organs for selected metals are indicated in Figure 4-5.

As demonstrated by the case history below, fume-forming processes that pose a significant hazard in confined space include welding and flame cutting, particularly if done on surfaces that have been galvanized (zinc coated); electroplated with heavy metals such as nickel, cadmium or chromium; or painted with lead-based paints.

Case 4-1: Cutting Cadmium Plated Bolts

Two utility workers working in an underground vault found it necessary to remove 26 cadmium plated bolts from a water meter. One of the workers cut the bolts off with an oxy-propane torch while the other assisted. Neither worker wore a respirator, and ventilation was not provided until after the job was completed—and then, only to exhaust the thick blue smoke which had filled the vault.

The worker who cut the bolts became nauseous shortly after finishing the job. A few days later he visited his physician and complained of fever, chest pains, cough and sore throat. On the third day following the exposure, he was feeling worse and was admitted to the hospital. He died two weeks later of massive coronary infarction.

The worker who assisted him also complained of chills, nausea, coughing and difficulty breathing. He was treated for pneumonia and eventually recovered. A reenactment of the job showed that airborne levels of cadmium fumes were well above permissible levels.

Mists

Mists are liquid droplets that are formed by splashing, foaming, atomizing or spraying. Mists may be generated when confined spaces are washed and steam-cleaned prior to entry. They may also be created when liquid surfaces are agitated or when coatings or adhesives are sprayed onto the walls of a space.

Vapors

Vapors are the gaseous phase of materials that usually exist in the solid or liquid state. Vapors may be converted to a liquid either by reducing the temperature or increasing the pressure. While vapors are primarily associated with solvents, some volatile solids like iodine, camphor and naphthalene can also produce vapors as they sublime (passing directly from the solid state to the vapor state without liquefying).

Vapors may remain in a confined space after volatile liquids have been drained off. They can also be formed during operations like solvent degreasing, spray painting and adhesive application. Since many solvent vapors have a depressant effect on the brain and central nervous system, an acute exposure may produce narcosis. Chronic exposure may result in a

METAL	G.I. Tract	Resp. Tract	CNS	Cardio-vascular	Liver	Skin	Blood	Kidney	Bone	Endocrine
Antimony	•	•		•	•	•				•
Arsenic	•	•	•		•	•	•			•
Beryllium		•				•			•	
Cadmium	•	•	•	•				•	•	
Chromium		•	•		•	•		•		
Copper	•						•			
Iron	•	•	•		•		•			
Lead	•		•				•	•		
Magnesium		•	•							
Mercury		•	•					•		
Nickel		•	•			•				
Selenium	•		•		•	•				
Strontium				•					•	
Thallium	•	•	•		•			•		•
Tin (organic)	•		•							
Titanium		•								
Zinc	•					•		•		

Figure 4-5. Target organs for selected metals. (Adapted from *Toxicology: The Basic Science for Poisons,* 1975, Macmillan Publishing Company.) A reenactment of the job showed that airborne levels of cadmium fumes were well above permissible limits.

wide range of systemic effects which include damage to the liver, kidneys and blood-forming system. General effects of exposure to selected, commonly encountered, solvent families are presented in Table 4-3.

It is often possible to see dusts, mists and fumes. Vapors, on the other hand, are invisible. While many vapors have a characteristic odor, individuals entering a confined space might be more likely to dismiss an odor than a contaminant that they could see with their own eyes. Hence, appropriate training in hazard recognition is vital. It is also important to note that not all vapors possess adequate odor warning properties. Some materials, such as toluene 2,4-diisocyanate, carbon tetrachloride and methanol may be hazardous at levels well below that which can be detected by the sense of smell.

The following case histories provide a variety of examples of how the presence of solvent vapors in confined spaces can produce serious consequences.

Case 4-2: Cleaning Airplane Fuel Tank with Trichloroethane

One of the processes at this facility required cleaning the inside of airplane fuel tanks with a solvent. Trichloroethylene had been previously used, but it was replaced with trichloroethane (methyl chloroform) which was believed to present less of a hazard.

Cleaning techniques varied from worker to worker. Some would moisten a pad with solvent and hand wipe the metal surfaces by reaching through an opening on the end of the tank. Others attached a pad to the end of a stick which they then pushed into the tank. Still others cleaned the tank by climbing into it.

One particular worker would saturate a pad with solvent, lower himself head first into the down-tilted tip of the tank, and clean as fast as possible. One day this worker was found unconscious with his legs sticking out of a 450 gallon tank. He was removed immediately and given artificial respiration, but was pronounced dead when a physician arrived.

Investigation of the incident revealed that methyl chloroform concentrations inside the tank could reach levels as high as 62,000 ppm. In addition, workers had generally assumed that since the new cleaning solvent was "safer" than the old one, they were in less danger when they used it. However, they didn't realize that methyl chloroform is a potent anesthetic at 30,000 ppm—half the concentration found in the tank.

Case 4-3: Paint Stripping Tank

Three laborers were assigned the job of removing sludge and other debris from the bottom of a paint-stripping dip tank measuring 10 feet long, 5 feet wide and 6 feet deep. The tank was drained of its methylene chloride stripper, but it was not ventilated and no atmospheric testing was performed.

Wearing splash suits and air-purifying respirators, the three laborers took turns working in the tank. They used crycilic acid to loosen the sludge, which was then scooped into buckets, passed up a ladder, and dumped into 55-gallon drums. After working the rotation for several hours, the worker who was inside the tank collapsed. His coworkers tried unsuccessfully to lift him out with a chain and in the process, one of them passed out and fell face-down into the sludge.

Arriving fire fighters pulled both workers out of the sludge and hosed them down to remove the acid from their clothing. The first labor received chemical burns over 40% of his body and died several days later of anoxia and methylene chloride exposure. The second worker suffered severe ventricular fibrillation and chemical burns to his face and upper-body, but eventually recovered. Four of the responding fire fighters also received chemical burns, but they were treated and released.

Case 4-4: Solvent Exposure in Sewer Regulator Chamber

An industrial waste treatment plant discharged its effluent into a sewer line that flowed a short distance into a regulator chamber. The chamber measured 4-feet wide by 8-feet long and 20-feet deep and was being modernized to operate under computer control. A superintendent and laborer momentarily entered the regulator area to inspect a railing. When they came out, the superintendent left the job site.

The laborer subsequently reentered the chamber and was overcome. He was discovered by a second worker who tied a lifeline to the top of the chamber and entered to attempt rescue. He too was overcome.

When the superintendent returned he found both workers unconscious and also attempted a rescue. He became dizzy and decided to go for help but fell backwards from about the fifth rung of the ladder, sustaining severe lacerations to his forehead and face. He recalled being unconscious for an unknown period of time, and when he came to he climbed back up the ladder, got into his truck, and went for help.

Table 4-3. Toxic effects of selected solvent families

Saturated Aliphatic Hydrocarbons

Straight or branched chains consisting of carbon and hydrogen

```
  H H H H H          H H H H H H          H H H H H H H
  | | | | |          | | | | | |          | | | | | | |
H-C-C-C-C-C-H      H-C-C-C-C-C-C-H      H-C-C-C-C-C-C-C-H
  | | | | |          | | | | | |          | | | | | | |
  H H H H H          H H H H H H          H H H H H H H

   Pentane             Hexane               Heptane
```

Least potentially toxic class of solvents in terms of acute toxicity. Vapors are only mildly irritating to mucous membranes at high concentrations. Liquids beginning with decane (C10) are fat solvents. They are primary irritants and can cause skin irritation upon prolonged or repeated contact. Chronic exposure to some aliphatics, notably hexane and heptane, has been associated with damage to the central nervous system.

Halogenated Hydrocarbons

Hydrocarbons where one or more hydrogen atoms has been replaced with a halogen (e.g., chlorine, fluorine, iodine or bromine)

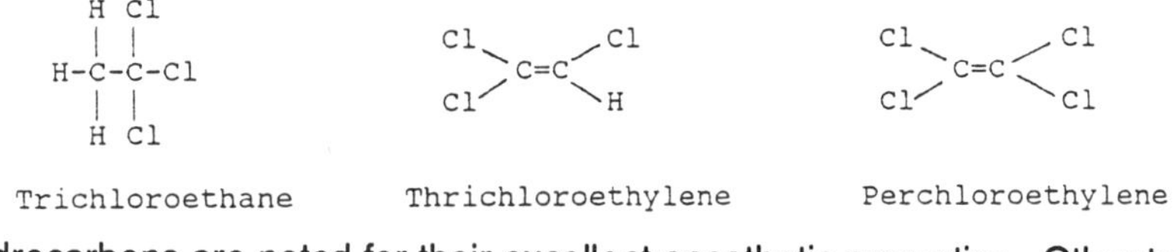

Trichloroethane Thrichloroethylene Perchloroethylene

Halogenated hydrocarbons are noted for their excellent anesthetic properties. Other toxic properties include potential for central nervous system, liver and kidney damage and can sensitize heart to adrenaline which can cause cardiac arrhythmias and cardiac arrest. Systemic toxicity increases with molecular size. The degree of chlorination increases the potential for CNS depression and liver and kidney injuries.

Alcohols

Hydrocarbons where a hydrogen atom is replaced with an -OH group

```
H-C-OH          H-C-C-OH        H-C-C-C-OH      H-C-C-C-C-OH

Methyl alcohol  Ethyl alcohol   Butyl alcohol   Propyl
alcohol
```

Alcohols are irritating to the mucous membranes, but irritation decreases with increasing molecular size. When compared to their aliphatic hydrocarbon analogs, they are more powerful CNS depressants. Toxicity varies, and tertiary alcohols are usually more potent then secondary alcohols, which are more potent than primary alcohols.

Cyclic Hydrocarbons

Ring structures consisting of hydrogen and carbon

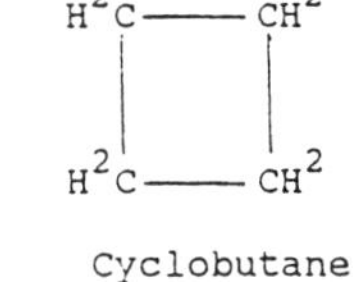

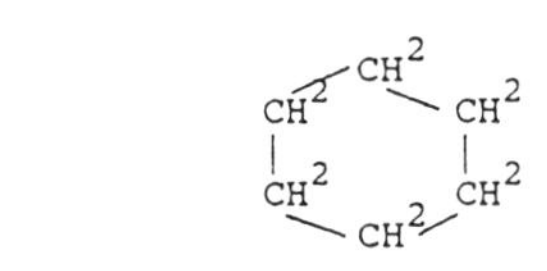

Cyclopropane Cyclobutane Cyclohexane

Industrial experience suggests that little to no chronic effects are observed after prolonged exposure to these compounds. Small alicyclics like cyclopropane have been used as medical anesthetics.

Table 4-3. Continued.

Aromatic Hydrocarbons

Six carbon ring molecules with one hydrogen per carbon atom

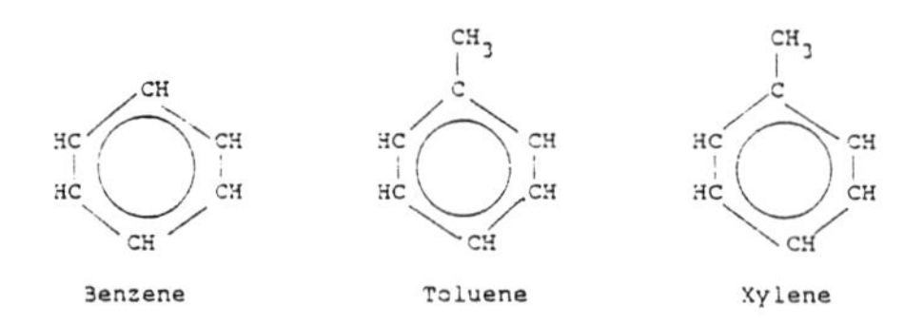

Aromatic compounds are far more irritating than aliphatics. They are primary irritants that can cause dermatitis. Prolonged or repeated exposure can severely defat the skin and lead to tissue injury or chemical burns. Chronic exposure to benzene has been linked to damage to the blood forming system. Substituted aromatics such as toluene and xylene are CNS depressants.

Aldehydes

Molecules with a C=O bonded to a hydrogen atom and a single functional (R) group

H
C = O
R

Aldehydes irritate the skin, eyes and respiratory tract. Although they can cause narcosis, narcotic effects are usually prevented because the irritating properties of most aldehydes provides a warning of exposure. Some aldehydes such as fluoracetaldehyde are converted metabolically into fluorinated acids which have a high level of systemic toxicity. These acids are toxic to all cells because they inhibit normal cellular metabolism.

Ketones

Molecules with a C=O bonded to two functional (R) groups

R′
C = O
R

Ketones are CNS depressants, but the vapor concentrations high enough to cause sedation are also irritating to the eyes and respiratory tract. However, low level concentrations may accumulate, impair judgement and lead to respiratory failure. Saturated ketones are less toxic than unsaturated species, and toxicity increases with molecular weight.

Ethers

Organic molecules which contain a C-O-C linkage

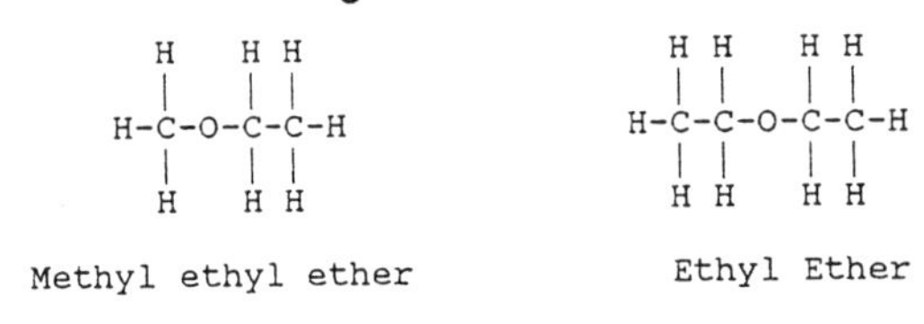

Ethers have excellent anesthetic properties that increase with the size of the molecules. Halogenated ethers can severely irritate the skin, eyes and lungs. Aromatic ethers are less irritating, less volatile and less toxic than straight chain ethers. Some chlorinated ethers are known carcinogens.

Table 4-3. Continued.

Esters

Compounds formed by the reaction of an acid and alcohol

$$R - C \begin{matrix} \nearrow O \\ \searrow OR' \end{matrix}$$

Esters are more potent anesthetics than ketones, alcohols and aldehydes, but weaker than halogenated hydrocarbons and ethers. Simple aliphatic esters are anesthetics. Halogenated species may cause severe lung irritation and lacrimation. Phosphate esters, which are often used as plasticizers, are very irritating and may act as depressants or convulsants. The toxicity of esters varies widely depending on the function groups attached.

In the interim, two consultant engineers who were making a routine visit to the chamber saw the two workers and flagged down a passing police car. Rescue units arrived shortly thereafter and retrieved the victims. One worker was pronounced dead on arrival at the hospital, the other was treated for exposure to toxic vapors and subsequently developed pneumonia. The supervisor was also treated for solvent vapor exposure and lacerations.

Test taken five hours after the incident showed the presence of high levels of toluene, xylene, tetrachloroethane and ethyl benzene. Apparently they had been illegally dumped into the sewer line.

Case 4-5: Tank Car Containing TDI Residue

A supervisor and two tank cleaners planned to clean two rail tank cars which previously contained toluene 2,4-diisocyanate (TDI). One worker entered the car and picked up solid chunks of TDI which were placed into a 35 gallon drum. The other worker outside used an “A” frame assembly mounted over the hatch-way to raise and lower the drum.

The workers cleaned the first car without incident and then broke for lunch. After the lunch break, they switched job duties and the man who previously operated the hoist entered the car. He worked for only a few minutes and suddenly collapsed. The worker outside yelled out to the supervisor for help and then climbed in to assist his coworker. The supervisor called for emergency services and returned to the car. Looking inside he saw the worker who had alerted him tying a rope around the unconscious man’s waist. The supervisor lowered the hook attached to the hoist and the worker inside attached it to the unconscious victim. However, as the victim was being pulled out, the rope slipped and he fell back to the bottom of the car.

They tried to pull the unconscious worker out three more times, but were unsuccessful. Emergency responders eventually arrived and were able to get the worker out of the car. He was then transported to the hospital but was pronounced dead on arrival. The cause of death was determined to be acute exposure to TDI.

Gases

Gases are formless fluids that completely fill any container into which they are introduced. Like vapors, gases can also be converted into liquids. But unlike vapors, which can be liquefied by either lowering the temperature or increasing the pressure, gases can be liquefied only by being both cooled and compressed. From an occupational health perspective, gases may generally be classified as anesthetic, irritating, asphyxiating or toxic.

Anesthetic gases are those that induce unconsciousness through narcotic effects on the brain and central nervous system. They include nitrous oxide, cyclopropane and halothane. Due to their limited use, they are likely to be found only in pharmaceutical manufacturing and specialty gas operations.

Irritant gases likely to be found in confined spaces include ammonia, chlorine, sulfur dioxide, ozone and nitrogen dioxide. While these gases have an adverse effect on the body, they are not considered to be systemic toxins because their effects are often local and reversible. These effects are usually characterized by irritation to the eyes, nose, mouth and respiratory system. These effects eventually disappear after the exposure ceases. In some situations, however, the effects may be delayed. Pulmonary edema, or the accumulation of fluid in the lungs, is one example of an effect that could manifest following exposure to an irritating environment.

Asphyxiant gases may be further divided into simple asphyxiants and chemical asphyxiants. As shown in Table 4-4, simple asphyxiants include gases like argon and nitrogen which, as explained in Chapter 2, can displace ambient air, creating an oxygen-deficient atmosphere. Chemical asphyxiants, on the other hand, include hydrogen sulfide, carbon monoxide and hydrogen cyanide. These gases cause asphyxiation through a biochemical reaction.

Table 4-4. Simple asphyxiants

• Acetylene	• Hydrogen
• Argon	• LP gas
• Ethane	• Methane
• Ethylene	• Neon
• Helium	• Nitrogen

Hydrogen Sulfide

Hydrogen sulfide is a colorless gas which frequently forms naturally in wastewater facilities and other areas where sulfur-containing materials can combine with hydrogen. It has a characteristic odor of rotten eggs, and exposure to excessive concentrations has been linked to innumerable confined space fatalities. While hydrogen sulfide has very good warning properties due to its low threshold of smell, prolonged exposure produces olfactory fatigue, that is, loss of sense of smell. Effects of exposure to selected concentrations of hydrogen sulfide are presented in Table 4-5.

The biochemical mechanism by which hydrogen sulfide causes asphyxiation is very complex and involves the inhibition of cytochrome oxidase enzymes which are needed for the exchange of oxygen on a cellular level. Once these enzymes are deactivated, oxygen can no longer be utilized by the cells, and the victim dies of oxygen starvation. Two cases involving fatal hydrogen sulfide poisoning are presented below.

Case 4-6: Sewer Pump Removal

A plumbing and heating contractor had the job of replacing a pump in a private sewer system, which was known to be hazardous from past experience. After the new pump was installed, an apprentice plumber entered the 16-foot deep manhole to open a closed valve. A short time later he was seen sitting unresponsive at the bottom of the manhole. An electrical contractor on the job entered to rescue him, but he was also overcome. When he collapsed, his head fell into the sewer water. Three other rescuers entered and were overcome, but they were able to be pulled out with a rope. Fire department personnel arrived and wearing self-contained breathing apparatus rescued the plumber and the electrical contractor who were now choking and spitting up blood. The plumber and one of the would-be rescuers were hospitalized. The other two rescuers were treated and released. Unfortunately, the electrical contractor died four days later. Subsequent investigation showed that all of the injured workers had inhaled hydrogen sulfide.

Case 4-7: Hydrogen Sulfide in Sewer Manhole

This manhole was located at the junction of four lines, including a sanitary sewer attached to a tannery settling tank. It was 7 feet deep and 4 feet in diameter with a 20-inch entrance. A maintenance man from the tannery entered it to clean the line and replace some loose tiles. Even though his foreman and a coworker were working just a few feet away in a trench, he did not notify them before he made his entry.

When the foreman and other worker finished their work, they went over to the manhole and discovered the maintenance man crumpled up at the bottom. The foreman entered and tried to lift him out, but was overcome in about 15–30 seconds. Three other workers who entered later were able to rescue both the foreman and maintenance man. They administered CPR to the maintenance man and oxygen to foreman until an ambulance arrived. Both of the injured workers were then transported to the hospital where the maintenance man was pronounced dead. The foreman was admitted and later recovered.

Hydrogen sulfide readings taken the next day showed levels between 50 and 960 ppm (the PEL and STEL are 10 and 15 ppm, respectively.) Interestingly, this cleaning operation had been performed every

Table 4-5. Signs and symptoms of exposure to hydrogen sulfide

Signs and Symptoms of Exposure	H_2S Level (ppm)	Exposure Time
Odor threshold	0.1	—
Moderate odor	5.0	—
OSHA PEL	20	8 hrs.
ACGIH TLV	10	8 hrs.
	15	15 min.
Tolerable, but strong, unpleasant odor	25	—
Eye irritation, coughing, loss of sense of smell	100	2 to 5 min.
Marked eye irritation and respiratory tract irritation	200–300	1 hr.
Loss of consciousness and possibly death	500–700	30 to 60 min.
Rapid unconsciousness, respiratory distress, death	700–1,000	minutes
Unconscious almost immediately. Respiration stops, death in a few minutes. Death may occur even if the victim is removed to fresh air.	1,000–2,000	—

Sources: ACGIH, ANSI, NIOSH.

two months for the previous thirty years without incident, providing an excellent example of how spaces that have historically been considered to be "safe" can develop deadly atmospheres without warning.

Carbon Monoxide

Carbon monoxide is an odorless, colorless gas with a density approximately the same as air. It is formed by burning carbon-containing fuels like wood, paper, oil and gasoline. The mechanism of carbon monoxide poisoning is well-known, and involves hemoglobin in blood. Effects of exposure to carbon monoxide at different levels are shown in Table 4-6.

Normally, oxygen in the air combines with hemoglobin and is transported to all parts of the body via the blood stream. Carbon monoxide, however, binds with hemoglobin much more readily than oxygen. As a result, it prevents the hemoglobin from carrying oxygen by attaching itself to the hemoglobin in place of oxygen. Without adequate oxygen, brain cells quickly die. One case of carbon monoxide poisoning in a confined space is summarized below.

Case 4-8: University Heating Plant Ash Silo

A foreman and boiler operator entered a 50-foot deep boiler ash silo to determine why its contents were not flowing out of the bottom outlet. After climbing 25 feet down a fixed ladder to the ash level, the boiler operator passed out and the foreman called for help. Another employee entered and attempted to carry the operator out, but he too became dizzy. Eventually, a rope was tied around the operator and he was pulled to safety. About five minutes later, when the foreman did not climb out, a fourth worker wearing a self-contained breathing apparatus was lowered in to rescue him. The foreman was dead on arrival at the hospital with the cause of death determined to be carbon monoxide poisoning and aspiration of ash. The operator was hospitalized overnight for carbon monoxide poisoning and later recovered.

Hydrogen Cyanide

Hydrogen cyanide is a colorless gas with a scent of bitter almonds. It is created whenever cyanide salts come in contact with acids. Cyanide solutions

Table 4-6. Signs and symptoms of exposure to carbon monoxide

Signs and Symptoms of Exposure	CO Level (ppm)	Exposure Time
OSHA PEL	50	8 hrs.
ACGIH TLV-TWA	25	8 hrs.
Possible mild frontal headache	200	2 to 3 hrs.
Frontal headache and nausea	400	1 to 2 hrs.
Occipital headache	400	2.5 to 3.5 hrs.
Headache, dizziness and nausea	800	20 min.
Collapse and possible death	800	2 hrs.
Headache, dizziness, nausea	1,600	20 min.
Collapse, possible death	1,600	2 hrs.
Headache and dizziness	3,200	5 to 10 min.
Unconsciousness, danger	3,200	10 to 15 min.
Immediate effect, unconsciousness, danger of death	128,000	1 to 3 min.

Sources: ACGIH, ANSI, NIOSH.

are employed extensively in the electroplating industry, and empty plating tanks may contain trace cyanide reside. If these tanks are not adequately flushed and rinsed prior to entry, the residue could react with the acids used to clean tank surfaces and produce cyanide gas. The mechanism of cyanide poisoning is similar to that of hydrogen sulfide in that hydrogen cyanide also inhibits the cytochrome oxidase enzymes necessary for oxygen transport.

Toxic Gases

For our purpose here, "toxic" gases are defined as gases other than anesthetics, irritants or asphyxiants that have a potential for causing systemic poisoning. Examples include arsine, ethylene oxide, phosphine and vinyl chloride.

Arsine

Arsine is a colorless gas with a slight garlic-like odor. It may be formed during metal pickling processes or when inorganic arsenic comes in contact with hydrogen. Arsine is an extremely toxic gas and early symptoms of exposure include malaise, giddiness, apprehension, headache, shivering and abdominal pain with vomiting.

Acute arsine poisoning is marked by three main effects. The first is discoloration of the urine with a hue up to that of port wine. Second is the onset of jaundice resulting in the body taking on a deep bronze color. Finally, there is a decrease in urine production, eventually resulting in complete suppression of the urinary function and severe kidney damage.

Phosphine

Phosphine is a colorless gas with an odor of decaying fish. Exposures may occur when acid or water mixes with metallic phosphides like aluminum phosphide and calcium phosphide. Phosphine exposure may also be encountered during cleaning of tanks that previously contained sulfuric acid. In addition to causing central nervous system depression and lung irritation, phosphine can also affect the liver. Other common effects of acute exposure include weakness, fatigue, headache, vertigo, nausea, vomiting, abdominal pain and difficulty swallowing.

Ethylene Oxide

Ethylene oxide is a colorless gas with a sweet odor. It is used as an intermediate in the manufacture of ethylene glycol, glycol ethers, ethanolamines, acrylonitrile, plastics and surfactants. It is also used as a medical and industrial sterilant. Exposure to ethylene oxide has been associated with reproductive effects, mutagenic changes, neurotoxic effects, sensitization and higher incidences of cancer in humans.

Vinyl Chloride

Vinyl chloride is a flammable gas with a pleasant ether-like odor. It is used both as a chemical intermediate and as the monomer in the manufacture of vinyl chloride and other resins. Vinyl chloride is a central nervous system depressant causing symptoms similar to mild alcohol intoxication. It is also considered to be a human carcinogen, affecting the lungs, liver, brain and kidneys.

CONCLUSIONS AND SUMMARY

Confined spaces atmospheres may contain an assortment of dusts, mists, fumes, vapors and gases. The widely varying nature of these contaminants dictates that a logical approach be used to identify those most likely to be present in a specific space. A good starting place is to determine what the space last contained since even traces of residue may create a hazardous atmosphere.

Next, attention can be directed to determining what types of planned or unplanned chemical reactions may have occurred in the space. Consideration should also be given to the nature and types of operations that might be performed in the space since processes like welding, painting, solvent cleaning and abrasive blasting can be significant sources of air contaminants. Finally, the environment around the space should be assessed for its potential to introduce contaminants into the space.

All materials are not equally toxic, and exposure standards like the ACGIH Threshold Limit Values (TLVs) and OSHA Permissible Exposure Limits (PELs) can be used as a frame of reference for determining the nature of hazard posed by specific air contaminants. However, it is important to remember that these exposure standards do not represent fine lines between safe and unsafe conditions.

After potential air contaminants are identified, they must be quantified by atmospheric testing that will explained in Chapter 9. In some cases, ventilation or respirators described in Chapters 8 and 11 may be required to protect entrants from atmospheric hazards.

But the control of atmospheric hazards is only one aspect of a confined space entry program. Another is the control of physical hazards posed by electrical equipment, machinery and fluids. These and other physical hazards in confined spaces will be described in the next chapter.

REFERENCES

Air Contaminants, Code of Federal Regulations, Vol. 29, Part 1910.1000.

Air Contaminants: Final Rule, Federal Register, January 19, 1989; 542332-2983.

Alpaugh, E.L., Particulates, in *Fundamentals of Industrial Hygiene,* B.A. Plogg, (ed.) Chicago, IL: National Safety Council, (1988).

Amos, David J., Labor's View On Confined Spaces, presented at the National Safety Council 1990 Congress and Exposition, Las Vegas NV.

Casarett, L.J. and M.C. Brice, Origin and Scope of Toxicology, in *Casarett and Doull's Toxicology,* J. Doull, C.D. Klaassen and M.O. Amdur, (eds.) New York, NY: Macmillan Publishing Co., Inc. (1980).

Castleman, B.I. and G.E. Zeim, Corporate influence on threshold limit values, *Am. J. Ind. Med.* 13:531–559.

Castleman, B.I. and G.E. Zeim, Threshold limit values: historical perspectives and current practice, *J. Occ. Med.* 31(11):910–818.

Cloe, W.W., Selected Occupational Fatalities and Asphyxiating Atmospheres in Confined Spaces as Found in Reports of OSHA Fatality/Catastrophe Investigations, Washington, DC, U.S. Department of Labor, Occupational Safety and Health Administration, (1985).

Cook, W.A., Maximum allowable concentrations of industrial atmospheric contaminants, *Ind. Med.* 11: 936–946.

Cornish, H.H., Solvents and vapors, in *Casarett and Doull's Toxicology,* J. Doull, C.D. Klaassen and M.O. Amdur, (eds.) New York, NY: Macmillan Publishing Co., Inc. (1980).

Criteria for a Recommended Standard, Occupational Exposure to Alkanes (C5-C8), National Institute for Occupational Safety and Health, NIOSH Publication No. 77-151 (1977).

Criteria for a Recommended Standard, Occupational Exposure to Carbon Monoxide, National Institute for Occupational Safety and Health, NIOSH Publication No. 73-11000 (1972).

Criteria for a Recommended Standard, Occupational Exposure to Hydrogen Cyanide and Cyanide Salts, National Institute for Occupational Safety and Health, NIOSH Publication No. 77-108 (1976).

Criteria for a Recommended Standard, Occupational Exposure to Hydrogen Sulfide, National Institute for

Occupational Safety and Health, NIOSH Publication No. 77-158 (1977).

Criteria for a Recommended Standard, Occupational Exposure to Ketones, National Institute for Occupational Safety and Health, NIOSH Publication No. 78-173 (1978).

Documentation of the Threshold Limit Values, Sixth Edition, Cincinnati, OH, American Conference of Industrial Hygienists.

Englund, A., K. Ringen and M.A. Mehlman, Occupational Health Hazards of Solvents, Princeton, NJ: Princeton Publishing Co., (1982).

Hammond, P.B. and R.P. Beliles, Metals, in *Casarett and Doull's Toxicology,* J. Doull, C.D. Klaassen and M.O. Amdur, (eds.) New York, NY: Macmillan Publishing Co., Inc. (1980).

James, R.C., Toxic effects of organic solvents, in *Industrial Toxicology: Safety and Health Applications in the Workplace,* P.L. Williams and J.L. Burson, (eds.) New York, NY: Van Nostrand Reinhold, (1985).

McFee, D.R. and P. Zavon, Solvents, in *Fundamentals of Industrial Hygiene,* B.A. Plogg, (ed.) Chicago, IL: National Safety Council, (1988).

NIOSH Pocket Guide to Chemical Hazards, National Institute for Occupational Safety and Health, NIOSH Publication No. 90-177, (1990).

Permit-Required Confined Spaces for General Industry, Code of Federal Regulations, Vol. 29, Part 1910.146.

Pettit, T.A., P.M. Gussey and R.S. Simmons, Criteria for a Recommended Standard: Working in Confined Spaces, National Institute for Occupational Safety and Health, DHEW/NIOSH Pub. 80-106) Government Printing Office, Washington, DC, (1980).

Pounds, J.G., The toxic effects of metals, in *Industrial Toxicology: Safety and Health Applications in the Workplace,* P.L. Williams and J.L. Burson, (eds.) New York, NY: Van Nostrand Reinhold, (1985).

Smith R.G. and J.B., Olishifski, Industrial toxicology, in *Fundamentals of Industrial Hygiene,* B.A. Plogg, (ed.) Chicago, IL: National Safety Council, (1988).

Tabershaw I.R., H.M.D. Utidjaim and B.L. Kawahara, Chemical hazards, in *Occupational Diseases: A Guide to their Recognition,* M.M. Key, A.F. Henschel, J. Butler, R.N. Ligo, and I.R. Tabershaw, (eds.) Cincinnati, OH: National Institute for Occupational Safety and Health (1977).

Threshold Limit Values for Chemical Substances and Physical Agents, Cincinnati, OH: American Conference of Industrial Hygienists, (1992).

CHAPTER 5

Physical Hazards

INTRODUCTION

In Chapter 1 we learned that confined space hazards could be broadly divided into two categories: atmospheric hazards and physical hazards. Since atmospheric hazards were discussed in the three preceding chapters, we can now turn our attention to the area of physical hazards.

For our purposes, physical hazards may be viewed as hazardous conditions that can cause physical stress to the body. Examples of physical hazards include energized conductors, moving machinery and temperature extremes. Unlike most atmospheric hazards which are invisible, many physical hazards can be detected by our senses. For example, we can see unguarded machinery and feel the effects of temperature extremes. Although we cannot actually see electricity we can infer potential electrical hazards from things like flexible power cords, switch gear and exposed electrical components.

Although it is easy to see physical conditions, the hazards they pose may not be apparent to every one. For instance, the casual observer might consider a handful of tools scattered around a street manhole to be nothing more than poor housekeeping. However, a safety professional will view the situation in quite a different light. In addition to poor housekeeping, the safety professional will see a potential slipping hazard and he will also realize that workers in the space could be severely injured if some of the tools accidentally rolled into the manhole opening. The goal of this chapter is to provide those in charge of planning confined space entries with an overview of the types of physical hazards they are likely to encounter. To that end, our focus will be on hazards posed by mechanical devices, electrical equipment, fluids and thermal conditions.

MECHANICAL HAZARDS

Mechanical hazards may be present in confined spaces both as fixed and portable equipment. Commonly encountered fixed equipment includes blenders, stirrers, mixers and agitators. Augers and conveyors may also be found in spaces where solid materials like coal, grain and fertilizer are handled, processed or stored. As demonstrated by the following three case histories, accidental activation of mechanical equipment can prove deadly to confined space entrants.

Case 5-1: Mixing Drum

A three man maintenance crew was in a mixing drum repairing a blade on one of the beaters. The machine was new and had been in service less than three months. The power to the machine had been turned off but not locked out. All three men were out of the operator's sight when he attempted to start the machine by throwing the main switch. When it didn't start he moved to a second control panel and threw another switch. When it still didn't start he climbed on to a platform surrounding the unit and threw a third switch. This time the beaters started.

When one of the men in the drum heard the sound of the switch being thrown he rushed out of the machine dragging a second man with him while shouting to the operator to turn the machine off. The third man did not escape in time and was fatally injured by the moving blades.

Case 5-2: Pulp Paper Mill

A mill worker was instructed by his supervisor to clean the screen on a pulp mill. He climbed into the machine and, as was common practice, stood on its blades to do the cleaning. Meanwhile, the supervisor began to troubleshoot an adjacent de-inking machine which had been shut down. As he was working through the de-inker's start-up procedure, he inadvertently activated the switch controlling the pulp mill. Recognizing his error, he immediately turned it off. Unfortunately, it was too late: the worker inside was killed during the momentary movement of the pulp mill's blades.

Case 5-3: Aluminum Plant Carbon Breaker

An aluminum plant used a rotating drum to break up large chunks of carbon used in the aluminum reduction process. Vertical steel plates and pieces of heavy linked chain were welded to the interior walls of the drum.

A welder was assigned the task of inspecting the drum and rewelding loose plates and chains as necessary. He locked-out the machine at the motor control station and went about his assigned work. When he finished he returned to the control station and removed his lock. It was customary to then operate the machine for about a half hour, after which the worker would reinspect the unit to determine if the welds held.

However, as he was headed back to the control station, three supervisors entered the breaker to inspect his work. They were still inside when he restored the power. All three were fatally injured.

While these three cases are quite dramatic, the inadvertent start-up of machinery and equipment is not the only mechanical hazard faced by entrants. Mechanical hazards may also be posed by portable tools such as drills, chipping hammers and grinders. Some of the more obvious hazards include exposed moving parts, flying chips and hot sparks.

A more subtle hazard is posed by high pressure air lines used on pneumatic tools. If an air-line coupling accidentally separates, the force of the escaping air will cause the line to fly wildly around the space striking occupants like a whip. This hazard can be controlled by securing all hose connections with a safety wire or chain.

In addition to moving equipment, some spaces may contain fixed obstructions like cable trays, pipe supports and baffle plates which project from side walls. These protrusions often have sharp or pointed edges which create bump, cut and abrasion hazards.

ELECTRICAL HAZARDS

Contact with energized conductors can result in involuntary muscular contractions, deep seated burns, respiratory paralysis and cardiac arrest. The consequences of electrical contact are the same regardless of whether it occurs inside or outside of a confined space. However, emergency personnel responding to an electrocution in a confined space will be confronted with obstacles that they are not likely to find in other environments.

For contrast, consider the situation where a worker is electrocuted while working on a piece of equipment in an industrial machine shop. Since the machine shop is open and unobstructed, emergency responders can take a variety of paths to reach the injured worker. The shop's openness also makes it relatively easy for responders to bring in stretchers and other lifesaving equipment. However, if the only way to reach a victim is through an 18 by 24 inch oval manhole located 10 feet above the floor, emergency treatment could be dangerously delayed.

Types of Electrical Injuries

A number of adverse consequences may result from contact with energized conductors. These include:

- *Burns* caused by heat generated as electricity flows through tissues in the body
- *Involuntary muscle contractions* which can prevent a victim from releasing an energized conductor until the flow of current is stopped
- *Ventricular fibrillation* (irregular heartbeat) when nerves controlling the heart muscles are disrupted by an electrical current; heart muscles quiver but fail to pump blood through the body
- *Cardiac arrest* (stoppage of the heart) resulting from paralysis of the nerve centers that regulate heart rhythm
- *Pulmonary arrest* (inability to breath) caused by paralysis of the nerve centers in the brain which control breathing

When oxygen ceases to be supplied to the brain because the heart stops pumping blood or the lungs stop moving air, death is almost certain to occur with in 4 to 6 minutes.

Nature of Electrical Injuries

The effect that electricity has on an individual largely depends on four factors:

1. The voltage of the circuit
2. The resistance of the person's body
3. The flow of current through the body
4. The circuit path through the body

Voltage

The dividing line between high-voltage and low-voltage is not well-defined, but it is generally agreed to be around 600 to 1,000 volts. Death from contact with low-voltage circuits usually results from ventricular fibrillation of the heart. High-voltage fatalities, on the other hand, result from electro-thermal burns. Although high-voltage transmission and distribution lines may be found in some electric utility vaults, it is more likely that entrants will be exposed to 110 volt circuits used to power tools and appliances that are taken into the space.

Body Resistance

Average body resistance measured hand-to-hand with dry skin is about 100,000 ohms. However, the resistance decreases as skin surfaces become moist through sweating or contact with water. Resistance also goes down as tissues are damaged during the uninterrupted flow of current through the body.

Current

The amount of current that flows through the body depends on the applied voltage and the body's resistance. The relationship between current, voltage and resistance is expressed by Ohm's law as indicated below:

$$\text{Current in amps} = \frac{\text{Voltage in volts}}{\text{Resistance in ohms}}$$

Using Ohm's law and assuming dry skin and an ordinary 110 volt circuit, we can calculate the current flow through the body as follows:

$$\frac{120 \text{ volts}}{100{,}000 \text{ ohms}} =$$

$$0.0012 \text{ amps or } 1.2 \text{ milliamps (ma)}$$

As indicated in Table 5-1, this level of current is just perceptible. However, any number of things can cause skin resistance to decrease, and Ohm's law tells us that more current flows as resistance decreases.

Light sweating, for example, may reduce the resistance to about 10,000 ohms. The current flowing though the body would then be

$$\frac{120 \text{ volts}}{10{,}000 \text{ ohms}} =$$

$$0.012 \text{ amps or } 12 \text{ milliamps (ma)}$$

As indicated by Table 5-1, this increase in current causes correspondingly severe consequences. A current of about 10–20 milliamps (ma) is frequently referred to as the "let-go current" because it is the level at which most people would be able to let go of an energized conductor. Above this level, the muscles in the hand contract so violently that it is impossible to release the conductor.

But as current flows through the body, the body's internal resistance decreases still further as cells and tissue are destroyed. If the resistance drops to say 1,000 ohms the new current flow would be

$$\frac{120 \text{ volts}}{1000 \text{ ohms}} = 0.12 \text{ amps or } 120 \text{ milliamps (ma)}$$

A continuous flow of current as low as 25 ma will cause ventricular fibrillation in a matter of minutes. As a frame of reference, a 25 watt light bulb draws 25 ma of current. However, as indicated by Table 5-2, stronger currents will produce ventricular fibrillation in a much shorter time.

Circuit Path Through the Body

The path that current takes through the body depends on the body parts that complete the circuit. The pathway is relatively easy to determine since electrical contact wounds are similar to bullet wounds. There is a discernible entry and exit point on the body, and the path between these points establishes the route the current followed. Interestingly, current does not always take the path of least resistance, it takes the shortest path to ground.

Accidental contact with *energized conductors* may occur in a number of ways, but some circuit paths are more dangerous than others. A circuit established between an elbow and finger on same hand, for example, is not as worrisome as a circuit established from hand-to-hand or hand-to-leg. The reason for this becomes clear from Figure 5-1. Note that current in each of the four circuits shown passes through organs that are essential for life such as the brain, heart and kidneys.

Secondary Electrical Effects

All cases of electrical contact are not fatal, some result in secondary, non-electrical injuries. For example, a current as low as 3 ma can cause involuntary

Table 5-1. Physiological effects of currents at different intensities

Current (milliamps)	Physiological Effect
<1	No sensation felt
3	Painful shock
10	About 2.5% of the population experiences sufficiently severe muscular contraction that they are unable to let go
15	About 50% of the population experiences sufficiently severe muscular contraction that they are unable to let go
30	Breathing difficult, may produce unconsciousness
50–100	Possible heart ventricular fibrillation
100–200	Certain heart ventricular fibrillation
>200	Severe burns and muscular contractions, heart ore likely to stop than to fibrillate

Adapted from *Electrical Hazards Course Manual,* OSHA Training Institute Course 100–17.

Table 5-2. Time and current intensity needed to produce ventricular fibrillation

Current Intensity (milliamps)	Time (sec.)
70–300	5
200–700	1
300–1,300	0.3
500–2,500	0.1
1,800–8,000	0.01

muscular reactions. The sudden jolt from this low level contact could cause an entrant to bang a hand on the side walls of a confined space so violently that it could be fractured. Similarly, an entrant working on a ladder or scaffold could be seriously injured by a fall caused by a mild shock.

Controlling Electrical Hazards

Electrical hazards may be controlled in a number of way including:

- Implementing an equipment inspection program
- Substituting pneumatic equipment
- Employing double-insulated tools
- Using properly grounded equipment or ground-fault circuit interrupters

Equipment Inspections

Perhaps the easiest way of minimizing electrical hazards is to inspect equipment, power hand tools and portable cords to assure that they are in good repair. Insulation on portable power tools and cord sets should be routinely examined for cracks, pinched areas and worn spots. Screw connections and contacts should be tight and free from corrosion since loose or corroded contacts can lead to arcing and shock hazards if they became detached.

Both the National Electrical Code and OSHA regulations prohibit splicing of portable cords. An exception is made for No. 14 gauge or greater, hard service cords if the splice provides the same insulating qualities, outer sheath properties and usage characteristics of the original cord.

As indicated by Table 5-3, the permissible working length of a cord set is determined by the size of wire gage and the current flowing through the cord.

Pneumatic Substitution

In some cases, the potential for electrical hazards can be eliminated by substituting pneumatic equipment for electrical hand tools. For example, air-driven grinders and sanders may be used instead of their electrical counterparts. However, if pneumatic tools are used, care must be exercised to assure that the exhausted air does not contain any hazardous contaminants which could adversely affect the space's atmosphere. The compressor system should be inspected to verify that the air inlet is located in uncontaminated air, not next to the exhaust stacks of furnaces, drying ovens, spray booths or other contaminant-producing processes. In addition, if the site has

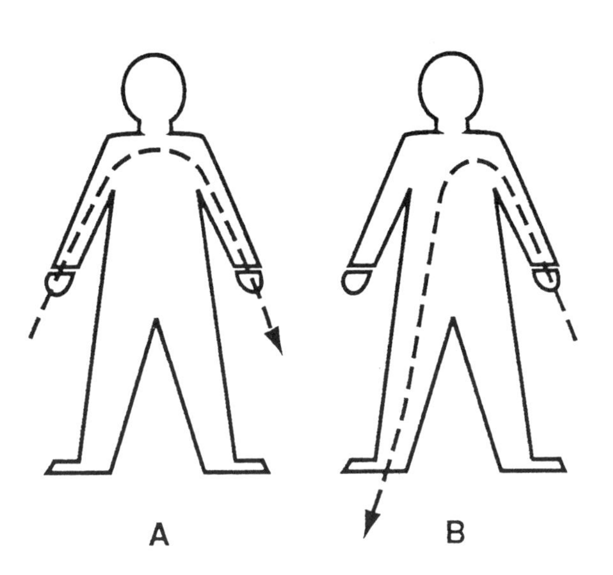

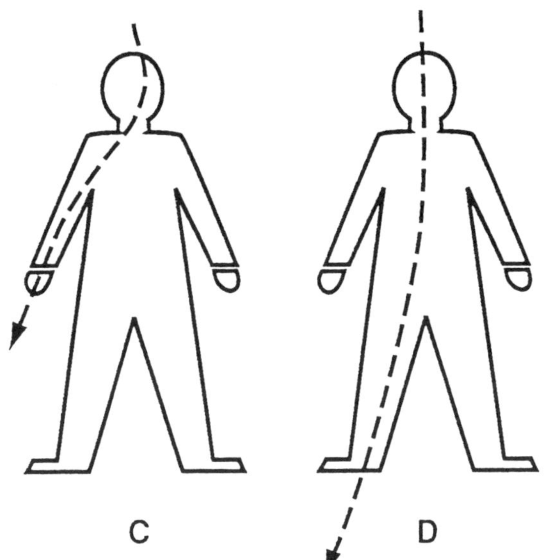

Figure 5-1. Paths of current. (Paul Trattner)

Table 5-3. Minimum recommended wire gauges for cord sets

Current Draw (amps)	Cord Length (feet)			
	25	50	100	150
0–6	18	16	16	14
10	18	16	14	12
10–12	16	16	14	12
12–16	14	12	(not recommended)	

Source: Power Tool Institute.

other utility lines containing gases like nitrogen, acetylene or oxygen, precautions must be taken to prevent pneumatic tools from being attached to those systems. For example, incompatible connection fittings could be installed to prevent the inadvertent use of another gas instead of air. As an additional precaution, different gas supply lines can be color-coded and labeled.

Double-Insulated Tools

Double-insulated tools are designed in such a way that an internal electrical fault will not energize the tool's case. Consumer-grade tools accomplish this by using a plastic case and a non-conductive on-off switch that prevents the operator from coming in contact any metal parts. Industrial-grade tools often require metal cases to provide the structural strength needed to withstand the added weight and greater torque imposed by heavier motors. For these tools, double-insulation is achieved by lining the inside of the metal case with a nonconductive material. Double-insulated tools can be easily identified because they are marked with the legend "double insulated" or "double insulation."

Equipment Grounding

Figures 5-2 and 5-3 graphically demonstrate the purpose of equipment grounding. As seen in Figure 5-2, a ground-fault that develops in a properly grounded tool will be directed back to the service entrance panel where it will blow fuse or trip a circuit breaker. However, if the ground return path is interrupted, for example, by cutting of the grounding pin, there is no place for the leakage current to go; therefore, the tool's case would be energized. It would still run and the operator would not be shocked until he became part of a ground loop. At that time, current would flow through him and back to the power source as seen previously in Figure 5-3. The severity of the shock the operator receives depends on the resistance of the ground loop. From Ohm's law we know that a resistance of as little as 10,000 ohms

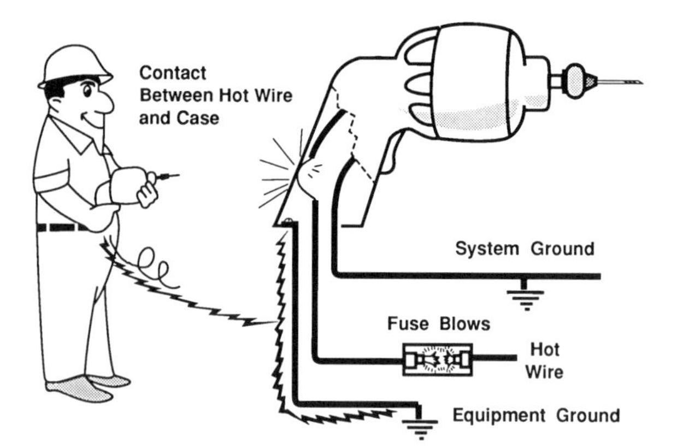

Figure 5-2. Properly grounded tools prevent shock. (Paul Trattner)

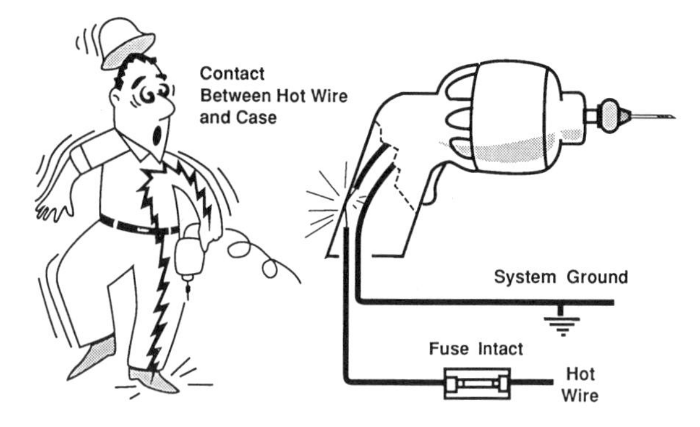

Figure 5-3. Open grounds can lead to shock hazard. (Paul Trattner)

could cause a 12 ma current to flow through, so even a slight lowering or body resistance could result in a severe, or perhaps even fatal, shock.

Ground-Fault Circuit Interrupters

A ground-fault circuit interrupter (GFCI) is essentially a fast-acting circuit breaker that electronically detects dangerous leakage currents and shuts off the flow of electricity before it becomes a hazard to personnel. As shown in Figure 5-4, the GFCI contains a differential transformer that monitors the flow of current on both legs of a circuit feeding a piece of electrical equipment. If the sensing element detects an imbalance, that is, if it sees more current flowing into the equipment on the "hot" side than it sees leaving on the neutral, it promptly opens a solid-state switch that interrupts the flow of current to the equipment.

GFCIs virtually assure employee protection from electric shock since the units can detect fault currents as low as 0.005 amp (+ 0.001 amp) and will trip open within 1/40 of a second (25 milliseconds) of detecting a current imbalance. It should be noted, however, that a GFCI will not provide protection in the case of line-to-line contact that results from someone touching the hot and neutral wire simultaneously with each hand.

As indicated by Figure 5-5, GFCIs are available in a number of styles that can be adapted to a wide variety of applications, including use in hazardous locations.

Low-Voltage Lighting

Lighting can be provided by low-voltage, isolated systems like that shown in Figure 5-6. These systems use an isolation transformer to reduce the 110-volt line supply to about 6 to 12 volts for powering the light bulb. As demonstrated by Figure 5-7, the transformer electrically isolates the lamp so that a fault on the supply side will not be conveyed to the light assembly. Of course, if portable lights are required for use in areas where flammable gases or vapors might be present, they must be approved for use in Class I hazardous locations as defined by Article 500 of the National Electrical Code® (Figure 5-8). The electrical code also stipulates that only extra-hard usage cords such as type S, SE, STO, STOO, SO or SOO be connected to portable lights used to hazardous locations.

HAZARDS POSED BY WELDING AND CUTTING

Welding and oxy-fuel gas cutting in confined spaces can present a host of hazards. Some of these hazards (such as toxic air contaminants and ignition of flammable gases and vapors) were touched on in previous chapters. However, hot work also poses a variety of physical hazards related to fire safety, equipment operation, work practices, materials handling and use of compressed gases. Methods for controlling these hazards are discussed below.

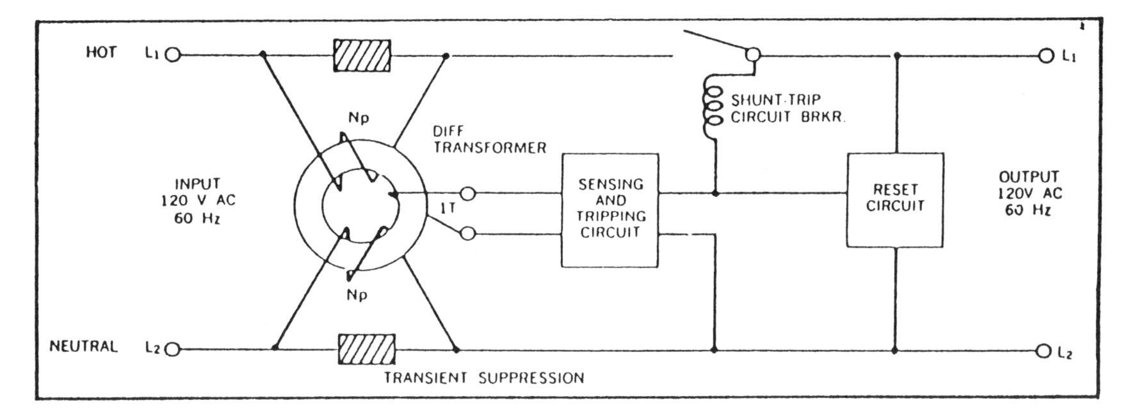

Figure 5-4. Ground-fault circuit interrupter. (National Safety Council)

Figure 5-5. In-line GFCI. (Hubble)

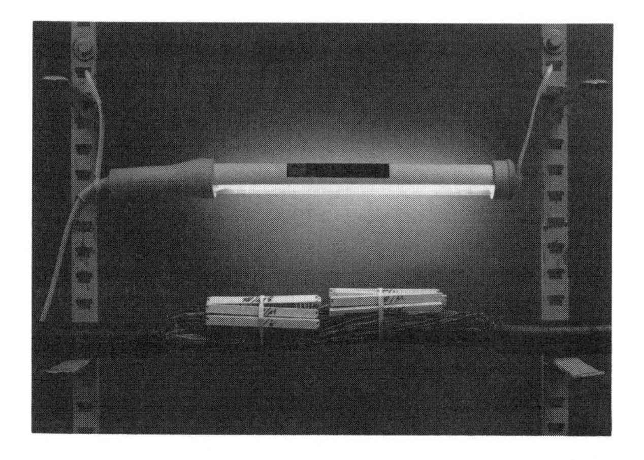

Figure 5-6. Low-voltage lights can reduce shock hazards. (Pelsue)

Identify Hot Surfaces

After welding operations are completed, the welder should mark the hot metal or provide some other means of warning workers of the thermal hazard.

Compressed Gas Cylinders

Compressed gas cylinders should be legibly marked to identify their contents. They should be kept away from heat sources and never be placed where they might become part of an electric circuit. Cylinder valves should be closed at the completion of the work, when cylinders are empty and before cylinders are moved. Valve protection caps should be in place and hand-tight except when cylinders are connected for use.

Cylinders must not be dropped or allowed to strike each other violently. Valve caps may not be used for lifting cylinders. Acetylene cylinder valves must not be opened more than one and one-half turns, but three-fourths of a turn is preferable. Cylinders that are not equipped with fixed hand wheels must have special keys, handles or nonadjustable wrenches on their valve stems while they are in service so the gas flow can be shut off in an emergency. If multiple cylinders are connected together, then at least one key or handle is required for each manifold.

OSHA welding standards prohibit gas cylinders and welding machines from being taken into a confined space because a leaking valve could quickly fill the space with oxygen or a flammable gas. Cylinders

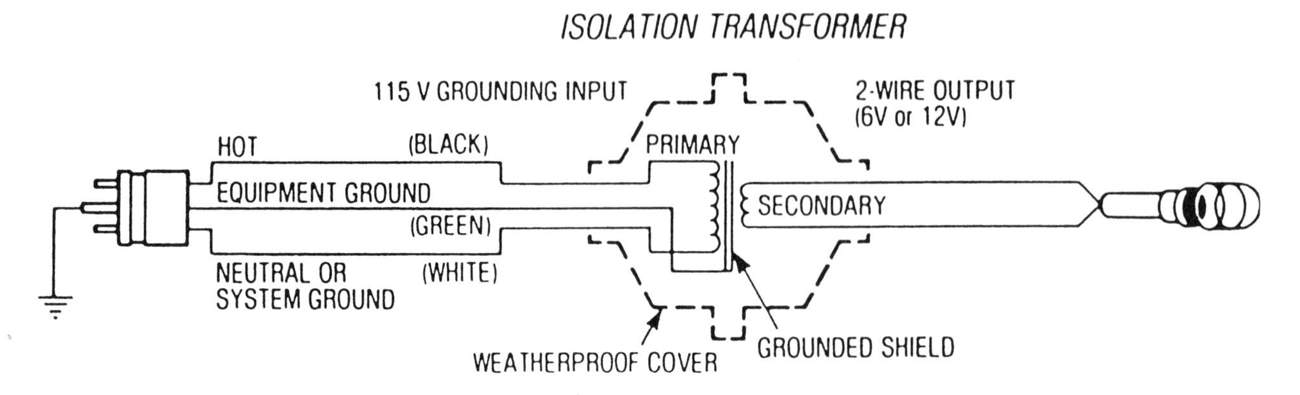

Figure 5-7. Isolation transformers prevent fault currents from being conducted to the lamp assembly. (Ericson Electric Co.)

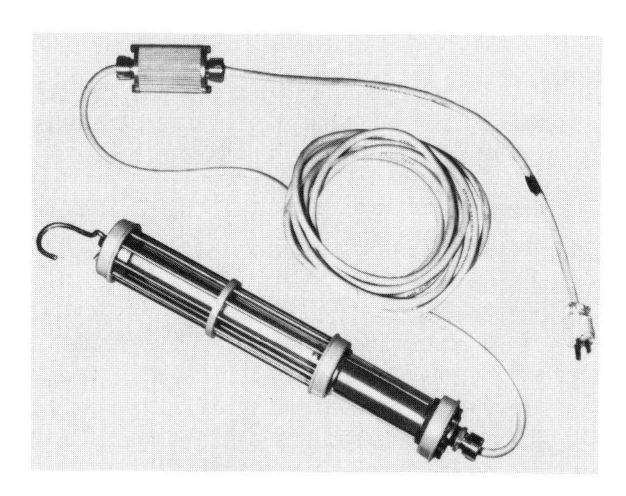

Figure 5-8. Low-voltage explosion-proof light approved for use in Class I hazardous location. (K.H. Industries)

positioned outside the space should be firmly attached to a welding cart (Figure 5-9) or otherwise secured from falling. Wheel-mounted equipment should be securely blocked to prevent accidental movement.

Since oxygen and oil can react explosively when combined under high pressure, oxygen cylinders or apparatus should never be handled with oily hands or with oil-impregnated gloves. Oxygen cylinders, valves, couplings, regulators, hoses and apparatus must also be kept free from oily or greasy substances. Jets of oxygen should not be allowed to strike an oily surface or greasy clothes.

Torches and Electrode Holders

Electrode holders and cables should be regularly inspected for cracks, worn spots or breaks in the insulation that could expose energized conductors. Torches, hoses, fittings and couplings should also be inspected for wear and damage.

Welding torches and electrode holders should not be taken into a space until they are ready to be used, and they should be removed as soon as the work requiring their use is completed.

When arc welding is suspended for a substantial period of time, such as during lunch or overnight, the welding machine should be disconnected from the power source and all electrodes should be removed from their holders. Unused electrode holders should be positioned to prevent accidental contact with people and conducting objects such as compressed gas cylinders.

Hoses and Welding Cable

Hoses and welding cables and other equipment should be arranged so that they do not block ladders and walkways. OSHA welding standards prohibit the use of cables with splices within 10 feet of the holder. Any nonconforming cables should be cut back to the spliced area and reconnected. It's also a good practice to prohibit welders from coiling or looping welding electrode cables around their bodies.

Arc Ray Hazards

Electric arc welding emits intense ultra-violet (UV) radiation that can damage the eyes and burn the

Figure 5-9. Welding gases secured on a cart. (John Rekus)

skin. Welders are aware of these hazards and ordinarily take precautions to protect themselves by using special protective clothing and equipment. However, non-welders working adjacent to welding operations may not be aware of the hazard and should be protected by an appropriate welding screen or shield.

Chlorinated solvents such as trichloroethylene and perchlorethylene are often used for surface cleaning and degreasing. The vapors of these solvents can decompose into hydrogen chloride and phosgene gas when exposed to UV light. Both of these gases are highly irritating to the lungs and respiratory tract.

Shielding Gas Hazard

Gas metal arc welding (GMAW) and gas tungsten arc welding (GTAW) use inert gases like argon, helium and carbon dioxide to shield fresh welds. If these shielding gases are allowed to accumulate in the space, they can displace the ambient air, resulting in an oxygen deficiency. Consequently, it is advisable to provide mechanical ventilation and to monitor the atmosphere continuously for oxygen. Alternatively, welders could wear positive-pressure air-line respirators with an appropriately sized escape cylinder.

Arc Welding Precautions

In addition to being sources of ignition, arc welding presents a potential for contact with energized conductors that may cause electrocution, thermal burns or shocks that can contribute to other injuries, such as falls from ladders or scaffolds. While many entrants may be alert to the hazards posed by 110-volt circuits, an easily overlooked source of electrical hazards is arc welding equipment. Unfortunately, as demonstrated by the following two case histories, this oversight can also prove to be deadly.

Case 5-4: Welding in Rail Car

A 10,000-gallon rail car which previously contained tallow had been steam cleaned in preparation for internal welding. The welder had been shocked several times by his equipment, and rather than checking the equipment he put on his gloves for protection. Late in the afternoon several workers heard a scream and ran to the tank. They found the welder laying on a coil of welding cable, still grasping the electrode holder. An employee who attempted to remove it received an electric shock. CPR was administered and continued all the way to the hospital, where the victim died shortly after arrival. The autopsy showed that he had several small burns on his arms and shoulder. The probable cause of death was determined to be electrocution.

Case 5-5: Welding in Tank Truck

An apprentice welder entered a 13-feet long by 64-inch diameter water tank truck to attach a baffle plate to the tank's bottom while another welder outside used a garden hose to apply cooling water to the tank's exterior shell. When the welder outside was informed of the arrival of a salesmen he was expecting, he ordered the apprentice out of the tank and left the area to talk with the salesman.

However, before he returned, the apprentice inexplicably reentered the tank. When the welder heard the apprentice scream he turned off the power to the welding generator and climbed up on the tank. Looking inside, he discovered the apprentice at the bottom of the tank under about 6 inches of water.

CPR was administered until rescue services arrived. The apprentice was transported to the hospital where he was pronounced dead. Autopsy reported the cause of death as cardiopulmonary arrest secondary to electrocution.

LIQUID AND GASEOUS HAZARDS

Piping systems containing flammable, corrosive or toxic materials may be connected to confined spaces. Unless adequate precautions are taken to disconnect, isolate and drain these lines prior to entry, workers could be inadvertently exposed to hazardous gases or liquids as a result of a leak or sudden release of material; for example, when a closed valve is accidentally opened. While isolation methods and techniques will be discussed in detail in Chapter 7, the following case histories clearly demonstrate why they are necessary.

Case 5-6: Sprayed with Sulfur Trioxide

An instrument mechanic and operations manager at a soap manufacturing plant entered a hot box to determine the cause of a pressure problem on a sulfur trioxide line. Once inside, they noticed that a valve was leaking and decided to disassemble it to find the cause. The manager then crawled out and went to the control room. A short time later he heard the mechanic scream and saw a white gas cloud pouring out of the hot box. He immediately shut down the flow of sulfur trioxide from the control room and ran over to rescue the mechanic. After two unsuccessful attempts, he was able to pull the mechanic out and get him under a safety shower keeping him there until an ambulance arrived. Even though prompt action was taken, the mechanic died several days later from chemical burns. Investigation showed that the sulfur trioxide line had not been shut down, nor had the line's internal pressure been relieved before work began on the valve.

Case 5-7: Kettles with a Common Drain

An operator entered a 12 to 14 feet high, 6 foot diameter reaction kettle through a 17½ inch topside manhole. His job was to chip hardened phenolic resin from its sides, a task he had performed for the previous four days without incident. After he had been inside for about two hours, a coworker discovered him unconscious at the bottom of the kettle. This worker entered and was also overcome. When another kettle operator did not hear any noise coming from inside the kettle, he too looked in and saw that the two other workers were unconscious. He notified other employees in the area, and then used a medical oxygen mask to enter the tank. Although he was able to recover one of the workers, he too was overcome just as he was climbing out. Fortunately, he was caught before he fell back into the reactor.

The shift foreman then donned the mask and entered to rescue the other worker. In the process of doing this, the mask was knocked off and he too was overcome. Another operator retrieved the mask, entered the kettle, and rescued the remaining worker and the foreman. All five employees were taken to the hospital where the kettle cleaner and first rescuer were pronounced dead. The three other rescuers were treated and released.

Investigators determined that the kettle being cleaned was connected to another kettle via a common drain line. Apparently, some extraneous material, including sulfur dioxide, back-flowed into the first kettle's receiving tank when the second kettle was drained. The sulfur dioxide then entered the occupied kettle, where it created a fatal atmosphere.

ENGULFMENT

Finely divided solid materials such as sawdust, grain and sugar can surround, trap and engulf a victim in a manner similar to that of quicksand.

As shown in Figure 5-10, cavities may be formed in material stored in bins and silos. Entrants who are unaware of this condition will be quickly buried when the seemingly solid surface suddenly gives way under their weight. Spaces free of voids are just as hazardous, but death comes much more slowly. As shown in Figure 5-11, a victim first begins to sink into the engulfing material, then as he struggles to escape he is drawn in deeper and deeper (Figure 5-12). In less than a minute he will be fully engulfed.

Another engulfment hazards results when material is drawn off from the bottom of storage bins. As material flows out of the bottom of the bin, a funnel-shaped depression forms over the outlet. As shown in Figure 5-13, material can move toward the center of the funnel so rapidly that a worker caught in its path will not be able not escape and will be drawn into the funnel's vortex.

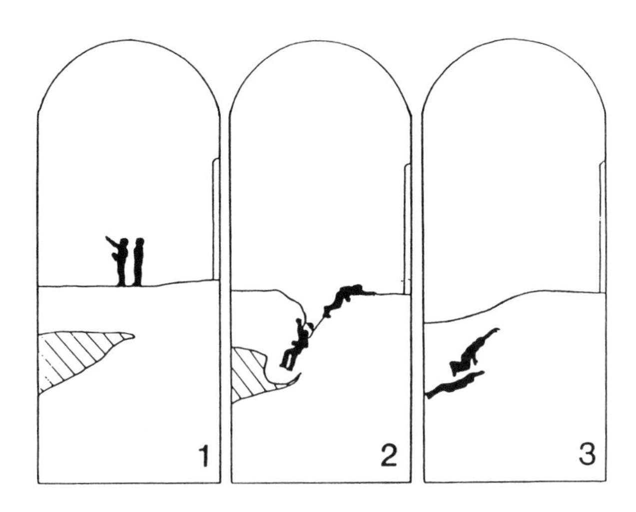

Figure 5-10. Cavity formed in finely divided material. (NIOSH)

Figure 5-12. In about 20 seconds victim may be buried. (Paul Trattner)

Figure 5-11. A victim may be trapped in a matter of seconds. (Paul Trattner)

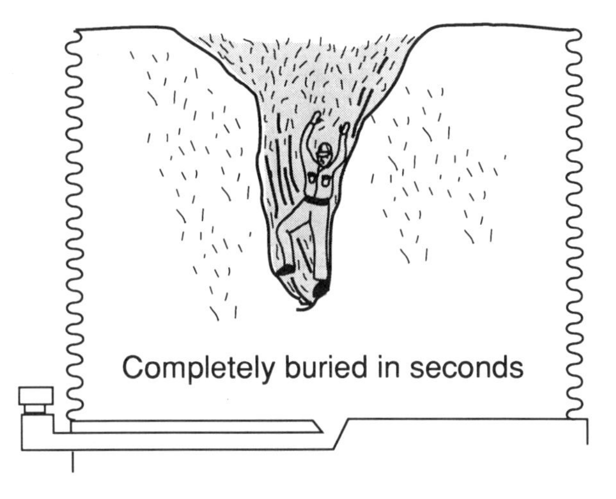

Figure 5-13. Workers may be trapped by material flowing out the bottom of a bin. (Paul Trattner)

Still another engulfment hazard is produced by a condition known as "bridging." Bridging is a condition that results when a layer of material extends across the cross-section of a bin, hopper or silo that has been emptied from below (Figure 5-14). While bridges often appear to be solid, they can give way without warning swallowing up workers who are standing on top of them. They can also break apart under their own weight and fall onto entrants working beneath them.

The diameter of the space and moisture content of the material are factors that contribute to bridging,

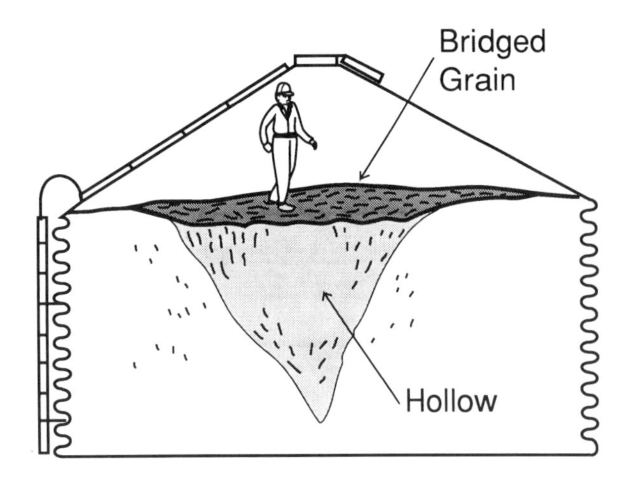

Figure 5-14. Bridging may result when material is unloaded from below. (Paul Trattner)

and it is most likely to occur in small diameter spaces filled with moist products. Bridging may occur in spaces containing ground agricultural products or other powdery materials like cement, limestone and sawdust. However, it is not likely to occur in spaces containing large particles such as unground grain, corn, oats and barley or because the individual kernels do not tend to stick to the sides of the container.

Regardless of how victims are engulfed, they are eventually surrounded by in-flowing material which exerts so much pressure on their chests that breathing becomes impossible. Once the body's supply of oxygen is cut off, permanent brain damage is inevitable in about 4 to 6 minutes.

Two typical case histories of fatal engulfment are presented below.

Case 5-8: Engulfment in Sawdust Silo

A wood window manufacturing plant stored its process-generated sawdust in a bin 20 feet in diameter and 40 feet high. One day, the bin's bottom outlet plugged up and the foreman decided to loosen the plug by running a blower-auger during the lunch hour. After lunch, 4 employees entered the nearly full bin through a 20 × 24 inch opening. As they worked on their hands and knees smoothing out the sawdust, one of them suddenly vanished into a hollow air pocket created by the auger. Even though the remaining operators summoned help immediately, it took 4 hours to recover their coworker, who was found under 20 feet of sawdust.

Curiously, this same operator was nearly buried two years earlier in a similar incident. Following that incident, bin workers were instructed to use a "safety rope," but the practice was viewed as inconvenient and quickly fell into disuse. The only rescue equipment in the silo at the time of the fatality was a 3/8-inch cotton clothesline which was worn and knotted.

Case 5-9: Engulfment in Soybean Mill

A mill worker at a feed-and-seed cooperative climbed on top of a 20 foot bin to loosen a jam so he could fill a customer's order for 800 pounds of soybean meal. In doing so, he climbed on beam supports that were spaced 32 inches apart. The jam was cleared and another employee filled the customer's order. When the employee who freed up the jam did not return to the mill floor, his coworker climbed up the bin to look for him. Suspecting the worst when he saw only a shovel, he immediately drained the bin and found his associate at the bottom. Other employees entered and were able to pull the victim out but efforts to revive him failed.

Investigators concluded that the employee either slipped from the widely spaced beams while breaking the feed loose, or fell into the bin after being overcome by an oxygen deficiency created by fermentation.

Case 5-10: Engulfed in Soy Meal

A 5 man crew was assigned the task of breaking up crusted soy meal inside a 100 foot silo so it could be removed from the silo's bottom hopper. The work level was located about 3 to 5 feet from the bottom cone leading to the hopper where the meal was augured into rail cars. The workers were individually lowered from the top of the silo using a boatswain's chair and they remained tethered to the surface with a lifeline attached to a body belt.

The men were standing on the hard crust, chipping at the product with air picks and spud hoes, when a large section of crust fell causing them to loose their balance. Aided by the lifelines, four of them were able to stabilize themselves, but one worker slipped out of his belt and fell into the product. He was covered with meal and suffocated.

Investigators noted that the ambient temperature was 100°F and concluded that sweat acted as a lubricant that allowed the safety belt to slip over the worker's hips.

THERMAL HAZARDS

Spaces located out of doors will be affected by changes in ambient temperatures. Depending on the area of the country, temperatures may range from below 0°F in winter to over 100°F in summer. Some spaces may also be located in plant areas that are subject to temperature extremes. For example, spaces may be located in refrigerated warehouses or adjacent to industrial ovens and furnaces. In some cases, attendants as well as entrants may be affected by environmental temperature extremes. For instance, attendants may be exposed to blustery cold winds as they maintain their vigil over workers in the space.

The effects of temperature extremes may be broadly divided into generalized and local effect. Generalized effects include hypothermia, heat stroke and heat exhaustion. Local effects include frostbite, heat cramps and burns.

Cold Stress

Shivering and pain in the extremities provide early warning signs of cold stress. Severe shivering which occurs when the body's core temperature falls to below 96.8°F signals hypothermia (Table 5-4).

Hypothermia results when the body looses heat faster than it generates it. The hands and feet are usually affected first, as blood vessels in the extremities contract to conserve heat.

Work in the space should be discontinued immediately with the onset of severe shivering because the continued loss of heat may result in a decrease in mental alertness, difficulty speaking and loss of manual dexterity.

Frostbite is a localized effect of cold that can occur without hypothermia. It is caused by fluids around cells and tissues that freeze because of inadequate circulation. One of the early warning signs of frostbite is pain in the nose, fingers, toes, cheeks and ears.

Unprotected skin should not be exposed continuously when wind chill temperature is less than 25°F. While wind chill will not be a problem for entrants shielded from the wind, it could be for attendants positioned outside the space.

Table 5-4. Progressive clinical signs of hypothermia

Core Temperature		Clinical Signs
°F	°C	
99.6	37.6	"Normal" rectal temperature
98.6	37.0	"Normal" oral temperature
96.8	36	Metabolic rate increases in an attempt to compensate for heat loss
95.0	35	Maximum shivering
93.2	34	Victim conscious and responsive with normal blood pressure
91.4	33	Severe hypothermia below this temperature
89.6	32	Consciousness clouded; muscular rigidity increases; pulse and blood pressure difficult to obtain; respiratory rate decreases
82.4	28	Ventricular fibrillation possible with myocardial irritability
80.6	27	Voluntary motion ceases; pupils nonreactive; deep tendon and superficial reflexes absent
78.8	26	Victim seldom conscious
75.2	24	Pulmonary edema
71.6	22	Maximum risk of ventricular fibrillation
69.8	21	
68.0	20	Cardiac standstill
64.4	18	Lowest accidental hypothermia victim to recover

Adapted from ACGIH Threshold Limit Values.

Heat Stress and Burns

Heat cramps are painful muscle spasms that result from drinking large amounts of water without adequate salt replacement. The once popular practice of consuming salt tablets to compensate for this imbalance has been replaced with a recommendation to increase dietary salt intake.

Heat exhaustion is characterized by fatigue, nausea, headache and giddiness. The victim's pulse and blood pressure are low, the skin is usually clammy and moist, and the complexion pale. Heat exhaustion victims should be removed to a cool environment where they can rest and drink cool liquids.

Heat stroke results when the victim's internal temperature regulating mechanism shuts down. The skin will appear red or mottled and usually warm and dry to the touch. Heat stress is a medical emergency and if not treated promptly can lead to death. The victim should be cooled as rapidly as possible by immersion in cold water or covering with ice packs.

Entrants may be burned when they touch surfaces that have been heated by processes like steam cleaning, welding or oxy-fuel gas cutting. Surfaces could also be heated by exothermic reactions, steam jackets or hot residue remaining in a space. Solar radiation beating down on outdoor vessels, tank cars and tank trucks can also produce hot surfaces, particularly in summer months.

While the burn itself may not be serious, the configuration of the confined space could affect an entrant's ability to self-rescue. Momentary contact with hot surfaces can also lead to secondary injuries such as those resulting from falls from ladders or scaffolds. In more serious burn cases, like those described below, entrants may be buried in hot material or scalded by steam.

Case 5-11: Burned By Hot Salt Cake

An operator and helper were assigned the task of chipping a 2-foot thick salt cake layer (70% sodium carbonate and 30% sodium sulfate) from the inside of a paper mill reactor vessel. The reactor, which was 30 feet high and 10 feet in diameter and normally operated between 1,285–1,310°F, had been shut down and ventilated for about 12 hours. The helper entered, chipped away at the 140 to 200°F salt cake with a shovel, and passed the loose material to the operator outside. After a few minutes, the entire layer broke free from the sides of the reactor and buried the helper up to his shoulders. He sustained first, second and third degree burns over 40 to 60% of his body and died a little over a month later.

Case 5-12: Scalded By Steam

Two steel mill maintenance workers and two blast furnace operators entered a pump vault to replace a leaking 4-inch water line. The operators isolated the 4-inch line by closing valves at both ends. However, they apparently neglected to open two other valves on an auxiliary line that supplied water to a vacuum level controller on the main steam line. Because these valves were not opened, the vacuum level in a 42-inch steam line dropped to a critical level. When this happened, one of the operators tried to compensate for the loss by starting a stand-by vacuum pump. This attempt was not only unsuccessful but also caused water to spill into the hot 42-inch steam line where it immediately vaporized. The increase in pressure from the vaporizing water ruptured the line and sprayed the workers with hot water. All four employees were hospitalized, and one died several weeks later.

STORED ENERGY AND GRAVITY

While not strictly physical hazards themselves, both gravity and the residual energy stored by some process equipment present hazardous situations. For example, the uncontrolled release of energy stored in pneumatic components, hydraulic lines, springs or electrical capacitors may be sufficient to rotate a mixing blade, forcefully eject a piston shaft or cause a fatal shock.

Gravity can also present a hazard to workers in a confined space if tools, parts or materials fall on top of them. Consequently, the area around vertically entered portals should be kept clear of all unnecessary materials. Entrants can be further protected by installing a manhole shield like that shown in Figure 5-15.

Figure 5-15. A manhole shield can keep stray materials from falling into the space. (Pelsue)

PEDESTRIAN AND TRAFFIC HAZARDS

Employees who work in street manholes are confronted with a variety of hazards posed by pedestrians and traffic. Personal accounts from municipal and utility workers suggest that employees who work in street manholes are subjected almost daily to a fusillade of bricks, bottles, cans, rocks and cigarette butts that is unleashed by adult and juvenile pedestrians. Partial protection from these hazards may be provided by surrounding spaces with protective railings, fences, high-visibility wire, rope or tape, and similar barricades that restrict access.

Protection from traffic hazards can be provided through the used of effective traffic control devices. As indicated by Table 5-5, traffic control devices can reduce the hazards posed by highway traffic, but they cannot eliminate them. Methods of protecting street manholes suggested by *The Traffic Control Devices Handbook* are shown in Figures 5-16 and 5-17.

SUMMARY AND CONCLUSIONS

Confined space entrants may encounter conditions that present a potential for electric shock, entanglement in mechanical equipment, exposure to hazardous liquids, burns and engulfment by finely divided solids. Fortunately, all of these hazards can be managed or eliminated.

Electrical hazards—like exposed current-carrying parts and worn or damaged insulation—can be detected through regular inspection of tools and cords. Prompt repair of these defects and the use of ground-fault circuit interrupters can greatly reduce the chance of employees being shocked.

Hazards posed by the accidental start-up of machinery or the inadvertent release of gases or liquids through pipes attached to a space can be addressed by a lockout-tagout program. Although it may not be possible to eliminate engulfment hazards, an education program that impresses employees with

Table 5-5. Summary of selected fatal accidents at manholes

Job Classification	Location	Description of Incident
Splicing foreman	Bridgeport, CT	A foreman visiting a splicing crew that was working at night was standing by a manhole. He was struck by a passenger car that crashed through the work area at about 3:00 am. Police charged the driver with driving under the influence of alcohol.
Apprentice splicer	Chicago, IL	At 4:30 am while working at a manhole on an around-the-clock splicing operation, the employee was struck by a hit and run motorist.
Cable splicer	Menands, NY	Employees were working at a manhole in the middle of a four-lane highway when a car crashed into the work area, pushing everything ahead of it like a bulldozer.
Lineman	Southern Bell	A drunk driver traveling at a fast rate of speed crashed through the manhole barricades and struck an employee before crashing into parked cars.
Cable splicer	Wisconsin	Employee was struck and fatally injured by a vehicle when he emerged from a manhole located in an intersection.
Cable splicer	Indiana Bell	A speeding vehicle crashed through protective equipment and demolished the work area, killing two splicers.
Cable splicer	Houston, TX	Employee was shot by unknown assailant as he was working above a manhole entrance.

Source: Bell Communication Research.

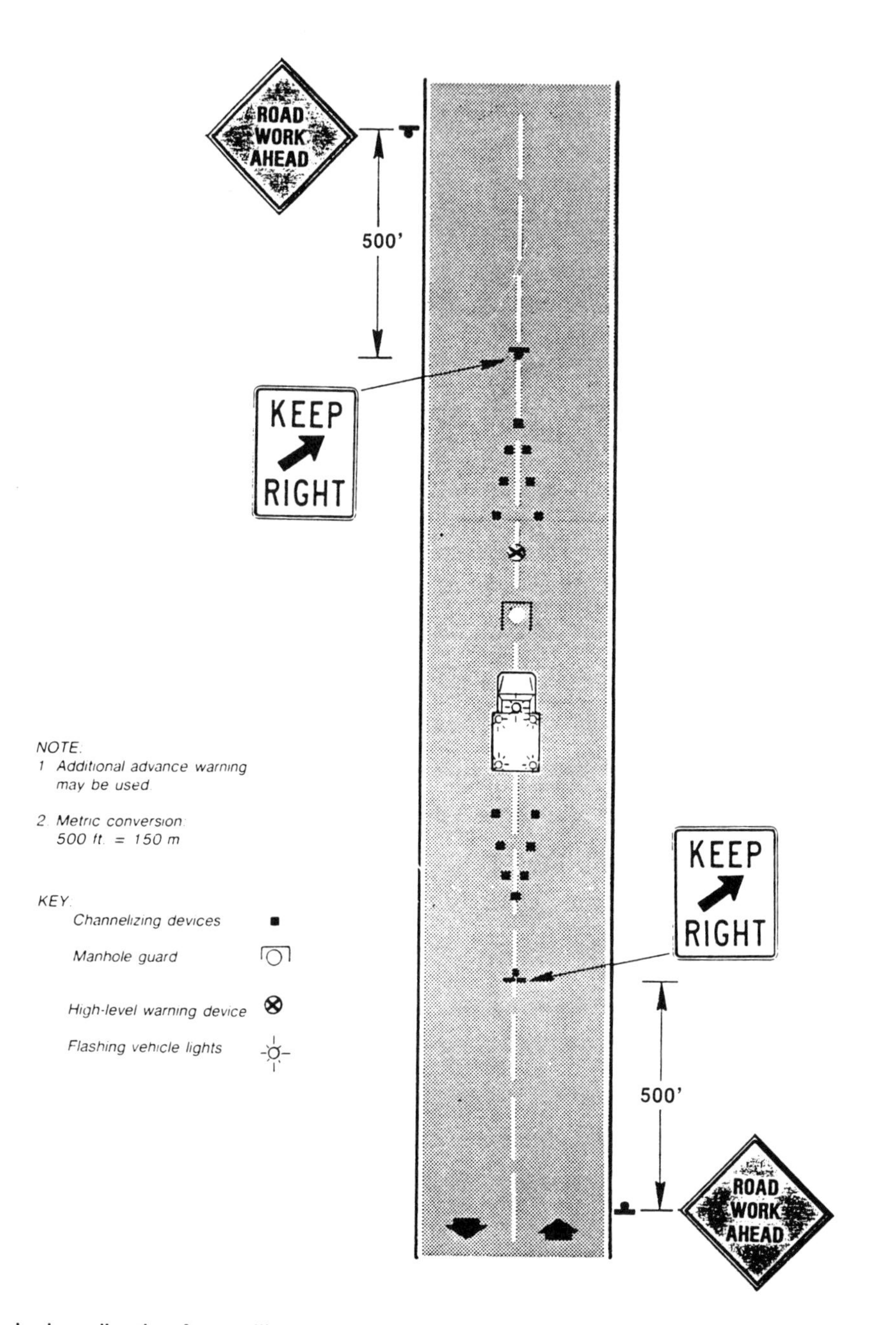

Figure 5-16. Typical application for a utility work zone on an urban location. (U.S. Department of Transportation)

the insidious nature of finely divided material may prevent accidents. Risk of engulfment can also be reduced through the use of appropriate protective equipment such as safety harnesses and lifelines.

The hazardous effects of heat and cold can be addressed by proper clothing and limiting the exposure time through the use of warm-up or cool-off breaks. Hazards posed by pedestrians and traffic can be controlled through the proper use of barriers and traffic control devices.

All of these hazard control methods must be integrated into a comprehensive confined space entry program which is described in the next chapter.

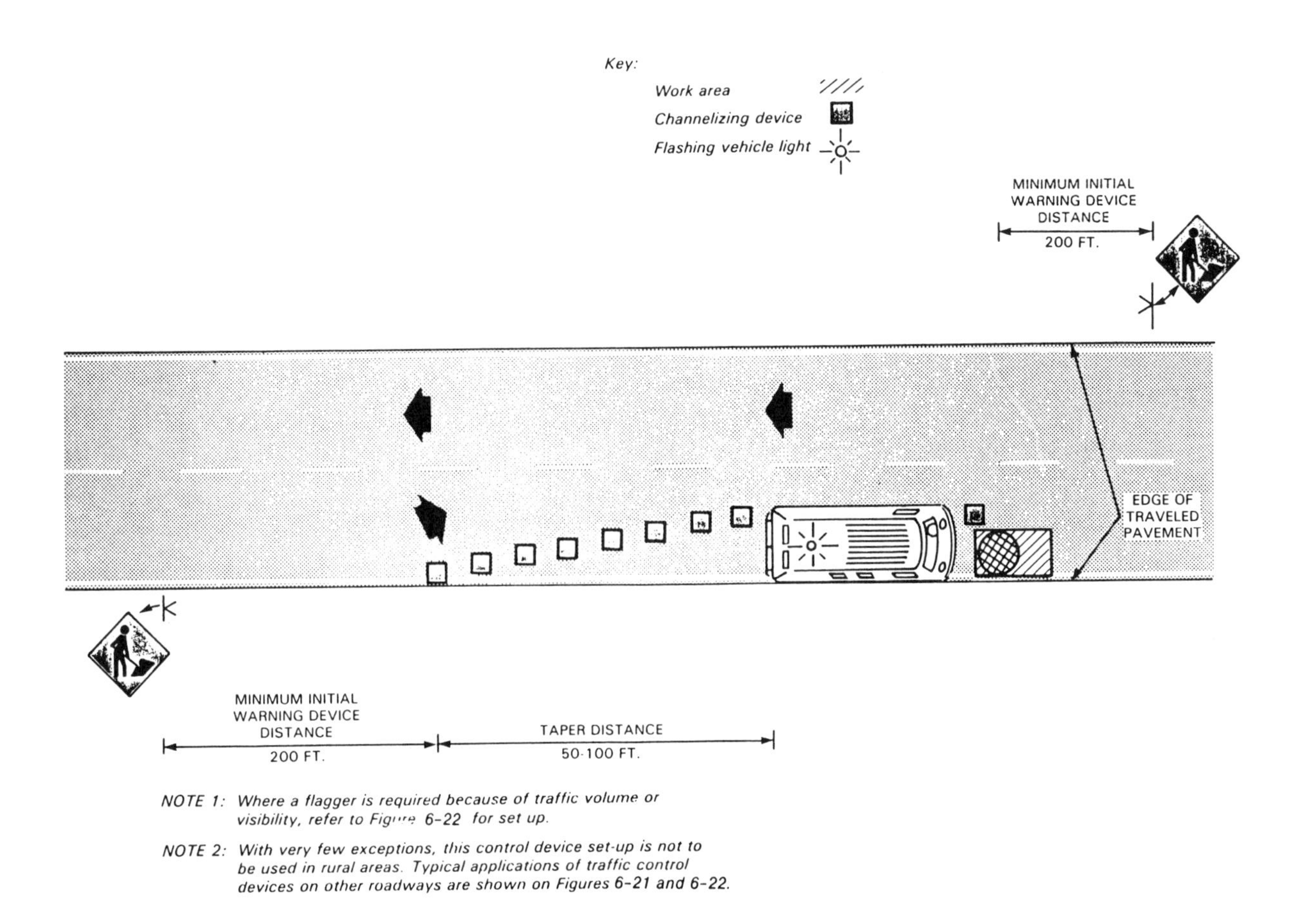

Figure 5-17. Typical application for a utility work zone on a two-lane residential street. (U.S. Department of Transportation)

REFERENCES

Amrhyn, O.C., M.H. Paret and D. Rainer, Testimony of Bellcor, Bell Communications Research, in letter dated September 15, 1989, letter to OSHA Docket S-109.

Arc Welding and Cutting, Code of Federal Regulations, Vol. 29, Part 1910.254.

Cloe, W.W., Selected Occupational Fatalities and Asphyxiating Atmospheres in Confined Spaces as Found in Reports of OSHA Fatality/Catastrophe Investigations, Washington, DC, U.S. Department of Labor, Occupational Safety and Health Administration, (1985).

Cloe, W.W., Selected Occupational Fatalities Related to Welding and Cutting as found in Reports of OSHA Fatality/Catastrophe Investigations, Washington, DC, U.S. Department of Labor, Occupational Safety and Health Administration, (1985).

Cloe, W.W., Selected Occupational Fatalities Related to Lockout/Tagout as Found in Reports of OSHA Fatality/Catastrophe Investigations, Washington, DC, U.S. Department of Labor, Occupational Safety and Health Administration, (1985).

Cloe, W.W., Selected Occupational Fatalities Related to Fixed Machinery Found in Reports of OSHA Fatality/Catastrophe Investigations, Washington, DC, U.S. Department of Labor, Occupational Safety and Health Administration, (1985).

Cloe, W.W., Selected Occupational Fatalities Related to Ship Building and Repairing as Found in Reports of OSHA Fatality/Catastrophe Investigations, Washington, DC, U.S. Department of Labor, Occupational Safety and Health Administration, (1985).

Control of Hazardous Energy, Code of Federal Regulations, Vol. 29, Part 1910.147.

Entry Into Grain Bins and Food Tanks, Data Sheet 663, Chicago, IL, National Safety Council, (1977).

Field, W.E. and R.W. Baily, Entrapment and Suffocations in Flowing Grain, presented at the 1979 Winter Meeting of the American Society of Agricultural Engineers December 11–14, New Orlando, LA, Paper No. 79-5535.

Grain Handling Facilities, Code of Federal Regulations, Vol. 29, Part 1910.272.

Hazardous (Classified) Locations, Code of Federal Regulations, Vol. 29, Part 1910.307

Machinery and Machine Guarding: General Requirements for All Machines, Code of Federal Regulations, Vol. 29, Part 1910.212.

Pettit, T.A., P.M. Gussey and R.S. Simmons, 1980, Criteria for a Recommended Standard: Working in Confined Spaces, National Institute for Occupational Safety and Health, (DHEW/NIOSH Pub. 80-106) Government Printing Office, Washington, DC, National Electrical Code, Quincy, MA: National Fire Protection Association, (1993).

NIOSH Alert, Request for Assistance in Preventing Entrapment and Suffocation Caused by the Unstable Surfaces of Stored Grain and other Materials, Cincinnati, OH: National Institute for Occupational Safety and Health, (1987).

Oxy-fuel Gas Welding and Cutting, Code of Federal Regulations, Vol. 29, Part 1910.253.

Permit-Required Confined Spaces for General Industry, Code of Federal Regulations, Vol. 29, Part 1910.146.

Traffic Control Device Handbook, Washington, DC: U.S. Department of Transportation, (1983).

CHAPTER 6

Entry Program Requirements

INTRODUCTION

One of the points emphasized in Chapter 1 is the wide diversity that exists among confined spaces. The same can be said of entry programs designed to control confined space hazards. These differences stem largely from variations in organizational structure, management philosophy and the level of cooperation that exists between staff members. While the regulatory mandate imposed on all facilities is identical, a program's philosophy and style are dictated by the organization's infrastructure. In one facility, a single crew may perform all activities, from isolating the space to doing the work inside it. In another facility, separate crafts may be responsible for performing specific procedures. For example, electricians disconnect motors, pipefitters detach and block fluid lines and industrial hygienists test the atmosphere in the space.

The preceeding chapters identified the types of atmospheric and physical hazards that may be encountered during confined space entry. This chapter builds on that foundation by explaining the OSHA standards requirements for a confined space entry program. Some program elements, like communications equipment and protection from external hazards, can be adequately addressed in a single section of this chapter. Other topics like ventilation, atmospheric testing, lockout-tagout and respiratory protection are much more complex. They will be introduced here but discussed more fully in succeeding chapters.

GENERAL PROGRAM REQUIREMENTS

The OSHA standard's legalistic wording suggests that a confined space program must address fourteen points. As a practical matter, these fourteen items can be distilled down to ten critical elements.

1. *Hazard Identification.* The hazards presented by permit-required confined spaces must be identified and evaluated prior to entry.
2. *Hazard control.* Procedures, methods and practices must be developed and implemented to eliminate or control confined space hazards. These precautions may include:
    - Implementing measures to prevent unauthorized entry
    - Specifying acceptable entry conditions
    - Isolating the space and controlling hazardous energy
    - Purging, inerting, flushing or ventilating the space
    - Providing pedestrian, vehicle and other barriers as necessary to protect entrants from external hazards
    - Verifying that all conditions remain acceptable for the duration of the authorized entry
3. *Permit system.* A written system for preparing, issuing, implementing and canceling entry permits must be developed. This system must also include provisions for closing off the permit space and returning it to service after the work is completed.
4. *Specialized equipment.* Any specialized equipment required for entry must be provided to employees at no cost, be maintained in effective condition and used properly. This equipment may include:
    - Air monitoring instruments
    - Ventilation equipment
    - Personal protective gear
    - Portable explosion-proof lighting
    - Communications equipment
    - Protective barriers and shields
    - Ladders and climbing devices
    - Emergency and rescue equipment, such as retrieval devices, portable fire extinguishers, first aid kits, etc.
5. *Employee designation.* Employees who play an active role in the entry such as authorized entrants, attendants and entry supervisors must be specifically designated. The program must also define the duties and responsibilities of these employees.
6. *Testing and monitoring.* Permit spaces must be tested and monitored to determine if conditions are suitable for entry and remain acceptable for the duration of the job.
7. *Outside Contractors.* Employers must coordinate the activities of any contractors who perform work in permit spaces.

8. *Emergency procedures.* Emergency procedures, including provision for rescue equipment, must be established and implemented. An attendant must be provided and remain available outside the space for the duration of the entry. If multiple spaces are to be monitored by a single attendant, the program must address how the attendant will respond to an emergency in one space without compromising supervision of the other spaces.
9. *Information and training.* Employees who serve as authorized entrants, attendants, entry supervisors or emergency responders must be trained in their respective duties and responsibilities.
10. *Program review.* Any problems encountered during the entry must be noted on the permit. At least once a year, the canceled permits must be reviewed to determine if there is a need to modify existing procedures in order to assure continued employee protection. Entry procedures must also be reviewed whenever there is reason to believe that they may not be fully protective. Program deficiencies must be corrected before future entries are authorized.

CONDUCTING AN INITIAL SURVEY

By April 15, 1993, you were supposed to have surveyed your facility to determine if there were any permit-required confined spaces. If there were, you were also supposed to have informed your employees of the existence of the spaces and hazards that they posed. Both the regulatory text and the decision tree provided in the standard's appendix imply that if the initial survey indicated the absence of permit-required spaces, nothing else had to be done.

Unfortunately, this thinking is flawed for two reasons. First, it overlooks the fact that existing *non-permit spaces* could become permit-required spaces at some point in their life cycle. Second, it fails to consider that equipment or process changes could introduce permit spaces at some future time.

In this light, I'd like to offer the following five step approach which not only addresses present conditions, but also anticipates changes in the future.

Step 1. Determine if your facility has *any* confined spaces. If it doesn't, there's nothing else you have to do *for the moment,* but bear in mind that process and facility changes should be continually evaluated to determine if they create permit spaces in the future.

Step 2. If your survey reveals that there are confined spaces, determine if they are permit-required spaces or non-permit spaces. If all of your spaces are non-permit spaces, you again don't have to do anything now. But you'll need to remember to monitor them to determine if they become permit spaces at some later date.

Step 3. If you determine that your facility has permit-spaces, you'll then have to do two things. First, you'll have to inform your employees of the locations of the spaces and the hazards they pose. Second, you'll have to institute measures to prevent entry by unauthorized people. One way to do this is by posting warning signs or notices with a legend that reads:

Danger Permit-required confined space Do not enter

However, the performance-based nature of the standard also allows you to warn workers through any other effective means such as training.

Step 4. Determine if any of your employees has to enter a permit space. If so, you'll have to develop a comprehensive written entry program.

Step 5. Determine if employees of another employer, for example, a contractor, will have to enter a permit-space. If so, you'll have to develop procedures to assure that the contractor abides by the confined space standard.

With this five step strategy in mind, let's now turn our attention to exploring each of the ten critical program elements.

ELEMENT 1: HAZARD IDENTIFICATION

The first and perhaps most critical aspect of an entry program is making an inventory of the potential hazards to which entrants may be exposed. The survey must be logical, systematic and comprehensive. Threats posed by seemingly insignificant hazards should not be ignored. After all, a condition that is dismissed because it appears trivial may be the one which triggers a chain of events that ends in catastrophe.

Although it is certainly not an all-inclusive list, Table 6-1 identifies some of the questions that should be asked as part of the hazard identification process. In some facilities, these questions may be answered

Table 6–1. Selected hazard inventory considerations

Characteristics of the Space

- What did the space last contain?
- Are there any hazards posed by residue?
- Does the configuration pose any unusual problems?
- Are interior surfaces potentially slippery?
- Are there any projections or objects that could cause cuts, bumps or abrasions?
- How large is the entry portal and where is it located?
- Is there anything around the portal that poses a hazard?
- Are there any potential hazards presented by adjacent processes or operations?

Atmospheric Hazards

- Could the atmosphere be deficient in oxygen?
- Could the atmosphere be oxygen enriched?
- What air contaminants might the space contain?
- Will air contaminants be introduced into the spaces by processes like welding, spray painting or solvent cleaning?
- Could the atmosphere be flammable?
- Does the atmosphere have the potential for becoming flammable?

Physical Hazards

- Does the space contain any mechanical equipment?
- Are there any fluid lines attached?
- Will any hazards be posed by portable equipment taken into the space?
- Is there a potential for engulfment?
- Are there any external hazards such as exposed electrical components, mechanical equipment, or vehicular or pedestrian traffic?

Other Considerations

- Will any noise-producing operations be performed?
- Are there any potential radiation hazards posed by thickness gage sources or X-ray equipment?
- Is there any potential for vermin or poisonous animals like spiders and snakes?

by a single person. In others, a number of individuals with different areas of expertise will have to be consulted. Regardless of the approach taken, the person who authorizes the entry is the one who ultimately assumes responsibility for verifying that all hazards have been identified and that appropriate protective measures have been implemented.

Since the hazard inventory reflects the interaction between the entrants and their environment, it must consider a number of factors such as the reason for entry, the space's physical characteristics and the operations to be performed in the space.

Determining the Reason for Entry

The hazard identification process should begin by determining the reason for entry. As mentioned previously, confined spaces may be entered for a variety of reasons, such as inspection, cleaning, maintenance and repair. Knowledge of why the space is being entered and the nature of work to be performed provides a frame of reference for relating entrants to their environment. It may also provide some initial insight to the types and magnitude of hazards that entrants could encounter.

For example, a worker who enters a water-meter vault merely to take a reading may be potentially exposed only to atmospheric hazards. On the other hand, maintenance workers who enter a chemical reactor to weld and adjust mechanical mixing blades could be exposed to a litany of hazards including accidental start-up of the mixer motor, inadvertent inflow of hazardous fluids, harmful residue, welding sparks, energized electrical components, ultraviolet light, hot surfaces and metal fumes.

In some situations, the hazard identification process might suggest an engineering change that could eliminate the need for future entries. For example, if workers have to enter a vault to operate a control valve, it might be possible to install a motor or chain drive so the job could be done from outside the space.

Physical Characteristics Survey

After determining the reason for entry, information should be collected on the space's characteristics. Some of the items that should be considered in making this assessment are summarized below.

Size and Configuration

Small, irregularly shaped spaces may restrict the movement of both entrants and rescuers. Large spaces are not immune from problems since they may contain internal obstructions like piping and side-wall baffle plates which can make ventilation and atmospheric testing difficult.

Space Mobility

Mobile spaces like tank trucks and rail tank cars can be moved inadvertently and should be chocked or otherwise secured to prevent unintended shifting.

Attached Equipment

Internal mechanical equipment such as mixers, blenders or stirrers and attached piping systems containing harmful fluids should be identified and isolated prior to entry.

Natural Lighting

Since limited lighting may contribute to some accidents, the availability and degree of interior lighting should be assessed. Spaces not adequately illuminated by natural light should be provided with artificial lighting.

Entry Portals

The number, size and location of entry portals should be noted. In some cases, portals will be located at ground level, making entry relatively easy. In other cases, portals may be located at an elevation that can only be reached by ladders, stairs or mobile work platforms. It is important to note that entry from one elevation to another—for example, entering a vertical tank from a roof portal—may introduce fall hazards when entrants climb down a ladder. Except for ladder wells meeting the dimensional requirements shown in Figure 6-1, OSHA standards require protective cages or ladder climbing devices be used on fixed ladders more than 20 feet long. Commercially available devices, like that shown in Figure 6-2, can also be attached to manhole ladders, making them easier to mount and dismount.

Previous Contents

If the space or any of its associated piping previously contained hazardous chemicals, the relevant material safety data sheets should be obtained and carefully reviewed for information on flammability, reactivity, health effects, signs and symptoms of exposure, emergency procedures, warning properties and any special precautions that entrants should take if they are exposed to the material. The potential for traces of hazardous materials remaining as residue or product sludge must also be considered since evaporating liquids may produce flammable or toxic atmospheres.

In situations where entry is possible through more than one portal, all possibilities should be considered and the portal which poses the least risk should be selected for entry.

Determine Operations to be Performed

The potential for other hazards will be suggested by the operations that will be performed. The use of portable electric tools, for example, suggests a potential shock hazard. Processes like abrasive blasting, welding, cutting and spray finishing may introduce atmospheric hazards like dusts, fumes and vapors. Similarly, air lines and welding cables lying in walkways pose a potential tripping hazard.

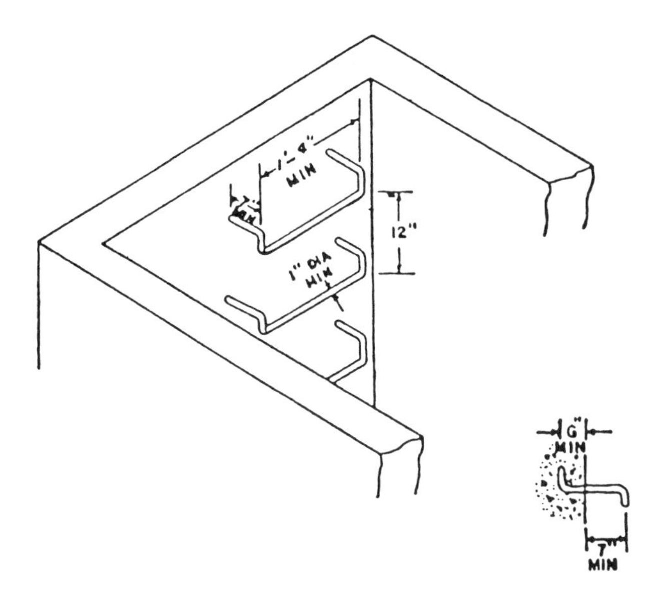

Figure 6-1. Dimensional requirements for ladder wells.

Figure 6-2. Manhole ladder attachments like this provide easier access. (Bilco)

Visual Inspection

At some point it will be necessary to visually inspect the interior of the space. This task is usually performed by viewing the interior through a manhole or other opening. However, since the standard stipulates that entry occurs when *any part* of a worker's body breaks the plane of an opening, employees must be careful not to breach any openings while making their visual inspection.

If it is not possible to inspect the space safely from outside, then the visual inspection should be delayed until all hazards have been characterized and controlled. Alternatively, the atmosphere may be presumed to be immediately dangerous to life or health, necessitating the use of positive-pressure atmosphere supplying respirators such as air-line respirators with escape bottles or self-contained breathing apparatus. This selection and use of this equipment will be discussed more fully in Chapter 11.

Hazards Outside the Space

Not all confined space hazards are located inside the space. An obvious example of an outside hazard is vehicular traffic around street manholes. But not all outside hazards may be as apparent. For example, both an unguarded belt and pulley and an exposed energized electrical conductor would be considered "guarded by location" if they were located more than seven feet above the floor. As a consequence, they would be viewed as posing very little hazard to personnel *under normal conditions.* On the other hand, if a vessel's entry portal was located adjacent to these elevated hazards, they would present a significant risk to entrants who climb into or out of the space.

ELEMENT 2: HAZARD CONTROL

Once all hazards have been inventoried, the next step is to eliminate or control them. Ideally, hazards should be eliminated whenever possible. For example, the hazards posed by explosive or oxygen-deficient atmospheres can be eliminated through ventilation. Similarly, hazards posed by toxic reside may be eliminated by draining and flushing the space. However, it is not always possible to eliminate hazards, and in those situations where they can not be eliminated, they must be controlled.

Ranking Hazards

Some spaces may present an array of hazards, each of which pose a different level of risk. Without a system that ranks hazards according to their potential risk, time and resources could be diverted to addressing low-risk hazards while more significant ones went unattended. Consequently, all hazards on the inventory should be ranked in descending order based on the level of risk they pose. The scope and direction of corrective action is then dictated by a particular hazard's position on the list. Those that present the greatest risk should be addressed first, followed in decreasing order by those which pose less of a threat.

One simple, semi-quantitative assessment technique considers both the severity of the hazard's consequences and probability that those consequences will occur. While this approach is highly subjective, it can be remarkably reliable when coupled with practical field experience and sound professional judgment.

Hazard Severity

Hazards may be divided into four categories depending on the seriousness of their adverse consequences.

Catastrophic hazards are those that may cause death (perhaps even multiple deaths); widespread or severe occupational illness; or loss of an entire production operation or operating process.

Critical hazards may also result in death, but additionally include situations that may produce serious injuries such as the loss of an eye or a limb; illness requiring medical treatment; or extensive property damage, including significant damage to operating or process equipment.

Marginal hazards may cause minor injury such as cuts and contusions; illness such as headache, nausea, or mild eye, skin or respiratory tract irritation; or slight damage to equipment or property.

Negligible hazards will not result in serious injury or illness and will cause little to no property or equipment damage.

Probability of Occurrence

The probability or likelihood that a hazard will result in an adverse consequence may be classified as:

- Frequent: likely to occur virtually every time
- Probable: likely to occur most of the time
- Occasional: likely to occur sometimes
- Remote: very unlikely to occur
- Improbable: negligible chance of occurring

Hazard Assessment Matrix

Once a hazard's severity and probability have been established, the relative risk it poses can be determined from the assessment matrix shown in Figure 6-3. This information can then be used to prepare a rank-listing of all the hazards

It's important to note, however, that both severity and probability may increase or decrease due to changes that occur in or around the space. For example, a change in weather that results in a sudden downpour could flood a below-ground space with rain water. Similarly, fugitive emissions, or a sudden uncontrolled release of hazardous materials from an adjacent process unit, may create atmospheric hazards which were not present initially. As a result, entrants and attendants must be ever vigilant for changes in the work environment that may affect conditions in the space—and, in the end, their safety.

General Control Strategies

Once all hazards have been ranked, the next step is to develop a strategy for controlling them. Again, the technical approach should be logical, systematic and thorough. Each of the rank listed items should be reviewed and possible controls identified. If there is more than one way of controlling a hazard, consider all the options and choose the most practical approach. Although confined spaces vary considerably, many of them share similar hazards. Thus, similar control measures may be applied in widely varying situations. Among the most commonly employed controls are:

- Implementing measures to prevent unauthorized entry
- Providing protection from external hazards
- Draining and flushing the space to remove residue
- Isolating the space and controlling hazardous energy
- Purging and ventilating
- Using protective equipment like respirators, hard hats, coveralls, gloves and safety glasses

Control of Unauthorized Entry

Analysis of case histories suggests that unauthorized entry principally occurs in two ways. First,

Frequency of Occurrence	HAZARD CATEGORIES			
	1 CATASTROPHIC	2 CRITICAL	3 MARGINAL	4 NEGLIGIBLE
(A) Frequent	1 A	2 A	3 A	4 A
(B) Probable	1 B	2 B	3 B	4 B
(C) Occasional	1 C	2 C	3 C	4 C
(D) Remote	1 D	2 D	3 D	4 D
(E) Improbable	1 E	2 E	3 E	4 E

Risk Assessment Code	Suggested Criteria
1A, 1B, 1C, 2A, 2B, 3A	Unacceptable
1D, 2C, 2D, 3B, 3C	Undesirable (management decision required)
1E, 2E, 3D, 3E, 4A, 4B	Acceptable with review by management
4C, 4D, 4E	Acceptable without review

Figure 6-3. Risk assessment matrix. (Adapted for MIL-STD-8828)

unqualified rescuers may enter a space in an attempt to aid unconscious or injured occupants. Second, employees who are unaware of the insidious nature of confined space hazards may enter for incidental reasons. For example, a maintenance mechanic might climb into a wastewater sump to recover an accidentally dropped part or tool. Not being aware of the potential for atmospheric hazards in the sump, he could be overcome by hydrogen sulfide or an oxygen-deficient atmosphere.

Because the potential hazards posed by confined spaces are not obvious to everyone, steps must be taken to inform employees of the possibly lethal consequences of entry. One way to do this is through the use of warning signs and labels. Another way is through employee education.

Although signs and labels cannot physically prevent an entry, they can alert employees to the existence of potentially dangerous environments that they might not otherwise recognize. As a result, permit spaces—other than those that cannot be entered without the use of special keys or tools—should be posted with warning signs which indicate the nature of the hazard and the prohibition against unauthorized entry. In some cases, labels may be affixed adjacent to or directly on hatch openings or access covers. In other cases, signs may be posted at entrances that lead to areas containing confined spaces.

Access to active confined spaces can be prevented by cordoning off the walkways with plastic safety cones, rope barriers, railings or other obstacles that will dissuade bystanders and casual observers from entering the work area. Entry by unqualified personnel, such as would-be rescuers, can be prevented by attendants who remain on duty outside the space.

External Hazard Protection

External hazards may be posed by moving equipment, energized electrical conductors, pedestrians and vehicular traffic. As mentioned earlier in this chapter, the area around entry portals should be examined for mechanical, electrical or thermal hazards that may threaten to entrants as they climb in and out of the space. Some specific hazards that should be considered are

- In-running roller nip points
- Exposed belts and pulleys
- Unguarded chains and sprockets
- Moving conveyors
- Energized electrical parts
- Exposed hot and cold surfaces
- Vibrating, rotating, oscillating or sliding mechanical components

Protection from pedestrians may be provided by surrounding spaces with protective railings, fences, high-visibility wire, rope or tape, and similar barricades. As noted in Chapter 5, specific guidance for protection from vehicle traffic can be found in *The Traffic Control Device Handbook.*

Draining and Flushing to Remove Residue

Tanks, reactors and process vessels should be drained of their contents and flushed from outside to remove sediment, sludge and traces of surface residue. In some cases, pre-entry cleaning procedures may require the use of caustic cleaners or gum-cutting solvents. If these materials are used, they too should be drained and interior surfaces rinsed to remove any remaining surface contamination.

Live steam is sometimes used to remove scale, tar and other substances that might coat side walls. Since steam cleaning creates hot surfaces that can burn unsuspecting entrants, sufficient cool off time must be allowed before entry. However, hot vessels should not be cooled so rapidly that their structural integrity is compromised. In one case, workers attempted to speed up the cooling of a freshly steamed chemical storage tank by spraying it with water from a plant fire hose. Unfortunately, the sudden change in temperature caused the tank shell to warp to the point of failure.

The nature and quantity of the residue in the space could be such that the steam condensate and flushing effluent may themselves be considered hazardous wastes. If so, they should be handled and disposed of in accordance with applicable environmental laws and regulations.

Lockout-Tagout Procedures

The accidental start-up of mechanical equipment in a space can be prevented by a lockout-tagout procedure. The regulatory requirements governing these procedures are rather complex and will be discussed in detail in the next chapter. However, a brief explanation of lockout-tagout is appropriate here.

In its most elementary form, lockout-tagout comprises three steps. First, an energy isolation device like an electrical disconnect switch or a valve is placed in the "off" or "safe" position. Next, a lock (lockout) or tag (tagout) is attached to the device to prevent it from being reactivated by anyone other than the person who attached the lock or tag (Figures 6-4 and 6-5). Finally, the lock or tag is removed by the person who applied it when the work is completed.

It is generally agreed that locks provide a more foolproof means of protection than tags since extraordinary force is required to remove them. Tags, on the other hand, can easily be by-passed or ignored. As will be seen later, these shortcomings can be partly resolved by taking additional precautions such as opening more than one electrical disconnect switch and increasing the level of employee training.

Disconnecting and Blanking Lines

Any lines attached to the space which contain harmful fluids should be disconnected or closed off with a blank flange. These devices, which are sometimes called "blinds," are circular metal plates which can be fitted over the bore of a pipe and are capable of withstanding the full head of pressure created by any materials upstream in the pipe.

Simply closing a valve does not provide adequate isolation for two reasons. First, unless the valve is locked out, it could be opened. Second, even if the valve was locked out it could leak. However, there is a configuration of valves known as "double-block-and-bleed" that can be used to achieve isolation simply by closing some valves and opening others. Double-block-and-bleed will be discussed more in the next chapter.

Purging and Ventilating

Atmospheric hazards in confined spaces are usually controlled by purging and ventilating.

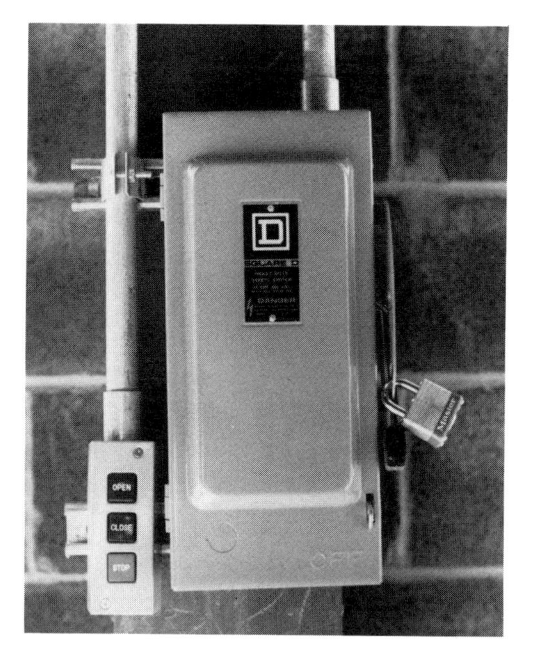

Figure 6-4. Lock-out used on electrical disconnect. (John Rekus)

Figure 6-5. Tagout-tag used on control valve. (John Rekus)

Purging in this context means the initial displacement of hazardous gases and vapors by air, steam or an inert gas forced into the space. The choice of purging agent depends on the previous contents of the space. Inert gases like nitrogen, carbon dioxide and argon are frequently used to purge spaces of flammable atmospheres, while steam and air are used to remove toxic air contaminants.

Ventilation may be defined as the process of continuously moving fresh air through the space. This continuous movement of air accomplishes three things. First, it maintains an adequate level of oxygen in the space. Second, it dilutes or removes toxic air contaminants that may be formed in the space. Third, it improves comfort by controlling temperature, humidity and nuisance odors.

The decision whether to force air into a space or exhaust it out depends on many factors including the size and configuration of the space, the nature of its previous contents and the number of openings in the space. In some situations it will be necessary to both blow air into the space and exhaust it out. For example, ventilation of a large storage tank may be accomplished using one fan to direct air in through a side-wall portal while a second fan exhausts it out the top (Figure 6-6).

Further information on ventilation concepts, principles and techniques will be presented in Chapter 8.

Protective Equipment

OSHA General Industry Standards require the use of personal protective equipment whenever there is a possibility of injury to any part of the body. But personal protective devices like gloves, hard hats, respirators and safety glasses do not remove the hazard, they merely place a barrier between it and the wearer.

Since no piece of personal protective equipment will be effective unless it is worn, employees must be informed when and how to wear it, and supervisors must be empowered with the authority to enforce its use. All equipment must retain its original level of protection, and regular inspections are necessary to identify defects such as cracks, breaks, tears and worn or missing parts which could reduce the equipment's effectiveness.

Additional information on the selection, use, care and maintenance of protective equipment will be provided in Chapters 10 (General Personal Protective Equipment) and 11 (Respiratory Protection).

ELEMENT 3: ENTRY PERMIT SYSTEM

Entry permits are written documents which certify that specific precautions such as isolation, atmospheric testing and ventilation have been taken before

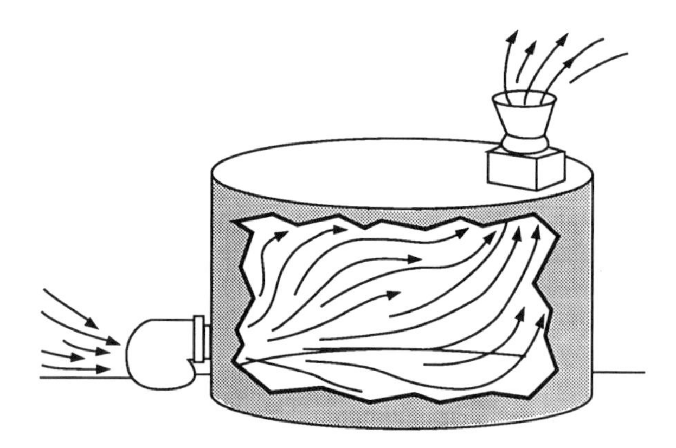

Figure 6-6. Large spaces may require blowing air in through a side-wall opening and exhausting it out the top. (Paul Trattner)

workers enter a confined space. To assure that all elements for an entry program are in place, there must be a written system for preparing, issuing, implementing and canceling entry permits. This permit system must also include provisions for closing off the permit space and returning it to service after work is completed.

Permit Elements

OSHA regulations do not prescribe a specific format for permits, but their content and layout should be standardized to facilitate easy use. At a minimum, permits are required to include the following information:

- The identification of the space to be entered
- The purpose of the entry
- The date and authorized duration of the entry
- Description of the hazards of the space
- Measures taken to isolate the space and manage the hazards
- The acceptable entry conditions
- Results of initial and periodic tests, including the name or initials of the tester and indication of when tests were made
- Communications procedures
- Special equipment required
- Identity of the authorized entrants and attendant
- The rescue and emergency services that can be summoned and the means for summoning them
- Any additional permits such as those required for hot work
- Any other relevant information
- The entry supervisor's signature or initials

Although OSHA provides three example permits in Appendix D to the confined standard, readers are cautioned not to be guided by them since they don't seem to comply with the standard's requirements.

Specifically, none of the examples has a provision for specifying communications procedures, and only one addressed emergency response-and-rescue procedures. One or more of the sample permits also lacked provisions for recording the purpose of the entry, the date and authorized duration of the entry, the acceptable entry conditions, a description of hazards in the space, results of tests and measurements with tester's name or initials and indication of when the tests were made.

Completing the Permit

People who authorize entries and complete permits must possess the knowledge required for the recognition, evaluation and control of confined space hazards. If "hot work" like welding or oxy-acetylene cutting will be performed, that fact must be prominently noted on either the entry permit or on a special "hot work permit" attached to the entry permit.

The permit must be signed by the individual authorizing the entry. However, his signature cannot be affixed until all of the actions and conditions required for safe entry have been met. The permit may be revoked by the authorizing individual whenever entry conditions become unacceptable. Finally, it is canceled after all entrants have exited the space following the completion of the job.

Any unusual conditions or problems encountered during the entry should be noted on the permit. These comments can then be reviewed to determine if program changes are warranted.

ELEMENT 4: SPECIALIZED EQUIPMENT

A variety of specialized equipment is required for evaluating and controlling confined space hazards. This equipment generally includes ventilation systems, atmospheric testing instruments, communications equipment, self-contained breathing apparatus and rescue devices. The work performed in some environments may also dictate the need for nonsparking tools, low-voltage lighting and ground-fault circuit interrupters.

As explained in Chapter 3, electrical or electronic equipment such as blowers, testing instruments and portable lighting which may be used in "classified locations" as defined by the National Electrical Code®—that is, locations where ignitable atmospheres may be present—must be intrinsically safe, explosion-proof or otherwise "approved" for use in the classified location.

Ventilation Systems

A wide variety of equipment is commercially available for ventilating confined spaces. Some models, such as that shown in Figure 6-7, are designed to be mounted directly onto a manhole opening. Other models, like that shown in Figure 6-8, consist of portable air movers attached to a length of flexible duct which is used to direct air into the space. Ventilation equipment may be driven by a variety of power sources including steam, compressed air (Figure 6-9), electric motors (Figure 6-10) and internal combustion engines (Figure 6-11). Typical air delivery rates range from about 500 to 17,000 cubic feet per minute (CFM), depending on equipment design, power source and ducting configuration.

Specific types of equipment as well as advantages and disadvantages of various approaches to confined space ventilation will be described more fully in Chapter 8.

Figure 6-7. Some ventilators like this are designed to mount directly on a tank manway. (Coppus)

Air Sampling Instruments

Portable air sampling instruments are needed to evaluate confined space atmospheres for oxygen, flammability and toxic contaminants. At one time, a separate instrument was required to evaluate each of these conditions. While it is still possible to purchase three individual instruments, the trend recently has been toward combination instruments that allow multiple atmospheric conditions to be evaluated with a single device. Many of these combination units incorporate both an oxygen and combustible gas meter (Figure 6-12). Some units also allow measurement of "toxic" gases like carbon monoxide and hydrogen sulfide (Figure 6-13).

Selection of the most appropriate equipment for a particular application depends on a number of factors including:

- The number and types of contaminants expected
- The desired degree of instrument sensitivity
- The level of portability required
- Ruggedness, stability and ease of operation
- Costs, including initial purchase price and operating expenses associated with expendables such as sensor elements and calibration gas
- The level of factory support available for calibration, maintenance and user training

Figure 6-8. A portable ventilator provided with flexible duct. (Coppus)

Figure 6-9. Venturi eductors like this are powered using compressed air or steam. (Coppus)

Figure 6-11. Blower powered by gasoline engine. (General)

Figure 6-10. Electric motor driven blower. (Pelsue)

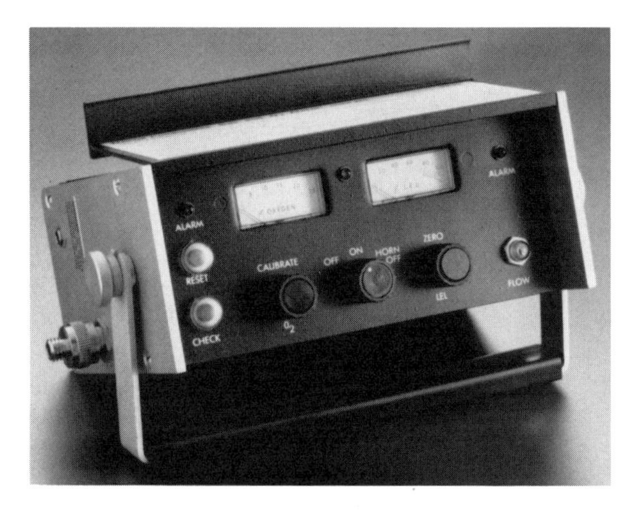

Figure 6-12. Combination oxygen and combustible meter. (MSA)

Some instruments incorporate features like microprocessors that store sampling data and allow subsequent printout as a permanent "hard copy." Others are equipped with audible alarms that sound warnings if pre-set contaminant levels are exceeded.

Recent developments in sensor technology have resulted in a new generation of instruments that can detect specific toxic contaminants in addition to oxygen and flammable atmospheres. As mentioned previously, the two sensors incorporated most commonly are carbon monoxide and hydrogen sulfide.

An instrument equipped with a carbon monoxide detector would be a good choice if measurements were routinely made around carbon monoxide-producing process equipment like ovens, furnaces, and boilers, or around street manholes where there is the danger of vehicle exhaust. Similarly, an instrument equipped with a hydrogen sulfide sensor would be an

Figure 6-13. Four gas instruments for oxygen, combustible gas, hydrogen sulfide and carbon monoxide. (Neotronics)

appropriate choice if measurements were made frequently around sanitary sewers, storm water drains, wastewater process equipment or other areas likely to contain hydrogen sulfide.

Instrument theory of operation, selection, calibration and use will be discussed in much more detail in Chapter 9.

Communications Equipment

Ordinary visual and voice communication between those inside and outside the space may be sufficient in many situations. However, in those cases where it is either inadequate or unreliable, it must be supplemented by other means such as two-way radios, sound powered headsets, field telephone or even tugs on a rope.

Rope Signals

In some situations a lifeline can serve double duty as a communications link if a prearranged set of signals has been established between the entrant and attendant. Table 6-2 outlines some typical rope-tug signals that can be used for two-way communication. While rope-signal systems have the advantage of being inexpensive, they are really only practical when a single entrant is involved.

Table 6-2. Example rope signals

Number of tugs	Signal	Meaning
1	Ok	When made from outside: "Is everything all right?"
		When made from inside: "Everything is all right."
2	Advance	When made from outside: "Go ahead."
		When made from inside: "Give me more line so I can advance."
3	Take up	When made from outside: "Come back."
		When made from inside: "Take up line. I am coming out or changing position."
4 or more	Help	When made from outside: "Come out immediately!"
		When made from inside: "Emergency. I need help."

Wire Systems

Wire communication links may be divided into two types: sound powered systems and electrically powered systems.

Sound-powered systems are a more sophisticated version of a child's "telephone" consisting of two tin cans connected by a piece of string. The units (Figure 6-14) consist of head- or hand-sets connected together by a pair of wires. When sound vibrations strike the mouthpiece, a diaphragm inside moves a coil of wire between the poles of a permanent magnet. As the coil moves in and out of the magnetic flux field, it creates an electrical current. This current then travels through the connecting wires to receiving earpieces where sound waves are recreated by the reverse process.

Electrical systems may be either telephones or solid state intercoms. These systems (Figure 6-15) are similar to their sound-powered counterparts in that they use head- or handsets connected together by wires. However, electrical energy is provided by either a battery or direct current power supply. This voltage powers a carbon microphone in the telephone, or a solid state audio-amplifier in the intercom. Both

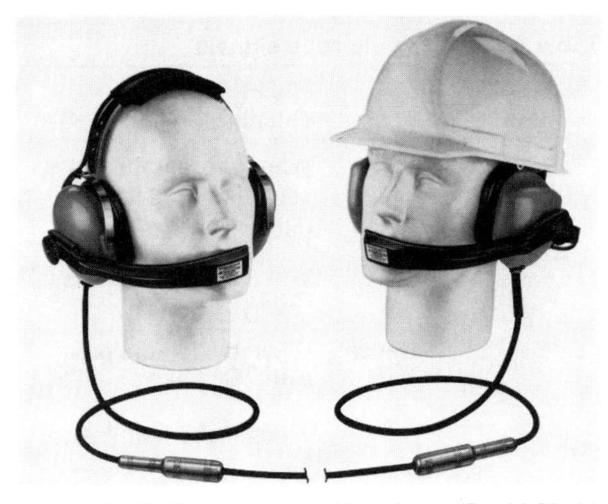

Figure 6-14. Sound powered head set. (David Clark)

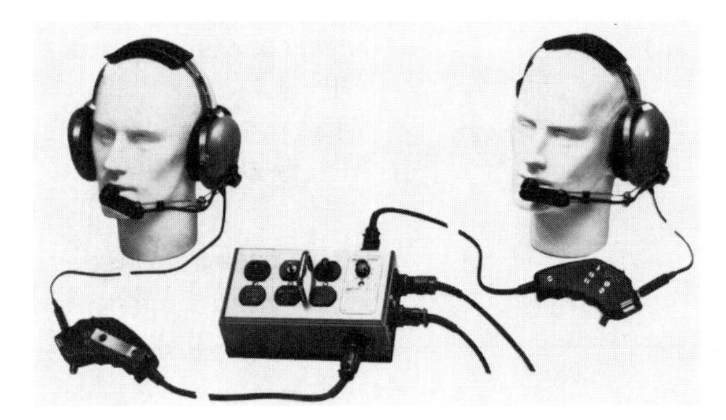

Figure 6-15. Intercom. (David Clark)

devices convert sound waves into electrical impulses which can then be transmitted over wires.

The earpiece volume of both the sound powered and telephone systems is not adjustable. The earpiece volume on intercom systems can be increased substantially. This feature is an important consideration in situations where devices will be used in high noise environments.

Hybrid Systems

One manufacturer offers a hybrid system consisting of both a lifeline and wire intercom (Figure 6-16). The system consists of both a 3/8-inch nylon rope which can be used for retrieval and an electrically powered headset system that can accommodate up to six users. The communications wires form the core of a rope whose ends terminate in quick-connect plug-in fittings for the headsets. The rope is attached to a full-body harness, the wearer plugs his headset into the termination connectors and he is set up for both communication and retrieval.

Wireless Systems

Wireless systems such as bullhorns and two-way radios offer two distinct advantages over wire-connected systems. First, there are no cables to string out, roll up, or to get cut, tangled or snagged. And second, radios can be equipped with either ear- or throat-microphones that can be used while wearing respirators. On the other hand, radios are more expensive to purchase and maintain than wire systems. Their operation also requires a license from the Federal Communications Commission (FCC).

While all radios must conform with certain performance specifications dictated by the FCC, there is a seemingly endless variety of makes and models available in the market place. Individual units differ in size, weight, battery life, operating features, accessories and output power. Of these, output power is one of four factors that influence the unit's effective range. The other three are the operating frequency, the antenna type and antenna location.

Almost all industrial two-way radio systems employ push-to-talk or "simplex" operation. Pushing the transmit button on the radio's case—or on an external microphone—turns the transmitter on, releasing it turns the receiver on. This means that an operator can *either* talk or listen, but can't do both at the same time—as can be done on a telephone.

Faraday Shielding. A problem encountered with entry into metal tanks and vessels is that the walls of the space act as a Faraday Shielding which blocks radio signals unless the two antennas are within line-of-sight. In other words, if the attendant and entrant can't see each other, they can't communicate. Faraday shielding may be overcome by using a "passive repeater" consisting of two portable whip antennas connected together with a short piece of coaxial cable. The antennas should be clamped to an entry portal so that one is inside the space and the other outside. This arrangement will passively reradiate the signals as long as both operators are in sight of the antennas, even if they are not in sight of each other.

Full-duplex Repeater. One manufacturer offers a system which not only overcomes Faraday Shielding

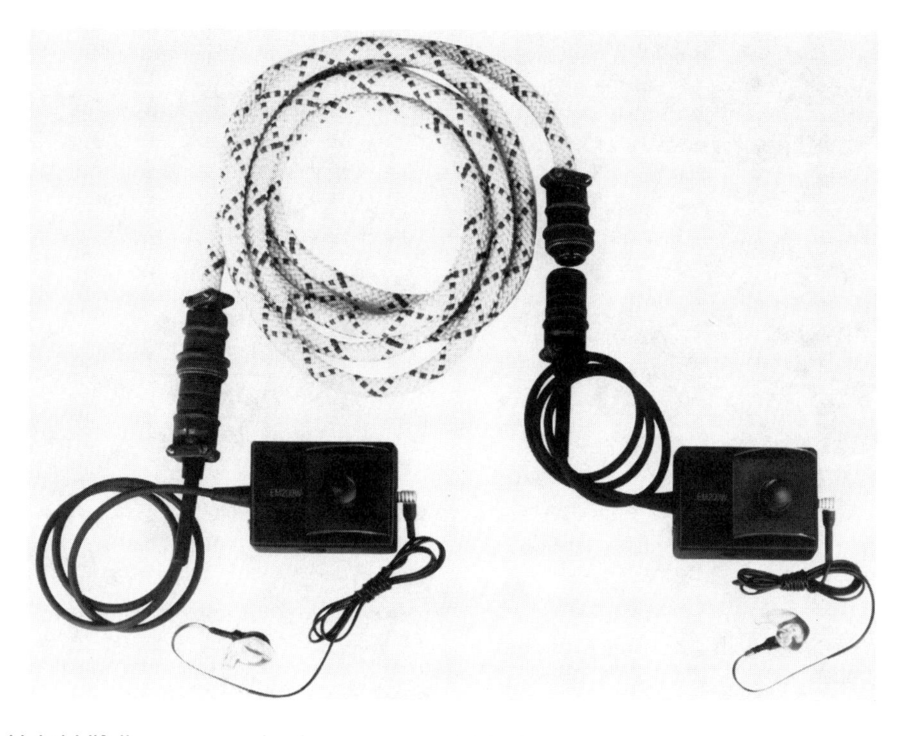

Figure 6-16. Hybrid lifeline communications system. (Telex)

but also provides full-duplex operation which allows simultaneous conversation in both directions. The system (Figure 6-17) consists of a battery powered module that is placed adjacent to an entry portal and self-contained radio headsets worn by entrants and attendants. The headset microphone may be either a throat mike as shown in Figure 6-18 or a boom mike like that shown in Figure 6-19.

Unlike the previously described passive repeater that rebroadcasts signals through induction, the active repeater receives the signal sent by the headsets and feeds them to a transmitter where they are rebroadcast. The headset's microphone is equipped with a voice-actuated switch which eliminates extraneous background noise, and the ear pieces can be attached to hard hats as shown in Figure 6-20.

Personal Alert Safety Systems (PASS). Personal Alert Safety Systems like those shown in Figure 6-21 were first developed for use in the fire service. The PASS senses a persons movements and if an injured or overcome wearer remains motionless for more than 30 seconds, the PASS alerts others by sounding an audible "person-down" alarm. PASS devices worn by confined space entrants can add an extra margin of safety both by providing almost immediate warning

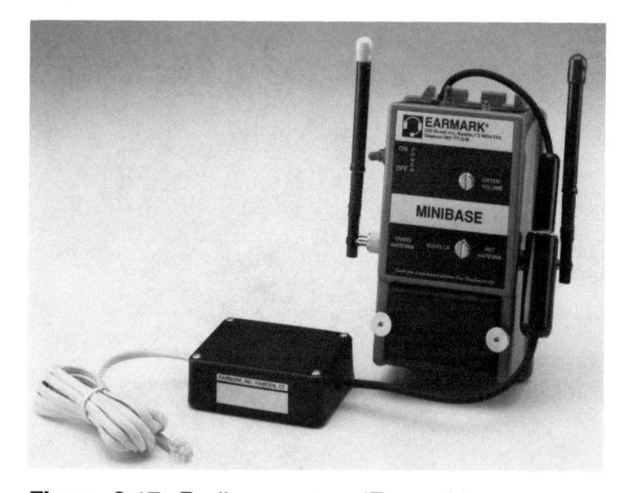

Figure 6-17. Radio repeater. (Earmark)

that a worker has become incapacitated and an audible signal that allows rescuers to more easily locate a downed victim (Figure 6-22).

The radio-PASS (Figure 6-23) works on the same principle but is designed for use on shared radio channels. Like its fire service counterpart, the radio-PASS monitors the wearer's motion and sounds a

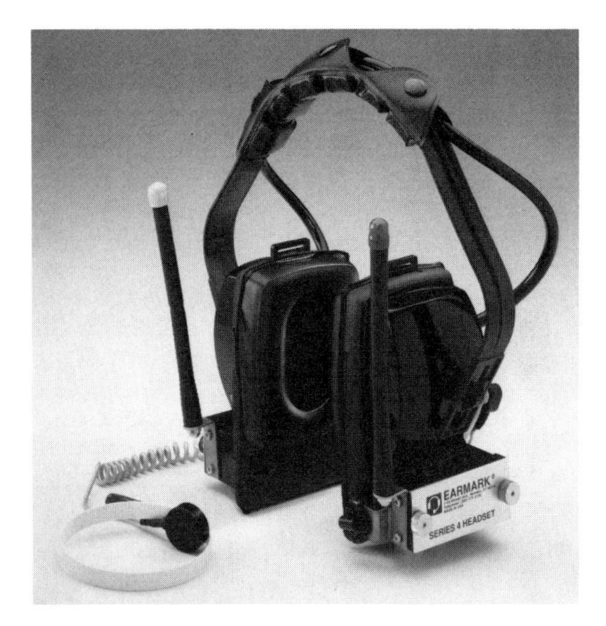

Figure 6-18. Radio headset with throat microphone. (Earmark)

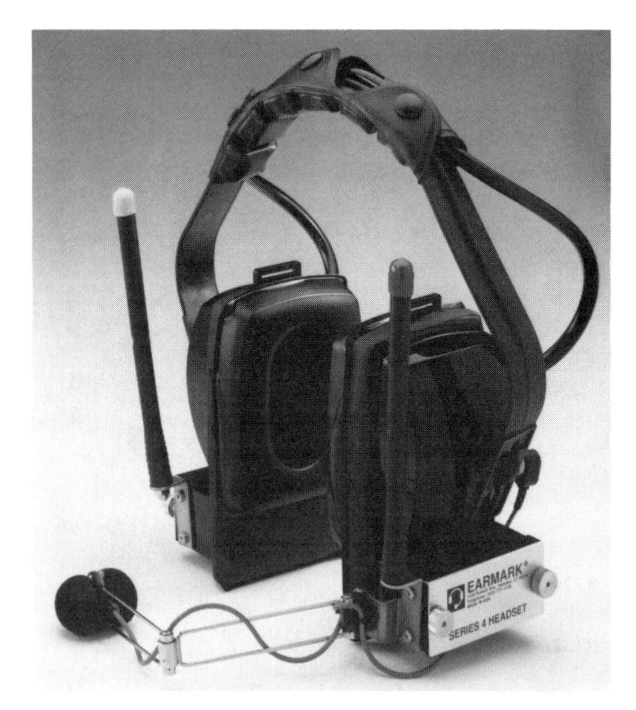

Figure 6-19. Radio headset with boom microphone. (Earmark)

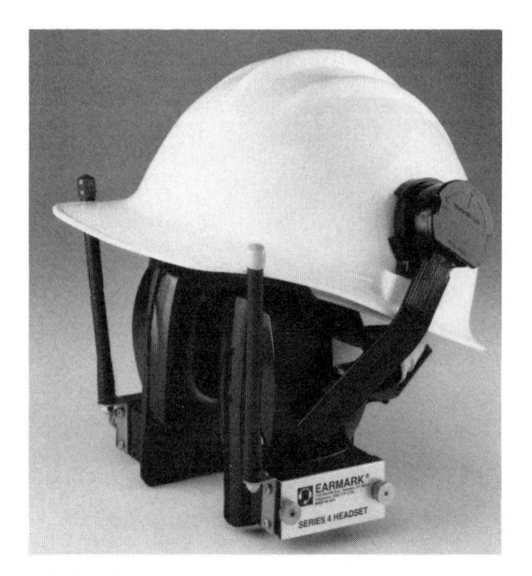

Figure 6-20. Radio headset fitted to hard hat. (Earmark)

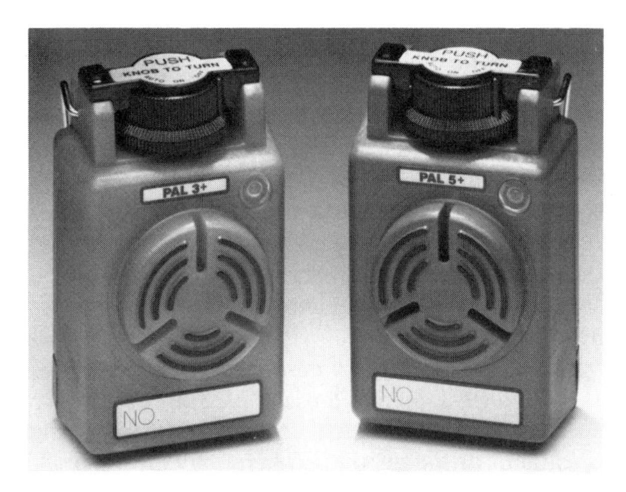

Figure 6-21. Pass devices. (Detex)

warning alert if no movement occurs for 20 seconds. The wearer then has 10 seconds to manually respond to this warning by pressing a reset button. If he does not, the unit alarms by transmitting a user identification number and an "alarm-condition" signal to a remote monitoring station. The monitoring station automatically acknowledges the PASS alarm signal and clears the radio channel for emergency response communications. This interactive approach minimizes the amount of time the PASS is on the air and without it, the PASS would lock-up the channel barring further communication.

Figure 6-22. Pass devices like this can help locate an incapacitated victim. (Detex)

Figure 6-23. The Pass on the left is a radio Pass. For comparison, a standard Pass is shown on the right. (Radio Systems, Inc.)

The radio-PASS also incorporates a high-amplitude audio-alarm which assists rescue personnel in locating the injured or immobilized victim. In situations where a number of people are working within audio range, the audible alarm may provide first warning that a co-worker is in trouble. In addition to the "person down" function, the radio-PASS contains a user-activated manual panic button that can be used in situations where the user can activate the alarm. When it is not in the alarm mode, the radio-PASS can be used for normal two-way communication.

Equipment Submersibility. Some manufacturers have designed their equipment to operate reliably even when fully submerged. This option may be an important consideration in situations where work is routinely performed in or around wet locations.

Component Substitution. Since radios and their accessories are tested and approved as a complete assembly, substitution of other than specified components will invalidate an intrinsic safety approval. That means that components from different manufacturers, makes and models may not be interchanged, even if they are all individually "intrinsically safe." For example, the battery, microphone or earpiece from one manufacturer's intrinsically safe radio cannot be used on another manufacturer's radio unless it has been specifically approved for use with that device.

An exception is provided for "entity" certified equipment. The entity concept allows the interconnection of intrinsically safe devices that have not been approved as a combination, provided that specific electrical compatibility criteria are met. The specification for each device will identify four parameters that must be considered to determine whether or not two devices may be interconnected. These parameters are

- Maximum Input Voltage V_{max}
- Maximum Input Current I_{max}
- Maximum Output Voltage V_{oc}
- Maximum Output Current I_{sc}

Apparatus may only be interconnected if both of the two following conditions are met

1. The maximum output voltage (V_{oc}) is not greater than maximum input voltage (V_{max}) and
2. The maximum output current (I_{sc}) is not greater than the maximum input current I_{max}

Bullhorns. Problems posed by hard-wire systems can also be overcome through the use of "bullhorns."

Bullhorns are battery-powered devices that incorporate a microphone, audio-amplifier and directional speaker all in a single handheld, self-contained package supported by a pistol grip handle. The user simply holds the bullhorn in front of his mouth, presses the handle-mounted push-to-talk switch and speaks into the microphone. His voice is amplified and directed out the speaker. Bullhorns are less expensive than radios and at least one manufacturer provides units approved for use in hazardous locations.

Air-Supplied Respirators

Although it is desirable to remove contaminants and increase the oxygen supply by engineering means like ventilation, it is not always possible. For example, in the case of a broken, hazardous material pipeline, the time taken to ventilate the pit and shut-off valve could result in a worse situation than if workers had entered the pit wearing self-contained breathing apparatus and stopped the uncontrolled release of material by closing the valve.

Air-supplied respirators may be broadly divided into three types: self-contained breathing apparatus (SCBA), air-line respirators and combination air-line/SCBA. The operating principle of all three is essentially the same in that the facepiece is connected to a supply of fresh breathable air. The air supply for the self-contained breathing apparatus is carried in a tank on the wearer's back (Figure 6-24) while air-line devices (Figure 6-25) are connected to an air compressor or compressed air cylinders located outside the space.

As their name suggests, combination units incorporate air-line and SCBA features into a single device. Primary breathing air is provided via an air-line and a small compressed air bottle worn on the worker's belt provides an escape supply that can be used in an emergency.

Other types of respirators and the requirements for a comprehensive respirator protection program will be discussed in Chapter 11.

ELEMENT 5: EMPLOYEE DESIGNATION

Employees who play an active role in the entry must be specifically designated. These employees include:

- *Authorized entrants* who are specially trained to work in confined spaces

Figure 6-24. Self-contained breathing apparatus provide users with an air supply in a cylinder worn on their back. (MSA)

- *Attendants* stationed outside the confined space who monitor entrants' activities and implement the rescue plan if necessary
- *Entry supervisors* who oversee and authorize the entry after determining that conditions are acceptable for entry

Authorized Entrants

An authorized entrant is an employee who is allowed to enter the space. These are the people who face the greatest risk of injury or death from exposure to the confined space hazards. Although the purpose of the permit program is to protect the entrants during entry operations, the entrants themselves must assume some responsibility for their personal safety. But to do this, they must know the hazards they might face during entry. They must also understand the consequences of exposure to those hazards.

Employees who can recognize hazards and who understand the consequences of exposure are more

Figure 6-25. Air-line respirators like this provide air from a source such as a compressor. (Scott)

likely to respond to a problem in time for a successful rescue. Properly trained entrants are those who know the modes, signs and symptoms of hazards to which they may be exposed and who follow safe work practices and procedures.

Authorized Attendants

One of the significant problems related to confined space entry operations is that an incapacitated entrant inside the space cannot normally be seen by people outside the space. For example, other employees working nearby might not notice an entrant who has succumbed to an oxygen deficiency or a toxic air contaminant inside a chemical reactor. As a result, the incapacitated employee would probably die before anyone even realized that something was wrong. In fact, a similarity shared by many confined space accidents is that incapacitated employees were not found until it was too late for rescue.

Assuring the presence of an attendant outside a permit space is a widely accepted method for assuring entrants' safety. Consequently, the OSHA standard requires an attendant to be stationed outside the space for the duration of the entry. From their vantage points, attendants should be able to monitor conditions in and around the space, track the activities of entrants and summon rescue services in an emergency.

The attendants' chief rescue responsibility is to summon emergency services. Thus, it's their ability to determine that entrants need help and to summon responders, not their proximity to the entrants that's critically important.

Acknowledging that electronic surveillance and communication equipment can assist attendants in carrying out their duties and responsibilities, the standard allows attendants to monitor more than one space. It is also interesting to note that attendants may also be stationed *anywhere* outside the space, provided they can effectively carry out their assigned duties. For example, attendants could be in control rooms that allow them to monitor entrants remotely using electronic tracking devices, closed-circuit television or public address systems.

But it is not sufficient for attendants to simply know how many entrants are inside the space. Instead, they must be able to track and *identify* the authorized entrants entering and exiting the space. The reason for having to track the identity of entrants is two-fold.

First, a tracking system assures that all entrants have exited the space at the conclusion of the job. Second, it provides rescuers with a positive means of identifying entrants.

This is critical during rescue operations, because with all the excitement, it becomes easy to loose track of exactly who exited the space. A system which only keeps count of the number of authorized entrants is not acceptable since it cannot ensure that all entrants have been rescued. Without a systematic approach for tracking employees, there is no way to account for entrants who failed to inform the attendant of their successful self-rescue from the space. Rescue personnel would then be exposed unnecessarily when they entered the space in an attempt to rescue entrants—who unknown to emergency responders—had already evacuated.

Another problem with the simple count method is that it does not address situations where unauthorized personnel may have entered the space. These people—who may have even caused the emergency in the first place—could easily be counted as "entrants" as they emerged from the space. The resulting miscount could lead the attendant to believe that every one had evacuated, when, in fact, some authorized entrants were still inside. These employees might then suffer further injury or death as a result of the mix up.

Clearly, it is essential during rescue operations to be able to determine whether or not all authorized entrants have evacuated. Any system for monitoring employee activity (badge-boards, sign-in/sign-out sheets, etc.) is acceptable as long as it identifies entrants inside the space.

Entry Supervisors

Many confined space accidents have occurred simply because the entry lacked oversight by someone with overall responsibility for enforcing workplace safety rules. As a result, the OSHA standard requires that all permit entries be directed by an entry supervisor who is directly responsible for the entrants' safety. Since these supervisors are responsible for all aspects of the entry operation, it is reasonable to expect them to know at least as much, if not more about confined space entry, than the entrants and attendants.

In general, entry supervisors are responsible for checking permits to assure that they are properly completed. They are also responsible for assuring that all required precautions such as lockout/tagout, isolation and ventilation have been taken prior to entry. Specific duties they are charged with include:

- Verifying the existence of acceptable entry conditions
- Verifying that rescue services are available, and that the means for summoning them is operable
- Revoking the permit if any conditions not allowed by the permit are observed
- Authorizing the entry by signing or initialing the permit
- Removing unauthorized entrants from the space
- Terminating the permit at the end of the job

ELEMENT 6: TESTING AND MONITORING

Permit space conditions must be tested and monitored to determine if the space is safe for entry and to assure that conditions remain acceptable for the duration of the job. This is accomplished through the use of testing instruments and visual inspections. Instruments are used to evaluate environmental conditions such as air contaminants, noise, radiation and temperature. Visual inspections are employed to ensure the continued effectiveness of barriers, lockout controls and isolation methods.

Initial tests should be performed from outside the space whenever possible. However, if the space must be entered to be evaluated, that entry must be done in accordance with all entry precautions. More specific information concerning testing and monitoring is provided in Chapter 9.

ELEMENT 7: COORDINATING OTHER EMPLOYER'S ENTRIES

In situations where an employer hires a contractor to work in a confined space, both the host employer and the contractor have certain responsibilities.

Host Employer Responsibilities

The host employer is obligated to do six things:

1. Advise contractors why the space is a permit space.
2. Share with the contractor any information concerning the host employer's experience with the space.
3. Inform contractors that they must comply with a permit-required confined space entry program which meets the requirements of the standard.
4. Inform the contractor of any special procedures or precautions the host employer takes to protect his employees who work in or around the space that the contractor will be working in.
5. Coordinate entry operations when employees of more than one employer will be working in the space simultaneously to prevent employees of different employers from endangering each other.
6. Conduct a debriefing session at the end of the job to discuss any hazards encountered during the entry.

Contractor Responsibilities

Contractors are obligated to:

1. Obtain from the host employer any available information regarding permit space hazards and entry operations.
2. Coordinate entry operations with the host employer when employees of both host employer and contractor will be working in or near a permit space.
3. Inform the host employer of the contents of the permit program that will be used to comply with the standard.
4. Advise the host employer of any hazards encountered in the permit space. This can be done either when the hazards arise, or at the debriefing session.

ELEMENT 8: EMERGENCY RESPONSE PROCEDURES

An effective confined space entry program attempts to identify and control all of the hazards to which entrants may be exposed. If everything goes as expected, work proceeds safely and nobody gets hurt. But we live in an imperfect world, things sometimes go wrong and emergencies arise. Clearly then, the time to plan for an emergency is before it happens, not when it happens.

Rescue may be necessary either because of extraordinary circumstances that arise suddenly and without warning, or because of some deficiency in the permit program. The standard, however, does not require employers to have on-scene rescue capability. Instead, it stipulates that employers develop and implement *procedures* for:

- Summoning rescue and emergency services
- Rescuing entrants from permit spaces
- Providing necessary emergency services to rescued employees
- Preventing unauthorized personnel from attempting rescue

Some employers may prefer to provide on-site emergency responders because off-site response may be slow, inadequate or ineffective. Other employers may prefer to rely on off-site rescue services perhaps because they believe that they do not have the resources to train employees to perform rescue, or because the ready availability of highly trained, quick-responding, off-site services makes on-site capability unnecessary.

Although emergency response procedures are discussed in detail in Chapter 12, two important regulatory issues related to program planning will be discussed here. They are the use of non-entry retrieval systems and specific entry requirements for emergency services personnel.

Non-Entry Retrieval

A rope tied around the waist or attached to a body belt will not be adequate for retrieval in an emergency. Instead, the standard requires that a non-entry retrieval system consisting of a rescue harness and lifeline be employed whenever an authorized entrant goes into a permit space. An exception is made for situations where the retrieval equipment itself would not contribute to the rescue effort, or would increase the overall risk of the entry.

One end of the retrieval line must be attached to the harness either above the head or at the center of the back near the shoulders. The other end of the line must be attached to a mechanical device or fixed point outside the space to facilitate prompt response.

Wristlets may be used in lieu of a harness if the employer can demonstrate that the harness use is infeasible or creates a greater hazard, and that wristlets provide the safest and most effective alternative.

If it is necessary to retrieve entrants from vertical permit spaces deeper than 5 feet, a *mechanical* lifting device must be available. In some cases, other equipment like emergency signal horns, two-way radios, stretchers and life-support oxygen may also be required.

Rescue Services

Rescue services may be provided by either on-site or off-site responders. On-site rescue teams have the advantage of being immediately available and intimately familiar with the facility. Team members may also have existing relationships with other craftsmen such as electricians, pipefitters and mechanics who may be called on in an emergency. Since response time is critical in any emergency, an on-site team reduces response time to a minimum. The principle disadvantage of on-site teams is the time and expense required for initial and annual training of team members.

The expense for municipal rescue services, on the other hand, is borne by the community. But not every community has the resources to train and equip a confined space rescue team. Assuming that a properly equipped municipal rescue squad does exist, the time spent in dispatch and travel can delay its arrival. Even with the best pre-emergency site familiarization and planning, response will not be as fast as that of an on-site team.

On-Site Rescue Service

If an employer chooses to use the on-site approach, then the rescue team members must be properly trained. At a minimum, the standard requires that they receive the same level of training required of authorized entrants. They must also be trained in the proper use of personal protective and rescue equipment.

Each member of the team must be trained in basic first-aid and in cardiopulmonary resuscitation (CPR), and at least one member must be *currently certified* in both.

Rescue team members must also participate in an annual hands-on drill. The drill should simulate rescue operations in which dummies, mannequins, or actual persons are removed from confined spaces. Either actual or simulated spaces may be used, provided the opening size, configuration, and degree of accessibility approximate the spaces from which rescue may be performed.

Off-Site Rescue Service

When an employer arranges for another employer's employees to perform rescue operations, the host employer must do two things.

First, he must inform the rescue service of the hazards that may be encountered when responding to an incident at the host employer's facility.

Second, the host employer must provide the rescue service with access to all permit spaces from which rescue may be necessary so that the rescue service can develop appropriate rescue plans and conduct practice operations.

Chemical Information

If an injured entrant is exposed to a substance for which there is written information such as a material safety data sheet, a procedure must be developed that assures that the information is provided to the medical facility treating the exposed employee.

ELEMENT 9: EMPLOYEE INFORMATION AND TRAINING

Virtually every aspect of confined space entry requires employee training, and the OSHA standard establishes specific training performance levels for entrants, attendants, entry supervisors and rescuers. The training, which must be conducted before employees are assigned confined space duties, needs to equip them with the knowledge and skills they need to do their jobs safely. While specific training requirements will be discussed more fully in Chapter 14, it should be noted that a comprehensive program must address a wide range of topics that include:

- Hazard recognition and control techniques
- Atmospheric testing equipment and methods
- Placement of ventilation equipment
- Lockout-tagout methods and procedures
- Line isolation, breaking and blanking
- Rescue procedures including first aid and CPR
- Operation and use of communications equipment
- Selection and use of personal protective equipment including self-contained breathing apparatus
- Selection and arrangement of barricades
- Method of completing the entry permit

Although it is the legal mandate to train employees whose work involves confined space entry, training should not be limited to *only* these employees. Recall that more than half of all confined space fatalities resulted from ill-fated rescue attempts made by workers who were not aware of confined space hazards. Consequently, *all employees* working in facilities where confined spaces are present should be instructed never to enter a space containing an unconscious coworker, but instead to immediately call for emergency services.

Entrants' Training

Entrants should receive general instruction on the types of hazards they are likely to encounter and the precautions they should take to protect themselves from those hazards. Training should also be trained in the methods used to communicate with attendants and in the operation of any specialized equipment entrants may be expected to use such as air monitoring instruments, communications devices and personal protective equipment. Since entrants may have to evacuate a space in an emergency they must also be trained in evacuation procedures.

Attendants' Training

Attendants must be trained to monitor activities inside and outside the space and be able to recognize conditions that might pose a hazard to the entrants. They must also be familiar with the types of hazards that entrants face, including information on the mode, signs, symptoms and consequences of exposure.

Attendants must be instructed on their communications responsibilities such as knowing when to order an evacuation. Attendants must be able to control access to the space and must be instructed to warn unauthorized people away from the space. If an unauthorized person does enter, the attendant should order the intruder out and inform the entrants and the entry supervisor. Attendants must know that they are to remain outside the space during entry unless they are relieved by another attendant. They must also be versed in non-entry rescue procedures including emergency communications and the use of any retrieval equipment.

Entry Supervisors' Training

Entry supervisors' training must be consistent with the tasks they are expected to perform. These include: determining if acceptable entry conditions are present, authorizing entry, overseeing entry operations and terminating the entry.

The training program must also develop skills which enable supervisors to verify that the permit has been completed properly, that all tests specified by the permit have been conducted, and that all procedures and equipment specified by the permit are in place before allowing entry to begin.

ELEMENT 10: ANNUAL PROGRAM REVIEW

At least once a year someone must review the entry program to ensure that it remains effective. At a minimum the canceled permits should be reviewed for trends, problems or inconsistencies which should be addressed and corrected.

SUMMARY

While confined space entry programs may vary widely in their complexity, they must include the following elements:

- Identification and control of hazards
- Establishing an entry permit system
- Selection and use of special equipment
- Designating employees who play an active role
- Testing and monitoring
- Establishing emergency procedures
- Coordinating multi-employer entries
- Providing employee information and training
- Conducting an annual program review

Hazards may be controlled through a variety of means such as locking and tagging out of mechanical equipment, blocking and blinding of fluid lines, and purging and ventilating spaces of hazardous atmospheres. Entry into some spaces may require highly specialized equipment like two-way radios, atmospheric testing instruments and portable lights approved for use in hazardous locations. Unauthorized entry into confined spaces must be prohibited, and those who are authorized to enter must be informed of potential hazard in and around the space and be trained in safe entry procedures. If outside contractors enter confined spaces, the "host" employer must assure that the contractor is aware of both the space's hazards and the steps that should be taken to abate them. Finally, an entry permit system must be established and used as a final checklist to verify that all necessary precautions have been taken prior to entry.

REFERENCES

Control of Hazardous Energy, Code of Federal Regulations, Vol. 29, Part 1910.147.

Fixed Ladders, Code of Federal Regulations, Vol. 29, Part 1910.27.

Hazardous (Classified) Locations, Code of Federal Regulations, Vol. 29, Part 1910.307.

Machinery and Machine Guarding: General Requirements for All Machines, Code of Federal Regulations, Vol. 29, Part 1910.212.

MagnaRope™ Intercom Line, Telex Technical Data, Minneapolis, MN, Telex Communications, (1990).

Modern Concepts in Communications, Ft. Lauderdale, FL, Radio Systems Inc.

Pathways to Lockout Compliance, New York, NY, Idesco, Personal Alert Safety Systems for Fire Fighters, NFPA 1982, Quincy MA: National Fire Protection Association (1988).

Personal Protective Equipment: General Requirements, Code of Federal Regulations, Vol. 29, Part 1910.133.

Permit-Required Confined Spaces for General Industry, Code of Federal Regulations, Vol. 29, Part 1910.146.

Portable Fire Extinguishers, Code of Federal Regulations, Vol. 29, Part 1910.157.

Portable Ventilators, Milbury, MA: Coppus Engineering, (1990). Model H3140 Headset with Throat Microphone, Worirseter, MA, David Clark National Electrical Code, Quincy, MA: National Fe Protection Association, (1993).

Systems Communications for Productivity, Hamden CN: Earmark.

System Safety Program Requirements, MIL-STD 882-B, Washington, DC: U.S. Department of Defense, (1984).

Traffic Control Devices Handbook, Washington, DC: U.S. Department of Transportation, (1983).

Valve Interlocking System for the Petrochem, Chemical and Process Industries, Erlanger, KY, Castel Safety Inc.

CHAPTER 7

Lockout/Tagout

INTRODUCTION

Some of the case histories presented in previous chapters revealed that confined space entrants can be killed or injured when hazardous liquids and gases leak into a space through attached pipes. Other cases discussed fatalities that resulted when mechanical equipment such as blenders, mixers and stirrers started unexpectedly.

Accidents like these can be prevented by a lockout/tagout program that assures that potentially hazardous equipment within the space cannot be accidentally activated. In its simplest form, a lockout/tagout procedure includes three steps. First, the hazardous equipment is isolated or deactivated, for example, by closing valves or opening electrical disconnect switches. Next a lock (lockout) or a tag (tagout) is attached to a valve or switch to prevent it from being operated. Finally, the lock or tag remains in place until the work is completed. It is then removed by the person who attached it after he or she determines that no hazardous conditions or exposures exist.

While the procedure outlined above generally describes the principle of lockout/tagout, the actual implementation of a comprehensive lockout/tagout program is far more complicated. In addition, specific provisions that lockout/tagout programs must include are dictated by a comprehensive OSHA standard. This chapter explains the requirements of that standard and describes how they may be integrated into a confined space entry program.

OSHA LOCKOUT/TAGOUT REQUIREMENTS

The OSHA lockout/tagout regulation, also known as *The Control of Hazardous Energy,* is contained in 29 CFR 1910.147. The regulation requires employers to establish a formal program for controlling hazardous energy (Table 7-1) which could be released unexpectedly during equipment servicing or maintenance. To be effective, a program must meet the following criteria. First, it must describe the specific procedures that will be used to control hazardous energy when equipment is being serviced or maintained. Second, it must provide employees with information and training on the energy control procedures. Third, it must be audited annually to assure that it remains effective.

Confined spaces are likely to have two systems that are subject to the lockout/tagout standard: electrical systems and piping systems. General isolation methods for both are described below.

ELECTRICAL SYSTEMS

Disconnect Switches

Electrical systems may be isolated by opening a main, branch or local disconnect switch (Figure 7-1). Opening a disconnect switch not only interrupts the flow of electrical energy to exposed conductors such as terminals, relay contacts and buss bar, but also stops motors that drive mixers, pumps and compressors. Start-stop buttons which are part of a control circuit should not be confused with energy-isolating devices.

In its simplest form, a control circuit consists of a pair of wires that connects the start-stop button to a relay (Figure 7-2). When the start button is pressed, the relay closes and power is provided to the equipment. However, the relay will also close if an electrical short occurs anywhere along the control circuit's path. For example, if bare spots on the control wires touched each other, the relay would close—even if the push button was in the "off" position. Since control switches do not provide complete isolation, they cannot be used as lockout/tagout points.

A disconnect knife switch, on the other hand, provides positive electrical isolation of the circuit. As shown in Figure 7-3 the blades of the switch are physically separated from the supply line when the switch handle is thrown into the "off" position.

Circuit Breakers

If the system is not equipped with a separate disconnect switch then it may be possible to isolate the equipment by opening a circuit breaker. At one time it was difficult to lock-out panel mounted breakers, but devices which are capable of locking out both

Table 7-1. Forms of hazardous energy

Electrical: electrical service feeds, branch circuits, exposed energized parts and circuit compounds
Mechanical: mixers, blenders, stirrers, agitators, belts, pulleys, gear trains in running nip, conveyors
Hydraulic: fluids under pressure in pipes, pistons, rams and cylinders
Pneumatic: air or other gases, e.g., carbon dioxide, argon, nitrogen under pressure in pipes, pistons, rams and cylinders
Thermal: hot and cold surfaces, e.g., hot water pipes, steam lines, cryogenic systems
Electromagnetic Radiation: ultraviolet light, visible light, infrared light, radio frequency energy, alpha- and beta- particles, gamma- and X-rays
Stored Energy: electrical energy stored in batteries, capacitors and power supplies; pneumatic and hydraulic stored in pressurized systems; mechanical energy in rotation parts, springs

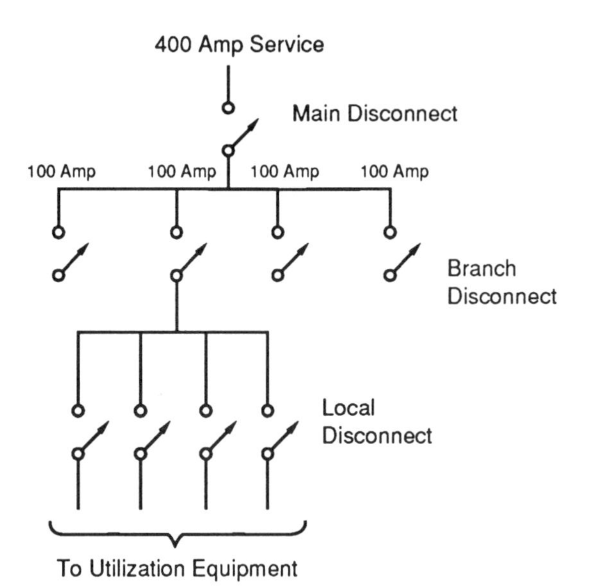

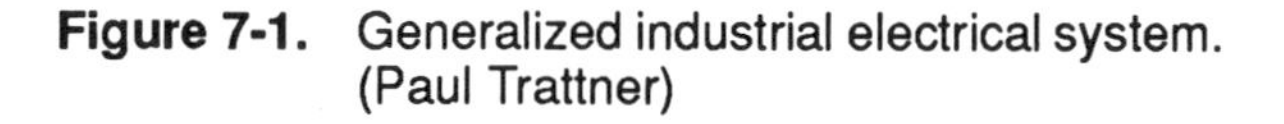

Figure 7-1. Generalized industrial electrical system. (Paul Trattner)

single-pole and double-pole breakers are now commercially available (Figure 7-4).

Cord and Plug Equipment

Cord and plug equipment is exempt from the provisions of lockout/tagout provided that two conditions are met. First, power to the equipment must be completely removed by unplugging. Second, the worker must have the plug under his or her *exclusive* control.

An internal tank agitator that was controlled by plugging and unplugging the motor would meet this criteria provided that the entrant could take the plug into the space. If not, then the plug would have to be locked-out using devices like those shown in Figure 7-5.

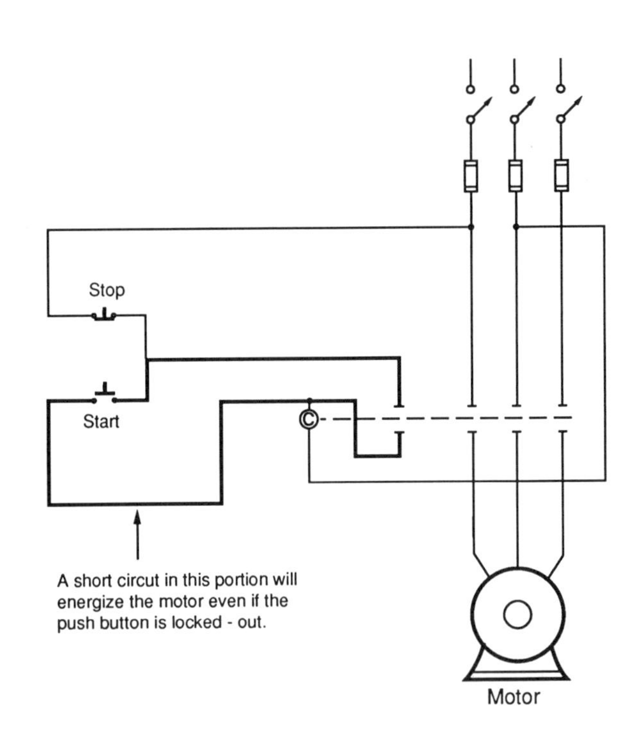

Figure 7-2. Simplified control circuit. (Paul Trattner)

PIPING SYSTEMS

Piping systems may be isolated in three ways: (1) by disconnecting them from the space; (2) by installing blank or blind flanges; or (3) by using a special arrangement of valves called double-block and bleed.

The initial separation or breaking of lines, frequently referred to as "the first break," presents some special hazards. For example, pressurized material which is trapped in the line may be forcefully ejected

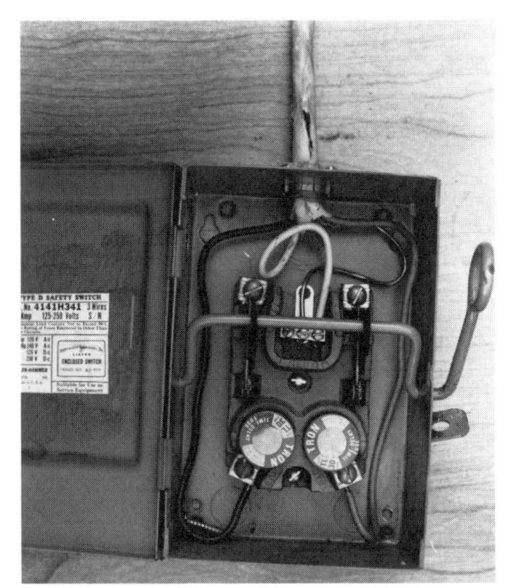

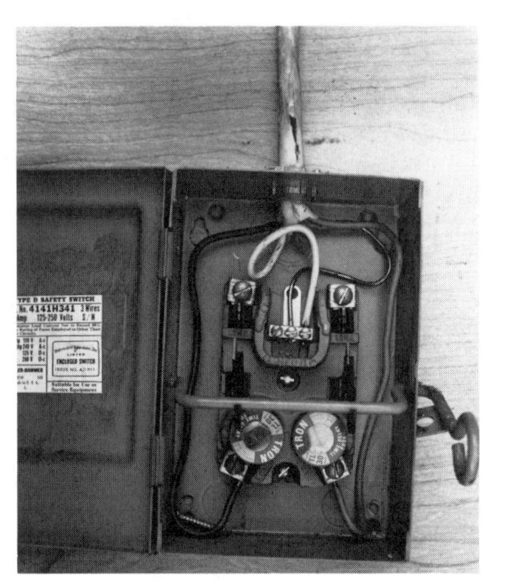

Figure 7-3. Disconnect switch with blades in the closed in the "on" or energized position in (a) and in open in the "off" or deenergized position in (b). (John Rekus)

Figure 7-4. Single- and double-pole circuit breaker lock-out. (Idesco)

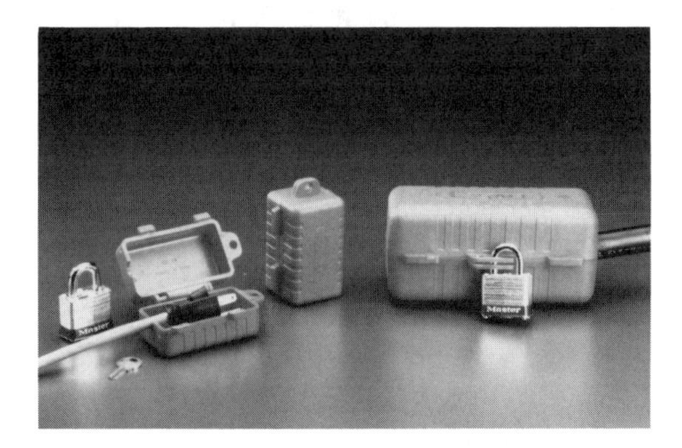

Figure 7-5. Plug lock-out box. (W.H. Brady)

when the line is opened. Consequently, "first-break" procedures should be established to control the hazards posed by the initial separation of the line. The material safety data sheets for any substances believed to be in line should be reviewed to determine the nature of any chemical hazards. An assessment should also be made to identify any physical hazards posed by liquids or gases at elevated temperatures or pressures. Any in-line valves that can be used to drain hazardous liquids, or to bleed-off stored hydraulic or pneumatic pressure, should also be identified.

If possible, lines that previously contained hazardous materials should be drained and flushed to remove any residue. In cases where this is not possible, precautions should be taken to prevent spills when lines are opened. For example, catch buckets can be placed below flanges and pipe unions to collect leaking material. Surfaces under joints that will be separated can be covered with plastic sheeting to prevent contamination and to facilitate clean-up in the event of a minor release.

Disconnecting Lines

Lines may be disconnected by removing the bolts that hold pipe flanges together or by loosening unions that connect threaded pipe sections. Disconnected sections of pipe can then be misaligned to prevent stray material from getting into the space.

Blanking and Blinding

Blanking or *blinding* is a method of closing off a pipe by installing a solid cap or plate so that the pipe's bore is completely covered. The plate's diameter is smaller than that of the flange so that the flange bolts can be reinserted and tightened to hold the blank in position. If the blind could be adversely affected by corrosive or reactive fluids in the line, it should be preceded with a layer of protective material. For example, steel blinds could be protected from sulfuric acid by a rubber or Teflon gasket.

The application of a special type of blind called a "spectacle blind"—because it looks like a pair of glasses—is shown in Figures 7-6 and 7-7. The open side of the spectacle visible in Figure 7-6 indicates the bore is covered by the spectacle's blank side. The reverse is shown in Figure 7-7, where the visible closed side of the spectacle indicates that the bore is open.

Figure 7-6. Spectacle blind indicating bore closed. (John Rekus)

Figure 7-7. Spectacle blind indicating bore open. (John Rekus)

Double-Block and Bleed

As an alternative to blanking or disconnecting lines, the standard allows the use of a valve system called "double-block and bleed." As shown in Figure 7-8, double-block and bleed consists of three valves arranged in a T-configuration. The two "block valves" that make up the arms of the "T" are locked in the "closed" position and the "bleed valve" positioned between them is locked open. The chance that both "blocked" valves will leak at exactly the same time is very small, and if the upstream valve does leak, any escaping fluid would be blocked by the downstream valve and vented through the open "bleed" valve.

LOCKOUT VS. TAGOUT

Lockout is the process of installing a lock on an energy isolating device such as a circuit breaker, disconnect switch or shut-off valve that prevents the device from being operated until the lock is removed by the person who applied it. Although they are not actually locks, blanks and blinds are also considered to be lock-out devices because extraordinary effort must be taken to remove them.

Tagout, on the other hand, is the process of attaching a sign, label or tag to an isolating device. The tag identifies specific equipment that is being used to achieve isolation and warns others not to operate it.

In general, the OSHA standard requires that when a device can be locked out, it must be locked out. If the device cannot be locked-out, then it has to be tagged-out. However, there is an exception to this rule.

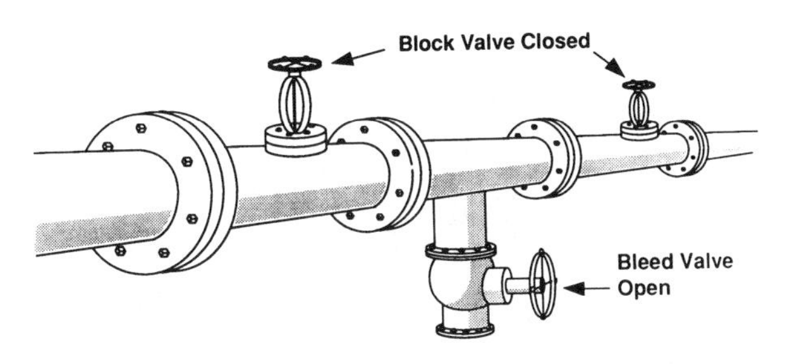

Figure 7-8. Double-block and bleed. (Paul Trattner)

If an employer can demonstrate that his tagout procedures provide employees with the same level of protection that is afforded by lockout, then tagout may be used instead. This may sound confusing since locks obviously provide a greater level of security than tags. However, an explanation is provided in the standard's preamble.

During OSHAs rule-making process, some commenters argued that neither locks nor tags alone prevented workplace accidents, but rather were only one element of a comprehensive *energy control program.* To be effective, this *program* also had to include provisions for employee training, education and auditing. These commenters went on to say that tagout could provide the same level of protection as lockout *provided* there was a rigorous training program and that tagout procedures were stringently followed and strictly enforced.

While OSHA was persuaded by these arguments, it insists that employers who choose to use tagout in lieu of lockout take additional steps to assure "full employee protection." Full employee protection means that additional precautions must be taken to make it more difficult to reactivate a system that has been tagged-out. Examples include removing a valve handle, opening additional disconnect switches or removing isolating circuit components such as fuses or critical printed circuit boards.

If an employer chooses to use tag-out, then the tags must be attached in the same location that a lock would be placed, for example, on a disconnect switch handle or valve hand-wheel. In addition, the employee training program must be expanded to inform employees of the following limitations posed by tags:

1. They do not provide a physical restraint.
2. They are warning devices and may evoke a false sense of security unless their meaning is clearly understood.
3. They must be legible and understandable by *all* employees.
4. Both tags and their means of attachment must be able to withstand hostile environmental conditions encountered in the workplace and must be securely attached so they cannot be inadvertently or accidentally detached.
5. Tags are not to be removed, by-passed, ignored or otherwise defeated without proper authorization.

EMPLOYEE CLASSIFICATIONS

The standard carefully differentiates between employees authorized to implement lockout/tagout and those who are affected by it.

Affected employees are those who meet either of two criteria: (1) their jobs require them to operate equipment that is subject to lockout/tagout or (2) their jobs require them to work in areas where lockout/tagout is used.

Authorized employees, on the other hand, are those who physically lock- or tagout a piece of equipment so that it may be serviced. In some cases, an affected and authorized employee may actually be the same person.

In addition to "authorized" and "affected employees," there may be other employees working in areas where equipment associated with a confined space has been locked out. These employees must also be aware of the existence of lockout/tagout procedures and understand that they should not remove lockout/tagout devices or tamper with the energy isolation device to which the lock or tag is attached.

COMPLETING THE PERMIT

In situations where each entrant is an "authorized employee" and only a few simple isolation devices are involved, all entrants would attach their personal locks or tags before the lockout portion of the entry permit could be completed. In other situations, there may be dozens of isolation points that can be identified only by specialized craftsmen such as electricians and pipefitters who are not part of the entry team. These craftsmen will then become key players in a group lockout procedure.

Using group lockout, an "authorized employee" attaches his lock or tag to the isolation device and assumes responsibility for the safety of all the entrants under his control. This authorized employee must also verify that all entrants were out of the space before removing any locks or tags.

Group lockout procedures are especially efficient for very complex processes that may utilize dozens of isolation points. Having each entrant affix a personal lock or tag to all these points would quickly become a management nightmare. In addition, the continual removal and replacement of locks and tags on many, perhaps widely separated, isolation devices increases the chance that one lockout will be overlooked. Missing just one isolation device could ultimately result in fatal consequences.

Regardless of whether individual or group lockout is performed, it is the responsibility of the person who authorizes the entry to assure that the space is isolated *before* the lockout/tagout section of the permit is checked off.

Figure 7-9. Nylon cable tie used to affix a tag. (John Rekus)

PROVIDING PROTECTIVE HARDWARE

Locks and any other equipment such as chains, blocks, pins and hasps required for energy control must be provided by the employer. Furthermore, these devices must be designated exclusively for lockout/tagout and may not be used for other purposes like securing tool boxes, lockers or truck cabs.

Tags must warn of specific hazardous conditions through the use of legends like "Do Not Operate," "Do Not Close," "Do Not Start," etc. The attachment device must also meet the following criteria:

1. Be able to be affixed by hand
2. Be non-reusable
3. Be self-locking
4. Have a minimum unlocking strength of 50 pounds

One device that meets all of these criteria is a one-piece nylon cable tie like that shown in Figure 7-9.

Locks and their associated attachment devices must be substantial enough to prevent removal without the use of special tools like cutting torches or bolt cutters. They must also be standardized by size, shape, or color. Similarly, tags must use a consistent typestyle, layout and format. Both locks and tags used outdoors or in hostile environments, like those found around acid pickling tanks, caustic cleaning vats or electro-plating baths, must be durable enough to survive the resulting wet, damp or corrosive atmosphere.

IDENTIFYING THE APPLICANT

The standard requires that all lockout/tagout devices be "singularly identified." This means that there must be a way of determining the identity of the person who applied them. Various methods for accomplishing this are described below.

Use of Hanging Tags

Hanging tags may be used alone as tagout devices, or they may be attached to locks to identity the key holder. Preprinted hanging tags (Figure 7-10) are available in a wide variety of legends. Some tags provide room to enter an employee's name and other relevant information such as the identity of the equipment being isolated, the reason for isolation, the time the tag was attached, etc.

They are usually manufactured from heavy gauge card stock. Some are coated or laminated with a plastic film that can be written on with a crayon or pencil and then wiped clean when the job is completed.

Photo Identification

One type of photo identification tagout device (Figure 7-11) is a hanging tag that uses a passport size photograph to identify the authorized employee

Figure 7-10. Preprinted hang tag. (Rockford Medical Systems)

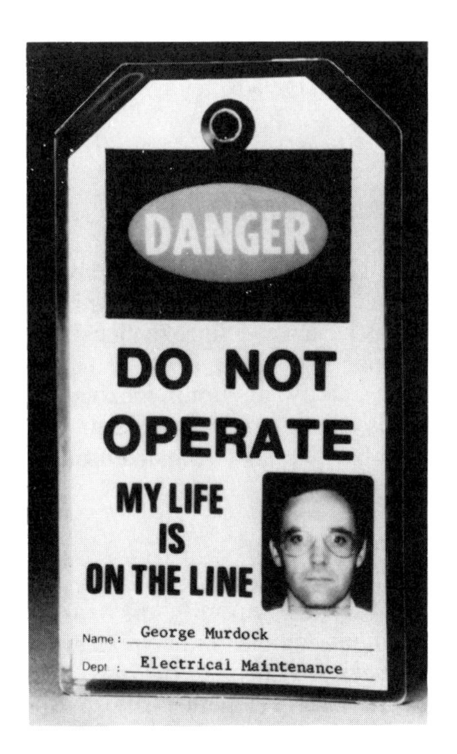

Figure 7-11. Photo identification tagout device. (Idesco Corp.)

who attached it. Another style uses an adhesive-backed photo-containing label that is wrapped around the body of a lock (Figure 7-12). Both are plastic laminated for durability.

Figure 7-12. Photo lock-wrap tag. (Idesco Corp.)

Color Coding

Some manufacturers anodize the bodies of their locks in as many as six colors. Others attach colored plastic bands to the bottom of the lock's body. Consequently, different colored locks may be assigned to different people, one given a blue, another a red, etc. Alternatively, as shown in Table 7-2, color coding may be used to identify different crafts who are involved in lockout/tagout procedures.

Physical Marking

Numbers or names can be painted, scribed, engraved or stamped into the body of the lock. Rather than physically marking the body of a lock, some manufacturer's offer metal tags that can be marked with an employee's name or identification number and then attached to the lock's shackle.

Establishing a pool of numbered locks provides a cost-effective alternative to issuing dozens, or perhaps even hundreds, of individual locks. Using the pooled approach, individually numbered locks are centralized at a readily accessible location such as a tool crib. Employees then use a sign-out log to requisition as many locks as they need for a particular project. When the job is completed, the borrowed locks are returned for reuse.

MULTILOCK HASPS

Multilock hasps like that shown in Figure 7-13 can be used when more than one employee must

Table 7-2. Color coding of locks by craft

Electricians	Green
Pipefitters	Blue
Millwrights	Red
Painters	Gold
Mechanics	Black
Welders	Silver
Carpenters	White
Laborers	Yellow

Figure 7-13. Multi-lock hasp. (Idesco Corp.)

attach a lock or tag to an isolation device. Adapters are available in different styles and shackle lengths, and most will accept up to six locks. If more than six authorized employees are involved in an entry, adapters may be connected together to accommodate as many locks as necessary. Although there are no limits as to how many adapters may be attached together, it is easier to use group lockout if more than ten people are involved.

VALVE LOCKOUT

Valves can be locked-out by threading a wire rope or a chain through the handwheel and around the valve body. If wire rope is used, its ends can be terminated with eye-splices that can be locked together. Alternatively, the ends can be joined to form a loop. The loop can then be folded in half and two ends locked together after being wrapped around the valve.

Valve Lockout Adapters

Plastic covers that can be placed over valve handwheels are also commercially available. The covers range in sizes from 2 to 24 inches (Figure 7-14) and are hinged so they can be easily slipped around the handwheel and locked in place (Figure 7-15). Similarly, ball valves may be locked-out using a chain or a commercially lockout adapter as shown in Figure 7-16.

Integral Locking Devices

Figures 7-17 and 7-18 illustrate valves that are equipped with a built-in locking mechanism which is operated by a special key. Since the locking mechanism is controlled by a single key, these devices lend themselves primarily to group lockout situations. Figure 7-19 is an example of pneumatic valves designed with a built-in padlock attachment.

TYPES OF PADLOCKS

Padlocks may be generally divided into three types: warded, pin-tumbler and combination.

Warded locks are operated by keys whose teeth are all the same length (Figure 7-20). Different key codes are established by varying the width and separation between the teeth. Warded locks have the advantages of being relatively inexpensive and may be used in locations where sand, ice and similar contaminants could jam other types of locks. On the other hand, they do not provide a very high degree of security and the number of key combinations is limited to only a few dozen.

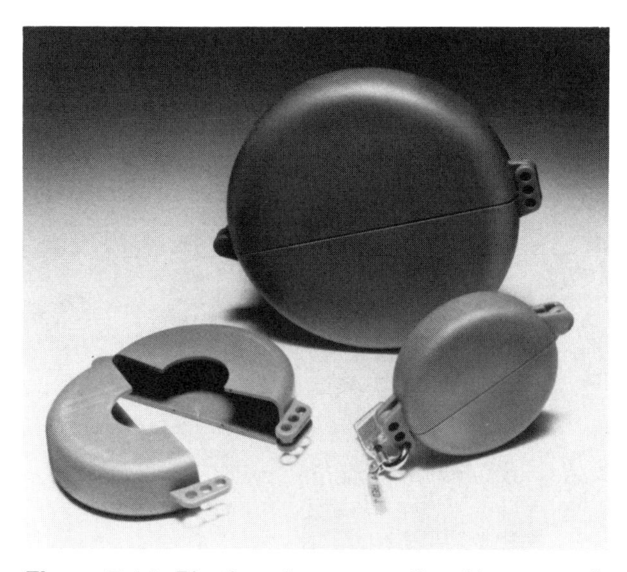

Figure 7-14. Plastic valve covers like this are available in a number of sizes up to 12 inches. (Rockford Medical Systems)

Figure 7-15. Valve covers locked in place. Note the cut out for the rising valve stem at the far left. (W.H. Brady)

Pin-tumbler locks employ a cylinder containing metal pins of varying length. The lock cannot be released unless the tops of all the pins are lined up at the same height. The number of key combinations is established by the number and length of pins in a cylinder. Pin tumbler locks typically employ between four to six pins and since pins are available in as many ten different lengths, millions of key codes are possible.

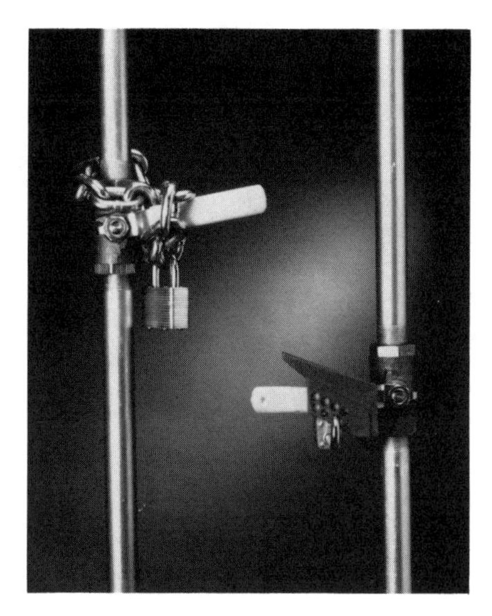

Figure 7-16. Ball valve locked out with chain and special adapters. (W.H. Brady)

Figure 7-17. Valve with integral lockout device. (Castell Safety)

The key code on some pin tumbler locks can be easily changed by the user. The old cylinder is removed with a hex wrench and replaced with one containing the new key code. Kits are also available that allow easy repinning of the original cylinder.

Figure 7-18. Built in lockout device on valve hand wheel. (Castell Safety)

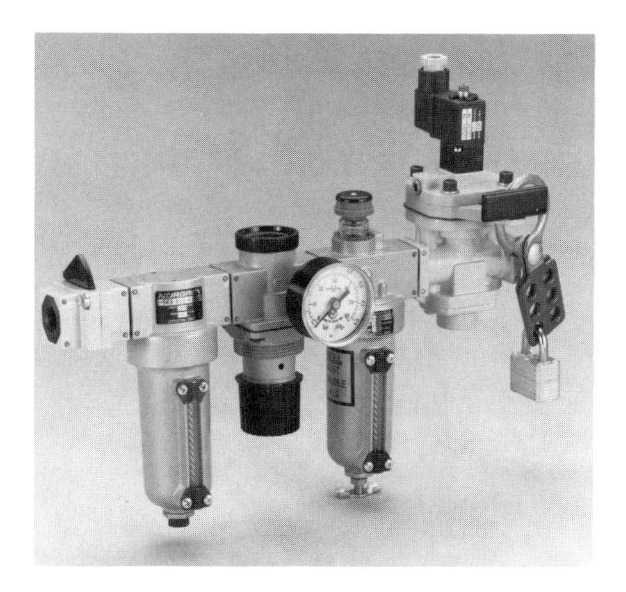

Figure 7-19. Pneumatic lockout devices. (Norgren)

Pin tumbler locks have the advantage of being keyed alike and master-keyed (Figure 7-21). A keyed-alike system allows all of the locks assigned to one person to be opened by a single key. This may be a convenient feature for entries that involve multiple isolation devices. For example, an entry involving three electrical disconnects and two double-block and bleeds would require nine locks. Rather than

Figure 7-20. Warded lock with keys. (John Rekus)

having to fumble with nine separate keys, a keyed-alike system would allow the entrant to open all nine locks with a single key.

Combination locks are available in both padlock and cable lock styles. Since they do not require a key for operation, they may be used in situations where handling a key could pose a hazard or inconvenience. The code on some thumb wheel combination locks can be reset by inserting a special pin into the lock's body and repositioning the thumb wheels to reflect the new combination. The combination is installed when the pin is removed. Since the combination is determined by the user, it is easy to remember and may be set on as many locks as necessary.

DEVELOPING AN ENERGY CONTROL PROGRAM

Lockout/tagout must be governed by an energy control program that includes two elements: control procedures and employee training. Although the standard specifies the level of performance that must be demonstrated, it allows a high degree of flexibility in selecting the means and methods used to achieve compliance. One important point that should be noted, though, is that any existing equipment which is not presently capable of being locked-out must be modified to accept lockout when it undergoes major repair or rehabilitation. Similarly, all new machinery or equipment that is installed must be capable of being locked-out.

Energy Control Procedures

Lockout/tagout procedures must include a general explanation of the reason why energy control is necessary. They must also clearly and specifically

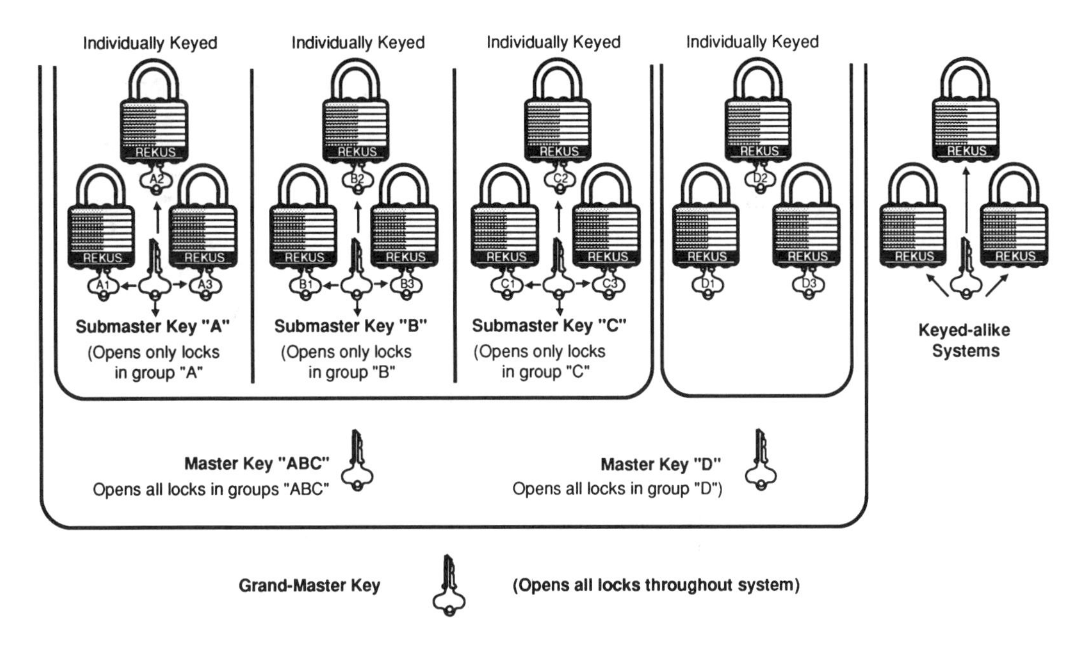

Figure 7-21. Master keying system. (Paul Trattner)

describe the scope, purpose, authorization rules and techniques that will be used to control hazardous energy. Finally, they must include an outline of the means by which compliance with the standard will be achieved. This outline must include the following:

- A specific statement of the intended use of the procedures
- Specific procedural steps for shutting down, isolating, blocking and securing machinery and equipment
- Specific procedures for placement, removal and transfer of lockout devices
- Specific requirements for testing a machine or piece of equipment to determine and verify the effectiveness of lockout or tagout devices

Implementing Lockout/Tagout

The standard stipulates that the following six step procedure be followed when implementing lockout or tagout.

Step 1: Preparation for Shutdown. In general, preparation for shutdown will involve an assortment of tasks including:

- Assuring that affected employees are informed that a piece of equipment is going to be locked or tagged will be in effect
- Identifying all isolation points and energy isolation devices
- Determining if there are any sources of stored energy such as pressurized air or oil in pneumatic and hydraulic reservoirs, springs in tension or compression, charged capacitors, etc.
- Selecting appropriate personal protective equipment
- Determining what tools and special equipment may be required
- Obtaining locks, tags and other hardware necessary for the task

It is essential that employees who perform lockout or tagout are knowledgeable concerning the types and magnitude of the energy present, the hazards of the energy to be controlled, and the methods and means of controlling that energy.

Step 2: Equipment Shutdown. Shutdown is the process of initially stopping the flow of energy to a piece of machinery or equipment, for example, pushing the stop button on a motor controller or closing a valve. Some shutdowns may be very simple and require only one or two steps. Others may be exceptionally complex and involve the operation of dozens of switches, valves and interlocks.

Step 3: Equipment Isolation. Once the equipment has been shut down, the next step is to isolate it. Isolation can be achieved by opening an electrical disconnect switch, breaking a line or establishing double-block and bleed.

Step 4: Apply Control. With the equipment shut down, locks or tags may then be attached to energy isolation devices.

Step 5: Dissipate Stored Energy. Energy may be stored in some machinery and equipment even after it has been shut down and isolated. For example, inertia permits flywheels and shafts to continue to rotate even after the motor controlling them has been turned off. Similarly, hydraulic and pneumatic lines may remain pressurized after pumps and compressors have been shut down. As a result, the potential for the following forms of stored energy should be considered.

Kinetic, or energy of motion, may be present in rotating flywheels, mixing blades and motor shafts. Depending on their mass and the friction they must overcome, some rotating parts could continue turning well after initial shutdown. Consequently, adequate time should be allowed for them to come to a complete stop.

Electrical energy stored in capacitors should be dissipated by grounding. If a potential hazard is posed by battery-powered circuits, they should be disconnected.

Potential energy may exist as pneumatic or hydraulic pressure stored in pipes, lines and tubes. This pressure can be relieved by opening a bleed valve or similar device. Hazardous potential energy may also be stored by springs in tension or compression, so they should be placed in neutral position whenever possible.

Step 6: Verify Effectiveness. Fatal accidents have occurred because a lock or tag was attached to the wrong energy isolation device. Thus, before any work begins on a piece of isolated equipment, the isolation must be confirmed by testing the operating controls to verify that the equipment will not function. While it is not directly related to confined space entry, the following case history demonstrates clearly the need for such verification procedures.

Case No. 7-1: Defective Disconnect Switch

Two maintenance men were assigned the task of servicing an overhead crane. They placed the disconnect switch in the "off" position and each attached a personal lock. They were on the job for about half an hour when one of the men touched a supposedly deenergized cable and was electrocuted.

Investigators confirmed that the locks had been placed on correct disconnect and that the switch handle was in the "off" position. However, when they removed the locks and inspected the switch, they found that the linkage between the outside handle and the knife blade rotating arm was broken. Although the switch handle was in the "off" position, the knife blades were connected to the supply side of the line.

Release From Lockout/Tagout

Release from lockout/tagout involves reversing the process of applying controls, but it incorporates only three steps.

Step 1: Work Area Inspection. The work area should be inspected to assure that it is safe to restart the machine. Loose parts and tools should be removed from the work area and any guards that were removed should be reinstalled.

Step 2: Notify Others in Area. Affected employees and other workers in the area should be informed that equipment is about to be placed back into service.

Step 3: Remove Lockout/Tagout Device. Finally, the lockout/tagout devices may be removed by the person who applied them.

Documentation of Procedures

Lockout procedures must be documented unless all of the following eight conditions are met:

1. There is no potential for energy to be stored or reaccumulated after shutdown.
2. There is a *single, readily identifiable,* source of energy isolation.
3. The isolation of *that single source* completely deenergizes the equipment.
4. The machine *is* locked-out during service.
5. Application of a *single* lockout device provides a complete lockout condition
6. The lockout device is under the exclusive control of the employee performing the work.
7. The maintenance or service does not itself create a hazard to other employees.
8. The employer has not had any accidents or incidents involving the unexpected activation of machines during service or maintenance.

One situation that would meet all these criteria is a chemical mixing vat where materials are stirred by a bottom-driven mixer whose motor is controlled by an adjacent disconnect switch. On those occasions where the vat must be entered, an entrant could simply open the disconnect switch and attach his lock without the need for a written program.

PROCEDURES FOR EMERGENCY REMOVAL OF LOCKS

Sooner or later someone is going to go home at the end of the day without removing a lock or tag from a completed project. When this occurs, an effort must be made to contact the errant employee, at home if necessary, and to have him return to the job site to remove his lock. If the employee cannot be reached and the equipment must be placed back in service before he is expected to return to the site, the lock or tag may be removed. However, removal must take place under the employer's direction and then only if specific training and procedures for the removal have been developed, documented and incorporated into the energy control program. Some facilities require at least two people be present to double check the procedure and witness the removal.

Finally, before the errant employees resume their duties at the facility, they must be informed both of their transgression and of the fact that their locks were removed.

EMPLOYEE TRAINING

Separate levels of training are required for "authorized employees," "affected employees," and other employees who might be in an area where lockout/tagout is used.

Authorized employees must be provided with an explanation of the methods and means by which hazardous energy will be isolated and controlled. They must also be equipped with the skills that allow them to recognize sources of hazardous energy.

Affected employees must be instructed on the purpose and use of the lockout/tagout procedures.

Other employees working in an energy control area need only be instructed on the procedures, not their use. However, they must be informed in no uncertain terms of prohibitions on trying to restart equipment that has been locked- or tagged-out.

PERIODIC AUDITS

An "authorized employee," other than the one actually using the procedures, must conduct an audit at least annually. This audit must review the entire program to assure that the energy control procedures are being followed. If tagout is used, then the auditor must also verify that the additional procedures established for assuring "full employee protection" are also being followed. The audit must be certified in writing, identifying the machinery or equipment involved, the date and name of the auditor.

ADDITIONAL REQUIREMENTS

The standard has four special provisions covering testing of equipment, outside contractors, group lockout and shift changes.

Testing and Positioning Equipment

The standard provides for situations where it may be necessary to temporarily remove a lockout device in order to test or position a piece of equipment. For example, a mixing paddle in a chemical reactor may have to be checked for clearance and balance after being repaired. In these situations, the area should be cleared of all unnecessary tools, equipment and employees. Next, the lockout devices are removed and the system is tested. When the test is complete, the lockout is reinstalled and the equipment is readjusted as necessary. The cycle of removal, testing, adjusting and reapplication of control is repeated until the equipment is functioning properly.

Outside Contractors

When isolation of a space is performed by an outside contractor, the contractor and host employer must inform each other of their respective lockout/tagout procedures. The on-site employer must also ensure that his employees understand and comply with the restrictions and prohibitions of the outside employer's energy control program.

Shift or Personnel Changes

Shift work presents an interesting situation since continuity of lockout/tagout must be assured at all times. Thus, specific procedures must be established to assure the orderly transfer of control devices between incoming and outgoing employees.

Group Lockout/Tagout

Group lockout, which was discussed earlier in this chapter, is a method whereby one person takes responsibility for performing lockout for a crew, craft or other group of employees. The standard permits group lockout provided that it assures the same level of protection as that afforded by a personal lockout program. Primary responsibility must be vested in the authorized employee who is responsible for the safety of a specified number of employees working under the protection of the group lockout. When more than one crew, craft or department is involved, an authorized employee must also be assigned the responsibility for coordinating the overall job-associated lockout/tagout control and ensuring the continuity of protection.

Authorized employees must also affix a personal lock or tag to the group lockout devices, group lock box or comparable device when they begin work, and remove it when they finish.

SUMMARY

Hazards posed by electrical and piping systems associated with confined spaces can be controlled by means of a lockout/tagout program. However, for the program to conform to OSHA requirements, it must include the following three elements: specific procedures for application of control and release from lockout, employee training on the lockout/tagout procedures and an annual audit to assure that the program remains effective. Lockout is the preferred method of hazard control and must be employed unless an employer can demonstrate that his tagout procedures provide a level of protection equal to lockout.

REFERENCES

American National Standard for Personnel Protection—Lockout/Tagout of Energy Sources—Minimum Safety Requirements, Z244.1-1982, New York, NY: American National Standards Institute, (1982).

An Introduction to Castell Interlocking, Erlanger, KY: Castell Safety, Inc.

Betz, G.M. and E.A. Eder, Locks and Keys—Sentinels of Plant Security, Plant Engineering, June, 1964.

Brady '92 Lockout/Tagout Product Line, Milwaukee, WI: W.H. Brady Co., (1992).

Cloe, W.W., Selected Occupational Fatalities Related to Welding and Cutting as found in Reports of OSHA Fatality/Catastrophe Investigations, Washington, DC, U.S. Department of Labor, Occupational Safety and Health Administration, (1985).

Cloe, W.W., Selected Occupational Fatalities Related to Lockout/Tagout as Found in Reports of OSHA Fatality/Catastrophe Investigations, Washington, DC, U.S. Department of Labor, Occupational Safety and Health Administration, (1985).

Cloe, W.W., Selected Occupational Fatalities Related to Fixed Machinery Found in Reports of OSHA Fatality/Catastrophe Investigations, Washington, DC, U.S. Department of Labor, Occupational Safety and Health Administration, (1985).

Control of Hazardous Energy, Code of Federal Regulations, Vol. 29, Part 1910.147.

Ed-CO Lockout and Padlocks, Solon Springs, WI: American Ed-CO, Inc.

Ellis Valve Interlocks, Erlanger, KY: Castell Safety, Inc.

Grain Handling Facilities, Code of Federal Regulations, Vol. 29, Part 1910.272.

Glasser, M.A., Data Regarding Fatal Injuries Concerned With Safety Lockout, Detroit, MI: United Automobile, Aerospace, & Agricultural Implement Workers.

Group Lockout Procedures for Single-Crew, Multi-crew and Multi-shift Jobs, New York, NY, Idesco, Corp., (1991).

Hand and Portable Power Tools and Equipment: General, Code of Federal Regulations, Vol. 29, Part 1910.242.

Heavy Duty Laminated Safety and Lock-out Devices, New York, NY, Idesco, Corp., (1991).

Interlocking Systems for Electrical Distribution Equipment, Erlanger, KY: Castell Safety, Inc.

L-O-X Valves for Pneumatic Energy Isolation, Bulletin 372, Troy, MI: Ross Operating Valve Co.

Machinery and Machine Guarding: General Requirements for All Machines, Code of Federal Regulations, Vol. 29, Part 1910.212.

New Safety Lockout System Makes OSHA Compliance as Easy as 1-2-3, Milwaukee, WI: Master Lock Company.

National Electrical Code, Quincy, MA: National Fire Protection Association, (1993).

NIOSH Alert, Request for Assistance in Preventing Entrapment and Suffocation Caused by the Unstable Surfaces of Stored Grain and other Materials, Cincinnati, OH: National Institute for Occupational Safety and Health, (1987).

Norgren "Safety Yellow" Lockout Products, Littleton, CO: Norgren (1989).

Padlock and Security Products, Crete, IL: American Lock Company.

Pathways to Lockout Compliance, New York, NY: Idesco, Corp.

Permit-Required Confined Spaces for General Industry, Code of Federal Regulations, Vol. 29, Part 1910.146.

Pettit, T.A., P.M. Gussey and R.S. Simmons, 1980, Criteria for a Recommended Standard: Working in Confined Spaces, National Institute for Occupational Safety and Health, DHEW/NIOSH Pub. 80-106, Government Printing Office, Washington, DC.

Pulppaper and Paperboard Mills, Code of Federal Regulations, Vol. 29, Part 1910.261.

Q-Tag System®, New York, NY: Idesco Corp., (1991).

Safety Bulletin No. 25 Crane & Rigging Safety Information, Subject: Tagout and Lockout Procedures, Orlando, FL: Crane Inspection & Certification Bureau.

Safety Interlocking in the Automated Factory, Westminster, MD: Telemecanique, Inc.

Textiles, Code of Federal Regulations, Vol. 29, Part 1910.262.

Valve Interlocking System for the Petrochem, Chemical and Process Industries, Erlanger, KY: Castel Safety, Inc.

CHAPTER 8

Ventilating Confined Spaces

INTRODUCTION

In Chapters 2, 3 and 4 we identified the four types of atmospheric hazards likely to be encountered in confined spaces. They were flammable materials, toxic substances, oxygen deficiency and oxygen enrichment. While there are a variety of methods for controlling these hazards, the best approach is to eliminate them through the use of ventilation.

But what's the best way to ventilate a confined space? Should you blow air in or exhaust it out? How many air changes are sufficient? How do you arrange ventilation equipment for maximum efficiency? Unfortunately, there are no simple answers to these and other questions related to confined space ventilation. Furthermore, since each space is unique, there is no single set of rules that applies to every situation.

As a result, identical ventilation equipment may be arranged one way when ventilating a street manhole and a different way when ventilating a chemical process vessel. As discussed in Chapter 1, some spaces, such as petroleum storage tanks, grain bins and open-topped vats, usually lack internal obstructions that could block the free flow of air. The opposite is true of other spaces such as tank truck trailers, process boilers and barges that often contain interior walls and baffle plates.

In some spaces, adequate ventilation may be provided by a single blower. Other spaces will require multiple blowers arranged so that some force air into the space, while others exhaust it out. Some tasks such as welding, cutting and brazing may also require the use of a local exhaust system that captures contaminants at their source. In other situations such as solvent cleaning, general dilution ventilation may be sufficient to control the hazard.

Recognizing that there is virtually an infinite variety of spaces and correspondingly infinite ways to ventilate them, this chapter will not focus on specific ventilation techniques. Instead, it describes some basic ventilation *principles and concepts* that may be applied to all spaces. These concepts and principles, when married with experience and sound professional judgment, will determine the best way of ventilating a specific space.

ACCIDENT CASE STUDIES

The following case studies set the stage for our discussion by providing examples of accidents that may have been prevented if those responsible for the entry had a working knowledge of basic ventilation principles and methods.

Case No. 8-1: Poor Procedures and Practices

A crew of untrained temporary day laborers was assigned the task of cleaning sludge from a chemical storage tank which measured about 30 feet high and 50 feet in diameter. The space was ventilated for 48 hours prior to the entry by a roof-mounted exhauster that drew make-up air in through a manway at the base. The manway was located on the same side of the tank as the exhauster.

Measurements were made for oxygen, combustible vapors and benzene, which was known to be present in the sludge. Since all were found to be within acceptable limits, an entry permit was issued and the laborers entered. They proceeded to squeegee sludge toward the manway so that it could be sucked through a hose into a waiting vacuum truck.

After a few hours, four of the workers became dizzy and crawled out of the space. One collapsed and was transported by ambulance to a nearby hospital, where he died without regaining consciousness. The cause of death was determined to be acute benzene poisoning.

Measurements taken during the investigation showed very high levels of benzene on the side of the tank opposite the manway, with concentrations becoming progressively lower as the manway was approached. This suggested that make-up air entering the manway was short-circuited and drawn directly up to the roof exhauster without mixing thoroughly with the atmosphere in the other areas of the space.

Although the atmosphere was tested prior to entry, the measurements were meaningless because they were made in the fresh airstream entering the tank.

Case 8-2: Wrong Blower Used

Two boatyard workers were using a flammable contact cement to attach soundproofing in a closed compartment below the rear deck of a 66-foot boat. A blower equipped with a length of flexible duct was available and could have been arranged to supply fresh air from outside, but the workers chose to use a non-explosion-proof fan which they took into the compartment.

Instead of diluting the flammable vapors given off by the contact cement, the fan simply recirculated them. When the concentration reached the lower explosive limit, the vapors were ignited by the fan's motor. Both workers were killed.

Case 8-3: Spray Painting Wing Tank

Two workers were assigned the task of painting the interior of a barge wing tank. A "competent person" checked the tank for flammable vapors and oxygen at the beginning of the shift and determined that the atmosphere was suitable for entry. No further testing was performed, even though the space was not continuously ventilated. A few hours later, flammable vapors in the space exploded. One worker was killed when he was struck by a section of tank wall that blew loose, the other died from burns suffered in the fire.

DIFFERENCE BETWEEN VENTILATION AND PURGING

Although the terms purging and ventilating are frequently used interchangeably, there is a difference between them.

Purging Defined

Purging is the process by which a space is initially cleared of contaminants by displacing the hazardous atmosphere with air, steam or an inert gas. The choice of purging medium depends largely on the nature of the contaminant in the space. To minimize the risk of fire, inert gases such as nitrogen and carbon dioxide are often used to purge spaces of flammable atmospheres. Air, on the other hand, is used to purge spaces of toxic atmospheres and gases such as carbon dioxide, nitrogen and argon which could create an oxygen deficiency. In some cases, more than one purging cycle may be required. For example, the fire hazard posed by flammable gases and vapors may be minimized by initially purging the space with nitrogen or carbon dioxide. The atmosphere would then be made breathable by purging with air.

Ventilation Defined

Ventilation is the process of *continuously* moving fresh, uncontaminated air through a space. This continuous air movement accomplishes three things.

First, it dilutes and displaces any air contaminants that may be present in the space. Second, it assures that an adequate supply of oxygen is delivered to the space and maintained during the entry. Third, it exhausts any contaminants formed by processes such as welding, oxy-fuel gas cutting, or abrasive blasting.

While not a primary consideration, ventilation may also be used to control unpleasant or irritating odors that may be a nuisance to the space's occupants. It can also increase the comfort level by either warming or cooling the ambient air.

DETERMINATION OF PURGE TIME

Perhaps the most thorough study of confined space purging times was performed in the early 1970s by Bell Telephone Laboratories. Bell engineers conducted a series of experiments to evaluate ventilation in each of the four following styles of manholes found in the Bell System:

- Conventional manholes ranging in size from 200 to 800 ft^3
- Off-set access manholes
- Deep neck manholes
- Irregularly shaped manholes

Representative samples of each manhole style were filled with smoke generated by smoke bombs. Ventilation blowers were turned on and the time necessary to clear the space of smoke was measured. Subsequent analysis of the empirical data by Bell engineers suggested that purge times could be estimated by the equation:

$$T = 7.5 \frac{V}{C}$$

where

T = purge time in minutes
V = the volume of the space in $ft.^3$
C = effective blower capacity in CFM

GENERAL VENTILATION AND LOCAL EXHAUST

Ventilation may be broadly divided into two categories: general ventilation and local exhaust.

General Ventilation

General ventilation is the process of introducing clean outside air into a space in such a way that it mixes with the inside atmosphere to dilute contaminants and restore oxygen. General ventilation is useful in providing comfort-cooling and for removing unpleasant odors; however, it is ineffective for controlling highly toxic contaminants, fumes formed by welding or oxy-fuel gas cutting and heavy dusts generated by grinding, chipping and abrasive blasting.

Local Exhaust

Local exhaust provides a positive means of removing contaminants at their source. Local exhaust systems are particularly well-suited for removing contaminants generated by point sources like arc welding and oxy-fuel gas cutting. They can also be used to remove vapors formed by the local application of solvents such as those encountered during dye penetrant testing or touch-up painting.

PURGING AND VENTILATION METHODS

Natural Ventilation

Natural ventilation, as its name suggests, uses the forces of nature to move air in and out of the space. The energy required for this movement is provided entirely by air currents, breezes, thermal gradients, and pressure differences which exist in and around the space.

There are both advantages and limitations associated with natural ventilation. The four principal advantages it offers are

- It is quiet since there are no moving parts.
- It is inexpensive since there are no costs associated with purchase or maintenance.
- It does not present a source of ignition.
- There are no electrical or mechanical parts to wear out or fail.

In some cases, the chimney effect resulting from temperature differentials inside large above-ground tanks can induce a strong draft of outside air. However, this draft, like all natural ventilation, is passive, variable and unreliable. Furthermore, the design and configuration of some spaces precludes them from being ventilated naturally. Two other disadvantages of natural ventilation are that it does not provide local exhaust, nor can it be readjusted to reflect changing conditions in the space.

Use of Steam

In some cases, steam may be used for purging spaces that contain low flash point hydrocarbon solvents, motor fuels and other petroleum products. Steam purging can also loosen scale, tars and viscous materials that may be stuck to side walls and seams. However, for obvious reasons, steam cannot be used for ventilation.

For steam to be an effective purging medium, it must be delivered to the space at a rate that raises the internal temperature to at least 170°F (77°C). Otherwise it will condense as fast as it is introduced, and contaminants will not be forced out. However, care must also be taken to assure that the vessel walls are not heated to the point where they buckle, warp or crack.

To take full advantage of the thermal head created by the rising hot air, top portals or hatches should be opened and steam should be introduced from the bottom of the space. The portals should also be left open during the entire process to prevent pressurization while steaming and vacuum formation during cooling. If steam vapor-freeing is used, sufficient vacuum relief must be provided to prevent a vacuum from being formed within the vessel as a result of rapid steam condensation.

Inert Gases

The fire hazard posed by flammable gases and vapors can be minimized by initially purging a space with inert gases such as carbon dioxide and nitrogen. As indicated in Table 8-1, inert gases can be used to lower the oxygen level to a point below which ignition will not occur. Carbon dioxide may be introduced either as a gas or as a solid in the form of dry ice. Nitrogen is normally introduced as a gas. Since static electricity may be generated when compressed gases are introduced, gas dispensing pipes and nozzles should be electrically bonded to the space.

Table 8-1. Maximum allowable oxygen level required to prevent ignition

	N_2-Air		CO_2-Air	
	O_2 % Above which Ignition can Take Place	Maximum Recommended O_2 %	O_2 % Above which Ignition can Take Place	Maximum Recommended O_2 %
Acetone	13.5	11	15.5	12.5
Benzene	11	9	14	11
Butane	12	9.5	14.5	10.5
Carbon disulfide	5	4	8	6.5
Cyclopropane	11.5	9	14	11
Ethyl alcohol	10.5	8.5	13	10.5
Gasoline				
73–100 Octane	12	9.5	15	12
100–130 Octane	12	9.5	15	12
115–145 Octane	12	9.5	14.5	11.5
Hexane	12	9.5	14.5	11.5
Hydrogen	5	4	6	5
JP-1 Fuel	10.5	8.5	14	11
JP-3 Fuel	12	9.5	14	11
JP-4 Fuel	11.5	9	14	11
Kerosene	11	9	14	11
Methane	12	9.5	14.5	11.5
Methyl alcohol	10	8	13.5	11
Propane	11.5	9	14	11

Adapted from NFPA-53M.

Mechanical Systems

Since mechanical systems use equipment like fans and blowers to force air through space, they can be used for both purging and ventilating. Mechanical systems also provide a high degree of reliability if they are properly serviced and maintained. They can also be used to provide local exhaust and can be adjusted as conditions in the space change.

The principal disadvantages posed by mechanical equipment include:

- Initial costs associated with equipment purchase and employee training
- Recurring cost for service and maintenance
- Ignition sources posed by metal parts and motors
- Electrical and mechanical hazards
- High noise levels which may interfere with communication between those working in or around the space

VENTILATION CONCEPTS AND PRINCIPLES

How Air Moves

Fans are devices that lower the atmospheric pressure inside a ventilation system. Since air, like any fluid, always flows from areas of high pressure to areas of low pressure, the higher pressure ambient air moves into the ventilation system. However, a pressure head is necessary to get air moving through the system. This pressure head has two components: static pressure (SP) and velocity pressure (VP). As shown in Figure 8-1, the sum of these two pressures is the total pressure (TP).

Static Pressure

Static pressure (SP) is the pressure inside a duct that tends to either suck the duct walls in or push them out. Negative static pressure is the force that causes a duct to collapse like a plugged up soda straw. Positive static pressure, on the other hand, is the pressure that tends to blow up a duct like a balloon. Static pressure energy is used to overcome the frictional resistance of the air on the duct walls as well as resistance offered by elbows, bends or filters.

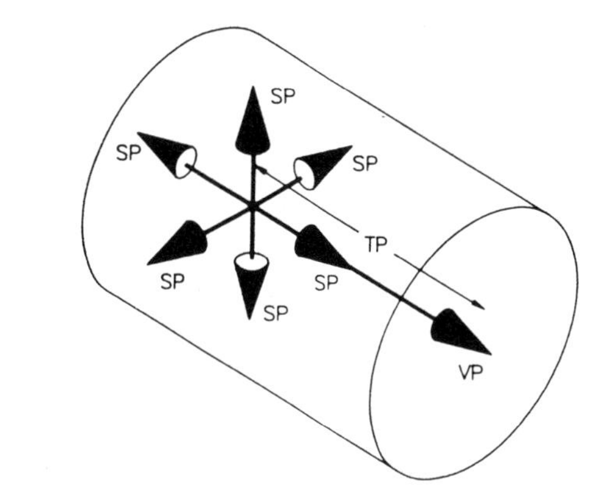

Figure 8-1. SP, VP and TP at a point. (From *Industrial Ventilation: A Manual of Recommended Practice* 20th Ed., ACGIH, Cincinnati, OH. Used with permission.)

Velocity Pressure

Velocity pressure (VP) is the pressure necessary to keep an air stream moving and is proportional to the kinetic energy of the air stream. More exactly, it is defined as the pressure required to accelerate air at rest (zero velocity) to some specified velocity (V). Velocity pressure is always positive and can be related to velocity as shown below:

$$V = 4005\sqrt{VP}$$

Volumetric Flow Rate

The volumetric flow rate, that is, the amount of air going past a point in the system can be calculated if the cross-sectional area of a duct and the velocity of the air stream flowing through the duct are known. The relationship between the volumetric flow rate, air velocity and duct area is shown below:

$$Q = A \times V$$

where

Q = the volumetric flow rate in cubic feet per minute
A = the cross-section area in square feet, and
V = the velocity in feet per minute

Since the air velocity and duct diameter can be measured, it is a simple task to calculate the volumetric flow rate. For example, the volumetric flow rate for a system where air is moving through an 8-inch round duct with a velocity of 1,000 fpm can be determined as follows.

First, determine the cross-sectional area of the 8-inch round duct using the formula

$$\text{Area} = \frac{\pi\, d^2}{4} = \frac{(3.14)(8)^2}{4} = 50.24 \text{ sq. in.}$$

Next, convert the cross-sectional area to square feet. Since there are 144 square inches in a square foot, the cross-sectional area can be converted to square feet by dividing it by 144, resulting in 0.349 square feet.

Finally, substituting the values for V and A, Q can be calculated as shown below:

$$\begin{aligned} Q &= A \times V \\ &= 0.349 \text{ ft}^2 \times 1{,}000 \text{ fpm} \\ &= 349 \text{ CFM} \end{aligned}$$

Supply and Exhaust Systems

Two other terms that need to be explained are *supply systems* and *exhaust systems*. Supply systems are those that force air into a space. Exhaust systems, on the other hand, remove air from a space through suction. Although either system may be used for purging and ventilating, selection of the most appropriate system depends on a number of factors including the nature of the contaminants, the configuration of the space and the work to be performed in the space.

A significant difference between supply and exhaust systems is that fans can exhaust or "blow" air much farther than they can capture or pull it in. In general, the ratio of exhausting to capture is 30:1 (Figure 8-2). That means that a fan that is capable of blowing air a distance of 30 feet would only be able to capture pull in contaminants that were within a one foot distance.

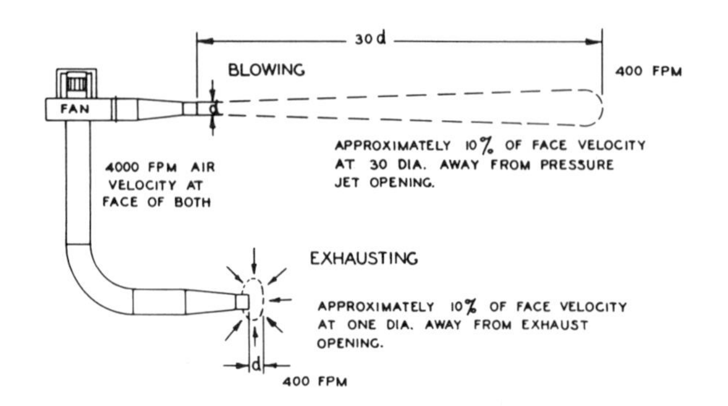

Figure 8-2. Blowing vs. suction range of fans. (From *Industrial Ventilation: A Manual of Recommended Practice* 20th ed., ACGIH, Cincinnati, OH. Used with permission.)

Rated vs. Actual Blower Capacity

Air flowing through a ventilation system is similar to water flowing through a pipe. The length of a duct, its interior surface roughness, and the number of bends, all increase resistance and decrease air flow. Thus, the *quantity* of air delivered at the end of a duct run will be less than the blower's rated capacity.

Free air delivery is the quantity of air delivered at the outlet of a blower without any outlet duct work attached. *Effective blower capacity,* on the other hand, is the actual quantity of air in cubic feet per minute (CFM) delivered at the end of a flexible duct connected to the blower. As a frame of reference, Table 8-2 illustrates the degree to which air flow is reduced with increasing lengths of duct. Further reductions would be produced by bends in the duct work.

Some blower manufacturers attach identification plates to their equipment indicating both free air delivery and the effective capacity under specified conditions, such as a stated length of duct with one and two 90 degree bends.

LOCAL EXHAUST SYSTEMS

As mentioned at the beginning of this chapter, local exhaust systems are used to capture contaminants at their source. A local exhaust system consists of four components:

1. Hoods for capturing the air contaminants
2. Rigid or flexible duct work for carrying the contaminants out of the space
3. Air cleaning devices that removing contaminants from the airstream
4. Fans that provide the static pressure (energy) needed to move the air through the system

Capture Velocity

For a local exhaust system to be effective, the air in front of the hood must be moving fast enough to overcome natural air currents so that the hood can capture contaminants and draw them into the duct. This *capture velocity* depends on three factors: the contaminant's physical state, its rate of generation and the turbulence of the surrounding air. In general, lower capture velocities are required for lighter materials (like gases and vapors) than for heavy materials (like dusts and metal fumes). Similarly, lower capture velocities are required for contaminants that are released at low rates or into relatively still air than for those released at high rates or into turbulent air. The approximate range of capture velocities for different contaminant states and conditions is provided in Table 8-3.

Transport Velocity

Transport, or duct velocity, is the minimum velocity required to keep a material from settling out in a duct as air flows through a system. Since gases and vapors are lighter and more diffuse than dusts and fumes, lower duct velocities are required to transport them than are required to move heavy materials like spent abrasive blasting shot and lead fume. Typical transport velocities are listed in Table 8-4.

Table 8-2. Air flow in cubic feet per minute (CFM) at 3,500 RPR motor speed

Horse Power	Inlet Duct Diameter (Inches)	Straight Run Length of Duct (Feet)				
		10	20	30	40	50
1/2	5	525	488	460	440	425
3/4	6	775	740	708	690	670
1	6	870	830	800	765	745
2	8	1,570	1,530	1,490	1,450	1,425

Adapted from *Portable Ventilators,* Coppus Engineering Corp., Worcester, MA (1989) pg.

Table 8-3. Range of capture velocities

Dispersion Condition	Examples	Capture Velocity feet/minute
Released with practically no velocity to still air	Solvent cleaning	50–100
Released at low velocity to moderately still air	Welding, cutting	100–200
Released at high initial velocity to zone of very rapid air movement	Grinding, abrasive blasting	500–2000

Adapted from *Industrial Ventilation: A Manual of Recommended Practice,* 20th Ed., ACGIH (1990).

Table 8-4. Transport or duct velocities

Nature of Contaminant	Design Velocity (feet/minute)
Vapors, gases, smoke	1000–1200
Welding fumes	1400–2000
Dry powders	2500–3000
Grinding dusts	3500–4000

Adapted from *Industrial Ventilation: A Manual of Recommended Practice,* 20th Ed., ACGIH (1990).

Hood Placement and Design

For a local exhaust system to be effective, the hood must be positioned as close as possible to the contaminant's source. Since the velocity of the air varies inversely with the square of the distance from the hood, doubling the distance from the hood will reduce the velocity by a factor of four. As illustrated in Figure 8-3, capture effectiveness can also be enhanced by the addition of a flange. While the velocity contours are similar, note that the flanged opening has a greater side capture range than the plain opening.

VENTILATION EQUIPMENT

Commercially available equipment used for purging and ventilating confined spaces may be broadly divided into two categories: fans and venturi eductors.

Venturi Eductors

Venturi eductors, (Figure 8-4) also known as air ejectors, air eductors and air horns, have no moving parts. Instead, they operate on the principle of the venturi effect in which the velocity of air moving through a tube increases as it passes through a constriction of smaller cross-sectional area.

As shown in Figure 8-5, compressed air or steam is fed into the eductor through a single side-inlet connector that leads to a nozzle chamber. As the air or steam rushes through the nozzle, it creates a venturi force which induces ambient air into the inlet. The induced air is then discharged at high velocity through a horn-shaped diffuser.

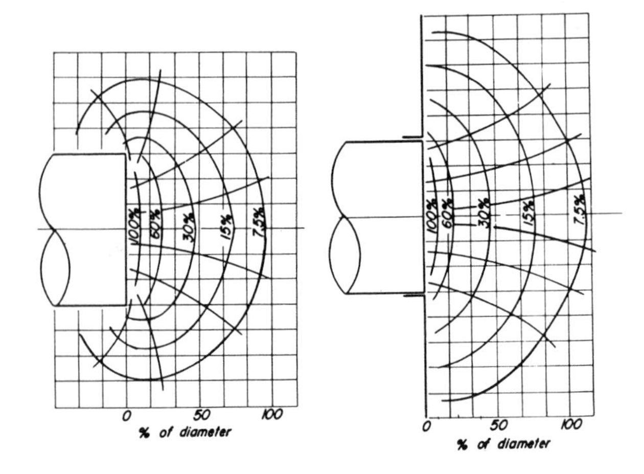

Figure 8-3. A flange can improve effective capture of a local exhaust system. (From *Industrial Ventilation: A Manual of Recommended Practice* 20th ed., ACGIH, Cincinnati, OH. Used with permission.)

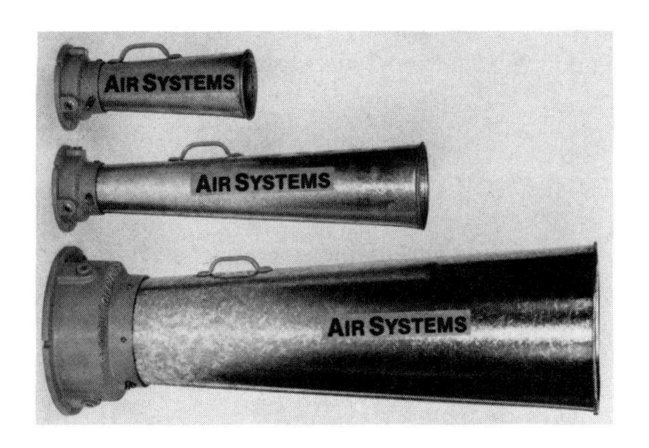

Figure 8-4. Venturi eductors. (Air Systems)

While eductors are unable to move large quantities of air, they do have the advantage of fitting into small openings. Another advantage is their ability to handle air containing corrosive or abrasive materials that would damage rotating fan blades. Air eductors should always be positioned firmly and inspected for internal obstructions before use. Otherwise, objects or debris trapped in the inlet could be ejected with tremendous force when the air or steam is turned on.

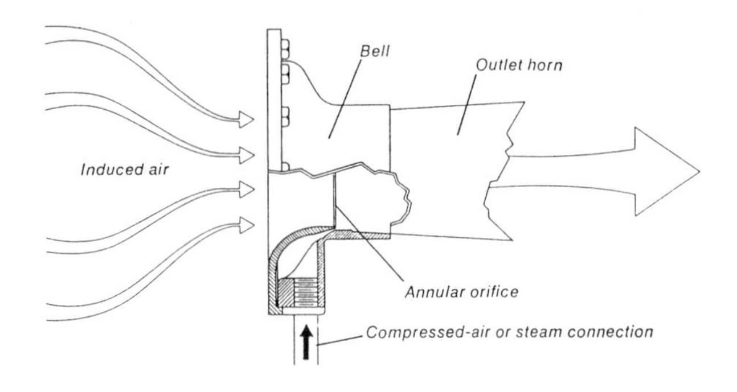

Figure 8-5. Venturi eductor principle of operation. (MSA)

Since the movement of high velocity air can generate significant quantities of static electricity, air eductors should always be electrically connected to the space with a bonding wire.

Fan Systems

Fans may be divided into two main classes: axial-flow or propeller types, in which the air flow is parallel to the axis of rotation (Figure 8-6) and centrifugal or radial-flow, in which the air flow is at right angles to the angle of rotation (Figure 8-7). Both axial flow and centrifugal flow blowers are commonly used to ventilate confined spaces. Axial flow units tend to be a little more compact but have limited ability to overcome system resistance. Noise levels associated with axial fans also tend to be quite high. Centrifugal fans, on the other hand, are much less sensitive to air flow resistance. They are also quieter than axial types. When fitted with appropriate ducting, both types of fans may be used for dilution ventilation and local exhaust.

Axial-Flow Fans

Axial-flow fans are those where air flow through the impellers is parallel to the shaft upon which the propeller is mounted. The airfoil shape of the propeller results in air being drawn in on the blade's leading edge and discharged from its trailing edge, forming a spiral pattern.

Axial-flow fans are further characterized by three designs: propeller types, tube-axial types and vane-axial types.

Figure 8-6. Axial-flow fan. (Pelsue)

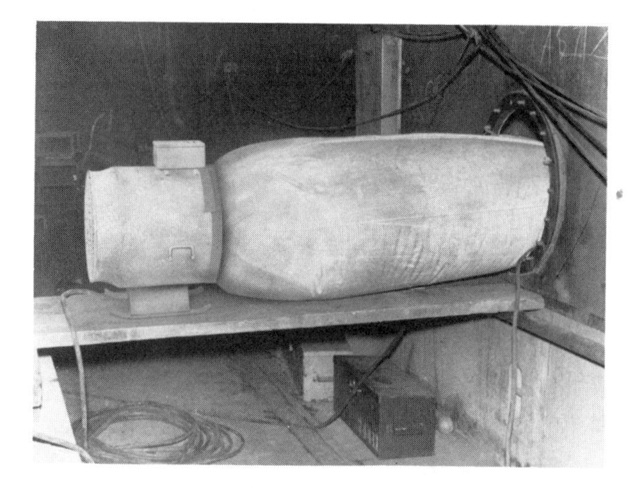

Figure 8-7. Tube-axial fan. (Coppus Engineering)

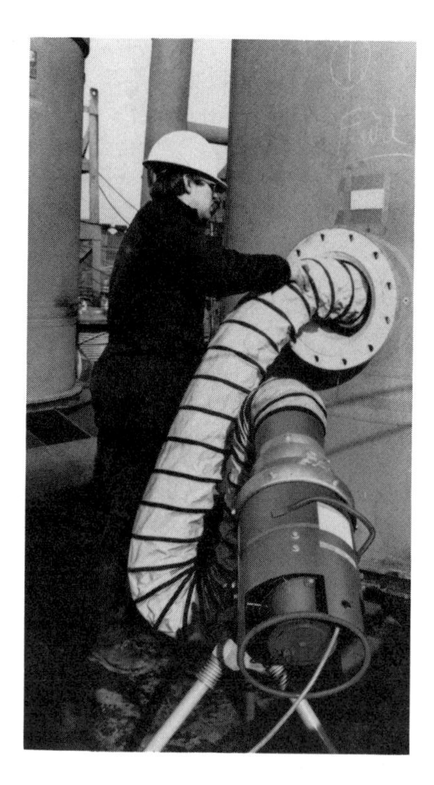

Figure 8-8. Vane-axial fan. (Coppus Engineering)

Propeller Fans. A familiar example of the propeller fan is the table-top oscillating fan used for personal cooling. Propeller fan efficiency is low and limited to low pressure applications. The impeller usually consists of two or more blades attached to a relatively small hub. Propeller fans can deliver high flow rates, but only against very low static pressures. Their maximum efficiency is reached near free delivery, and they are not well-suited for connections to duct work. The air discharge pattern is circular and air stream swirls because of the action of the blades and absence of straightening vanes.

Tube-axial Fans. Tube-axial fans (Figure 8-7) provide significant improvement over propeller fans because they are more efficient and capable of operating at higher static pressures. The blades of a tube axial fan are mounted inside a steel tube or cylinder. The tube fits closely around the blades so that there is very little clearance between the tube and the tips of the blades. Like the propeller fan, the air discharge pattern is circular and whirling because of absence of air straightening vanes.

Vane-axial Fans. Vane-axial fans (Figure 8-8) are similar in design to tube axial fans, but are equipped with a set of guide vanes on the upstream and downstream of the propeller. The vanes straighten out the circular air flow of the rotating blades and improve the pressure characteristics and efficiency of the fan.

Centrifugal-Flow Fans

Centrifugal fans (Figure 8-9) consist of a wheel or rotor that is mounted on a shaft which rotates inside a scroll-shaped housing. Air enters the center of the rotor and makes a right angle turn as it is forced through the rotor blades and into the housing by centrifugal force. The centrifugal force imparts static pressure to the air and the diverging shape of the scroll also converts a portion of the velocity pressure into static pressure.

Figure 8-9. Centrifugal-flow fan. (Coppus Engineering)

Figure 8-10. Self-contained ventilator LP gas. (Air Systems)

COMMERCIAL EQUIPMENT OPTIONS

Commercially available ventilation systems may be generally divided into three categories: self-contained ventilators, motor driven blowers and units driven by compressed air.

Self-Contained Ventilators

Self-contained ventilators use either a small LP-gas or gasoline engine to directly drive the fan blades (Figures 8-10 and 8-11). These units are highly portable and are available in a number of sizes. Typical air delivery rate range between about 600 to 4,000 CFM. Some units like that shown in Figure 8-12 may also be fitted with an electric generator that can be used to operate auxiliary equipment such as lights and power tools.

Motor Driven Blowers

Motor driven blowers (Figure 8-13) depend on an external source of electrical power provided either by fixed commercial wiring or by a portable generator. Smaller units which offer the same degree of portability as self-contained ventilators can deliver about 1,500 to 3,000 CFM. Larger units, which because of their weight and size cannot be moved as easily, can deliver more than 20,000 CFM. As an extra option, some manufacturers will install explosion-proof motors to allow use in "classified" locations.

Figure 8-11. Self-contained gasoline. (Air Systems)

Compressed Air Driven Units

Compressed air driven units may be subdivided into three types: air eductors, turbines and reactor fans.

Air eductors (air horns) which were discussed earlier, use compressed air to draw large volumes of ambient air into one end of a venturi chamber and exhaust it out the other. Air delivery rates range between about 1,000 to 8,000 CFM. The amount of air an eductor delivers depends on its size, the inlet air pressure and the length of duct attached. Typical values under selected operating conditions are listed in Table 8-5.

Turbine units use compressed air to drive an impeller similar to that on a water pump, which in turn

Figure 8-12. Self-contained gasoline with built in generator. (Pelsue)

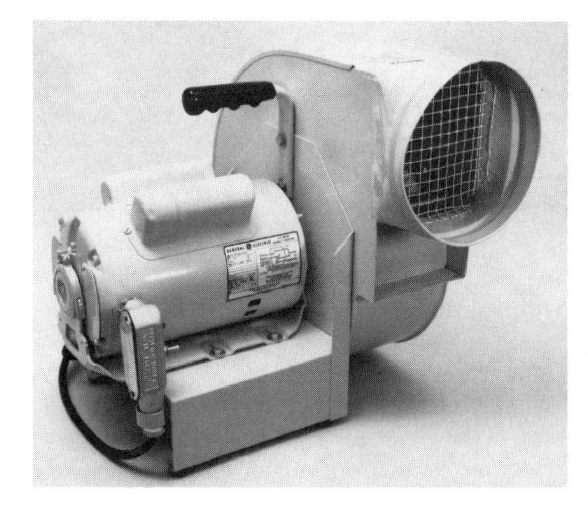

Figure 8-13. Motor-driven blower. (Pelsue)

rotates the fan blade. Volumetric capacity ranges between 7,000 to 16,000 CFM, depending on the pressure of the inlet air.

Reactor fan blades are hollow and designed with a small outlet nozzle on the tip of the blade (Figure 8-14). The force of air exhausted out the hole turns the blade. Small, portable units designed to be used with ductwork (Figure 8-15) have delivery rates of about 1,400 to 1,500 CFM. Larger models that are designed to be attached directly to a space (Figure 8-16) have delivery rates of about 11,000 to 17,000 CFM.

A comparison of free air delivery rates for selected sizes of venturi eductors and reactor fans is provided in Table 8-6.

EQUIPMENT MAINTENANCE

All blowers should be checked and cleaned periodically. Improper operation resulting from worn, loose or sticking bearings, shafts, belts and couplings can reduce a blower's speed and lower its air delivery. Since air delivery rates of electrically powered blowers depend on the correct voltage and line frequency, both must be maintained as specified in the operating instructions. Similarly, air delivered by blowers powered by internal combustion engines will vary appreciably if the engine is not properly maintained and adjusted to operate at the manufacturer's specified engine speed.

Flexible ducts should be visually inspected for defects such as tears, cracks, obstructions, separations and kinks which can reduce air delivery. Ends of ducts should fit snugly over blower attachment points, and if two or more duct sections are joined together, the coupling sleeve should also be inspected to assure a tight fit.

FAN NOISE

Fans can generate noise levels high enough to interfere with communication inside the space. Since noise levels decrease with increasing distance from the source, increasing the distance between entrants and the fan will reduce their exposure to noise. As a general rule, doubling the distance will reduce the sound pressure level by about 6 dBA. Typical noise levels based on manufacturers' data for different types of ventilation equipment are provided in Table 8-7.

PLANNING CONSIDERATION

The introduction to this chapter pointed out that since every confined space is unique, it is impossible to provide a single set of ventilation guidelines that will apply in all situations. Nevertheless, there are some broad-based concepts and considerations that can establish a starting point and frame of reference for arriving at the best solution to a specific ventilation problem.

Let us first ask ourselves the question, what is the purpose for ventilating a confined space? The ultimate answer, of course, is to provide an atmosphere in which entrants may work safely. To achieve this goal, we need to accomplish two things. First, we must maintain the level of oxygen within a range

Table 8-5. Eductor performance: air flow through flexible duct cubic feet per minute (CFM) at 80 psig inlet pressure

Inlet Diameter and Length (Inches)	Duct Diameter (Inches)	Straight Run Duct Length in Feet					
		None	10	20	30	40	50
7.3 × 16.5	6	1,400	1,200	1,040	930	850	780
7.3 × 33.0	8	1,540	1,520	1,432	1,385	1,335	1,280
11.5 × 44.2	12	3,860	3,337	3,580	3,450	3,335	3,225
14.3 × 46.0	14	6,250	5,800	5,550	5,280	5,050	4,850
16.6 × 46.0	14	8,000	7,300	6,850	6,550	6,250	6,000

Adapted from *Portable Ventilators,* Coppus Engineering Corp., Worcester, MA (1989).

Figure 8-14. Detail of reactor fan blade. (Coppus Engineering)

Figure 8-15. Portable reactor fan with duct. (Coppus Engineering)

of 19.5 to 23.5%. Second, we must control potentially harmful or explosive air contaminants.

If we focus our attention on how to best achieve these two objectives, we can logically establish the best approach to ventilating a particular space. Some specific items that should be considered during the decision making process are discussed below.

Previous Contents

If the space is a process or storage vessel, has it been drained and flushed? The more residue that can be removed through external means like flushing the less effort will be required for ventilation.

Internal Obstructions

Are there baffles or obstructions that will adversely affect air flow? If so, it may be necessary to thread lengths of flexible duct through and around them in order to assure that contaminants are removed from pockets and dead ends.

Existing Openings

The number, location and relative position of all openings in the space should be noted during the

Figure 8-16. Reactor fan mounted directly on manhold. (Coppus Engineering)

initial hazard assessment. Small diameter pipe connections should not be overlooked because they might be arranged in a way that allows them to be used either for exhausting contaminants out of the space or for drawing make-up air in.

Natural Drafts

Determine if there are any natural drafts flowing through openings or portals. Some large, above ground spaces may have strong natural drafts resulting from thermal gradients. If so, it may be possible to take advantage of these drafts. Draft patterns through various openings in the space will also influence the positioning of air moving equipment. For example, it might be better to place a blower at an opening where it could take advantage of the forces posed by a natural draft rather than working against them.

Supplying vs. Exhausting Air

The decision whether to force air into the space or exhaust it out depends largely on the nature of the space and the contaminants that are likely to be present. As shown in Figure 8-17, forcing air into a space tends to create turbulence which can agitate gases and vapors and evaporate volatile residue. Positive ventilation is an appropriate choice in situations where contaminant levels are relatively low and the air displaced from the space will not pose a hazard to others. It is commonly used in utility operations because the atmospheric hazards most likely to be encountered are oxygen deficiency, methane and hydrogen sulfide which, when diluted and dispersed, are not likely to cause any adverse consequences.

On the other hand, if the space is a process vessel or storage tank that still contains a volatile product, the uncontrolled release of vapors resulting from air being forced into the space as shown in Figure 8-18 may indeed pose serious hazards. For example, displaced materials that are within the flammable range could be ignited by nearby ignition sources. Alternatively, exhausted air may contain contaminants at levels in excess of permissible health standards which could adversely affect employees working in adjacent areas. Clearly, these possibilities must be considered and addressed.

In some cases, exhaust outlets may be arranged so that potentially flammable or toxic atmospheres are vented to a location where they may be adequately dispersed and diluted. In other situations, it may be possible to vent contaminants through an opening at the top of the vessel where they will be less likely to encounter ignition sources or pose a hazard to workers.

Vapor Density Considerations

Although contaminant gases and vapors diffuse throughout a space, they will have a tendency to stratify on the basis of their vapor density. Those lighter than air tend to rise to the top of the space, while those heavier than air will sink to the bottom. Temperature increases resulting from solar loading will also hasten evaporation and increase the convective rates with which gases and vapors diffuse and rise to upper regions of a space.

The arrangement and the placement of exhaust and make-up air inlets should take these characteristics into account. When contaminants are heavier than air, it may be advantageous to provide exhaust at the bottom of the space while drawing make-up air in through openings at the top (Figure 8-19). When contaminants are lighter than air, or when elevated temperatures are encountered, the system should be reversed with exhaust provided at the top of the space and make-up air provided at the bottom (Figure 8-20).

Table 8-6. Comparison of free air delivery rates for selected sizes of venturi eductors and reactor fans

Inlet Diameter and Length (Inches)	INLET PRESSURE 40 psig		60 psig		80 psig	
	Total Air Flow (CFM)	SCFM of Air Consumed	Total Air Flow (CFM)	SCFM of Air Consumed	Total Air Flow (CFM)	SCFM of Air Consumed
Eductors						
7.3 x 16.5	970	38	1,150	49	1,400	63
7.3 x 33.0	1,100	37	1,280	52	1,540	63
11.5 x 44.2	2,840	73	3,440	97	3,960	132
14.3 x 46.0	4,270	120	5,350	178	6,250	233
16.8 x 46.0	5,530	194	6,730	265	8,000	332
Reactor Fan (Diameter, in.)						
12	900	17	1,240	23	1,400	29
16	2,490	53	3,130	73	3,810	93
20	7,000	125	9,500	9,500	11,000	210
24	11,700	230	14,600	14,600	16,900	400

Adapted from *Portable Ventilators,* Coppus Engineering Corp., Worcester, MA (1989).

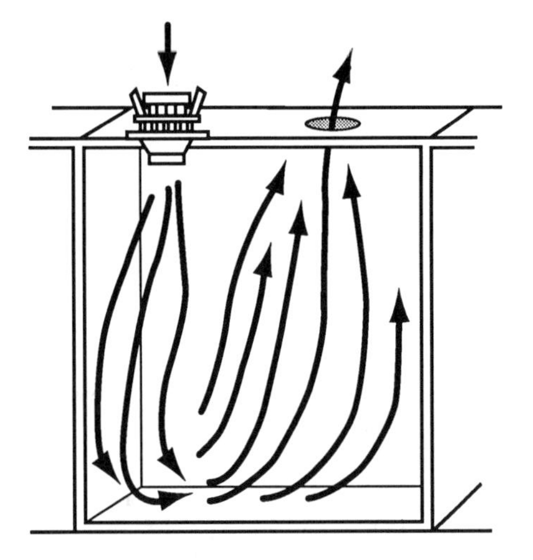

Figure 8-17. Turbulence created by blowing air into a space can cause evaporation of volatile reside. (Paul Trattner)

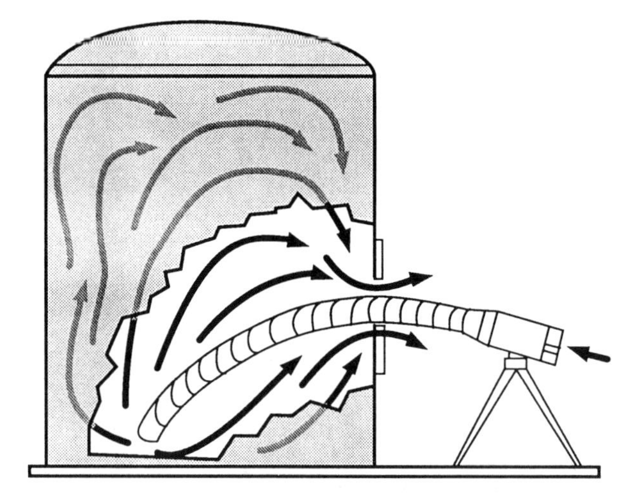

Figure 8-18. Blowing air into a space may result in the uncontrolled release of hazardous gases and vapors. (Paul Trattner)

These arrangements also minimize the dispersal of gases and vapors throughout the space because they capture them where they are present in the greatest concentration and remove them in a highly controlled fashion.

Operations in the Space

Work that is performed in the space may also introduce certain atmospheric hazards that need to be controlled by ventilation.

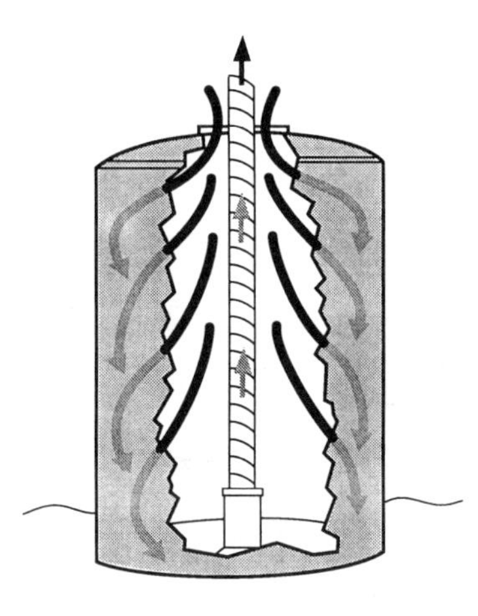

Figure 8-19. Gases and vapor that are heavier than air should be drawn off from the bottom while providing make up air from the top. (Paul Trattner)

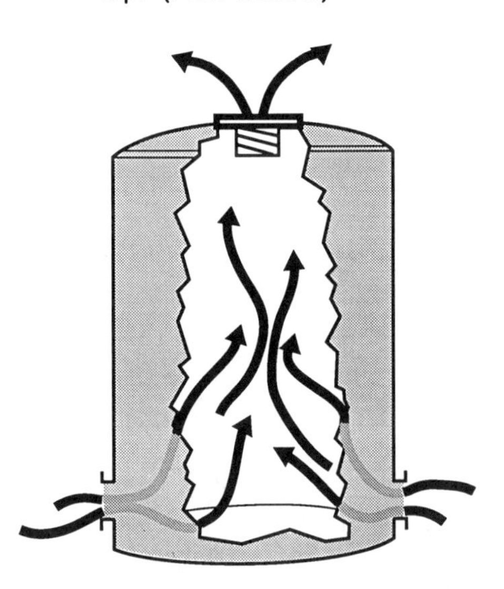

Figure 8-20. Lighter than air gases should be drawn off from the top and make up air provided from the bottom. (Paul Trattner)

Welding, cutting and burning operations will all produce metal fumes and gases such as carbon monoxide, ozone and oxides of nitrogen. These contaminants can easily be controlled by a local exhaust system that captures them at their source (Figure 8-21). Typical system requirements are listed in Table 8-7.

Solvent based products such as paints, preservative coatings, adhesives, cleaning agents and similar materials are frequently toxic or flammable. Processes that utilize these materials, such as spray paint-

Figure 8-21. Local exhaust used to control welding fume at its source. (Paul Trattner)

Table 8-7. Fan noise levels

Electric

Vaneaxial		Centrifugal	
HP	dB_A	HP	dB_A
1/2	84	1	84
3/4	87	2	88
1	90	5	94
2	91	15	100

Pneumatic

Equipment Type Diameter in.	Inlet Pressure (psig) 80	60	40
Air Ejector	Noise Level dB_A		
3	88	85	81
6	92	89	85
8	94	91	87
9	95	92	88
Blower/ Exhauster			
12	94	92	90
16	94	92	88
20	108	106	103
24	111	109	104

Adapted from *Portable Ventilators,* Coppus Engineering Corp., Worcester, MA (1989) pg.

ing and degreasing, tend to disperse contaminants over a large area rather than at a single point. Consequently, local exhaust may not be effective for controlling vapors except for low volume spot processes such as dye penetrant testing.

Sludge removal and spray application of flammable materials pose particularly thorny problems. In many cases, even the best possible ventilation will only reduce contaminant levels to below 10% of the lower explosive level. While this is acceptable from a fire and explosion perspective, health standards for most materials will be exceeded at this level.

As a result, two additional precautions should be taken. First, entrants should be provided with air line respirators with escape cylinders which are described in detail in Chapter 11. Second, the atmosphere should be monitored continuously, and entrants must stop spraying and leave the space whenever levels begin to exceed 10% of the lower explosive level.

Abrasive blasting contaminants cannot generally be controlled by dilution ventilation, and blasters who use open-air methods will have to be protected by air-supplied abrasive blasting helmets. However, dilution ventilation can be used to improve visibility in the space by removing suspended dust particles. Guidelines based on maritime experience suggest that air flow rates of about 15,000 CFM per blast nozzle, or 80 CFM per square foot of floor area, will be required. Transport velocities should be at least 4,500 fpm to prevent heavy particles from settling out in the exhaust ducts.

Recent advances in abrasive blasting technology have created an attractive alternative to open-air blasting. A vacuum blasting system consists of a blasting nozzle equipped with an integral local exhaust system (Figure 8-22). Since the local exhaust captures both fines and spent shot, very little dust is released into the space.

Vacuum blasting equipment may be fitted with corner attachments like those shown in Figure 8-23 and 8-24 that allow access in tight spots. They can also be connected to separation and recovery devices which recycle usable shot back into the system and minimize the quantity of hazardous waste that is generated. This last feature can be a real cost saver on many jobs.

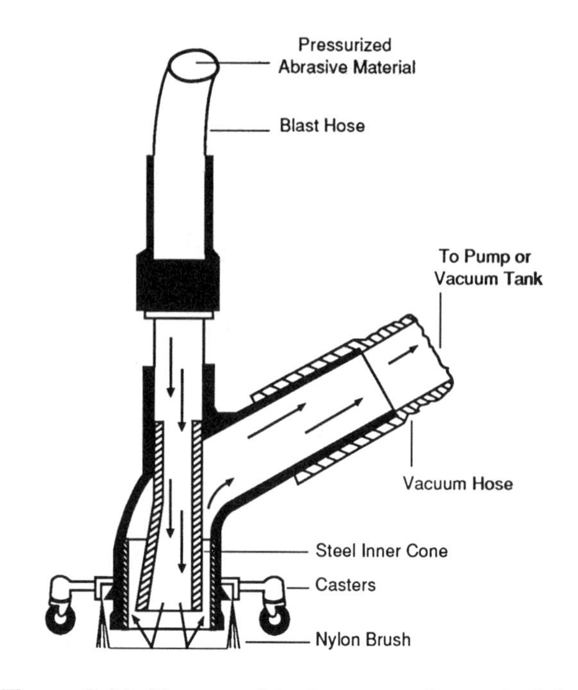

Figure 8-22. Vacuum blaster operating principle. (Paul Trattner)

Figure 8-23. Inside corner cutter. (LTC Corp.)

Contaminant Reentrainment

If the blower is powered by an internal combustion engine, care should be taken to assure that exhaust gases are not drawn into the blower's fresh air intake. While the blower can be oriented to take advantage of prevailing winds, a more effective approach is to vent exhaust gases away from the blower's air inlet. The flexible metallic tubing shown attached to the exhaust of the blower shown in Figure 8-25 can be extended away from the unit to take advantage of natural dilution.

Figure 8-24. Outside corner cutter. (LTC Corp.)

Figure 8-25. Blower with flexible exhaust pipe. (Air Systems)

Blower inlets should also be kept away from sources of contamination such as vehicle exhaust pipes, or contaminant-producing operations or processes such as welding, spray painting and solvent degreasing. If local exhaust systems are used in the space, care must be taken to assure that the outlets from these systems are positioned away from air inlets of general ventilation systems. Otherwise, exhausted contaminants may be captured by the general ventilation system and directed back into the space.

Short-Circuiting

A ventilation system may be short-circuited if openings used for exhausting and supplying air are too close together as shown in Figure 8-26. This condition can be corrected by installing a length of duct work as shown in Figure 8-27.

Short circuiting may also occur when a roof-mounted exhauster draws make-up air in through a ground-level, side-wall manway. If the manway and exhauster are located on the same side of the tank as shown in Figure 8-28, air that enters the manway will be drawn straight up to the roof by the exhauster, and contaminants on the opposite side of the tank will not be removed. A far better approach is to provide exhaust and make-up air through openings on opposite sides of the tank so that air currents can flow throughout the entire space to dilute and remove contaminants (Figure 8-29).

The decision whether to place the exhauster at the roof with make-up air provided through a manway in the side wall, or vice versa, depends on vapor density and thermal considerations discussed previously. If vapors tend to accumulate near the top of the space, the former would be a better choice; if they tend to accumulate near the bottom, the latter method should be chosen.

Portal Obstructions

In some cases the ductwork used to ventilate a space must be placed through the same portal that is used for entry and exit. If the portal is relatively small, the space taken up by the duct can restrict access and exit. This problem can be solved through the use of a Saddle Vent™ adapter like that shown in Figure 8-30, which reduces the amount of the portal area needed to accommodate the duct.

Atmospheric Testing

While it is theoretically possible to calculate how much time is required to purge a space of contaminants, the type and number of variables generally makes calculations impractical. Among the factors that need to be considered are:

- The magnitude of any air contaminants or extent of a deficiency in oxygen
- The size and geometry of the space, including obstructions that would limit air flow

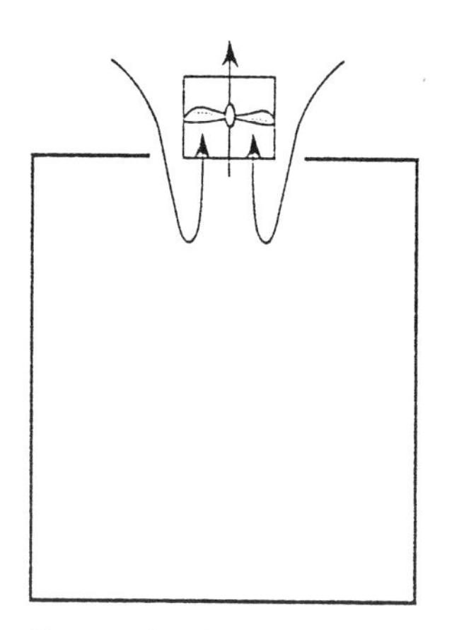

Figure 8-26. Short circuiting in a manhole. (Paul Trattner)

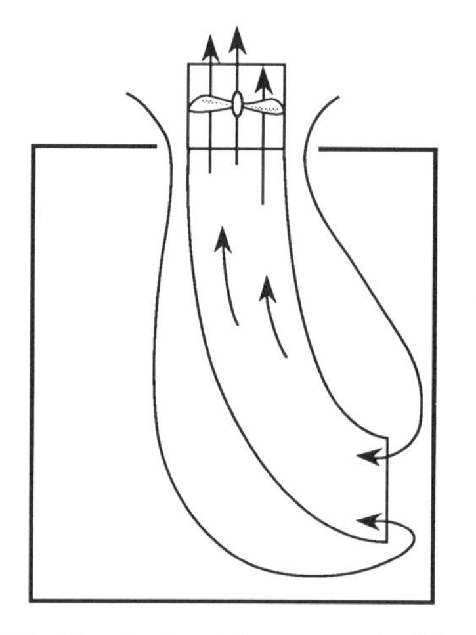

Figure 8-27. Short circuiting corrected by adding a length of duct. (Paul Trattner)

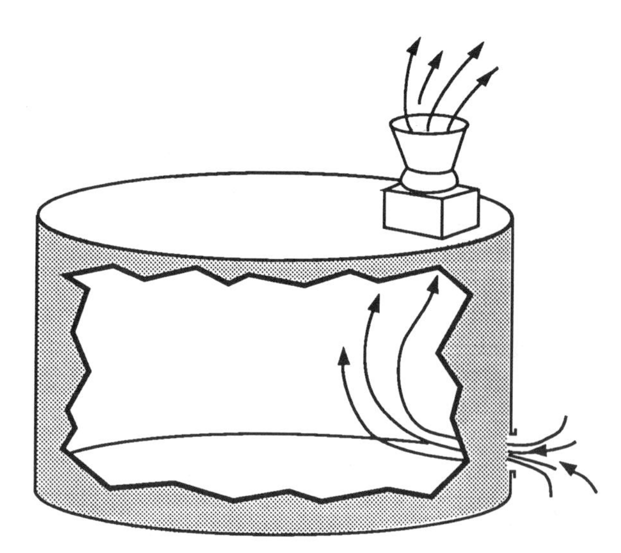

Figure 8-28. Short circuiting in a vessel. (Paul Trattner)

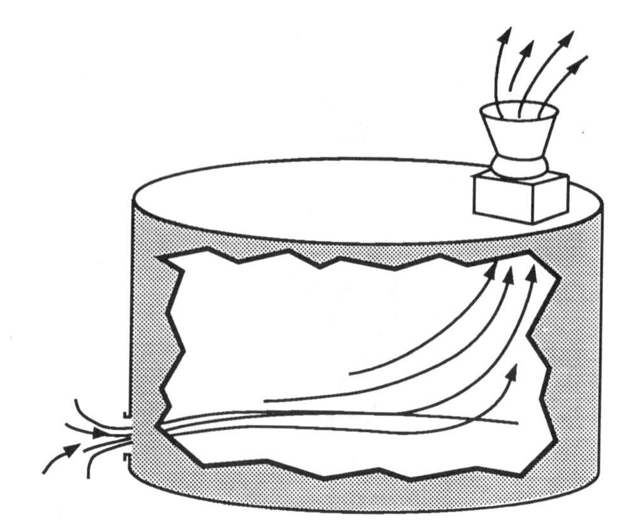

Figure 8-29. Short circuiting in a vessel corrected by using a different air inlet. (Paul Trattner)

Figure 8-30. Saddle vent adapter. (Air Systems)

- Effective blower capacity after compensating for friction losses imposed by length of the duct, as well as twists, turns and elbows
- The degree of air mixing that occurs
- The number, position and size of any openings

As a result, virtually the only way to assure that a space has been adequately purged is to test the atmosphere. The sampling strategy, of course, will depend on the anticipated hazards. Some spaces will only need to be tested for oxygen, others will also require an evaluation of combustible gases and vapors and toxic air contaminants.

While atmospheric testing may be used to verify the adequacy of ventilation, it should not be relied on to determine if purging or ventilation is necessary. An atmosphere that is initially deemed to be suitable for entry could change over time. For example, hazardous materials could enter accidentally or be introduced by operations performed in the space. Continuous ventilation provides an added margin of safety against unexpected events.

ALTERNATIVE ENTRY PROCEDURES

Streamlined entry procedures are allowed for permit-spaces that meet the following two-prong test:

1. The *only hazard* posed by the permit space is an actual or potentially hazardous atmosphere.
2. Continuous forced air ventilation alone is sufficient to maintain the permit space safe for entry.

Documentation of Supporting Data

The first thing you have to do under the alternative entry provisions is to gather monitoring and inspection data that support the two conditions cited above. If an initial entry of the permit space is necessary to obtain the data, that entry must be performed in compliance with a full entry program as described in Chapter 6. The determinations and supporting data must be documented and made available to each employee who enters the permit space.

Cover Removal and Opening Protection

Any conditions that make it unsafe to remove an entrance cover must be eliminated before it is removed. Once the cover is removed, the opening must be promptly guarded by a railing, temporary cover or other equally effective barrier similar to that shown in Figure 8-31. The guard must be sufficient to prevent people and foreign object from accidentally falling into the space through the opening.

Provision for Atmospheric Testing

Before an employee enters the space, the internal atmosphere must first be tested with a *calibrated direct-reading instrument*. The proper sequence for testing is oxygen content, then flammable gases and vapors, and finally toxic air contaminants.

Figure 8-31. Manhole protected by guard rail. (Pelsue)

Continuous Forced Ventilation

Continuous forced air ventilation must be used and employees may not enter the space until the ventilation has eliminated any hazardous atmosphere. The air supply must be from a clean source and may not increase the hazards in the space and there may be no hazardous atmosphere within the space whenever any employee is inside the space.

The system must be arranged to ventilate the immediate areas where employees are working and it must continue to operate until all employees have left the space. The standard stipulates that the space atmosphere be tested "periodically" to ensure that the ventilation is preventing the accumulation of a hazardous atmosphere. However, it is more prudent to test *continuously* with an instrument that will alarm if a hazardous atmosphere is detected.

If a hazardous atmosphere occurs during entry, employees must leave the space immediately and conditions must be evaluated to determine how the hazardous atmosphere developed. Additional measures must then be implemented to protect employees from the hazardous atmosphere before any subsequent entry takes place.

Certification Required

A certificate indicating that the space is safe for entry must also be prepared and made available to entrants prior to entry. The certificate must contain the date and the location of the space as well as the signature of the person who prepared it.

WHEN VENTILATION IS NOT POSSIBLE

There are some situations where ventilation may not be possible. For example, in an emergency it may be necessary to enter a valve pit to shut off the flow of a hazardous material, which, if left unchecked, could result in a catastrophic release. Since time is of the essence, ventilating and testing prior to entry would be counterproductive and the time lag in shutting off the valve could lead to disastrous consequences.

In cases like this, it is more prudent to presume that the atmosphere in the vault is immediately dangerous to life or health and to rely on a self-contained breathing apparatus in lieu of ventilation.

SUMMARY

The goal of confined space ventilation is to provide an atmosphere in which entrants can work safely. This goal can be achieved if the level of oxygen is maintained within prescribed limits and if potentially harmful or explosive contaminants are controlled.

Since confined spaces vary in size, shape and function it is impossible to develop a single set of ventilation guidelines that will apply in every situation. However, decisions for selecting the most appropriate method to ventilate a particular space should be based on some general considerations. Among these are:

- The geometry of the space
- Its previous contents
- The existence of natural drafts, the number and location of any openings, and
- The nature of any contaminant-producing tasks that may be performed in the space.

Although some spaces can be ventilated by natural air movement, mechanical ventilation is preferred because it is more reliable and versatile. Mechanical systems can also be readjusted to compensate for changing conditions in the space as well as arranged to provide local exhaust that removes contaminants at their source.

The only practical way to determine if a space has been ventilated adequately is to test the atmosphere. Air sampling instruments as well as measuring methods and techniques will be discussed in the next chapter.

REFERENCES

American National Standard for Safety Requirements for Confined Spaces, ANSI Z117.1-1989, New York, NY: American National Standards Institute, (1989).

Arc Welding and Cutting, Code of Federal Regulations, Vol. 29, Part 1910.254.

Amos, D.J., *Labor's View on Confined Space,* paper presented at the National Safety Council Congress and Exposition, Las Vegas, NV, October, 1990.

Bell Laboratories Report, Manhole Ventilating Practices, Case No. 37974-129, August 29, 1974.

Brief, R.S., L.W. Raymond and W.H. Meyer, Better ventilation for close-quarter work spaces, Air Conditioning, *Heating and Ventilation,* 58:74–88, September, 1961.

Burton, D.J., *Industrial Ventilation Work Book,* Salt Lake City, UT: DJBA Publishing, (1989).

Cleaning Petroleum Storage Tanks, API Publication 2015, Washington, DC: American Petroleum Institute, (1985).

Cleaning Open-top and Covered Floating Roof Tanks, API Publication 2015B, Washington, DC: American Petroleum Institute, (1981).

Cloe, W.W., *Selected Occupational Fatalities Related to Ship Building and Repairing as Found in Reports of OSHA Fatality/Catastrophe Investigations,* Washington, DC: U.S. Department of Labor, Occupational Safety and Health Administration, (1985).

Cloe, W.W., *Selected Occupational Fatalities and Asphyxiating Atmospheres in Confined Spaces as Found in Reports of OSHA Fatality/Catastrophe Investigations,* Washington, DC: U.S. Department of Labor, Occupational Safety and Health Administration, (1985).

Coppus Vano Blower/Exhauster Models 175 & 250, Operating and Maintenance Instructions, Worcester, MA: Coppus Engineering, (1985).

Coppus C-12AW & C-15AW Tank Ventilator, Operating and Maintenance Instructions, Worcester, MA: Coppus Engineering, (1989).

Coppus Air-Driven Blower-Exhauster Type RF-20, Operating and Maintenance Instructions, Worcester, MA: Coppus Engineering, (1983).

Coppus RF-12 Air-Driven Axial Blower-Exhauster, Operating and Maintenance Instructions, Worcester, MA: Coppus Engineering, (1985).

Coppus RF-16 Air-Driven Blower-Exhauster, Operating and Maintenance Instructions, Worcester, MA: Coppus Engineering, (1984).

Coppus Ventair Blower "TM" electric motor, & "TE" gas engine, Operating and Maintenance Instructions, Worcester, MA: Coppus Engineering, (1987).

Coppus Jet Air High Performance Types 3S-HP, 3-HP, and 6-HP, Operating and Maintenance Instructions, Worcester, MA: Coppus Engineering, (1988).

Garrison, R.P. and D.R. McFee, Confined spaces—a case for ventilation, *Am. Ind. Hyg. J.* (47) A-708-A-714, November, 1986.

Guidelines for Confined Space Work in the Petroleum Industry, API Publication 2217, Washington, DC: American Petroleum Institute, (1984).

Guide for Controlling the Lead Hazard Associated with Tank Entry and Cleaning, API Publication 2015A, Washington, DC: American Petroleum Institute, (1982).

Hot Work, Code of Federal Regulations, Vol. 29, Part 1926.14.

Industrial Ventilation: A Manual of Recommended Practice, 20th Ed., Cincinnati, OH: American Conference of Governmental Industrial Hygienists.

Mutchler, J.E., Principles of ventilation, in *The Industrial Environment its Evaluation and Control,* Cincinnati, OH: National Institute for Occupational Safety and Health, (1973).

May, J.W., *Physics of Air,* Louisville, KY: American Air Filter, (1970).

Naval Sea Systems Command, Gas Free Engineering Program, Naval Sea Systems Command, NAVSEA Publication S6470-AA-SAF-010.

Oxy-Fuel Gas Welding and Cutting, Code of Federal Regulations, Vol. 29, Part 1910.253.

Pacific Bell, Testimony to U.S. Department of Labor, Docket No. S-019, in letter of September 29, 1989 from Marty Kaplan.

Park, C. and R.P. Garrison, Multicellular model for contaminant dispersion and ventilation effectiveness with application for oxygen deficiency in a confined space, *Am. Ind. Hyg. J.* (51):70–78, February, 1990.

Pelsue, Heaters, Blowers and Generators, Englewood, CO: T.A. Pelsue Co.

Pelsue Model 1590 Heater Blower, Product Data Sheet, Englewood, CO: T.A. Pelsue Co.

Pelsue 1600 LP Gas Heaters, Product Data Sheet, Englewood, CO: T.A. Pelsue Co.

Pelsue Model 1000 Series PEL-PORT™ Portable Blowers, Product Data Sheet, Englewood, CO: T.A. Pelsue Co.

Pelsue Model 1240 2-speed Blower, Product Data Sheet, Englewood, CO: T.A. Pelsue Co.

Pelsue Model 1450 "PEL-PORT G.O." Blower, Product Data Sheet, Englewood, CO: T.A. Pelsue Co.

Pelsue Model 2000 Series Ventilation Hose, Product Data Sheet, Englewood, CO: T.A. Pelsue Co.

Pelsue Model 1325/1400B Axial Blower, Product Data Sheet, Englewood, CO: T.A. Pelsue Co.

Pelsue Model 1900A, TAP19™, Product Data Sheet, Englewood, CO: T.A. Pelsue Co.

Permit-Required Confined Spaces for General Industry, Code of Federal Regulations, Vol. 29, Part 1910.146.

Pettit, T.A., P.M. Gussey and R.S. Simmons, *Criteria for a Recommended Standard: Working in Confined Spaces,* Cincinnati, OH: National Institute for Occupational Safety and Health, DHEW/NIOSH Publication 80-106, (1979).

Pettit, T.A. and H. Linn, *A Guide to Safety in Confined Spaces,* Cincinnati, OH: National Institute for Occupational Safety and Health, DHHS/NIOSH Publication No. 87-113, (1987).

Precautions Before Entering Compartments and or Spaces, Code of Federal Regulations, Vol. 29, Part 1915.12.

Soule, R.D., Industrial hygiene controls, in *Patty's Industrial Hygiene and Toxicology,* 3rd. Rev. Ed., G.D and F.E. Clayton, (eds), New York: NY, John Wiley and Sons, (1978).

Testing and Ventilating Manholes, Bell System Practices, Section 620-140-501sv, Issue A, May 1989.

Technologically Advanced Air Systems, Catalog 592, Chesapeake, VA: Air Systems International, (1992).

Ventilation, Code of Federal Regulations, Vol. 29, Part 1910.94.

Welding, Cutting and Heating in Way of Preservative Coatings, 1926.354.

Welding Smoke and Fume Control, Moorestown, NJ: Securus.

CHAPTER 9

Atmospheric Testing

INTRODUCTION

The oil lamp may have been the first instrument used to detect the presence of hazardous atmospheres. While ancient miners probably did not know that naturally occurring methane gas was being ignited by the naked flames of their lamps, they were certainly aware of the subsequent explosion! Some may have also inferred that a weakening of the flame indicated another problem with the air. Today, we would call that problem an oxygen deficiency.

As time progressed, miners became more sophisticated and for a while they used caged canaries to detect the presence of carbon monoxide. The theory was that the canaries would be affected by carbon monoxide before the miners and when miners saw that the canaries were being adversely affected, they could evacuate before they too were overcome. While the idea was good in theory, it failed in practice. Canaries, it seems, provided a reliable indication of high levels of carbon monoxide and were affected well before the miners. However, because of the way canaries metabolized carbon monoxide, they were unaffected by lower concentrations which proved fatal to the miners.

Perhaps the first significant advance in gas detection instrumentation came in 1855 with the invention of the miner's safety lamp by Sir Humphry Davy. The Davy lamp not only provided miners with illumination, but it also served as a combination oxygen and explosive gas detector.

As shown in Figure 9-1, the wick of Davy's lantern was shrouded by two layers of fine metal mesh which acted as a flame arrestor, permitting air to flow into the wick area, while at the same time preventing the open flame from igniting combustible gases that may have been present in the mine. By observing the flame height, miners could also obtain information about the air quality. A shortening of the flame indicated a decrease in oxygen, and a lengthening indicated the presence of a flammable gas.

While the Davy lamp may have been state-of-the art in its day, modern advances in electronics have progressed to the point where it is no longer necessary to watch the height of a flame to evaluate atmospheric conditions. Modern field-portable instruments have become more sophisticated and rely heavily on electrical and chemical methods to detect the presence of literally hundreds of substances. However, the increasing sophistication of instruments has resulted in a corresponding need to increase the level of expertise of instrument operators.

Consequently, anyone who is going to use gas detection instruments to evaluate confined space atmospheres must possess four essential skills. They must be able to:

1. Select the most appropriate instrumentation for accurately determining the anticipated atmospheric hazards
2. Check the instrument to verify that it's functioning properly and reading correctly
3. Use the instrument in a manner that assures that the atmosphere in the space is thoroughly evaluated
4. Correctly interpret the measurement results

Because instrument manufacturers are constantly refining their existing models and regularly introducing new ones, this chapter cannot describe all of the makes, models and styles of available instruments. However, since much of the equipment is constructed along similar lines, it is possible to describe general types of instruments and to explain their theory of operation and use. It is also possible to identify special features offered by some manufacturers and to discuss purchase options and considerations which may provide an advantage in confined space operations.

CLASSES OF INSTRUMENTS

Rapid advances in the field of electronics have spawned a proliferation of instrumentation that can be used to evaluate confined space atmospheric hazards. However, it can all ultimately be divided into two categories: direct-reading instruments and indirect-reading instruments.

Direct-Reading Instruments

Direct-reading instruments are those where the substance of interest is collected and analyzed within the testing instrument. Instruments typically used for

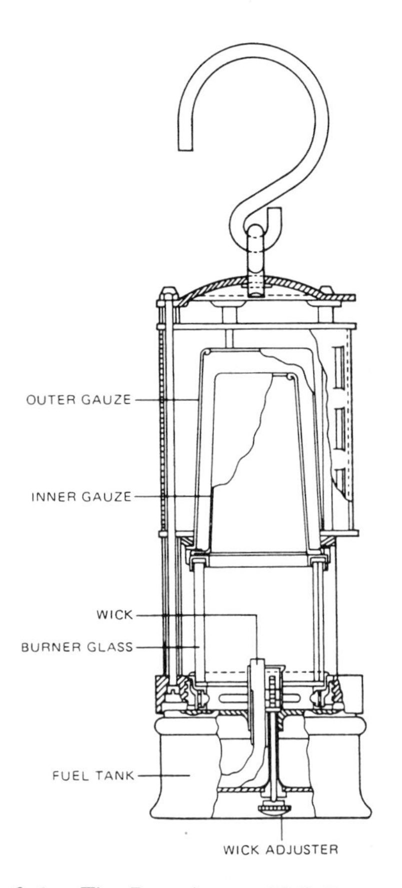

Figure 9-1. The Davy Lamp. (U.S. Bureau of Mines)

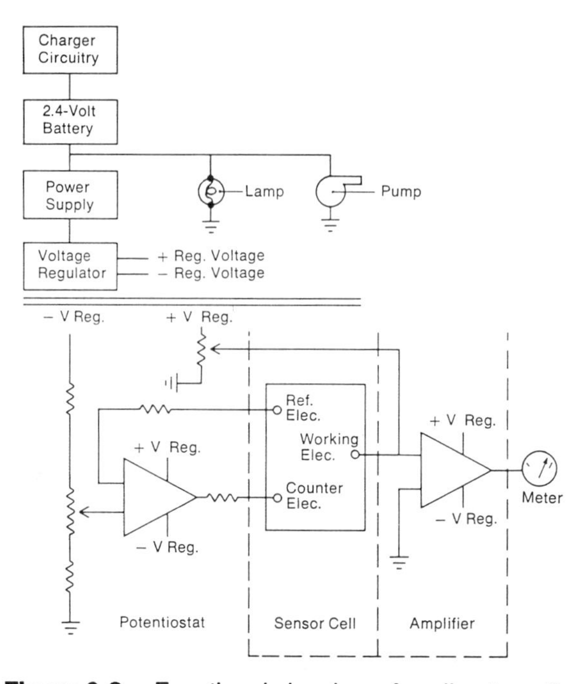

Figure 9-2. Functional drawing of a direct-reading instrument. (MSA)

evaluating confined spaces employ sensors which detect the presence of specific gases. Signals produced by the sensors are processed electronically, and measurement results are reported on a meter or digital display (Figure 9-2). Sensor-based instruments enjoy wide popularity because they are highly portable and are able to evaluate some of the more frequently encountered atmospheric hazards. Only a limited number of toxic contaminants can be evaluated with existing sensor technology, but this limitation may be overcome by other instruments such as colorimetric indicating tubes, infrared analyzers, and portable gas chromatographs, all of which will be discussed later.

Indirect-Reading Instruments

Indirect-reading instruments require two separate steps to measure atmospheric contamination. First, the contaminant of interest must be collected, then the collected sample must be subjected to laboratory analysis. Indirect-reading instruments typically use a small, battery-powered vacuum pump to draw contaminated air through collection media such as filters, solid sorbents or liquids as shown in Figure 9-3. The collection media traps the contaminant of interest and preserves it for subsequent laboratory analysis. The selection of the collection media, flow rate and sampling time are largely a function of the laboratory analytical method. However, in general, filters are used to collect particulate contaminants such as fumes and dusts, and solid sorbents and liquid media are used for gases and vapors.

Because conditions in the space may change between the time the sample is collected and receipt of the analytical results from the laboratory, indirect reading instruments are rarely used to evaluate confined space atmospheres prior to entry. On the other hand, they are particularly well-suited for use in evaluating employee exposures to contaminants resulting from operations performed in the space.

INSTRUMENTS FOR EVALUATING CONFINED SPACES

The pre-entry evaluation of a confined space is generally performed using direct-reading portable

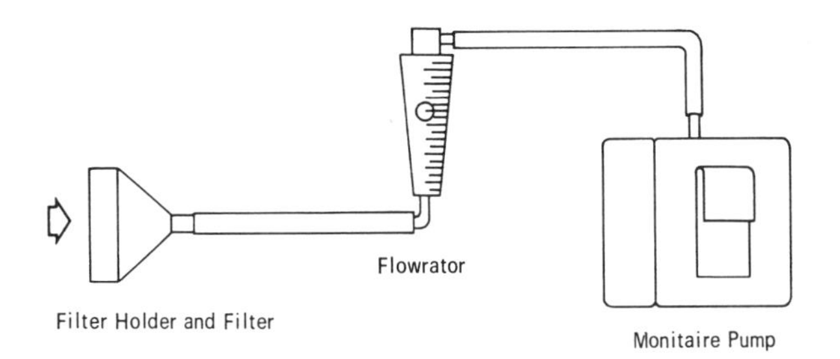

Figure 9-3. Indirect-reading instrument sampling train. (MSA)

instruments. These instruments may be categorized as either single gas or multi-gas devices. As might be expected, single gas instruments are available for oxygen, combustibles and a variety of toxic materials. Multi-gas instruments most often incorporate detectors for both oxygen and combustible gases, but many also include sensors for one or more toxic gases. While sensors for hydrogen sulfide and carbon monoxide are most frequently incorporated, some manufacturers offer other choices such as chlorine, sulfur dioxide and nitrogen dioxide.

A sample line or probe attached to the instrument is usually inserted into the space and air is drawn into the instrument as shown in Figure 9-4. The operator then reads the measurement results directly from a meter or digital display (Figure 9-5). If measurements indicate that levels of oxygen, combustibles and toxic materials are within acceptable limits, the entry may proceed. If the atmosphere is deemed to be unacceptable, entry must be delayed until the space can be more thoroughly ventilated, or until entrants can be provided with suitable respiratory protection.

Although handheld electronic devices are widely used for evaluating oxygen and combustibles, the number of toxic gases that can be measured with these instruments is limited to less than a dozen (Table 9-1). Consequently, other instruments such as colorimetric detector tubes, portable chromatographs and portable infrared spectrophotometers may be needed to evaluate toxic atmospheres.

Regardless of the type of instrument, users must be aware of three important concepts: interference from materials other than the substance of interest, the effects of instrument response on measurements and the difference between accuracy and precision.

Figure 9-4. Direct-reading instrument sampling line is inserted into the space. (MSA)

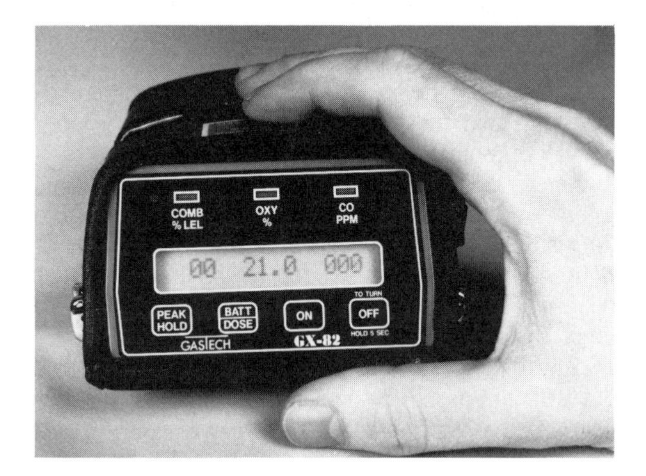

Figure 9-5. Instrument digital display reading for combustible gas, oxygen and carbon monoxide. (Gastec)

Interferences

All gas detection systems will be affected to some degree by contaminants other than that for which the system is calibrated. For example, an instrument calibrated for acetone may be affected by other ketones such as methyl ethyl ketone and methyl isobutyl ketone. Contaminants other than the species of interest that influence the value of the measurement are called interferences.

Table 9-1. Contaminants for which electrochemical sensors are commercially available

- Ammonia
- Carbon monoxide
- Chlorine
- Hydrogen sulfide
- Nitric oxide
- Nitrogen dioxide
- Ozone
- Sulfur dioxide

Interferent materials that result in the instrument reading higher than the actual concentration are called positive interferents, while those that cause a lower than actual reading are called negative interferents.

The biggest problem with interferent gases is that their presence can lead to misleading conclusions about the measurement results. For example, negative interferents can mask a hazardous atmosphere by causing an instrument to indicate a lower airborne contaminant concentration than actually present. This could lead the person making the measurement to conclude that the space is suitable for entry even though it contains a hazardous atmosphere.

Interferents may also trigger false alarms in instruments equipped with an alarm feature. For example, an instrument set to alarm at a level of 10 ppm of hydrogen sulfide may be affected by low levels of other sulfur-containing contaminants such as mercaptans. If nuisance tripping persists, employees may decide to ignore the alarm or turn it off. In either case, the alarm will be defeated and entrants will not be warned should an actual toxic gas hazard arise.

Instrument Response Time

Air sampling instruments do not provide measurement results instantaneously. Instead, there is a slight delay encountered as the air sample makes its way to the sensor. Once it arrives, there is another delay as the sensor reacts to the substance of interest. The period between the time the measurement starts and the time when a reliable reading can be obtained is called the response time.

Response times for many portable instruments are on the order of thirty to sixty seconds, and some may be as long as two or three minutes. That means that an operator must sample the atmosphere for the period established by the instrument's response time before he can be confident in the measurement result. In other words, measurements made by inserting an instrument into a space for a few seconds—as is often done by improperly trained technicians—will be misleading and possibly dangerous.

Accuracy vs. Precision

The terms accuracy and precision are often incorrectly used interchangeably.

Accuracy may be defined as how close a measured value is to the true value. For example, if the actual concentration of particular gas or vapors is 200 ppm, a measurement indicating 201 ppm would be more accurate than a measurement of 190 ppm.

Precision, on the other hand, is a term used to describe the level of variability that exists between repeated measurements. A high degree of precision exists when repeated measurements produce the same result. Since precision is a measure of variability, it is possible for measurements to be very precise (reproducible) without being accurate (correct). This apparent paradox is demonstrated graphically in Figure 9-6.

OXYGEN METERS

Operational Theory

The earliest oxygen measuring instruments used a wet chemical process that was based on the principle of absorption of oxygen in a solution of chromous chloride. Although these instruments were employed widely in the maritime industry, they were cumbersome and required a high degree of manual dexterity. In addition, the oxygen absorbing solution was corrosive, and splashes could irritate the skin and eyes. While wet chemical instruments still find some use in industry, they have largely been replaced by electronic devices.

Present day handheld oxygen meters employ electro-chemical or "wet-cell" sensors (Figure 9-7). The cell consists of a plastic cylinder that houses a lead-sensing electrode and gold-counter electrode which are surrounded by an electrolyte solution of potassium hydroxide and water. One end of the cylinder is covered with a Teflon membrane that is permeable to gases, but impermeable to the potassium hydroxide solution. Oxygen molecules diffuse through the membrane and dissolve in the solution where they enter into an electro-chemical reaction that causes electrons to flow from the gold cathode to the lead anode. The reactions that occur at each electrode are:

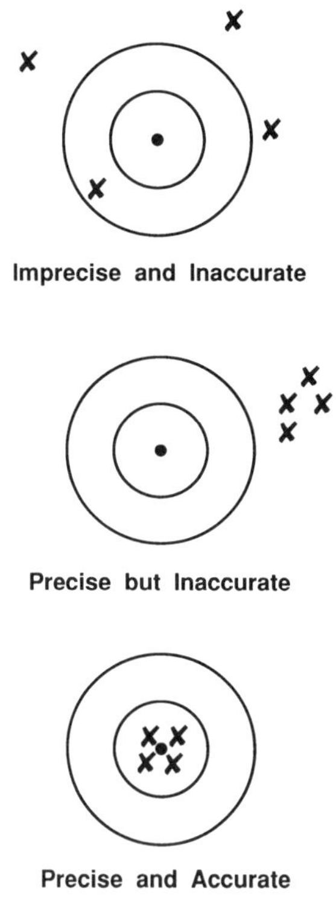

Imprecise and Inaccurate

Precise but Inaccurate

Precise and Accurate

Figure 9-6. Accuracy vs. precision. (Paul Trattner)

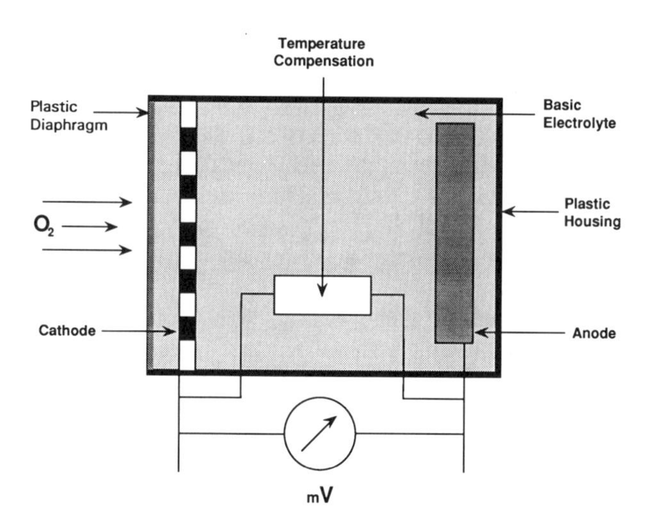

Figure 9-7. Wet-cell oxygen sensor. (Paul Trattner)

Gold cathode reaction $4e^- + 2H_2O + O_2 \rightarrow 4OH^-$

Lead anode reaction $2OH^- + Pb \rightarrow PbO + H_2O + 2e^-$

Although the current produced by the flow of electrons is directly proportional to the concentration of oxygen, it should be noted that the instrument is actually detecting the partial pressure of oxygen. This means that a unit calibrated at sea level will indicate a *lower percentage* of oxygen at higher altitudes. (Refer to Chapter 2 for a discussion of oxygen partial pressures.)

Field Calibration Check

Oxygen sensors are checked in the field by exposing them to atmospheric air (Figure 9-8). Since the concentration of atmospheric oxygen is generally agreed to be 20.95%, a reading of about 21% in "clean" air verifies that the instrument is operating properly.

The exact field calibration checkpoint varies among instrument manufacturers. Some specify 20.8%, allowing 0.15% for ambient water vapor, others specify 20.95, and still others round the checkpoint off to 21%.

The electronics in some instruments automatically set the meter to a specified field calibration checkpoint when they are turned on in clean air. Other instruments are equipped with a calibration control that the user adjusts until the display reads the specified value.

High- and low-level alarms are another feature incorporated into some oxygen meters. To provide warning of oxygen deficiency and oxygen enrichment, the factory will often set the low-level alarm point at 19.5% and the high-level alarm point at either 22 or 23.5%. The alarm function of these units can be checked by introducing special gas mixtures available from the instrument's manufacturer.

Care and Maintenance

Oxygen sensors are virtually maintenance free and while their useful life depends on the conditions under which they are used, most should last for about a year. However, service life can be substantially reduced if they are exposed to extremes of temperature. Literature provided by one manufacturer also indicates that oxygen sensor life may be reduced significantly by exposure to high levels of carbon dioxide (Table 9-2). Thus, sensors that are used in atmospheres rich in carbon dioxide, such as those found in brewing and wine-making industries or in spaces where carbon dioxide has been used as an

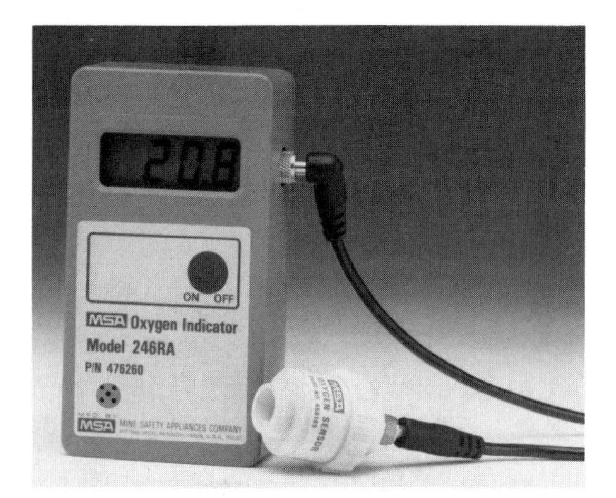

Figure 9-8. Oxygen meter reading adjusted in clean air. (MSA)

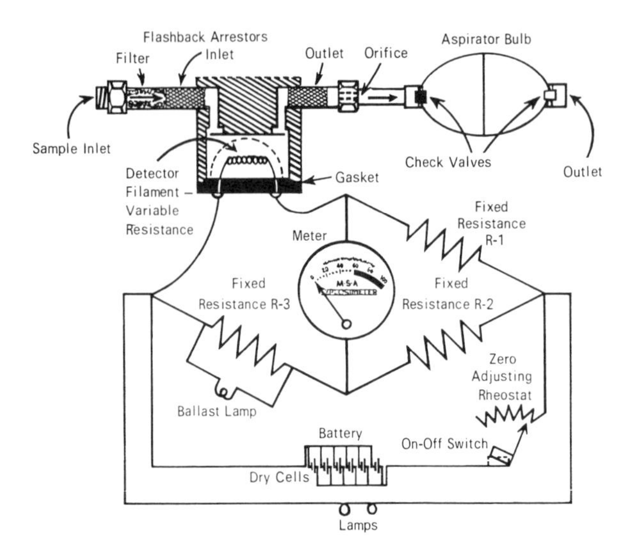

Figure 9-9. Functional drawing of a combustible gas meter. (MSA)

Table 9-2. Effect of carbon dioxide on oxygen sensor life

Carbon Dioxide Concentration (%)	Sensor Life (Days)
100	3
5	50
1	100

Source: MSA, Pittsburgh PA.

inerting agent for fire control or hot-work, may need to be replaced more frequently than sensors used in other areas.

COMBUSTIBLE GAS MONITORS

Operational Theory

Electronic combustible gas meters have been in existence since about the 1930s and their operating principle has changed little since that time. Most operate on the principle of detecting the change in resistance that occurs when combustible gases and vapors react with a hot catalyst-coated filament, but a few devices employ metal oxide semiconductor detectors.

A functional diagram of a typical heated-filament type of combustible gas meter is shown in Figure 9-9. The meter employs a Wheatstone Bridge circuit which consists of four resistors arranged in a diamond configuration. Voltage is applied across two opposite points of the diamond and a meter is connected across the remaining two points. If the resistance of all four legs of circuit is the same, the meter reads zero. However, a change in the resistance in any leg causes an up-scale meter deflection that is proportional to the degree of resistive change.

In the combustible gas meter, a heated filament is substituted for one of the resistors. The earliest units used a bare filament; however, later designs incorporated a catalyst-coated filament or catalytic bead (Figure 9-10) which provides the same degree of combustion efficiency at lower operating temperatures. The catalytic sensor (Figure 9-11) actually employs two of these beads. Only one actually senses combustible gases, the other is used to compensate for changes in ambient temperature.

The instrument is first electrically zeroed to compensate for the initial resistance offered by the heated filament, and air is then drawn into the filament chamber. Flammable gases and vapors in the air are ignited on the bead and the resulting increase in filament resistance produces an upscale meter reading. This reading is proportional to the concentration of the flammable gas or vapor present in the air.

While the instruments are capable of detecting a wide variety of flammable gases and vapors, they are non-specific. This should not be surprising since all combustible gases produce heat when they burn, and it is this heat of combustion that causes the upscale

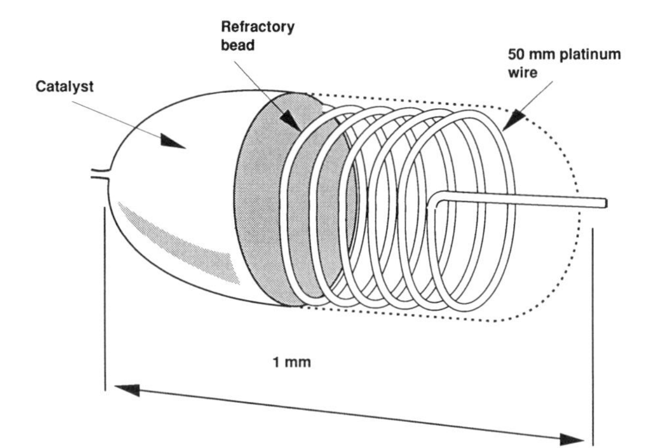

Figure 9-10. Catalytic filament bead. (Paul Trattner)

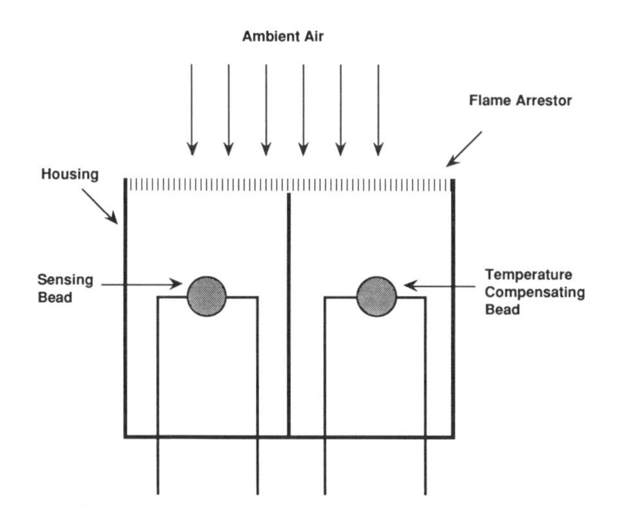

Figure 9-11. Catalytic combustion dual-bead sensor. (Paul Trattner)

meter deflection. For this reason, combustible gas meters are unable to differentiate between contaminants in atmospheres that contain a mixture of combustible gases and vapors.

Since combustible gas meters are factory-calibrated for a single gas, usually methane or pentane, the meter will only be accurate when measuring the specific gas for which it was calibrated. For example, consider a meter calibrated for pentane. Since pentane has a lower explosive limit (LEL) of 1.5%, the meter will read full-scale (100% of the LEL) in atmospheres containing 1.5% pentane. Pentane concentrations of 0.15% and 0.75%, which are 0.1 and 0.5 of the LEL, would produce meter readings of 10 and 50% of the LEL, respectively.

But if measurements were being made for another contaminant, say xylene, then the meter reading would not necessarily indicate the true percentage of the LEL. In fact, significant differences may be observed. It is possible to compensate for these differences by using calibration correction charts or tables provided by some manufacturers in their instruction manuals.

Consider, for example, the family of curves shown in Figure 9-12. Note that a meter reading of 10% LEL for ethyl alcohol (substance 1) indicates an actual concentration of only about 7.6%. In this case, the measurement errs in the direction of safety since the meter indicates a higher concentration than actually present.

But contrast this situation with that of JP4 (substance 6). In this case, the instrument reads only 6% when the actual concentration is 10%! Since entry into atmospheres of less than 10% of the LEL are acceptable, workers could enter a hazardous atmosphere thinking it was safe.

Calibration corrections may also be made by using a set of correction factors like those shown in Table 9-3. If the contaminant of interest is isobutyl acetate a correction factor of 1.5 is used. To determine the actual concentration of isobutyl acetate, the operator would multiply the reading on the meter by 1.5. In other words, a meter reading of 7% would be multiplied by 1.5 to obtain an actual concentration of 10.5%.

Note that again in this case the meter reading indicates a level that is "safe" for entry from a fire and explosion perspective. However, the actual level of isobutyl acetate exceeds 10% of the LEL. It should also be noted that the factors listed in Table 9-3 are for a *specific manufacturer's* instrument and cannot be applied to instruments made by other manufacturers.

Because the detection principle is based on combustion, there are two other limitations that users must be aware of when using these instruments. First, there must be enough oxygen in the space to support combustion. Second, the meter may not be accurate in atmospheres above the upper explosive level (UEL).

Meters that use the catalytic-combustion method of detection will not provide an accurate reading in atmospheres containing less than 10% oxygen. In fact, the readings obtained in oxygen-deficient atmospheres will be lower than actual because combustion in a low oxygen atmosphere will result in less heat

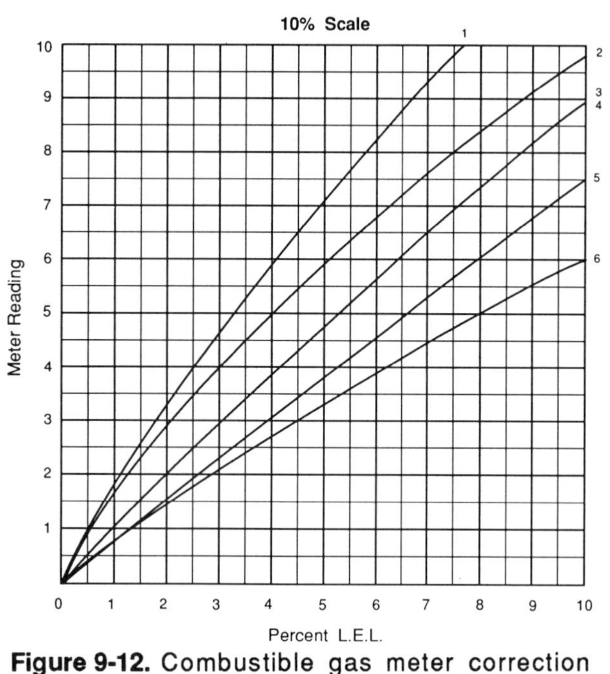

No.	Name	Formula	L.E.L.
1	Ethyl Alcohol	C_2H_6O	3.3
2	Styrene	C_8H_8O	1.1
3	Pentane	$C_5H_{12}O$	1.5
4	Gasoline Leaded - Use Pt. No. 47740 Inhibitor Filter		1.4
5	Acetone	C_3H_6O	2.6
6	JP4 Fuel		1.3

Figure 9-12. Combustible gas meter correction curves. (Paul Trattner)

being produced. Oxygen-enriched atmospheres pose a different problem. An increase in oxygen may produce combustion temperatures that exceed the safe working limits of the flame arrestor and the meter itself could become an ignition source. These problems can addressed by always testing for oxygen first. Once it is established that sufficient oxygen is present, the atmosphere can be evaluated for combustibles.

Measurements in atmospheres above the upper explosive level may also produce misleading results. As shown in Figure 9-13, measurements in atmospheres above the upper explosive level result in the meter rising, going off scale, and then returning to zero or close to zero. The initial rise is caused by the combustible gas entering and mixing with oxygen present in the sensor block. However, as the gas level equilibrates at the upper explosive level, combustion ceases because the mixture is too rich to burn. As a result, the display falls to zero.

It is essential that operators be aware of this phenomenon and constantly monitor the reading as it is displayed. If an operator's attention is diverted, he may miss the rise and fall of the meter and mistakenly determine that a space whose atmosphere is above the upper explosive level is "safe" for entry. As air diffuses into the space over time, the atmosphere will enter the explosive range. Should a source of ignition be present, the gas will ignite explosively.

Table 9-3. Combustible gas meter correction factors

Substance	Correction Factor
Acetone	1.1
Acetylene	0.7
Acrylonitrile	0.8
Benzene	1.1
1,3-Butadiene	0.9
n-Butane	1.0
n-Butanol	1.8
Carbon disulfide	2.2
Cyclohexane	1.1
2,2 Dimethylbutane	1.2
2,3 Dimethylpentane	1.2
Ethane	0.7
Ethyl acetate	1.2
Ethyl alcohol	0.8
Ethylene	0.7
Formaldehyde	0.5
Gasoline	1.3
Heptane	1.1
Hexane	1.3
Hydrogen	0.5
Isobutane	0.9
Isobutyl acetate	1.5
Isopropanol	1.1
Isooctane	1.1
Methane	0.5
Methanol	1.1
Methylcyclohexane	1.1
Methyl ethyl ketone	1.1
Methyl isobutyl ketone	1.2
Mineral spirits	1.1
VM&P naphtha	1.6
n-Pentane	1.1
Propane	1.0
Propylene	0.8
Styrene	1.9
Tetrahydrofuran	1.1
Toluene	1.1
Vinyl acetate	0.9
Xylene	1.2

Adapted from Microguard Instruction Manual (MSA).

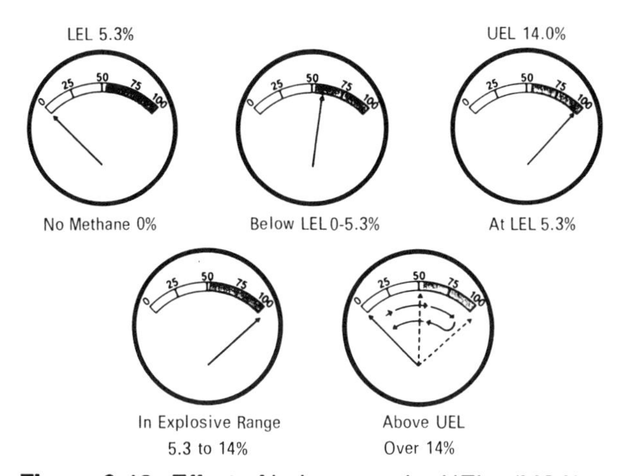

Figure 9-13. Effect of being over the UEL. (MSA)

Health Hazard Considerations

To provide a margin of safety from potential fire and explosion hazards, spaces containing more than 10% of the LEL must not be entered. However, since the LEL for many materials is well above the concentration which could pose a health hazard, exposure to as little as 10% of the LEL could have serious health-related consequences. In other words, a toxic atmosphere may exist at vapor concentrations that are considered "safe" from a fire and explosion perspective. This can be clearly seen by reviewing the data presented in Table 9-4.

Using carbon disulfide (CS_2) as an example, we find that the lower explosive level is 1.3%. Since 1% by volume is the same as 10,000 parts per million (ppm), 1.3% equates to 13,000 ppm. Thus, 10% of the 13,000 ppm LEL is the same as 1,300 ppm. However, OSHAs health standard for employee exposure to CS_2 is only 20 ppm. That means that at 10% of the LEL, the air would exceed the OSHA permissible exposure limit (PEL) by 62 times.

Field Calibration Check

The calibration of combustible gas meters may be verified in the field by introducing a gas of known concentration and checking the upscale deflection of the meter. Small disposable cylinders of calibration gases (Figure 9-14) ranging from about 20 to 60% of the LEL are available from instrument manufacturers and specialty gas supply houses. While methane and pentane are most frequently used calibration gases, suppliers of specialty gases can blend mixtures of other materials.

Care and Maintenance

Because of the simplicity of their design, combustible gas meters tend to be rather rugged. However, the filament wires can break if the instrument is subjected to a sharp blow, such as being dropped. Repeated exposure either to atmospheres within the explosive range or to oxygen-enriched atmospheres can shorten the filament life. The catalyst applied to combustible gas meter filaments can also be "poisoned" or inactivated by a variety of materials such as lead, phosphorous, silicates and silicones. Once poisoned, the detector filament loses sensitivity and will not function properly, resulting in inaccurate measurements.

Care should be taken whenever catalytic combustion instruments are used in areas where catalyst poisons may be encountered. Silicates and phosphorous compounds may be found in hydraulic fluids and lubricating greases. Silicones are used in some floor waxes, furniture polishes and products used in the rubber and plastics industry. Airborne lead may exist as an artifact of hot-work performed on surfaces coated with lead-based paint, and in gasolines containing tetraethyl lead anti-knock agents.

INSTRUMENTS FOR TOXIC GAS AND VAPORS

For our purposes, toxic gases and vapors are those other than simple asphyxiants, which have an adverse affect on the body. At a minimum, they include materials that have an OSHA PEL or an ACGIH Threshold Limit Value. Some chemical manufacturers have also established their own in-house exposure standards for toxic materials that are not listed by ACGIH or OSHA. Since many chemicals are hazardous at concentrations that cannot be detected by smell, instruments play a vital role in identifying hazardous conditions that might otherwise go unnoticed.

The two types of instruments most commonly employed in evaluating toxic atmospheres in confined spaces are colorimetric indicating tubes and direct-reading, sensor-based gas monitors. However, portable infrared analyzers and gas chromatographs may also be used in some circumstances.

Table 9-4. Comparison of lower explosive level with the OSHA permissible exposure limit

Contaminant	LEL % Vol	ppm LEL	ppm 10% LEL	ppm PEL	Times over the PEL
Carbon disulfide	1.3	13,000	1,300	20	62
Cyclohexane	1.3	13,000	1,300	300	4
Ethyl alcohol	3.3	33,800	3,380	1,000	3.3
n-Hexane	1.1	11,000	1,100	50	22
Toluene	1.2	12,000	1,200	100	12
Xylene	1.1	11,000	1,100	100	11

PEL is the OSHA Permissible Exposure Limit expressed as an 8-hour, time-weighted average.

Figure 9-14. Calibration field-check of combustible gas meter. (MSA)

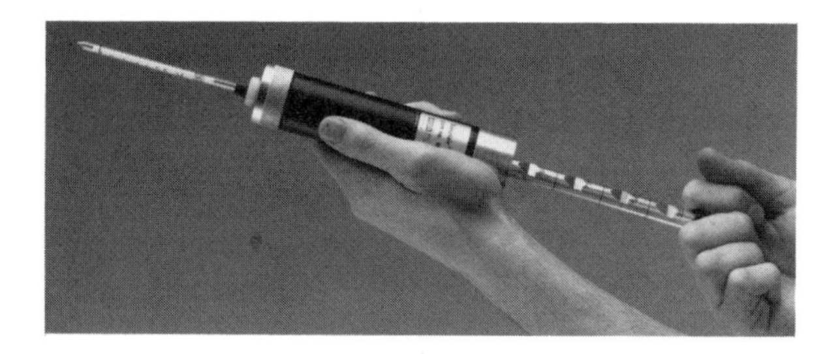

Figure 9-15. Piston-type detector tube pump. (MSA)

DETECTOR TUBES

Detector tubes are glass tubes about five inches long and an eighth of an inch in diameter that are filled with a granular chemical or chemically treated material. The sealed ends of the tube are broken off and air is drawn through the packing by means of a manually operated piston (Figure 9-15) or bellows pump (Figure 9-16). The contaminant of interest reacts with the solid sorbent in the tube to produce a color change whose length is directly proportional to the contaminant concentration (Figure 9-17). In other words, the greater the contaminant's concentration, the longer the length of the color stain. Since the tubes are scribed with gradations, the degree of contamination can be determined by noting the length of the stain. The gradations marked on most tubes allow the concentration to be read directly (Figure 9-18). However, some tubes are marked with a millimeter scale that requires the user to interpret the concentration using a table on the tube's instruction sheet.

Detector tubes can be used to measure about 200 different contaminants, and a partial listing of materials for which tubes are available is provided in Table 9-5. The principal advantages of detector tubes are that they require little skill to use, are relatively inexpensive and provide a quick means of quantification. However, their apparent simplicity can be very deceptive, and unless users are fully aware of a tube's specific operating limitations, the measurements obtained may not be correct. Specific limitations for

Figure 9-16. Bellows-type detector tube pump. (National Dreager)

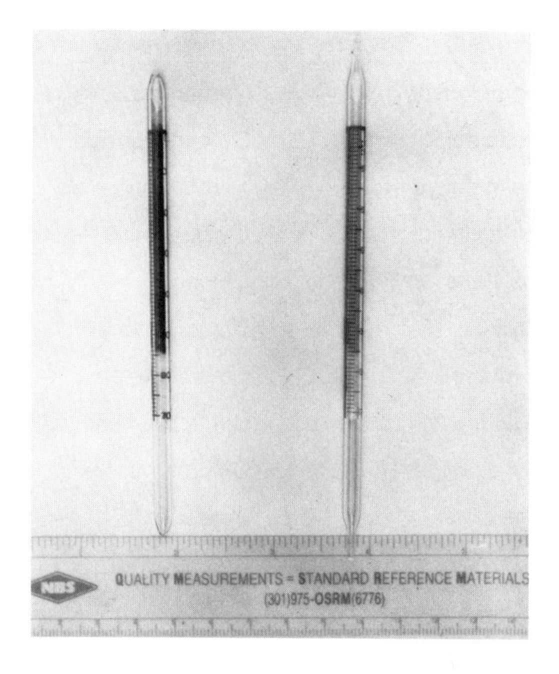

Figure 9-18. Comparison of new and used detector tubes. Note that the packing on the used tube on the left is much darker than the unused tube on the right. (John Rekus)

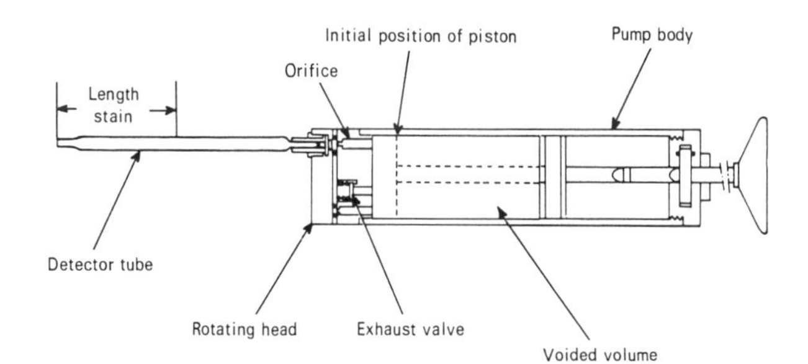

Figure 9-17. Operation of detector tube system. (MSA)

each substance are discussed in great detail on individual instruction sheets enclosed with each box of tubes.

Detector Tube Certification

At one time, NIOSH certified detector tubes for about two dozen materials. Certified tubes were required to have an accuracy of ±25% when tested at 1–5 times the permissible exposure limit, and ±35% when tested at levels of one-half the standard. However, in 1982, NIOSH announced that it was suspending the certification program and had no intention of reinstating it. The resulting gap in quality assurance was filled in 1987 when the Safety Equipment Institute, a non-profit organization funded by manufacturers of safety equipment, started its own certification program. Testing is performed by AIHA-accredited labs using the NIOSH protocols, and quality assurance audits are made of participating members' manufacturing facilities. A current list of all certified detector tubes can be obtained from the Safety Equipment Institute, 1901 North Moore Street, Arlington, VA 22209, phone (703) 528-2148.

Use and Limitations

Detector tubes are usually packaged ten to a box. Included with each box is a detailed instruction sheet similar to those shown in Figures 9-19 and 9-20. The instruction sheet includes essential information that must be reviewed to determine if there are any special conditions that influence the use of the equipment. For example, some pumps use different size orifices to establish the flow rate required for a specific tube. In order to obtain the proper flow rate, and hence an accurate reading, the correct orifice setting must be selected.

Environmental conditions such as temperature, pressure and relative humidity may also affect measurement accuracy, so instruction sheets often contain

Table 9-5. Selected substances for which detector tubes are available

Acetaldehyde	Carbon dioxide	Ethanolamine	Methylene chloride
Acetic acid	Carbon disulfide	2-Ethoxyethanol	Methyl ethyl ketone
Acetic anhydride	Carbon monoxide	Ethylacetate	Methyl hydrazine
Acetonitrile	Carbon tetrachloride	Ethylacralate	Methyl isobutyl ketone
Acetylene	Chlorine	Ethyl amine	Methyl mercaptan
Acrolein	Chlorine dioxide	Ethyl benzene	Methyl methacrylate
Acrylonitrile	Chlorobenzene	Ethyl bromide	Morpholine
Allyl alcohol	Chlorobromomethane	Ethylchloride	Nitric acid
Allyl chloride	Chloroform	Ethylenediamine	Nitrogen oxides
Ammonia	Cresol	Ethylene glycol	Octane
n-Amyl acetate	Cumene	Ethyleneimine	Ozone
n-Amyl alcohol	Cyanide	Ethylene oxide	Pentachloroethane
Amyl butrate	Cyclohexane	Ethyl ether	Pentane
Amyl chloride	Cyclohexanol	Ethyl mercaptan	Phosgene
Amyl formate	Cyclohexanone	Formaldehyde	Phosphine
Amyl propionate	Cyclohexene	Formic acid	*n*-Propyl alcohol
Aniline	Cyclohexylamine	Furfural	*n*-Propyl acetate
o-Anisidine	Diborane	*n*-Heptane	Propyleneimine
p-Anisidine	Diacetone alcohol	*n*-Hexane	Propylene oxide
Benzene	Diborane	Hexone	Pyridine
Benzoic acid	*o*-Dichlorobenzene	Hydrazine	Styrene
Benzyl acetate	*p*-Dichlorobenzene	Hydrogen bromide	Sulfur dioxide
Benzyl alcohol	Dichlorodifluoromethane	Hydrogen sulfide	Toluene
Benzyl chloride	1,1-Dichloroethane	Isoamyl acetate	1,2,4-Trichlorobenzene
Bromine	1,2-Dichloroethane	Isobutyl acetate	1,1,1-Trichloroethane
Butadiene	1,2-Dichloroethylene	Isobutyl alcohol	1,1,2-Trichloroethane
1-Butanol	Diethanolamine	Isophorone	Trichlorofluoromethane
2-Butanol	Diethylamine	Isopropyl alcohol	Trimethylamine
2-Butanone	Diethylenetriamine	Methyl alcohol	Vinyl acetate
2-Butoxyethanol	Diethyl ether	Methyl acrylate	Vinyl chloride
n-Butylacetate	Epichlorohydrin	Methyl bromide	Vinylidene chloride
n-Butyl amine	Ethyl alcohol	Methyl cyclohexanol	Xylene

GASTEC

HALOGENATED HYDROCARBON TUBE CALIBRATED FOR TRICHLOROETHYLENE AND PERCHLOROETHYLENE (TETRACHLOROETHYLENE) HIGH RANGE TUBE NO. 132HA

The Gastec Detector Tube No. 132HA provides a rapid and fully quantitative analysis of the concentration of TRICHLOROETHYLENE in air with an accuracy tolerance of ±25% at 1, 2, and 5 times TLV and ±35% at 1/2 TLV-TWA utilyzing the Gastec Multi-Stroke Gas Sampling Pump.

PERFORMANCE:

Calibration Scale	50—500 ppm (based on 1 pump stroke)		
Measuring Range	20—200 ppm	50—500 ppm	130—1300 ppm
Number of Pump Strokes	2	1	1/2
Multiplication Factor	Tube reading ÷ 2.5	Tube reading × 1	Tube reading × 2.6
Detecting Limit*	4 ppm	—	—
Sampling Time	30 seconds per pump stroke		
Color Change	Yellow — Purple		
Shelf Life	2 years		

*The minimum detectable concentration.

MEASUREMENT PROCEDURE:

1. Break tips off a fresh detector tube by bending each tube end in the tube tip breaker of the pump.
2. Insert the tube securely into the rubber inlet of the pump with the arrow on the tube pointing toward the pump.
3. Make certain the pump handle is all the way in. Align the guide marks on the shaft and pump body.
4. Pull the handle all the way out until it locks on 1 pump stroke (100 ml). Wait 30 seconds until staining stops.
5. Read concentration at the interface of the stained-to-unstained reagent.
6. If the stain exceeds the highest calibration mark by 1 pump stroke sampling, use 1/2 stroke sampling (50 ml) in which case the true concentration is obtained by multiplying the tube reading by 2.6.
7. If the stain does not attain the first calibrtation mark or for more accurate measurement up to 100 ppm, repeat one more pump stroke and devide the tube reading by 2.5.

CORRECTION FOR TEMPERATURE, HUMIDITY AND PRESSURE:

Calibration of the Gastec Detector Tube No. 132HA is based on a tube temperature of 20°C (68°F) and not the temperature of the gas being sampled, approximately 50% relative humidity, and normal atmospheric pressure.

1. For temperature other than 20°C (68°F), tube reading must be corrected according to the following Temperature Correction Factor:

Temperature	0°C (32°F)	10°C (50°F)	20°C (68°F)	30°C (86°F)	40°C (104°F)
Factors	1.4	1.15	1	0.8	0.65

2. No correction is required for relative humidity range of 0～90%.
3. Tube reading is proportional to absolute pressure. To correct for pressure, multiply by

$$\frac{760}{\text{Atmospheric Pressure (mmHg)}}$$

DETECTION PRINCIPLE:

Trichloroethylene is decomposed by nasent oxygen generated by oxidizing agent to liberate hydrogen chloride, which discolors Hammett indicator (4-phenylazodiphenilamine) to purple.

$$CHCl{:}CCl_2 \xrightarrow[H_2SO_4]{PbO_3} HCl$$

$$HCl + \text{Indicator} \longrightarrow \text{Violet Reaction Product}$$

INTERFERENCES:

Interferents	Result
Perchloroethylene, Chloroprene, Dichloroethylene, Methyl bromide and other Halogenated hydrocarbons, Hydrogen chloride, Hydrogen fluoride, Nitric acid and other acid gases and vapors.	Produce similar stain by themselves and if coexisted they give plus error on tube reading.

DANGEROUS AND HAZARDOUS PROPERTIES:

Threshold Limit Value-Time Weighted Average by ACGIH (1986): 50 ppm (7—8 hours)
Threshold Limit Value-Short Term Exposure Limit by ACGIH (1986): 200 ppm (15 minutes)

APPLICATION FOR OTHER GASES:

Gastec Detector Tube No. 132HA can also be used for detection of PERCHLOROETHYLENE (TETRACHLOROETHYLENE) in air. Measureing ranges and correction factors are as folows:

Measuring Range	20—200 ppm	50—500 ppm	150—1500 ppm
Number of Pump Strokes	2	1	1/2
Multiplication Factor	Tube reading ÷ 2.5	Tube reading × 1	Tube reading × 3
Detecting Limit*	4 ppm	—	

Temperature Correction Factor

Temperature	0°C (32°F)	10°C (50°F)	20°C (68°F)	30°C (86°F)	40°C (104°F)
Factors	1.95	1.35	1	0.75	0.65

STORE TUBES BELOW 0°C, OTHERWISE SHELF LIFE WILL BE SHORTENED.
SEE OPERATING INSTRUCTIONS INCLUDED WITH THE GASTEC MULTI-STROKE GAS SAMPLING PUMP. PUMP.

Manufacturer: Gastec Corporation, Yokohama, Japan
86J-132HA-1

Printed in Japan

Figure 9-19. Detector tube instruction sheet. (Gastec)

MSA Detector Tubes

instructions for detecting Alcohols using detector tube part no. 95097 Lot No. 53

1. **WARNING.** *These instructions are applicable for use with the Universal Tester™ Pump Pt. No. 83499 or the Samplair® Pump Pt. No. 463998 only. When using these tubes with any other MSA® sampling device, use the instructions and calibration values provided with that device. These tubes are not calibrated for use with other than MSA sampling devices.*

2. **Storage.** Do not store at temperatures above 90°F. Protect from prolonged exposure to light. Either of these conditions could cause a change in the sensitivity and accuracy of the tubes.

3. **Shelf Life.** If stored properly, these tubes can be used up to 30 months from date of manufacture. They must be used prior to the expiration date stamped on the box.

4. **Chemical Reaction and Color Change.** This detector tube provides a quantitative method for estimating the concentration of alcohols in air. Indication is based on the reduction of potassium dichromate by the alcohols. The color change is from yellow to green. Furfuryl alcohol generally results in a partly dark brown, partly green stain.

5. **Test Procedure.**
   - 5.1 Check pump for leaks in accordance with the instruction manual for the pump.
   - 5.2 When using the Universal Tester Pump, set the rotating pump head on #2 index mark. With the Samplair Pump, no indexing is necessary.
   - 5.3 Remove one detector tube from box and break off both tips using the tube breaker hole in the head of the pump.
   - 5.4 Insert the detector tube into the rubber tube holder of the pump, with the arrow on the tube pointing toward the pump.
   - 5.5 Refer to the calibration chart and decide on the number of pump strokes to be taken. This is dependent on the compound of interest and the expected concentration.
   - 5.6 Align the index marks on the handle and back plate of the pump.
   - 5.7 Face the mounted tube into the atmosphere to be tested. Take an appropriate volume of air to be sampled by pulling the pump handle out the required number of strokes. Wait a minimum of 50 seconds for each full pump stroke for the evacuated pump to fill. If alcohol is present, a greenish stain develops in the detector chemical. The stain may have a light end point when first formed, but in a few minutes intensity increases producing a distinct end point. As mentioned, furfuryl alcohol gives a partly brown, partly green stain.
   - 5.8 To determine the concentration, read the length of stain in millimeters directly from the tube. If the end of stain should be uneven, read at the position of the average length of stain. Then refer to the Calibration Chart. Read opposite the compound of interest and the number of pump strokes used, the concentration in ppm corresponding to the measured length of stain in millimeters.

NOTE: If the stain length is too short, additional pump strokes may be taken immediately and the concentration read opposite the compound of interest and the **total** number of pump strokes taken.

6. **Limitations and Corrections.**
   - 6.1 Interferences. Other common gases in the range of their allowable limits do not interfere with accurate measurements, although strong reducing agents could interfere by increasing the length of stain.
   - 6.2 Temperature. Where required, temperature correction factors are given on the Calibration Chart. In order to avoid correction factors at test conditions colder than 60°F., it is recommended that the tube be kept in an inner pocket prior to use and warmed with the hand during test.
   - 6.3 Relative Humidity. Moist air causes an increase in stain length. At 80% relative humidity, multiply the stain length by 0.8 and then convert this corrected stain length to concentrations in ppm.
   - 6.4 Pressure. Calibrations are made at 740 mm mercury. A pressure correction must be used for pressures greatly different from this. To correct for pressure, use the following formula:

$$\text{corrected reading, ppm} = \text{actual reading, ppm} \times \frac{740}{\text{test pressure in mm mercury}}$$

7. **Measurement range.** This tube measures the alcohols listed in the Calibration Chart over the concentration ranges listed there.

8. **Calibration and Accuracy.** Each lot of tubes is separately calibrated. This instruction sheet should be used only with the designated lot of detector tubes.

9. **Remote Sampling Procedure.**
   - 9.1 Remove the metal connecting tube from the spare parts vial and insert one end into the sampling line and the other end into the rubber tube holder on the pump.
   - 9.2 Follow the directions under Paragraph 5, Test Procedure, inserting for step 5.4: "Insert the detector tube into the other end of the sampling line, with the arrow on the tube pointing in the direction of the pump."

10. **Lot No.** 53

CALIBRATION CHART

For the various temperatures, where a correction is required, multiply the test stain length by the correction factor and then convert this corrected length of stain to ppm concentration.

	Number of Strokes	Light Figures Below Indicate Length of Stain in Millimeters. Bold Figures Below Indicate Concentration in Parts Per Million.									°F. Temperature Correction Factors
n-Amyl Alcohol	5	0 mm **0**	4 **50**	6 **100**	9 **200**	12 **400**	14 **600**	17 **1000**	20 **2000**		@ 110° x 0.9
	10	0 mm **0**	5 **25**	6 **50**	10 **100**	14 **200**	18 **400**	21 **600**	24 **1000**	29 **2000**	
Iso-Amyl Alcohol	5	0 mm **0**	5.5 **50**	7 **100**	10 **200**	12 **300**	14 **400**	16 **500**	18 **750**	20 **1000**	
	10	0 mm **0**	8 **50**	11 **100**	16 **200**	19 **300**	23 **400**	27 **500**	32 **750**	37 **1000**	
sec-Amyl Alcohol	5	0 mm **0**	4 **25**	5.5 **50**	7 **100**	10 **200**	15 **400**	18 **600**	25 **1000**	33 **2000**	
	10	0 mm **0**	5 **25**	7 **50**	11 **100**	16 **200**	19 **300**	23 **400**	29 **600**	40 **1000**	

Figure 9-20. Detector tube instruction sheet. (MSA)

tert. Amyl Alcohol	5	0 mm 0	4 50	6 100	10 200	13 300	15 400	19 800	27 1000	41 2000	
	10	0 mm 0	5.5 25	7 50	11 100	16 200	20 300	24 400	29 600	41 1000	
2-Butoxy Ethanol (butyl cellosolve)	10	0 mm 0	4 30	5 50	5.5 100	6 175	8 375	10 650	11 900		
n-Butyl Alcohol	3	0 mm 0	5 50	6 100	8 200	11 400	14 600	17 1000	22 2000	29 4000	@ 110° x 0.9
iso-Butyl Alcohol	3	0 mm 0	5 50	6 100	9 200	13 400	16 600	20 1000	28 2000	37 4000	@ 110° x 0.8
sec-Butyl Alcohol	3	0 mm 0	4 50	6 100	8 200	13 400	17 700	20 1000	28 2000	37 4000	
tert.-Butyl Alcohol	10	0 mm 0	13 100	19 200	24 300	28 400	34 600	40 800	46 1000		@ 110° x 0.9
Cyclohexanol	10	0 mm 0	4 25	5 50	7 100	10 200	13 300	16 500	18 750	20 1000	@ 110° x 0.9
2-Ethoxyethanol (cellosolve)	10	0 mm 0	1.5 50	3 100	4 200	5.5 400	6.5 700	7 1000			@ 110° x 0.8
Ethyl Alcohol	3	0 mm 0	8 200	12 500	15 1000	20 2000	29 5000	40 10000			@ 110° x 0.9
Furfuryl Alcohol	10	0 mm 0	3.5 25	4.5 50	6.5 100	8 200	9 300	11 400	12 500		
Methyl Alcohol	3	0 mm 0	4.5 100	8 200	11 400	15 1000	19 2000	28 5000	37 10000		@ 110° x 0.8
2-Methyl Cyclohexanol	5	0 mm 0	2 25	3 50	5.5 100	8 200	12 400	14 600	15 800		
	10	0 mm 0	3 25	6 50	10 100	15 200	19 400	21 600	22 800		
Methyl Isobutyl Carbinol (methyl amyl alcohol)	5	0 mm 0	4 25	5 50	7 100	9 200	11 300	15 500	19 700	24 1000	
	10	0 mm 0	5.5 25	7 50	11 100	15 200	18 300	26 500	33 700	43 1000	
n-Propyl Alcohol	3	0 mm 0	6 100	9 200	13 400	17 700	18 1000	23 2000	33 5000	41 10000	@ 110° x 0.9
iso-Propyl Alcohol	3	0 mm 0	6 100	8 200	11 400	15 700	17 1000	22 2000	33 5000	40 10000	

TAL 185 (L) REV. 3 　　　　 894872

Figure 9-20. Continued.

correction factors that must be applied when sampling conditions exceed certain limits. While instruction sheet formats vary among manufacturers, most will discuss the following information.

Component Substitution

Detectors tubes and pumps are designed, manufactured and calibrated as a unit. Since the accuracy of the system depends on the suction characteristics of the pump, tubes from one manufacturer cannot be used with another's pump.

Accuracy

Even SEI certified detector tubes do not provide a high level of accuracy, and field measurements may vary by as much as 25% for contaminant concentrations at the threshold limit value (TLV) or PEL. That means that measurements of a contaminant whose actual concentration is 100 ppm could be off by much as 25 ppm.

Failure to consider this variation could mislead a user to believe that the atmosphere was suitable for entry when it was actually hazardous.

Shelf-Life

Detector tubes have a limited shelf-life that begins at the date of manufacture. Shelf-life typically ranges from about 12 to 36 months and boxes are marked with an expiration date. Tubes that have passed their expiration date may be used for classroom training and demonstrations, but they must not be used for actual field measurements.

Storage Recommendations

The useful life of detector tubes may be reduced if they are exposed to high temperatures, and tube manufacturers generally specify an acceptable ambient temperature range for storage. In a few cases they recommend that specific tubes be stored under refrigerated conditions. For example, note that the tube manufacturer whose trichloroethylene instruction sheet is shown in Figure 9-19 recommends that they be stored at 0°C.

Measurement Range

The range of contaminant concentrations over which tubes may be used is usually reported in parts per million. The measurement range may also be a function of the number of pump strokes. For example, notice that the halogenated hydrocarbon tube instruction sheet (Figure 9-19) indicates three measurement ranges—20–200 ppm, 50–500 ppm and 130–1,300 ppm—are possible depending on the number of pump strokes.

Pump Strokes

The length of the color stain is affected by the volume of contaminated air that is drawn through the tube. In order to accurately relate the stain length to the concentration, a specified air volume must be sampled. Since pumps draw a preset volume with each stroke, increasing the number of strokes increases the air volume drawn through the tube. Since different air volumes may be required for different chemical reactions, the number of strokes may also vary from tube to tube or substance to substance, as indicated by the alcohol instruction sheet shown in Figure 9-20.

Time Per Stroke

Since each pump stroke draws a specified volume of air through the tube, sufficient time must be allowed to assure that air completely fills the evacuated pump. If too short a time is allowed, the required volume of air will not be drawn through the tube and the reading obtained will be lower than actual. The exact sampling time per stroke is a function of the pump orifice size and the resistance to air flow offered by the tube packing, and sampling times frequently range from 30 and 60 seconds per stroke.

Note that the sampling time in the two examples are 30 seconds for the trichloroethylenes and 50 seconds for the alcohols. As a convenience, some pumps have self-contained end-of-stroke indicators that visually signal when the next stroke may be made.

Reaction Principle

This section describes the chemical reaction that is employed to produce the color change. It also usually indicates the original color of the tube and the expected color change. Note that in our examples the instructions indicate a change from yellow to green for the alcohol tube and from yellow to purple for the trichloroethylenes. Other color changes could easily be from white to blue, yellow to brown, red to black and so forth.

Correction Factors

Since detector tubes are calibrated under specified conditions of pressure, temperature and relative humidity, use under conditions different from those for which the tubes were calibrated may produce inaccurate results. Tubes that are affected by environmental variations will be accompanied by tables of correction factors that are used to adjust the field reading up or down. For example, a correction factor of 1.2 may be applied when a specific tube is used at temperatures between 80 and 90°F. If the reading on the tube indicated a concentration of 100 ppm, the actual concentration would be 1.2 times higher than that indicated, or 120 ppm.

In our examples, correction factors ranging from 1.4 at 0°C (32°F) to 0.65 at 40°C (104°F) must be applied to the halogenated hydrocarbon tube, and at temperatures greater than 110°F some alcohol readings must be corrected by 0.8 to 0.9. Note too that both of our examples indicate that corrections are necessary for changes in atmospheric pressure that vary from 760 mm Hg for the Gastech tube and 740 mm Hg for MSAs.

Interferences

Detector tubes may be subject to interference from materials other than the species of interest. The presence of these materials may result in a lengthening or shortening of the stain length, or cause the color reaction to fade. Tubes may also be cross-sensitive to other compounds, resulting in a masking effect.

For example, the instruction for Gastech's trichloroethylene tubes list chloroprene, dichloroethylene, methyl bromide, hydrogen chloride and nitric acid among the interferents which produce a stain similar to that of trichloroethylene.

Leak and Flow Testing

Samples of solvent vapors, acid gases and oxidizers that are drawn into a pump may cause wear and leakage by damaging elastomeric parts. For this reason, detector tube manufacturers recommend that all pumps be leak-tested before each use. While the specific leak-testing procedure varies depending on the type of pump, it typically involves operating the pump while blocking the air inlet with an unopened tube. After a prescribed period of time the pump is checked for signs of leakage. The handles of piston-type pumps should return to their approximate starting position when released, and squeeze-bulb on bellows-type pumps should remain collapsed.

In addition to the pre-use leak check, manufacturers also recommend that their pumps be periodically flow-tested to assure that the air flow through the pump is maintained at the specified rate. The flow check is usually made using a soap bubble flow meter. Failure to perform either the pre-use leakage check or periodic flow-rate check can lead to incorrect results in the field.

Field Use

It should be clear from the discussion above that while there is a generic similarity among detector tube equipment, there are many differences between the individual tubes. It is critical that anyone who intends to use detector tubes to evaluate confined spaces, reads and understands all of the manufacturer's instructions for the specific detector tube system that they are using. Users should also be cautious about being lulled into a false sense of security that may result from the misguided belief that because they are familiar with one style of pump, they can use others. Each system has unique features and limitations that must be understood in order to obtain valid results.

Remote Sampling

Most manufacturers have an accessory that allows remote sampling of confined spaces. Accessory kits typically include an adapter that allows a piece of plastic tubing to be attached to the end of the pump. A detector tube which is fitted into the end of the tubing can then be inserted into the space via a manhole or other opening. When using remote sampling attachments, it is critical that the detector tube be placed on the end of the accessory tubing, rather than the other way around. If the flexible tubing is attached ahead of the detector tube, it will trap some of the air and the required sample volume will not be drawn through the tubes. However, if the detector tube is placed on the end of the plastic tubing, the full volume of air will be drawn through the tube.

Sampling Protocol

The following protocol outlines one approach that can be used when sampling with detector tubes.

1. Identify what contaminants are likely to be present in the space. Don't forget to consider trace materials or impurities that may be present in addition to the principal substances of interest.
2. Consult the instruction sheets for the substances of interest and determine if there are any interferences or environmental correction factors that must be considered.
3. Evaluate the environmental conditions in the space and determine if sampling conditions will necessitate the use of correction factors. If more than one contaminant is present, determine whether or not interference will occur.
4. Perform the leak test in accordance with the pump manufacturer's recommendation.
5. Break off the ends of the sample tube and insert it into the pump. Detector tubes are marked with an arrow that indicates the direction of air flow, and they must be inserted into the pump in the proper direction.
6. Operate the pump, allowing the full time specified on the instruction sheet for each pump stroke.
7. Apply any correction factors and consider the limits of accuracy in interpreting the sampling results. Remember that the actual concentration of gases and vapors in the space may vary between 25–35% of the reading indicated by the tube.

SENSOR BASED INSTRUMENTS

Sensor based instruments are available in dozens of makes and models, but all are functionally similar. As shown in Figure 9-21, air enters the instrument either passively by diffusion or actively by means by a small battery-powered vacuum pump. Next, the

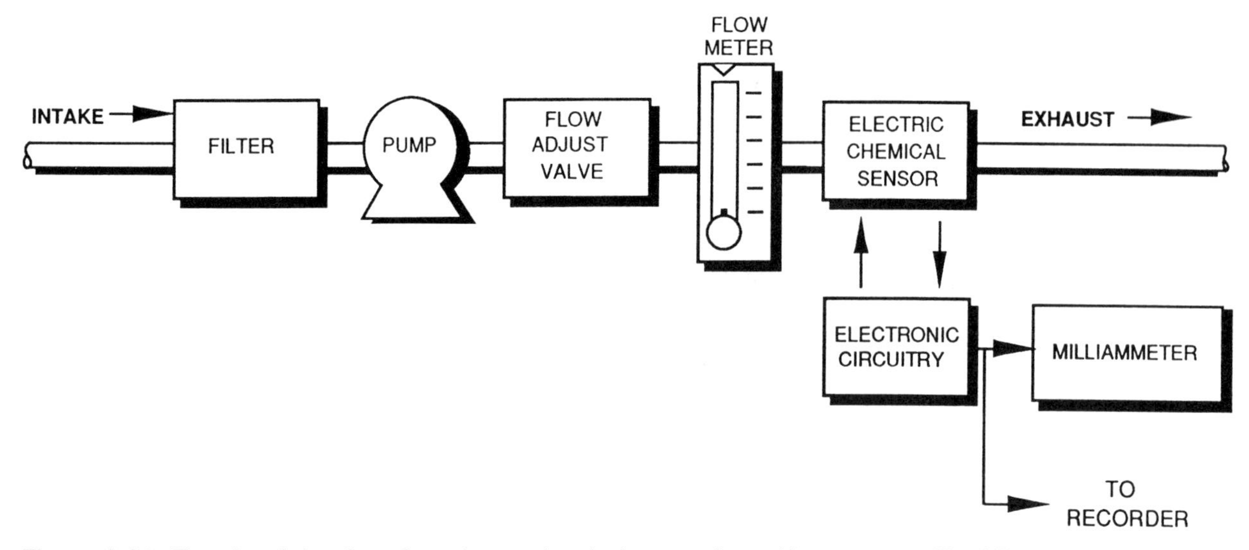

Figure 9-21. Functional drawing of an electro-chemical sensor based instrument. (Paul Trattner)

gases of interest interact with sensor elements which produce an electrical signal that is proportional to the gas concentration. Finally, the signal is processed and displayed directly on a meter or digital readout.

Instruments are available in either single gas or multiple gas styles. Dual-gas devices that allowed measurement of both oxygen and combustible gases have been available for many years (Figure 9-22). More recently, however, the capabilities of these units have expanded to include measurement of toxic gases (Figure 9-23). The two gases most commonly chosen are carbon monoxide (CO) and hydrogen sulfide, (H_2S), but some instrument manufacturers offer sensors that can measure other materials.

Electro-Chemical Sensors

The galvanic oxygen cell previously described is one example of an electro-chemical sensor. However, sensor elements are also available for measuring toxic gases such as carbon monoxide, hydrogen sulfide, nitrogen dioxide, sulfur dioxide and chlorine. Electro-chemical cells are usually assembled in small plastic cylinders that are opened on one end (Figure 9-24). Inside the cylinder are two electrodes that are surrounded either by a liquid, a gel-like fluid, or a porous, liquid-impregnated solid. After assembly, the open end of the cell is covered with a gas-permeable membrane. A polarizing voltage is established between the two electrodes so that one is negatively charged and the other positively charged. When the contaminant gas of interest flows into the sensor, it produces an oxidation-reduction reaction which causes electrons to flow between the two electrodes. The flow of electrons establishes a current that is proportional to the toxic gas concentration. Thus, higher gas concentrations produce greater currents. The current flow is subsequently converted to a voltage which can be read on a meter or digital display.

Metal Oxide Semiconductor Sensors

Metal oxide semiconductor (MOS) sensors are solid state devices (Figure 9-25) that consist of a pellet or film of metal oxides, usually a proprietary mixture of oxides of iron, tin and zinc, which is embedded with a noble metal heating wire. The heating wire raises the temperature of the semiconductor material and electrodes in contact with the surface of the MOS material measure its resistance. An electrical resistance baseline is established once oxygen absorbed on the MOS surface equilibrates with oxygen in the oxide mix. However, when contaminant gases react with absorbed oxygen on the sensor's surface, they change the surface oxygen concentration, lowering the surface electron charge. This lowering of surface charge greatly affects the resistance of the semiconductor, and changes of only a few parts per million of many reducing gases will cause a significant lowering in resistance. Absolute selectivity is unattainable in semiconductor sensors since they respond to a variety of gases. However, sensor fabrication techniques, coupled with the carefully formulated oxide blends and appropriately selected operating

Figure 9-22. Dual-gas instrument for oxygen and combustibles. (MSA)

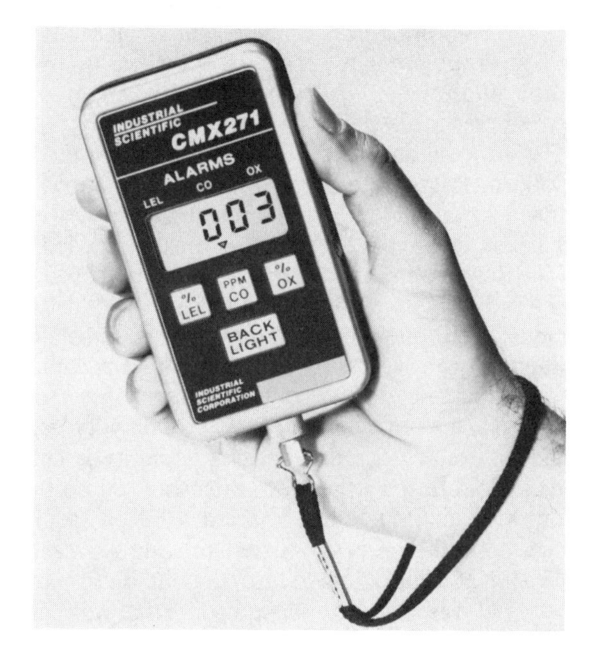

Figure 9-23. Three-gas instrument for oxygen, combustibles and a toxic gas such as CO or H_2S. (Industrial Scientific)

temperatures, can enhance the response of some gases over others, producing an acceptably low response to interferences in many applications.

Some manufacturers use MOS devices as the sensing element for combustible gas meters. MOS sensors in this application have an advantage over catalytic filaments in that they operate at lower oxygen levels. MOS sensors are very reliable and much less prone to being poisoned by lead and silicon containing materials than are catalytic filament detectors.

Figure 9-24. Typical electro-chemical cell. (Paul Trattner)

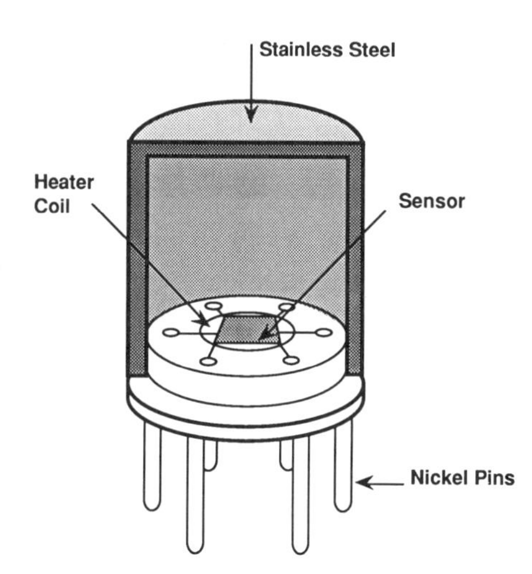

Figure 9-25. Metal oxide semiconductor. (Paul Trattner)

Field Calibration Check of Sensor Based Instruments

Field calibration checks of sensor based instruments are usually performed by first "zeroing" the instrument in uncontaminated air, then introducing a calibration gas of known concentration. The upscale meter deflection is adjusted to the value of the calibration gas with a "span" control and the instrument is then rezeroed. This procedure of zeroing and spanning is repeated until the meter reads correctly at both zero and the upscale calibration checkpoint.

The principal advantage that electro-chemical sensors offer over detector tubes is that they can monitor an environment continuously and indicate changes as they occur. Some sensor based instruments are also equipped with alarms that sound at a preset contaminant level, while others contain internal memory circuits that electronically record the sampling results. The stored data can be downloaded to a computer or printed out as hard copy.

This feature can greatly reduce the time needed to prepare an entry permit since testers don't have to perform the laborious task of transcribing dozens of measurements onto the permit. They can simply download the instrument's internal memory and attach a printout of the data to the permit.

PORTABLE GAS ANALYZERS

While it is theoretically possible to design sensors that respond to any toxic gas, commercially available devices are limited to only a handful of contaminants. However, toxic gases can also be measured with portable direct-reading instruments that incorporate infrared spectrophotometry and gas chromatography. Both of these methods have been used for many years in analytical laboratories to identify and quantify chemical agents. Now, thanks to advances in miniaturization and solid state circuitry, instruments which formerly occupied an entire laboratory bench have been scaled down to battery-powered units that can easily be carried into the field.

Although the initial purchase price of these instruments is higher than sensor-based instruments, they are capable of measuring hundreds of contaminants. This versatility makes them very attractive choices for certain applications. For example, a contract tank-cleaning firm that encounters different toxic materials every day, may find that an instrument capable of near real-time measurement of many hazardous materials provides a level of flexibility that permits the contractor to respond more quickly to customer needs. Similarly, chemical plants or petroleum refineries where confined space entries are performed on a regular basis may find these instruments to be cost effective because they can be used for other purposes such as routine industrial hygiene surveys and leak detection around flanges, valves and fittings.

Infrared Analyzers

Infrared analyzers are capable of detecting most organic compounds and a few inorganic materials. They operate on the principle that gases and vapors characteristically absorb infrared energy between specific wavelengths. For example, as shown in Table 9-6 alcohols strongly absorb infrared energy between 2.8 to 3.1 micrometers, while aromatic compounds like benzene, toluene and xylene will absorb more strongly between 6.15 to 6.35 micrometers.

Since small variations exist in the absorption spectra of similar compounds of the same chemical family, selectivity can be achieved by choosing an analytical wavelength corresponding to a strong characteristic absorption peak of the contaminant of interest. The desired sensitivity can be achieved by panel control adjustments of the optical path length and electronic response time.

Early model analyzers suffered from three drawbacks which severely limited their application in confined space work. First, the selection of operating parameters required a high degree of user knowledge and skill. At a minimum, instrument operators had to have a rudimentary understanding of the principles of infrared spectrophotometry. They also had to be able to interpret various charts and tables that specified operating parameters and control settings.

Second, instruments had to be individually calibrated for each compound that was going to be sampled. Calibration involved injecting microliter volumes of the compound of interest into a special closed-loop calibration system and relating the resulting vapor concentration to the amount of infrared energy that was absorbed—a procedure that was both complicated and time consuming.

Third, since the analyzers were not approved for use in hazardous locations, they could not be used in places where ignitable concentrations of gases and vapors were likely to exist. This precluded their use in many petroleum and chemical operations.

Today, however, there is at least one commercially available infrared instrument that overcomes these obstacles (Figure 9-26). The unit incorporates microprocessor technology which removes many of the previously encountered operator burdens. Rather than consulting tables and manually setting instrument controls, the operator simply uses a touch sensitive keyboard to enter a six character alpha-numeric code corresponding to the contaminant of interest. For example, the substance codes for acetone, isopropyl alcohol and sulfur dioxide would be entered as "ACETON," "IPA" and SO2. Once the contaminant code is entered, the microprocessor automatically sets

Table 9-6. Specific infrared absorption bands

Functional Group	Absorption Band (micrometers)
Alkanes	3.35 to 3.65
Alkenes	3.25 to 3.45
Alkynes	3.05 to 3.25
Aromatic (hydrocarbons)	3.25 to 3.35
Aromatic (substituted benzenes)	6.15 to 6.35
Alcohols	2.80 to 3.10
Organic Acids	5.75 to 6.00
Aldehydes	6.60 to 6.90
Ketones	5.60 to 5.90
Esters	5.75 to 6.00
Chlorinated compounds	12.80 to 15.50

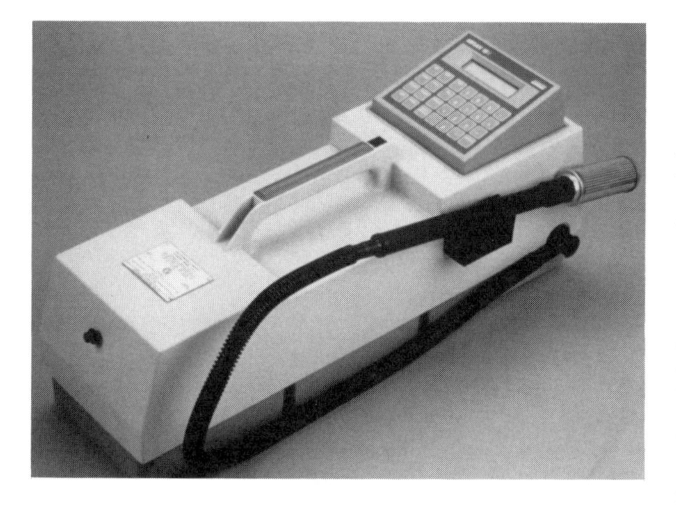

Figure 9-26. Infrared spectrophotometry. (Foxboro)

the sampling parameters for wavelength and path length based on information contained in an internal library of chemicals.

The internal library contains information on 116 materials for which there is an OSHA Permissible Exposure Limit or an ACGIH Threshold Limit Value. The library also contains spare storage space that accommodates up to 10 additional user-selected substances. The specific analytical wavelengths selected for use in the library were generally chosen to maximize analytical sensitivity for the substance of interest, while at the same time minimizing interferences from other vapors that are commonly present in the air. For example, ambient air always contains water vapor and carbon dioxide so whenever possible, the analytical wavelength was selected to minimize interference from these materials.

Factory calibration for materials in the internal chemical library provides an accuracy of ±15%, but better accuracy can be achieved by custom calibration by the user. The required calibration equipment is available from the manufacturer as an accessory.

The instrument has been performance tested and approved for use in Class I Division 1 (Groups B, C, D) hazardous locations. That means it can be used in atmospheres containing ignitable concentrations of any gas except acetylene.

The unit can operate for one hour on its internal battery pack, and an external battery is available that provides up to four hours of use. The analyzer contains no user-serviceable parts and does not require routine cleaning or lubrication. Except for the battery pack, all parts have an indefinite storage life.

Gas Chromatographs

Gas chromatographs (Figure 9-27) operate on the principle that gases and vapors of different chemical species will travel through a semi-solid medium at different speeds. The heart of the chromatograph is a piece of tubing filled with a special packing. The tubing is wound into a coil referred to as the column.

A sample of air is introduced at one end of the column and the contaminant gases and vapors are propelled through the solid packing by a carrier gas like hydrogen or nitrogen. As air moves through the column, some contaminants travel faster than others. As a result, they begin to separate from each other much like runners in a marathon (Figure 9-28).

Faster molecules reach the end of the column sooner than slower molecules, and when they arrive a detector notes their presence. Since the retention time on the column is characteristic for each contaminant, different materials can be identified by the elapsed time from injection to detection. The quantity of the material is determined by the magnitude of an electric current generated by the detector.

Figure 9-29 indicates that portable chromatographs like that shown in Figure 9-30 employ the same operating principles as fixed laboratory bench units. Although they are transportable, many have not yet reached the same level of operational simplicity as infrared analyzers, and in some cases, their use still requires a high level of technical knowledge

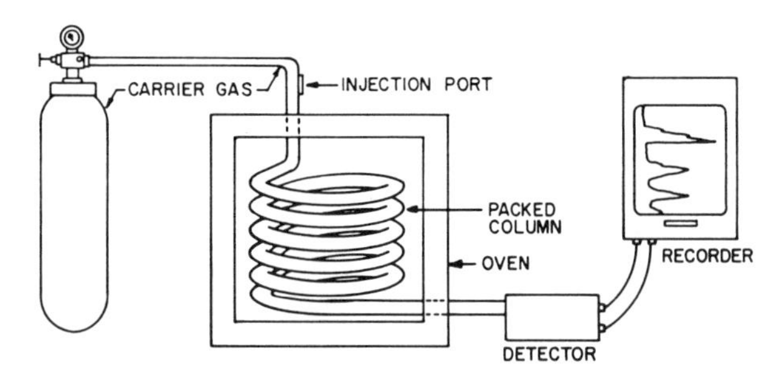

Figure 9-27. Functional drawing of a gas chromatograph. (NIOSH)

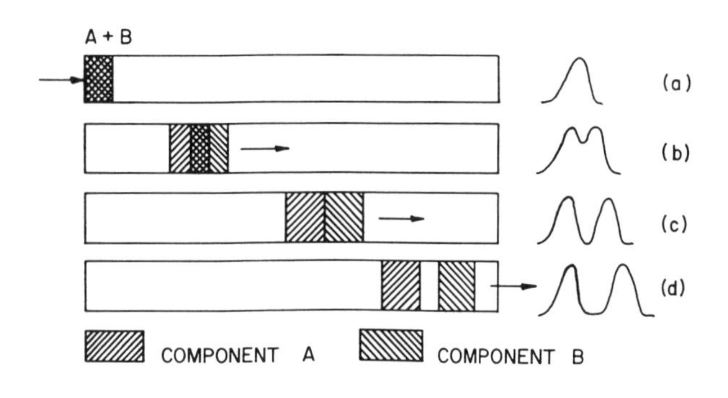

Figure 9-28. Separation of components within the column. (NIOSH)

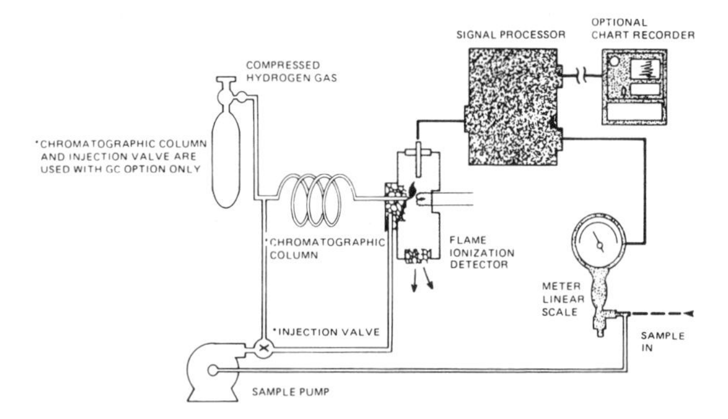

Figure 9-29. Functional drawing of a field-portable chromatograph. (Foxboro)

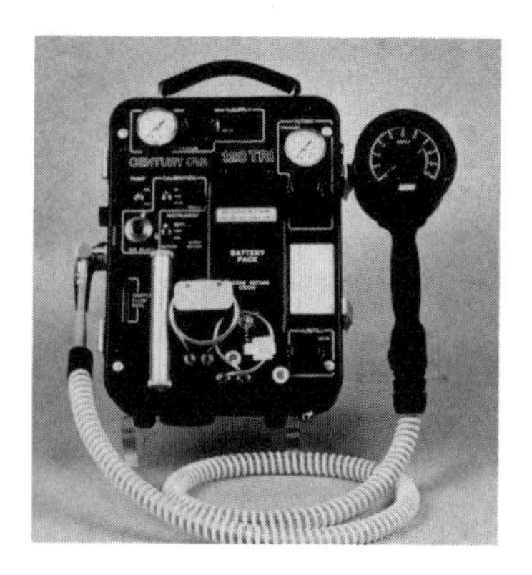

Figure 9-30. Field-portable chromatograph. (Foxboro)

Figure 9-31. Hand-held portable gas chromatograph. (Photovac)

and skill. One exception is a handheld unit which has only recently become commercially available (Figure 9-31). The unit weighs only a few pounds and is intrinsically safe. The unit is fairly easy to operate but requires a different operating module for different classes of compounds such as alcohols, aromatics, ketones, etc.

Most chromatographs also require regular maintenance that includes cleaning filters and refilling an internal compressed gas cylinders. Accuracy may vary between 10 and 20% depending on the operating temperature and the scale selected for measurement.

UNDERSTANDING INSTRUMENT SPECIFICATIONS

Technical specifications outlined in instruction manuals and product literature describe an instrument's capabilities and limitations. Although specifications may appear confusing and cryptic, it is essential for you to understand them in order to select the most appropriate instrument for your needs. For example, if you plan on making measurements in steam tunnels where temperatures usually exceed 110°F, you better select an instrument that is capable of operating at that temperature.

At a minimum, purchasers should evaluate the following eight areas of performance:

- Accuracy of reading
- Environmental operating range
- Intrinsic safety
- Interferences
- Warm-up time
- Response time
- Zero and span drift
- Detectable limits

An explanation of these specifications is provided below, but not all manufacturers publish this information in their product literature or instrument manuals. That means that you may have to *specifically* request it. You should also carefully consider the professional and legal implications of purchasing instruments for which you do not obtain detailed specifications.

Instrument Accuracy

One of the most important considerations in selecting any instrument is its accuracy. Recall that accuracy is a measure of how close the instrument's reading is to the actual contaminant concentration. For example, a measurement of 100 ppm made with an instrument with an accuracy of ±10% could be off 10 ppm in either direction. In other words, the *actual concentration* could be as high as 110 ppm or as low as 90 ppm. For our purposes, this 20 ppm spread between the high and low value will be called the margin of error.

Methods of Reporting Accuracy

Specifications for most direct-reading instruments report accuracy as a function of the full-scale reading. For example, the accuracy of an oxygen meter may be reported as ±2% of full scale, usually abbreviated as FS. Some manufacturers, however, report accuracy as a function of the meter reading, or as a function of the calibration range. The differences between these reporting methods may be significant, and a careful review of the specifications may suggest that a particular instrument may not be suitable for some applications.

Interpreting Accuracy Specifications

Interestingly, two instruments with the same specified accuracy may have different margins of error. Consider, for example, two carbon monoxide meters, both of which have an accuracy of 3%; however, the full-scale limits are 100 ppm for one, and 500 ppm for the other. Even though both instruments have the same level of *accuracy,* 3%, the one with the 100 ppm scale has a *margin of error* of 6 ppm, vs. 30 ppm for the one with the 500 ppm scale. This smaller margin of error provides the advantage of reducing the level of uncertainty about the results of measurements.

Using Accuracy Specifications

Careful reading of an instrument's specifications may suggest that it may not be suitable for confined space applications. For example, published specifications for at least two commercially available multigas handheld carbon monoxide instruments indicate a full scale range 0–1,000 ppm with an accuracy of ±5% FS. In other words, instrument readings may vary as much as ±50 ppm from the true value. This is particularly interesting since the PEL and TLV for CO are 50 ppm and 25 ppm, respectively. This means that at a CO concentration of 50 ppm, the meter could indicate 0 ppm even though the CO level exceeded both legal (OSHA) and recommended (ACGIH) exposure standards!

Resolution

Resolution is the smallest number that can be presented by an instruments display. Although resolution is often reported in the same format as accuracy, for example "Range of 0 to 999 ppm Resolution ± 1 ppm," resolution is not the same as accuracy. To understand why, consider the odometer in your car. The smallest number that be displayed by most odometers 0.1 miles, so the odometer has a resolution of 0.1 mile. But this doesn't tell you anything about the accuracy of the measurements made by the odometer. For example,if you have to drive 0.3 of a mile before the odometer rolls over, your measurement is not very accurate even though your display shows distances down to 0.1 of a mile.

Operating Environment

Instruments are affected by environmental conditions and they will only provide an accurate reading if they are operated within prescribed limits for temperature and humidity.

Operating Temperature

As shown in Table 9-7, operating temperatures vary among manufacturers and the selection of an appropriate instrument depends on the temperatures to which it may be subjected during use. It should also be noted that even though an instrument may function at low temperatures, liquid crystal displays will be slower to respond and battery life will be reduced.

Relative Humidity

As shown in Table 9-7 some instrument specifications identify operating range in terms of relative humidity (RH). However, relative humidity is just that—relative—*and it is relative to the ambient temperature.*

Since warm air can hold more water vapor than cold air, air with a 95% RH at 120°F contains more moisture than air with a 95% RH at 32°F. Since relative humidity specifications are based on room temperature, humidity effects become an important consideration when measurements are made in steam tunnels or hot humid process areas. Measurements made outdoors may also merit consideration, especially in areas that are very "dry" such as northern or southwestern states, or very "humid" such along the southern seaboard and gulf coast.

Intrinsic Safety

The National Electrical Code® (NEC) defines "hazardous" or "classified" locations as those where ignitable atmospheres are likely to be present. It also requires that any electrical equipment used in a Class I hazardous location be explosion-proof, purged-and-pressurized, or intrinsically safe. As a practical matter, the intrinsically safe approach is generally the method of choice for portable instruments.

Classes of Locations

Hazardous atmospheres are divided into three classes on the basis of physical properties of the contaminant. Class I locations contain flammable gases and vapors, Class II combustible dusts, and Class III ignitable fibers and flyings.

Divisions Within Classes

Each group is further divided into two divisions. Divisions for Class I and II materials are established on the basis of whether the hazardous condition exists under normal conditions (Division 1), or exist under "unusual" conditions (Division 2). Divisions for Class III substances are determined on the basis of whether fibers and flyings are in suspension—Division 1, or stored Division 2.

Groups Within Classes

Class I and II materials are also placed into groups based on similarities of their physical characteristics and flammable properties. As shown in Table 9-8, Class I atmospheres are divided into four groups A, B, C and D.

The principal factor used to establish group assignment for Class I materials is the Maximum Experimental Safe Gap (MESG) for the material of interest. The MESG is the gap between two plates of metal, placed edge to edge, below which a burning mixture on one side of the plates will not propagate to the other side. If the experimental safe gap is greater than that of gasoline (0.029 inches), the material is classified as Group D. Similarly, if the experimental gap falls between 0.029 and 0.012 inches—the MESG for ethyl ether—it is classified as Group C. Materials with MESGs between 0.0012 and 0.003 are not as easy to classify and the decision to place them in Group A or Group B depends on their explosion pressure. Those with an explosion pressure similar to hydrogen are placed in Group B. However, as a practical matter, acetylene is the only material in Group A.

Class II combustible dusts are classified as Groups E, F or G depending on their ignition temperature and conductivity.

Group E includes combustible metal dust—such as aluminum, magnesium or zirconium regardless of their resistivity—and other combustible dusts that have a resistivity of less than 10^2 ohm-cm.

Group F atmospheres are those that contain carbon black, charcoal, coal and coke dusts which have more than 8% total volatile material as determined by ASTM 3175-82. Atmospheres containing dusts with a resistivity between 10^2 and 10^8 ohm-cm which are sensitized by other materials so that they present an explosion hazard are also classified as Group F.

Table 9-7. Environmental operating ranges for selected multi-gas, handheld instruments

	Operating Temperature Range		% Relative Humidity (non-condensing)
	°C	°F	
Manufacturer No. 1	-15 to 50	5 to 122	0 to 99
Manufacturer No. 2	-10 to 120		0 to 95
Manufacturer No. 3			15 to 90
Continuous	-15 to 40	5 to 104	
Intermittent	-20 to 55	-4 to 131	
Manufacturer No. 4			10 to 90
Combustibles	-35 to 55	-31 to 131	
Oxygen	0 to 40	32 to 104	
When calibrated at temperature of use:	- Low limit: 18°C (0°F) - High limit: 50°C (122°F)		

Group G atmospheres are those containing combustible dusts having a resistivity greater than 10^8 ohm-cm.

Class III ignitable fibers and flyings such as lint, cotton dust and wood chips are not divided into groups.

Intrinsically Safe Equipment

An intrinsically safe device is one which will not release sufficient energy either in normal operation or under fault conditions to ignite the most easily ignitable concentration of a particular gas or vapor. In evaluating a specific instrument's claim of intrinsic safety, it is important to differentiate between equipment that is intrinsically safe *by design* and equipment that is intrinsically safe *by certification.*

Intrinsically safe by design means that the manufacturer claims that the instrument incorporates certain construction methods that will limit the release of heat and electrical energy. On the other hand, equipment *certified intrinsically safe* has been submitted to an independent testing laboratory that evaluates the instrument against established criteria and subjects its circuits to proof-tests in specified atmospheres. Following the evaluation and test, the instrument is deemed to be acceptable for use in specified hazardous atmospheres. The atmospheres for which the instrument is approved are indicated on a label like that shown Figure 9-32.

Instruments may be used only in those atmospheres for which they are certified, so an instrument approved only for use in atmospheres containing Groups C and D materials could not be used in atmospheres containing acetylene, butadiene and hydrogen. Intrinsic safety approval is also based on a design that incorporates specific components, and any circuit modifications or substitution of parts automatically voids the approval. For example, rechargeable batteries could not be substituted for alkaline batteries unless approval was granted for both.

It is also important to note that while an instrument itself may be intrinsically safe, accessories such as battery chargers and external sample-draw pumps may not be. If they are not, then they cannot be used in hazardous locations.

Finally, instruments approved for use in hazardous locations must be marked with the safe operating temperature or a code identification number as shown in Table 9-9.

Interferences

All gas detection systems will be affected to some degree by contaminants other than that for which the system is calibrated. Contaminants other than the species of interest that influence the value of the measurement are called interferences. As noted earlier in this chapter, interferents that cause an instrument reading to be higher than actual are called positive interferents, while those that cause readings lower than actual are called negative interferents.

Interferents may be reported in a variety of ways, two examples are provided in Tables 9-10 and 9-11. When choosing an instrument, look for a low degree of interference from other substances whose presence is anticipated in the confined space.

Warm Up Time

The warm up time is the elapsed time necessary for an instrument to meet its performance specifications after being turned off for at least 24 hours.

Table 9-8. NEC Group Classification

Group A	Group D
Acetylene	Acetone
	Acronitrile
	Benzene
	Butane
Group B	Cyclohexane
Butadiene	Ethyl alcohol
Ethylene oxide	Gasoline
Hydrogen	Heptane
Propylene oxide	Hexane
	Isopropyl acetate
	Isopropyl ether
	Methane
Group C	Methanol
Acetaldehyde	Methyl ethyl ketone
Allyl alcohol	Octane
n-Butyl aldehyde	Pentane
Crotonaldehyde	Pentanol
Cyclopropane	1-Pentane
Diethyl ether	Propane
Diethylamine	Propanol
Epichlorohydrin	Propylene
Ethylene	Styrene
Ethylenimine	Toluene
Methyl ether	Turpentine
2-Nitropropane	Vinyl chloride
Nitromethane	Xylene

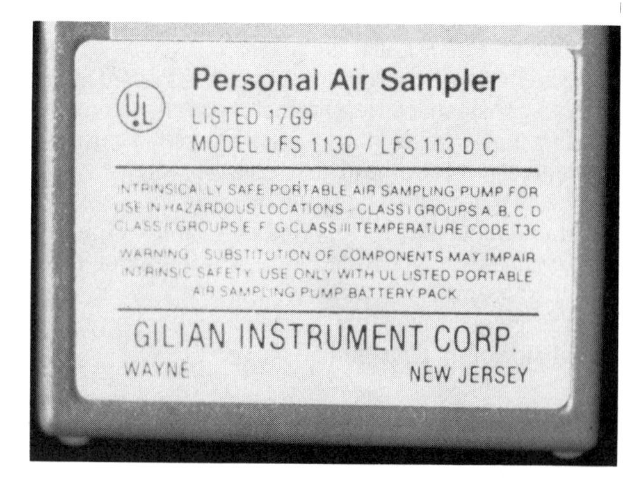

Figure 9-32. This label indicates that the instrument is certified as intrinsically safe. (John Rekus)

Table 9-9. Temperature identification

Identification Number	Maximum (°C)	Temperature (°F)
T1	450	842
T2	300	572
T2A	280	536
T2B	260	500
T2C	230	446
T2D	215	419
T3	200	392
T3A	180	356
T3B	165	329
T3C	160	320
T4	135	275
T4A	120	248
T5	100	212
T6	85	185

Source: NFPA-70

Instruments with short warm up times offer the practical advantage of minimizing the time an operator must wait before making a measurement.

Response Time

Air sampling instruments do not provide measurement results instantaneously. Instead, there is a slight delay encountered as the air sample makes its way to the sensor. Then once it arrives, there is another delay as the sensor reacts to the substance of interest. The elapsed time from the start of the measurement until a reliable reading can be obtained is called the response time. If an instrument has a 30-second response time, the operator must wait at least that long before he can be confident that the reading is correct. Instruments with fast response times can offer long-term financial savings because they minimize the time needed to evaluate a space.

It is important to note that different sensors in the same instrument may have different response times. For example, an oxygen sensor could have a response time of 10 seconds, while the combustible gas and toxic gas sensors in the same instrument may have response times of 30 and 45 seconds, respectively. Typical response times for various types of sensors are shown in Figure 9-33.

Instrument Drift

Drift is a measure of how stable an instrument is over time. Instruments with a low level of drift are

Table 9-10. Cross-sensitivities expressed as a percentage of primary sensitivity when exposed to 100 ppm of selected interfering gases

Test Gas	Sensor				
	CO	H_2S	NO_2	SO_2	Cl_2
CO	100	<7	0	0	0
H_2S	<10	100	-20	125	-20
SO_2	<10	<20	<-0.5	100	-0.5
NO	<30	0	0	0	0
NO_2	<15	-20	100	-125	120
Cl_2	N/D	-20	90	-60	100
H_2	<40	<5	0	0	0
HCN	N/D	0	<1	N/D	<1
HCl	N/D	N/D	N/D	N/D	N/D
C_2H_2	N/D	0	0	<1	N/D

Adapted from Industrial Scientific Product Specification for TMX 410 multi-gas monitor.

more stable and consequently require less frequent re-adjustment. 2 types of drift are generally specified.

Zero drift is the change in a meter's output response over a stated period of unadjusted continuous operation when the input gas concentration is zero.

Span drift is the change in a meter's output response over a stated period of unadjusted continuous operation when the input gas concentration is about 50–75% of the meter's full-scale range.

Detectable Limit

The detectable limit provides an indication of how sensitive an instrument is by specifying the lowest gas concentration that can be measured reliably.

INSTRUMENT FEATURES AND OPTIONS

Although there is a generic similarity among handheld, direct-reading gas meters, different instrument manufacturers offer a variety of features from which to choose. Some of the generally available options are discussed below. It should be noted, though, that all options are not offered by every manufacturer.

Sample-Draw vs. Diffusion

Portable direct-reading instruments may operate in either the sample-draw or diffusion mode. Sample-draw devices use either a squeeze-bulb aspirator or a battery-powered pump to mechanically induce air into sensors (Figure 9-34). On the other hand, sensors of instruments operating in the diffusion mode are exposed passively to air (Figure 9-35).

Some diffusion style instruments can be fitted with an accessory that also allows them to operate in the sample-draw mode. The accessory consists of an aspirator bulb or battery-powered pump that is connected to a cover which attaches over the sensor's protective grill. A piece of tubing can be affixed to the bulb or pump and inserted into the space, allowing air to be drawn into the sensors.

When sample-draw device accessories are used, they must be attached tightly enough to prevent leakage. The instrument's instructions should also be consulted to determine the number of aspirator bulb strokes that are required to draw the sample into the sensing element. In some cases as many as 20 or 30 strokes may be required before a reading can be obtained. If the number of pump strokes is not specified, a general rule of thumb is to allow one aspirator squeeze for every foot of sample line attached.

Response times are influenced not only by whether the instrument is operated in the diffusion or sample-draw mode, but also on whether sample-draw is provided by an aspirator bulb or motor-driven pump. An illustration of response time variations in different modes for one manufacturer's instrument is provided by Table 9-12.

Table 9-11. Effects of selected interferences on H_2S and CO sensors

Interferent Gas	Interferent Conc. (ppm)	Sensor Response (ppm)	
		H_2S	CO
Ammonia	100	0	-4
Benzene	17.7	0	0
Carbon dioxide	5,000	0	-4
Carbon disulfide	14.5	0	2
Carbon monoxide	100	0	0
Chlorine	5	0	2
Dimethyl sulfide	4.5	0	100
Ethylene	50	0	-2
Freon 12	1,000	0	-2
Hexane	500	0	70
Hydrogen	500	1	30
Hydrogen cyanide	42	0	170
Isopropyl alcohol	50	0	40
Ethyl mercaptan	44	3	6
Methyl mercaptan	5	5	7
Methane	50,000	0	-3
Methanol	50	0	130
Nitric oxide	100	2	260
Nitrogen dioxide	100	-8	80
Sulfur dioxide	150	5	30

Adapted from MSA Product Specification for Models 360 and 361.

Filters and Liquid Traps

Suspended dusts or liquid residues found in some spaces can damage sensor elements on contact. Instruments can be protected from these insults by installing accessory in-line filters and liquid traps which prevent harmful materials from getting into the sensors.

Sample Lines and Probes

Lengths of flexible tubing and rigid extender probes like those shown in Figure 9-36 may be attached to sample-draw devices to allow testing of remote areas. Some tubing materials may absorb gases and vapors. Others may be adversely affected by certain chemicals and will deteriorate over time. To avoid these problems, only sample lines made of materials recommended by the instrument's manufacturer should be used.

Since long lines increase the time required for the air sample to reach the sensors, tubing lengths should be kept as short as practical. The temperature of the tubing should also be at or above the temperature in the space. If it is cooler than the atmosphere being tested, vapors may condense in the line resulting in inaccurate measurements.

Alarm Function

Some instruments are equipped with audible alarms that will sound at pre-set levels. Alarm set-points can often be selected by the user; however. typical factory settings are:

- 10% of the LEL for combustibles
- 23.5 and 19.5% oxygen (high- and low-level)
- The PEL or TLV for specific toxic gases

Alarms may be either latching or non-latching. Non-latching alarms are activated only as long as the set-point concentration is reached. Latching alarms, on the other hand, stay on until they are manually reset by the operator.

Multi-Gas Sensing

As mentioned previously, multi-gas instruments are typically designed to sense oxygen, combustibles, and one or more toxic gases. The method used to select gas modes varies among instruments. Some require the user to manually select a specific gas function by using switches or a touch sensitive pad. The principal drawback to these devices is that they permit only a single gas to be monitored at a time. Other instrument designs permit measurement of multiple gas functions simultaneously. Some of these instruments provide a continuous display of all gas

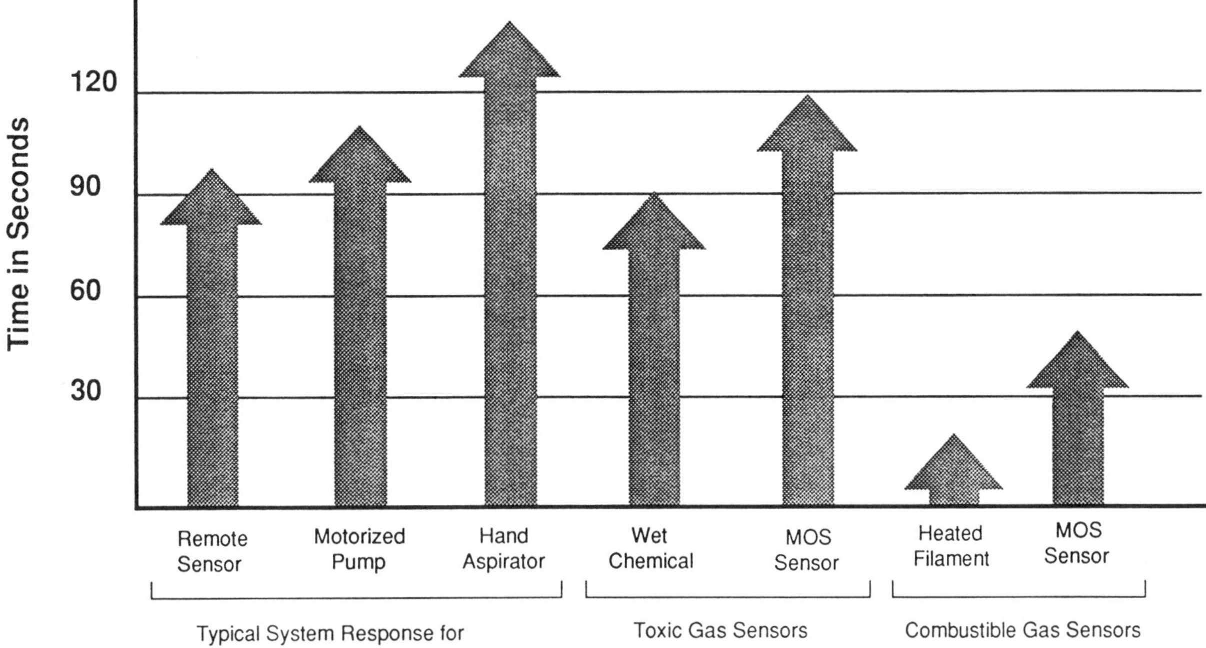

Figure 9-33. Typical response times for selected gas sensors. (Paul Trattner)

Figure 9-34. The sample draw pump is shown attached to the top of this handheld, multi-gas instrument. (Industrial Scientific)

concentrations. Others display only a single gas level at a time, but will alarm if any of the set points is reached.

Data Logging Function

A few instruments are equipped with microprocessors that electronically log the sample results. The

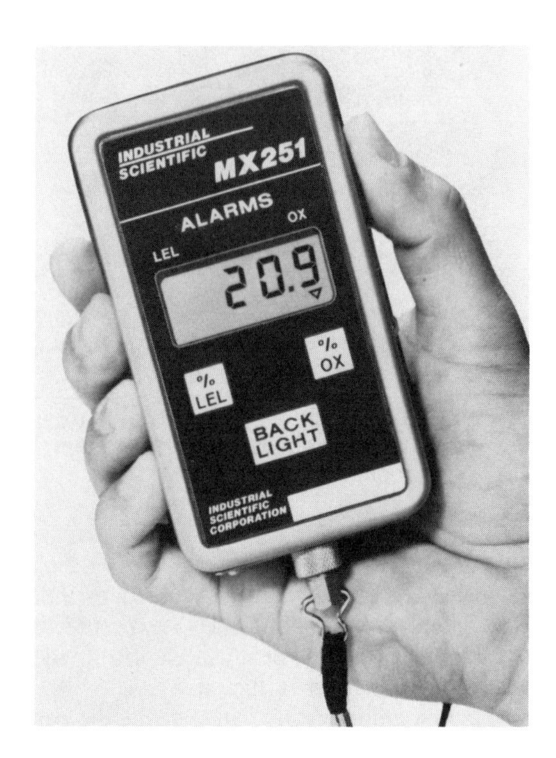

Figure 9-35. This instrument operating in the diffusion mode indicates an oxygen concentration of 20.9%. (Industrial Scientific)

Table 9-12. Comparison between sample draw and diffusion mode

	Response Time (Seconds)		
Gas	**Motor Pump**	**Aspirator Bulb**	**Diffusion Mode**
Oxygen	<10	<30	<30
Combustibles	<10	<30	<60
Carbon monoxide	<30	<50	<70
Hydrogen sulfide	<20	<40	<50

Source: Bacharach, Sentinel 4, Personal Gas Monitor, Instruction Manual.

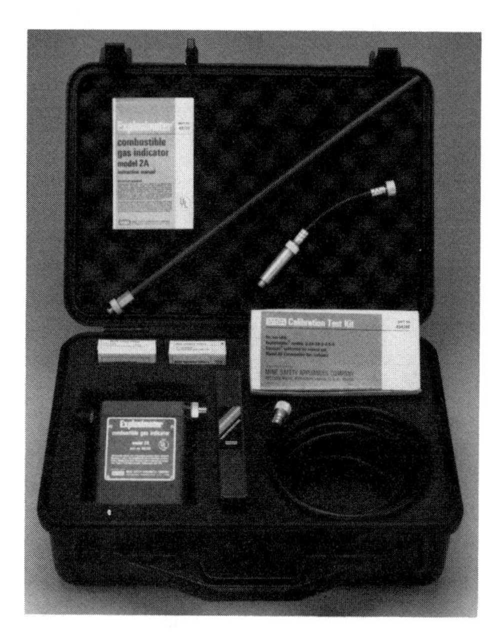

Figure 9-36. A rigid sample probe like this can be used for remote sampling through small openings. (MSA)

stored data can be internally processed to indicate time-weighted averages, short-term exposure levels and peak concentrations, or it can be downloaded to a personal computer for further analysis.

The data logging feature can expedite the permitting process because if frees the tester from the tedious task of having to write down the results of each measurement made. Once the space has been deemed suitable for entry, hard copy of the sampling results can be printed out and attached to the permit.

Radio Frequency Shielding

Radio frequency (RF) energy emitted from broadcast band, HF, VHF and UHF transmitters can induce stray currents into instruments. These currents can trigger false alarms or cause erratic fluctuations in the instrument's display. The effects of spurious currents can be minimized by instrument design that incorporate RF shielding.

The potential for interference exists whenever instruments are used in areas where radio frequency energy is likely to be present (for example, around airports, maritime facilities, military reservations, railroad yards and television and radio broadcasting stations). The degree to which interference will pose a problem is a function of a number of variables including the operating frequency, transmitter power and distance from the antenna.

A simple field-test can determine if an instrument is susceptible to RF interference from portable two-way radios used for communication between entrants and attendants. Key the transmitter and move the antenna around the instrument case while observing the display (Figure 9-37). Anomalous readings on the display are an indication of RF interference. It is important to check all gas modes, since some sensing circuits may be more affected by RF energy than others. While this method can be used to check for interference from handheld job site radios, it will not indicate RF susceptibility from other sources such as radio and TV broadcasting stations, cellular phones, radar sets or microwave towers, all of which may be found in RF-rich cities such as Los Angeles, Washington and New York.

CALIBRATION VS. FIELD-CHECKING

Calibration is a laboratory or bench-top process, usually conducted by the factory, a factory service center or a factory trained technician. It involves a multi-point check of an instrument's response to a variety of test gas concentrations and may also include a check of the alarm points and response time.

A field-check, on the other hand, is an expedient method used to verify that the instrument is responding properly in the field. It is essentially a "go/no-go" test of functional response that involves subjecting the instrument to a test gas and checking the instrument's reading to determine if it is operating within prescribed limits.

Some manufacturers instruction manuals recommend field-checking an instrument every thirty days, but this is a questionable practice from a practical

Figure 9-37. RF shielding can be tested by moving the antenna of a two-way radio around the instrument while keying the transmitter. (John Rekus)

perspective. Investigation revealed that this recommendation is based on laboratory tests which show some instruments to be stable for 30 days or more. While this may true for instruments tested under laboratory conditions, it is not necessarily true for instruments used in the field. Field instruments are often subject to rough handling and environmental conditions that may affect the instruments' operation. For example, they may be subjected to shock and vibration when being transported in the back of a pickup truck, or be exposed to wide swings in ambient temperature when stored in vehicles and job trailers.

INSTRUMENT OPERATIONAL CHECKS

All instruments should be checked before use in accordance with the instructions provided in the operator's manual. While not intended as a substitute for those instructions, the summary below identifies items that should be checked as a matter of good practice.

The general mechanical condition of all instruments should be checked for any abnormal conditions such as cracked, loose or missing parts that may adversely affect the instrument's performance.

Protective grills covering sensor elements should be free of debris, coatings or other obstructions that could restrict the flow of air.

Connections for sample-draw lines should be tight. If threaded connections are used, they should not be bent, damaged, cross-threaded or filled with dirt which could cause leakage.

Instrument response should be verified by conducting both a zero check and span check before and after the instrument is used.

Batteries should be fully charged to provide maximum operating life. Rechargeable batteries may require 16 hours of charging to obtain 8 hours of use.

Prefilters should be installed if measurements will be made in spaces containing dusts or liquids that could damage the sensors.

SAMPLING STRATEGIES

Any confined space sampling strategy must address three issues. First, the atmosphere in the space must be tested prior to entry to determine if it is acceptable. Second, the atmosphere should be checked during the entry to verify that it remains acceptable. Third, in some situations it will be necessary to evaluate workers' exposure to air contaminants generated by processes performed during the entry.

Pre-Entry Evaluation

Pre-entry evaluation typically employs direct-reading, sample-draw instruments to evaluate the level of oxygen, combustible materials and toxic contaminants. As shown in Figure 9-38, a probe or piece of flexible tubing attached to the instrument is inserted into the space through an opening such as an entry portal; after waiting the specified response time, the concentration of the gas or vapor of interest is noted. If the measurements indicate that the atmosphere is within acceptable limits, entry may proceed. If not, additional ventilation or respiratory protective equipment must be provided.

An important consideration of pre-entry testing is to assure that the entire space is evaluated. The size and shape of the space will dictate how many measurements will be required and where they should be made. Because it is possible for contaminants to stratify at different levels, the top, middle and bottom of a space must all be evaluated as indicated by Figure 9-39. Horizontal spaces such as rail tank cars should also be evaluated at representative points along their length, for example, at the middle and at both ends.

Many spaces can be tested adequately from the same portal that will be used for entry. For instance, an underground valve pit could be fully evaluated by dropping a probe in through the surface manhole used to access the valves. On the other hand, it may not

Figure 9-38. Initial testing should be performed from outside the space by inserting a probe or piece of flexible tubing. (Industrial Scientific)

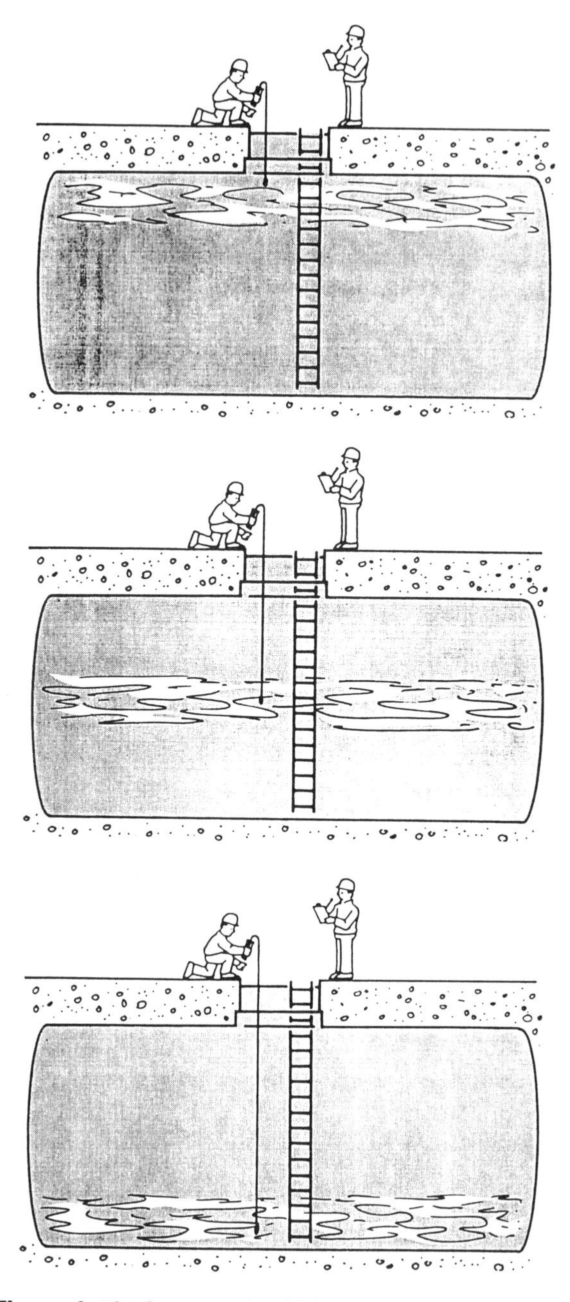

Figure 9-39. Spaces should be checked at top, middle, and bottom levels because of stratification of gases and vapors. (Industrial Scientific)

be possible to completely evaluate spaces such as petroleum storage tanks or large process vessels from a single point, and sampling lines may have to be inserted through vents, gauging hatch covers, pipe connections and other openings. This task can be made easier by using telescoping extension probes like the one shown in Figure 9-40.

Monitoring During Entry

Although pre-entry testing indicates whether or not the atmosphere in a space is initially acceptable for entry, atmospheric conditions can change over time. An oxygen enrichment could be produced by a leaking cutting torch, air could be displaced by an inert gas, and toxic contaminants could seep in from outside. Since serious consequences could result if these conditions go unnoticed, a policy should be developed that assures that the atmosphere is retested during entry. Periodic retesting is one alternative, but the determination of how frequently testing should be done is subject to the vagaries of professional judgment.

A better alternative is to employ personal, direct-reading instruments that continuously monitor the atmosphere (Figure 9-41). This approach has the advantage of being able to quickly detect even minor atmospheric changes. Since many continuous monitors can also be equipped with alarms, entrants can be warned of an impending hazardous condition before it becomes critical.

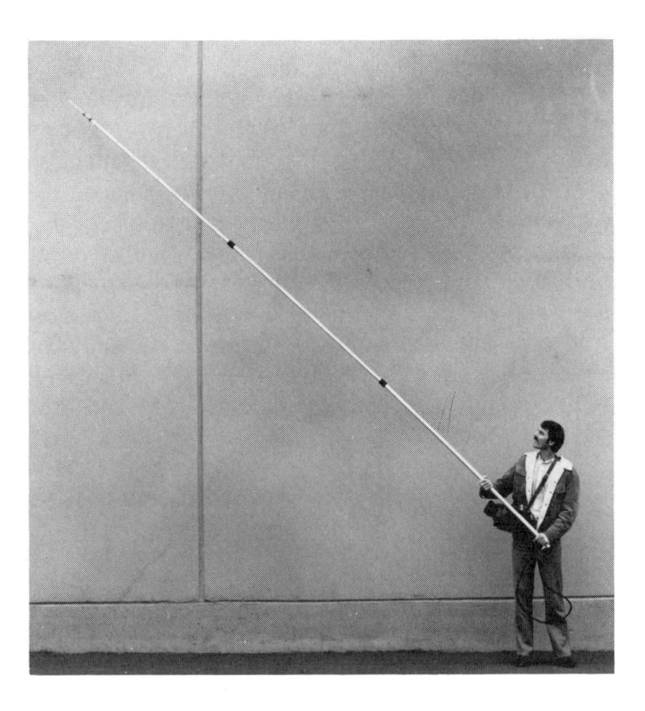

Figure 9-40. Telescoping extension probes like this can be used to reach distant areas in a large space. (Gastech)

Figure 9-41. Continuous monitors provide constant surveillance of atmospheric conditions in a space. This instrument can be used to initially test the space and then worn by a worker when he or she enters. (Industrial Scientific)

Personal Sampling

Personal sampling is an industrial hygiene method that is employed to measure employee exposures to air contaminants. The employee whose exposure is being evaluated wears a lightweight, battery-powered vacuum pump on his belt and a sample collection device on his collar (Figure 9-42). The pump and collection device are connected together with a piece of plastic tubing. As air is drawn through the collection device, contaminants of interest are trapped for subsequent laboratory analysis. The analytical results indicate the employee's level of exposure during the sampling period.

Personal sampling is principally performed to evaluate employee exposures during specific tasks or operations. The sampling results can be used to determine the effectiveness of ventilation systems, to provide a historical document of exposure and to comply with specific OSHA health standards.

There are presently 12 OSHA standards that specifically mandate personal sampling. If the entry involves exposure to any of the materials listed in Table 9-13, initial air sampling must be performed to determine the level of employee exposure. The process of lining acid tanks is a good example of a situation requiring monitoring since employees would be exposed to lead fumes when lead rods and lead sheeting are melted with oxy-fuel gas torches. These standards also prescribe certain protective measures such as employee training, medical surveillance and work practices that are predicated on specified exposure levels. If the entry exposes employees to any of the materials listed in Table 9-13, the full text of the standard for that material should be consulted to determine what, if any, requirements apply.

Personal sampling results are also useful for documenting employee exposures in a more general way. For instance, personal samples taken during cleaning of a gasoline storage tank can be used to evaluate exposure to aromatic and aliphatic compounds such as benzene, toluene, hexane and octane. This information can be shared with employees to demonstrate that adequate precautions were taken during the project. It could also prove valuable in the event of future litigation.

PERSONAL SAMPLING EQUIPMENT

Sampling Pumps

Personal sampling pumps may be generally divided into two categories: high-flow (1–3 L per minute

Figure 9-42. Sampling train for evaluating worker exposures. (MSA)

Table 9-13. OSHA health standards requiring personal monitoring

Substance	Regulation
Asbestos	29 CFR 1910.1001
Acrylonitrile	29 CFR 1910.1045
Arsenic (inorganic)	29 CFR 1920.1018
Benzene	29 CFR 1910.1028
Cadmium	29 CFR 1910.1027
Coke oven emissions	29 CFR 1910.1029
Cotton dust	29 CFR 1910.1043
Formaldehyde	29 CFR 1910.1048
Lead	29 CFR 1910.1025
Methylenedianiline	29 CFR 1910.1058
Vinyl chloride	29 CFR 1910.1017

minute) and low-flow (5–1,000 mL per minute). Some pumps, however, are capable of operating in both the high- and low-flow modes. Flow rates for specific contaminants depend on the collection medium, the contaminant concentration and the sensitivity of the analytical method. However, as a practical matter, lower flow rates are generally used when sampling gases, and vapors and higher flow rates are used when sampling particulates.

The highly portable nature of these instruments allows them to be worn by workers wherever they go over an entire shift. Since they require little attention from the investigator, it is possible to evaluate the exposure of a number of employees simultaneously.

The earliest models of pumps were rather unsophisticated and required manual flow adjustment as the battery wore down or as the collection medium loaded up with contaminants. However, recent technological advances have eliminated this need through electronic circuitry which maintains a constant flow rate.

Another feature found on some pumps is a clock which not only displays the elapsed operating time, but can also be programmed to start and stop the pump at pre-set intervals.

Collection Media

Collection media can be divided into three major categories: filters, solid sorbent tubes and liquid absorbers. Each has its own particular application.

Filters

Filters are very thin disks of a semi-permeable plastic membranes. They collect particulates like dusts, welding fumes and paint-spray mists by mechanically separating them from the air stream.

Sorbent Tubes

Solid sorbent tubes are glass tubes that contain activated charcoal, silica gel or a proprietary granular material. Air is drawn through the tube and contaminant gases and vapors are trapped on the surfaces of the solid sorbent. The tubes consist of two sections of sorbent with the sample collected primarily on the first sections while the second serves as a back-up to trap any material escaping from the first. This "break-through," as it is called, results when the primary section becomes saturated and thus unable to retain any more of the contaminant. Without the back-up, the portion breaking through would go undetected, and analysis would indicate a lower concentration than that actually present in the environment.

The contaminant is desorbed in the laboratory with a solvent, and the front and back sections of the tube are analyzed separately. If no material is present in the "back-up," then that in the front section alone indicates the quantity of the contaminant present in the environment.

Small quantities of break-through material may be added to that found in the front section. However, if the break-through quantities exceed a certain level established by the lab, then all that can be said about the environment is that it contained at least the amount reported, but it may *actually* contain much more.

Liquid Media

Liquid media are used to collect some inorganic gases like ammonia, ozone and hydrogen cyanide which are difficult to collect on solid sorbents. The collection device is a glass vial containing about 15–20 mL of water, an organic solvent or a reagent solution. Air is bubbled through the liquid and the contaminant either dissolves or reacts with the reagent to form a stable chemical species. The contaminant-containing liquid can then be analyzed using standard laboratory methods like polarography, spectrophotometry and potentiometry.

Sample Analysis

Many labs have the equipment necessary for the analysis of air samples, but their personnel may be unfamiliar with industrial hygiene analytical techniques. They may also lack appropriate quality control programs to assure the accuracy of results. These deficiencies may unduly delay laboratory reports. Worse yet, they could precipitate faulty analysis, yielding incorrect results.

Therefore, users who contract for laboratory services should give careful consideration to employing only those laboratories that are accredited by the American Industrial Hygiene Association (AIHA). A listing of accredited labs may be obtained from AIHA's offices in Fairfax, VA, or by consulting a recent edition of the AIHA Journal.

SUMMARY

Air monitoring equipment has come a long way since the Davy lamp of a century and half ago. Today, dozens of instruments are available to measure hundreds of contaminants. Even though there has been a proliferation of air sampling instruments, they can all be classified as either direct-reading or indirect-reading devices. Direct-reading instruments include oxygen meters, combustible gas meters, detector tubes and toxic gas monitors which allow the operator to directly measure the concentration of the substance of interest. Indirect-reading instruments, on the other hand, use a battery-powered vacuum pump to draw air through collection media such as filters, liquids or solid sorbent tubes which are subsequently analyzed in a laboratory. Direct-reading instruments are most commonly used to evaluate confined spaces prior to entry, and indirect-reading instruments are used to evaluate employee exposures.

Prior to entering a confined space, the atmosphere should be tested for oxygen, combustible gases and vapors and toxic contaminants.

Oxygen meters employ galvanic cells which are highly specific and virtually maintenance free. Sensors typically last for about a year, but their life can be shortened by exposure to high levels of carbon dioxide.

Combustible gas meters most frequently operate on the principle of catalytic combustion on a heated filament, but some instruments employ metal oxide semiconductors. Combustible gas meters are nonspecific and cannot differentiate between atmospheres containing a single gas and those containing a mixture. In addition, they are only accurate when used to measure the gas for which they are calibrated. Meters operating on the catalytic combustion process provide inaccurate results when oxygen concentrations are below 10%, so the oxygen level should always be verified before testing for combustibles.

Toxic gases may be evaluated by a variety of devices including sensor-based instruments, detector tubes, gas chromatographs and infrared analyzers. Only a handful of contaminants can be evaluated with sensor-based instruments, but this difficulty may be overcome by using colorimetric detector tubes which can quantify hundreds of gases and vapors. Although detectors tubes are easy to use, their accuracy is limited and may be affected by variations in temperature, pressure and humidity. Portable gas chromatographs and infrared spectrophotometers may also be used to evaluate toxic gases and vapors, but they are expensive and require a greater degree of user sophistication than either sensor-based instruments or detector tubes.

The National Electrical Code® divides atmospheres containing ignitable materials into various classes, divisions and groups. If an instrument will be used in a hazardous location as defined by the code, it must be "approved" for use in that location. Even though an instrument may be "approved," accessories used with it, such as battery chargers and sample draw pumps, may not be. Some "approved" devices are not suitable for use in all atmospheres and can only be used in atmospheres containing gases and vapors for which they are approved.

Finally, personnel responsible for evaluating confined spaces must possess certain skills. They must be able to select the appropriate instrument for the task, determine if the instrument is functioning properly, develop a sampling strategy that assures that the entire space is evaluated and correctly interpret the measurement results.

REFERENCES

American National Standard for Safety Requirements for Confined Spaces, ANSI Z117.1-1989, New York, NY: American National Standards Institute, (1989).

Bevington, P.R., *Data Reduction and Data Analysis for the Physical Sciences,* New York: McGraw Hill, (1969).

CGM 929 Instruction Manual, Ann Arbor, MI: Dynamation, Inc.

CGS-20 and CGS-20M Instruction Manual, Ann Arbor, MI: Enmet Corporation.

Combo 434 Instruction Manual, Ann Arbor, MI: Dynamation, Inc.

Cleaning Petroleum Storage Tanks, API Publication 2015, Washington, DC: American Petroleum Institute, (1985).

Cleaning Open-top and Covered Floating Roof Tanks, API Publication 2015B, Washington, DC: American Petroleum Institute, (1981).

Cloe, W.W., *Selected Occupational Fatalities and Asphyxiating Atmospheres in Confined Spaces as Found in Reports of OSHA Fatality/Catastrophe Investigations,* Washington, DC: U.S. Department of Labor, Occupational Safety and Health Administration, (1985).

Ecotox 60/75 Multi-Gas Monitor, Product Data Sheet, Gainesville, GA: Neotronics of North America, (1992).

Guidelines for Confined Space Work in the Petroleum Industry, API Publication 2217, Washington, DC: American Petroleum Institute, (1984).

Guide for Controlling the Lead Hazard Associated with Tank Entry and Cleaning, API Publication 2015A, Washington, DC: American Petroleum Institute, (1982).

Hazardous Location Equipment Directory, Northbrook, IL: Underwriters Laboratories (1990).

Installation of Intrinsically Safe Systems for Hazardous (Classified) Locations, Research Triangle Park, NC: Instrument Society of America, (1987).

Intrinsically Safe Apparatus and Associated Apparatus for use in Class I, II, and III Division 1 Hazardous (Classified) Locations, ANSI/UL 913-1988, Northbrook, IL: Underwriters Laboratories, (1988).

Intrinsically Safe Apparatus and Associated Apparatus for Use in Class I, II and III Division 1 Hazardous Locations, NFPA-493-1978, Quincy, MA: National Fire Protection Association, (1978).

Installation Operation and Maintenance of Combustible Gas Detection Instruments, ISA-RP12.13 Part II, Research Triangle Park, NC: Instrument Society of America, (1987).

Keith, L.H., *Principles of Environmental Air Sampling,* Washington, DC: American Chemical Society, (1988).

Korolkoff, N.O., Survey of toxic gas sensors and monitoring systems, *Solid State Technology,* December, 1989.

Manual of Recommended Practice for Combustible Gas Indicators and Portable Direct Reading Hydrocarbon Detectors, Fairfax, VA: American Industrial Hygiene Association, (1980).

MicroGuard Instruction Manual, Pittsburgh, PA: MSA.

Miran 1B2 and 1BX, Portable Ambient Air Analyzer Instruction Manual, Foxboro, MA: Foxboro Company.

Model 70 Carbon Monoxide Indicator Instruction Manual, Pittsburgh, PA: MSA.

Model 260 Instruction Manual, Pittsburgh, PA: MSA.

Model 261 Instruction Manual, Pittsburgh, PA: MSA.

Model 361 Instruction Manual, Pittsburgh, PA: MSA.

Models 1338 and 1314 Castechtor Instruction Manual, Newark, CA: GasTech, Inc.

Model GX-83 Three-Gas Personal Monitor Instruction Manual, Newark, CA: GasTech, Inc.

Model GX-86 Four-Gas Personal Monitor Instruction Manual, Newark, CA: GasTech, Inc.

Model GX-91 Personal Four-Alarm Instruction Manual, Newark, CA: GasTech, Inc.

Model HMX 271 Instruction Manual, Oakdale, PA: Industrial Scientific, (1992).

Model OVA 128 Organic Vapor Analyzer Instruction Manual, Foxboro, MA: Foxboro Corp.

Model pm-7440 Multi-Gas Monitor Instruction Manual, Rochester, NY: Metrosonics.

Mueller, P.K., Comments on advances in the analysis of air contaminants, *J. Air Poll. Control Assoc.* 30:998 (1980).

MSA Detector Tube and Dosimeter Handbook, Pittsburgh, PA: Mine Safety Appliances (1988).

National Electrical Code,® NFPA-70, Quincy, MA: National Fire Protection Association, (1993).

National Institute for Safety and Health: Certification of Gas Detector Tube Units, *Federal Register* 38:11458 May 8, 1973.

Oyab, T., T. Kurobe and T. Hidai, Development of Tin Oxide Sensor and Monitoring Equipment, *Proceedings of the International Meeting on Chemical Sensors,* Furuoka, Japan, September 19-22, 1983.

Permit-Required Confined Spaces, Code of Federal Regulations, Vol. 29, Part 1910.146.

Pettit, T.A., P.M. Gussey and R.S. Simmons, *Criteria for a Recommended Standard: Working in Confined Spaces,* Cincinnati, OH: National Institute for Occupational Safety and Health, DHEW/NIOSH Publication 80-106, (1979).

Pettit, T.A. and H. Linn, *A Guide to Safety in Confined Spaces,* Cincinnati, OH: National Institute for Occupational Safety and Health, DHHS/NIOSH Publication No. 87-113, (1987).

Precautions Before Entering Compartments and or Spaces, Code of Federal Regulations, Vol. 29, Part 1915.12.

Shaw, M., *More Straight Talk About Toxic Gas Monitors,* Chatsworths CA: Interscan Corporation.

Sentinel 4® Instruction Manual, Pittsburgh, PA: Bacharach Instruments, Inc., 1984.

Sniffer 302 Instruction Manual, Pittsburgh, PA: Bacharach Instruments, Inc., 1984.

CHAPTER 10

Personal Protective Equipment

INTRODUCTION

Personal protective equipment such as hard hats, safety glasses, cover-alls and gloves may be required to protect against hazards that may be encountered during some entries. While the use of personal protective equipment plays a critical role in protecting employees, prudent practice dictates that hazards be eliminated or controlled whenever possible. Some hazards can be eliminated through engineering controls such as ventilation or isolation of lines containing toxic materials; others hazards can be controlled by substituting different methods, materials or work practices. In those cases where hazards cannot be eliminated, employees should be required to use protective equipment.

It is important to remember that protective equipment does not eliminate hazards such as flying chips, hot sparks, harmful liquids, sharp surfaces or contaminated air. Instead, it places a physical barrier between the worker and the hazard. But equipment will only be effective if it is properly selected, used and maintained. Chemical-resistant gloves, for instance, won't provide any protection if they are riddled with holes. Similarly, a respirator with a broken headband, distorted valve or a depleted chemical cartridge will provide little protection.

OSHA General Industry Standards require that personal protective equipment be worn whenever its use could prevent injury. But merely making protective equipment available to employees is not sufficient. A good personal protective program must also include provisions for employee training, plus ongoing workplace surveillance to assure that the equipment is properly worn and maintained.

This chapter discusses the general types of protective equipment—except for respirators, which will be discussed in Chapter 11—that might be required during a confined space entry. It also explains how to establish an effective personal protective equipment program.

ANSI SPECIFICATIONS

Specifications for safety eye wear, head protection and foot protection have been established by the American National Standards Institute (ANSI). ANSI is a non-profit, non-governmental standards-setting organization whose membership is made up of representatives from industry, organized labor, government and education. The ANSI standards-setting process is voluntary and its publications express the consensus of the members who sit on its technical committees.

ANSI standards establish minimum performance specifications that protective equipment must meet. Hard hats, for example, must meet certain criteria for fire resistance, electrical conductivity and penetration by sharp objects. Similarly, eye protection must meet specifications for optical performance, corrosion resistance and impact.

Equipment that is designed to meet ANSI specifications can also be submitted to the Safety Equipment Institute (SEI) for performance certification. The SEI is a private, not-for-profit organization that operates a third-party, non-government certification program which physically tests representative samples of equipment to verify that they do in fact meet ANSI specifications. The quality assurance of various testing protocols is also monitored by an independent auditor. The type of products tested and laboratories that conduct the tests are listed in Table 10-1.

Use of only SEI certified equipment is encouraged since there is a greater degree of assurance that the equipment meets ANSI performance criteria. A current list of certified equipment may be obtained by writing SEI at 1901 North Moore Street, Arlington, VA, or by calling (703) 525-3354.

CONTROL HIERARCHY

Safety and health professionals have historically taken a tiered approach to managing workplace hazards. The first step is to determine if the exposure can be eliminated by means of engineering changes or process modifications. For example, rather than having employees enter an underground steam vault to operate a control valve, it might be possible to move the valve to a more accessible location above ground. Alternatively, it might be possible to equip the valve with a pull chain, a remotely controlled motor or a chuck-type fitting that could be operated by a rod inserted from grade level.

Table 10-1. Safety equipment institute tested products

Product Category	Applicable Standard	Testing Laboratory
Eye and face protection	ANSI Z 87.1-1989	Professional Services Industries, Inc.
Head protection	ANSI Z 89.1-1986	United States Testing Company
Liquid splash protective suits	NFPA-1992 (1990 ed.)	Texas Research Institute
Vapor protective suits	NFPA-1991 (1990 ed.)	Texas Research Institute

Source: Safety Equipment Institute.

If engineering controls are not feasible, the next step is to determine if less hazardous materials, methods or techniques can be substituted. For example, exposure to paint vapors might be controlled by substituting water-based for solvent-based materials. If the hazard cannot be controlled by engineering methods, materials substitution or changes in work practices, then personal protective equipment provides the final alternative.

OSHA STANDARDS

For the most part, OSHA standards do not identify specific operations for which protective equipment is required, but there are some exceptions. OSHA standards 29 CFR 1910.252(b)(2) and (b)(3), for example, require welders and their helpers to wear special protective clothing and equipment. But a much broader mandate is given in 29 CFR 1910.132 which requires protective equipment to be provided, used and maintained whenever it necessary because of hazards or environmental conditions. This means that it's up to employers to evaluate their operations and to determine what, if any, protective equipment is required for specific operations, processes or activities.

Since OSHA personal protective equipment standards are broad-based, complying with them requires a working knowledge of the hazards to which employees are exposed. Selection of the appropriate protection is then a matter of simply matching the right equipment to the job.

Except for respirators—which *must be* provided by the employer—employees may provide their own protective equipment. However, it is the employer's responsibility to assure that *appropriate* equipment is selected and that it is properly maintained. If an employee chooses the wrong equipment, or fails to take care of it, OSHA will penalize the employer, not the worker.

An overview of the most commonly used types of personal protective equipment is provided below.

HEAD PROTECTION

Hard hats are intended primarily to protect workers' heads from impact and penetration by small falling objects such as hand-tools, debris, and loose parts that may accidentally fall into a space; however, they also provide limited protection against splashes by hazardous liquids and contact with electrically energized conductors. Hard hats also protect workers' heads from bumps on overhangs, pipes, ducts, steam lines and cables that may be present in the space. Consequently, protective headwear should be worn whenever a potential for head injury exists.

ANSI Classifications

Protective headwear meeting ANSI Z89.1 specifications is divided into three classes and two types. A helmet's type is established by the shape of its brim, while its class is determined on the basis of specific performance tests. All helmets meeting ANSI performance criteria are identified by a label affixed to the shell's interior.

Type 1 helmets (Figure 10-1) have a full brim that extends around the entire circumference of the helmet. This wide brim provides extra protection from sun glare, splash and rain.

Type 2 helmets (Figure 10-2) have a partial brim—technically called a "peak"—that extends only over the eyes.

Class A helmets are suitable for general service use and are intended primarily to reduce the force of impact of falling objects. Their level of electrical protection is limited to contact with *low-voltage* conductors, and representative helmet shells are proof-tested to only 2,200 volts. Class A helmets may not have metal parts in direct contact with the head, and their shells must be free of holes except for those

Figure 10-1. ANSI Type 1 full-brim helmet. (Bullard)

Figure 10-2. ANSI Type 2 Peaked brim helmet. (Bullard)

needed to attach the suspension and accessories such as ear muffs and face shields.

Class B hats are designed to provide protection from both falling objects and contact with *high-voltage* electrical circuits, and their shells are proof-tested to 20,000 volts.

Class C helmets offer no protection from electrical contact and are intended only to provide protection from impact and penetration of falling objects. As a result, they must not be worn in spaces containing exposed electrical conductors. Type C helmets made of aluminum should not be worn in areas where they may be subjected to splash from acids, alkalis or other corrosive materials which could react chemically with the shell.

Adjustments and Accessories

The outer shell of the hard hat is supported by an inner webbed suspension that functions like a shock absorber to cushion impacts by distributing the load over the entire head. To be effective, the suspension must be adjusted so that it is about an inch away from the helmet's shell. Hard hats should fit snugly, but not so tightly that they are uncomfortable. Appropriate sizing can be determined from Table 10-2.

Some hard hats are also fitted with hardware that allows easy mounting of accessories such as ear muffs and face shields (Figure 10-3).

Care and Maintenance

Helmet shells may be cleaned by scrubbing with mild detergent and water. To prevent damage to plastic and elastomeric parts, wash water temperatures should not exceed 140°F. Hard hats should not be cleaned with solvents since they can attack and damage shell polymers. Shells, headbands, sweat bands and attached accessories should also be visually inspected daily for dents, cracks, wear and evidence of penetration or impact which could compromise a helmet's effectiveness.

Table 10-2. Hard hat sizing guide

Headband	Head Circumference	
	Inches	Centimeters
6-1/2	20-1/2	52.07
6-5/8	20-7/8	53.02
6-3/4	21-1/4	53.98
6-7/8	21-5/8	54.93
7	22	55.88
7-1/8	22-3/8	56.83
7-1/4	22-3/4	57.79
7-3/8	23-1/8	58.74
7-1/2	23-1/2	59.69
7-5/8	23-7/8	60.64
7-3/4	24-1/4	61.59
7-7/8	24-5/8	62.55
8	25	63.50

Adapted from ANSI Z89.1 1986.

Figure 10-3. Helmets can be fitted with ear muffs and face shields. This one has the face shield attached. (Cabot Safety)

Storage

As a general rule, hard hats should not be stored in areas where they would be subject to contact with solvents or corrosives which could damage the shell or its suspension or where surfaces might be contaminated with toxic materials that could be absorbed through the skin when the hat was handled or worn. Hard hats should also not be stored on a car's rear window shelf because in the event of an accident, or a sudden stop, the hat could fly forward and injure the passengers. In addition, constant exposure to heat and ultraviolet rays from sunlight can weaken the polymers used to fabricate some shells. Early warning signs of UV-degradation are loss of surface gloss, chalking, powdering and flaking. Any hats that severely exhibit these conditions should be removed from service.

EYE AND FACE PROTECTION

ANSI Classifications

ANSI standard Z-87.1 divides eye and face protection into two types: primary protectors and secondary protectors. Primary protectors are devices which may either be worn alone or with secondary protectors. Secondary protectors, on the other hand, are devices that *must be worn* with a primary device.

Safety spectacles are classified as primary protectors and are used to shield the eyes from a variety of hazards including radiation and flying particles such as chips, sparks and splinters. Spectacles may be provided with or without side-shields (Figure 10-4), but even those with side-shields offer only limited splash protection since liquids can flow around the shield.

Goggles are also classified as primary protectors. They are available in two styles: the eyecup style (Figure 10-5) which covers only the eye sockets and the cover-type which protects the eyes and part of the face around the eyes. Some goggles are ventilated, others are not (Figure 10-6). Ventilated goggles are usually more comfortable because they are cooler.

Directly ventilated goggles (Figure 10-7) have vents that block large particles like wood chips, but permit the passage of very fine powders and liquids.

Indirectly ventilated goggles (Figure 10-8) overcome this difficulty by using baffled vents to form a maze that prevents dust and liquid from getting inside.

Face shields are used to protect the wearer's face and neck. It should be noted, though, that ANSI classifies face shields as secondary protectors. That means they must be worn with primary protectors such as safety spectacles or goggles.

Welding helmets are a special type of face shield that protects the eyes and face from flying particles and UV radiation created by welding processes. Since electrode size influences the intensity of the arc, welding helmets are designed to accept filters of varying optical densities.

Darker filters are required for more intense arcs. The optical densities required by OSHA Standard 29 CFR 1910.252(h) for specific welding and burning processes are listed in Table 10-3.

In some cases a combination of protection may be required. Welders, for example, may wear both safety glasses and a welding shield (Figure 10-9). The

Figure 10-4. Safety spectacles with side-shields. (Cabot Safety)

Figure 10-5. Eyecup goggles cover the eye sockets. (OSHA)

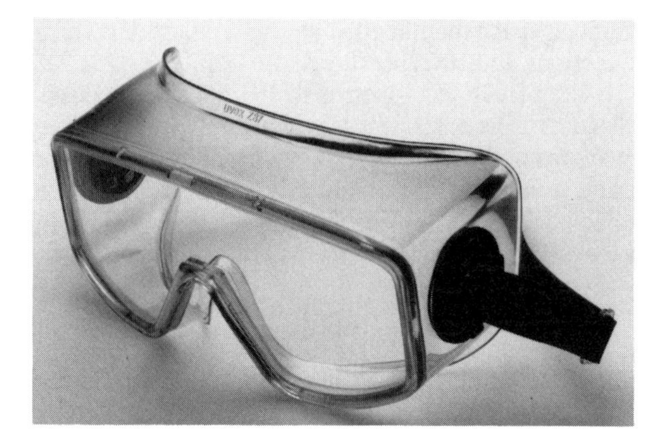

Figure 10-6. Unvented cover goggles. (Uvex)

Figure 10-7. Directly ventilated cover goggles. (Uvex)

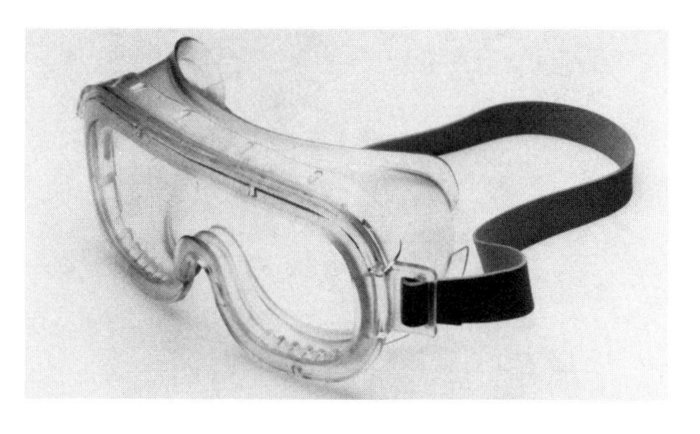

Figure 10-8. Indirectly ventilated cover goggles. (Uvex)

Table 10-3. Lens shades required for selected welding processes

Welding Operations	Lens Shade
Shielded metal-arc welding 1/16-, 3/32-, 1/8-, 5/32-inch electrodes	10
Gas-shielded arc welding (non ferrous) 1/16, 3/32, 1/8, 5/32 inch electrodes	11
Gas-shielded arc welding (ferrous) 1/16, 3/32, 1/8, 5/32 inch electrodes	12
Shield metal-arc welding	
3/16, 7/32, 1/4-inch electrodes	12
5/16, 3/8-inch electrodes	14
Atomic hydrogen welding	10 - 14
Carbon arc welding	14
Soldering	2
Torch brazing	3 - 4
Light cutting (up to 1-inch)	3 - 4
Medium cutting (1 to 6 inches)	4 - 5
Heavy cutting (over 6 inches)	5 - 6
Light gas welding (up to 1/8 inch)	4 - 5
Medium gas welding (1/8 to 1/2 inch)	5 - 6
Heavy gas welding (over 1/2 inch)	6 - 8

Source: 29 CFR 1910.252.

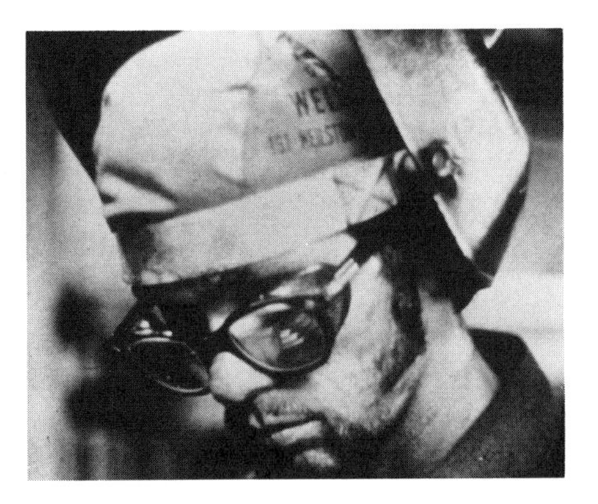

Figure 10-9. Welder with glasses under welding shield. (OSHA)

welding shield provides protection against the harmful ultraviolet rays created by the electric arc, while the safety glasses protect the welder's eyes from flying particles when he chips flux from a fresh weld.

Eye Protection: Selection and Fitting

The eyewear selection process is highly subjective because of the infinite variety of situations which might require protection. ANSI recognizes this difficulty and has developed guidelines (Figure 10-10) to aid in selecting the appropriate type of protection for a specific application.

Accident data collected by the Bureau of Labor Statistics (BLS) indicate that eye injuries may still occur even if safety glasses are worn. However, further analysis of the BLS reports suggests that most of the injuries that did occur could have been prevented if the glasses had been fitted with side-shields. Consequently, ANSI encourages that side-shields be used whenever they are practical.

Eye protection that slides off the face, pinches the nose or is otherwise uncomfortable is not likely to be worn.

Fortunately, the wide variety of makes, models and styles of eye protection offers so many choices that it is virtually impossible not to find a device that will be acceptable to every wearer. To assure a proper fit, however, eye protection should always be sized by a qualified person.

Care, Maintenance and Training

Eye and face protection should be cleaned regularly and maintained in a sanitary and usable condition. Any equipment that is damaged, worn or badly scratched should be replaced. Employees should be trained in use, maintenance, storage and limitations of any eye protection they are required to use. For example, they should know that ordinary tinted lenses will not provide protection from harmful optical radiation sources such as welding arcs or lasers.

HEARING PROTECTION

Noise, which is generally defined as unwanted sound, can be a hazard to workers who chip, grind, hammer or sandblast in a confined space. In fact, noise created by these processes will be louder in a confined space than in open air due to interior surface reflections and reverberations.

Under free-field (open space) conditions, acoustical energy dissipates rapidly as sound waves move away from their source. However, in a confined space, sound waves which strike interior walls are reflected back, in effect amplifying the sound. In some cases, these reflections may produce dangerously high noise levels.

Noise is generally described in terms of its frequency, or pitch, measured in hertz (Hz) and its amplitude, or loudness, measured in decibels (dB).

Effects of Noise Exposure

Extra-Auditory Effects

While most people know that prolonged exposure to loud noise may cause a hearing loss, many are unaware of the subtle *extra-auditory* effects of noise. Selected adverse extra-auditory effects which could compromise worker safety during confined space entry are discussed below:

- Constriction of small blood vessels in the extremities which may increase arterial-blood pressure as the heart tries to overcome the increased resistance to blood flow created by the constriction
- Increase in muscle reflex activity that produces a general tightening of muscles and results in greater tension levels
- The startle-response triggered by exposure to a sudden unexpected sound may contribute to secondary injury such as bumps, scrapes or contusions

- Changes in vision, including narrowing of the visual field and modification of color perception, particularly a partial deficiency in the ability to perceive red
- Loss of sensitivity to touch or pain which can rob the perceiver of early warning signs of impending danger

Taken either individually or as a group, these otherwise inconsequential effects could significantly increase the level of risk posed to employees working in a confined space.

Auditory Effects

An occupational hearing loss results when small hair cells in the inner ear are damaged from repeated exposure to noise. These hair cells are similar to blades of grass, which when stepped on, will flatten and then spring back. However, if the grass is stepped on repeatedly, some blades will eventually be stressed to the point where they no longer bounce back. The hair cells in the inner ear respond in much the same way, and when exposed to repeated noise, they too lose some of their resilience.

This type of hearing loss is referred to as sensorineural loss. People with sensorineural losses cannot hear high frequencies and voices are distorted so that they sound as if they were being heard under water. Sensorineural loss is permanent and cannot be corrected with a hearing aid.

The OSHA Noise Standard

OSHA has a comprehensive standard governing exposure to noise. The standard, which is contained in 29 CFR 1910.95, considers two factors in establishing the permissible level of exposure: (1) the amplitude of the noise and (2) the length of time the worker is exposed it. As shown in Table 10-4, the louder the noise level, the less time an unprotected employee may be exposed to it.

The first step of a noise control program is to measure the noise using either a sound level meter or a personal noise dosimeter. The electronics of the dosimeter integrates both the sound level and length of exposure and automatically calculates the exposure dose, which is expressed as a percentage. A dose over 100% indicates that the employee has been over exposed to noise levels permitted by the standard.

While there is no shortage of instrumentation for measuring noise, a quick and dirty rule of thumb is that if you have to raise your voice significantly to be heard by someone about foot away, the sound level is probably above OSHA's 90 dB_A criterion.

Except for major turn-arounds in the petroleum industry, which may take weeks, or even months, most confined space activities can be completed in a few hours, or at most, a couple of days. Given the short duration of these entries, and the intermittent nature of the work, most jobs probably won't exceed noise exposure limits.

But compliance with the noise standard is not the only consideration. In addition to the extra-auditory effects discussed previously, high noise levels can affect an employee's ability to communicate or to hear verbal warnings or audible alarms.

Engineering Controls for Noise

Like all hazards, noise is best controlled by engineering methods. For example, mufflers can be used to control the loud whistle created when air is exhausted from a pneumatic tool. Until recently, noise created by chipping, grinding or abrasive blasting inside a steel tank was thought to be impossible to control. However, with the advent of noise cancellation technology, even these sources can be controlled.

Commercially available noise cancellation equipment uses a microphone to pick up sound waves emanating from the noise source. The resulting electrical impulse is fed to a signal processor which generates an identical wave form 180° out-of-phase with the original sound. The out-of-phase signal is then amplified and directed back toward the noise source via a speaker system. When the two sound waves combine, they cancel each other out, greatly reducing the noise.

Hearing Conservation Program

OSHA requires employers to establish hearing conservation programs to protect those workers exposed to excessive noise levels. These programs must include provisions for audiometric testing, use of protective equipment and employee training. It should be noted, though, that the OSHA standard establishes only *minimum requirements* for hearing conservation, and some workers will suffer a hearing loss even when exposed to noise levels permitted by the regulation.

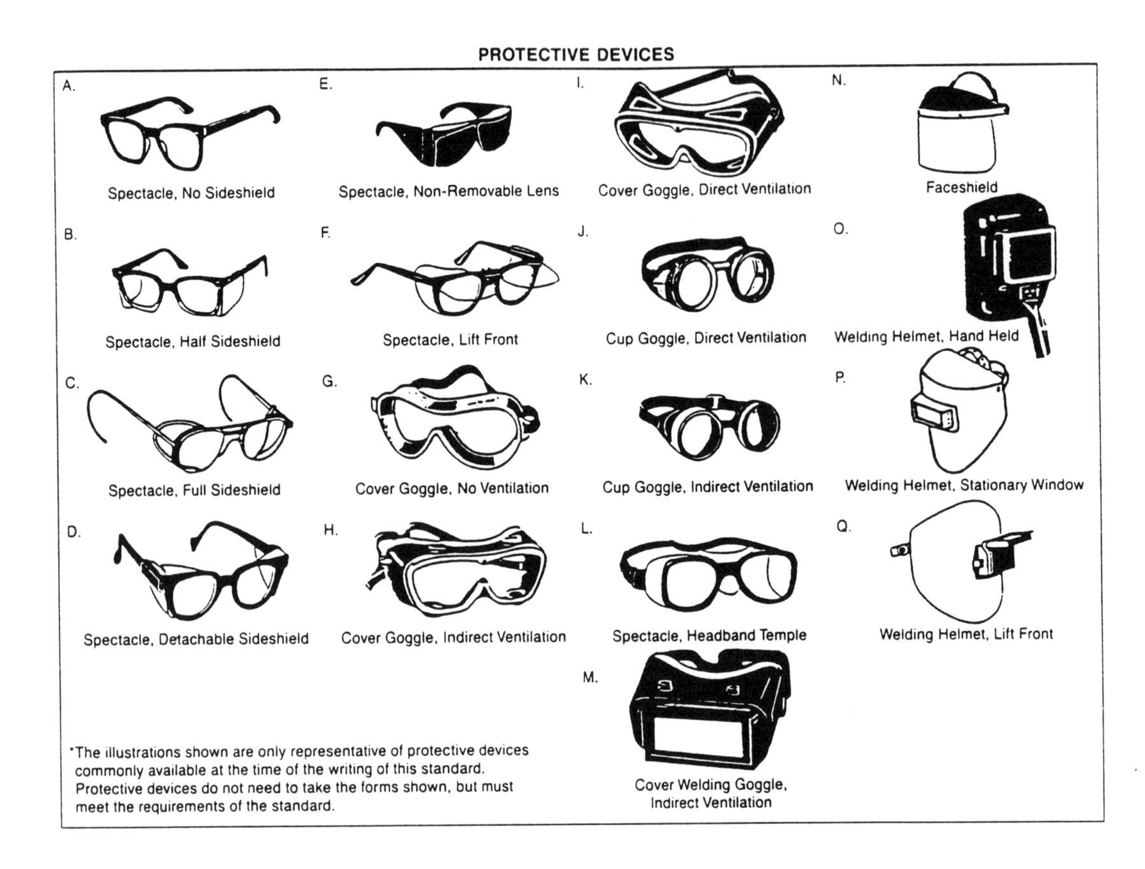

NOTES:

(1) Care shall be taken to recognize the possibility of multiple and simultaneous exposure to a variety of hazards. Adequate protection against the highest level of each of the hazards must be provided.

(2) Operations involving heat may also involve optical radiation. Protection from both hazards shall be provided.

(3) Faceshields shall only be worn over primary eye protection.

(4) Filter lenses shall meet the requirements for shade designations in Table 1.

(5) Persons whose vision requires the use of prescription (Rx) lenses shall wear either protective devices fitted with prescription (Rx) lenses or protective devices designed to be worn over regular prescription (Rx) eyewear.

(6) Wearers of contact lenses shall also be required to wear appropriate covering eye and face protection devices in a hazardous environment. It should be recognized that dusty and/or chemical environments may represent an additional hazard to contact lens wearers.

(7) Caution should be exercised in the use of metal frame protective devices in electrical hazard areas.

(8) Refer to Section 6.5, Special Purpose Lenses.

(9) Welding helmets or handshields shall be used only over primary eye protection.

(10) Non-sideshield spectacles are available for frontal protection only.

Figure 10-10. ANSI Selection Chart. (This material is reproduced with permission from the American National Standard Z 87.1, copyright 1989 by the American National Standards Institute. Copies may be purchased from the American National Standards Institute at 11 West 42nd Street, New York, NY 10036.)

AMERICAN NATIONAL STANDARD Z87.1-1989

SELECTION CHART — **PROTECTORS**

		ASSESSMENT SEE NOTE (1)	PROTECTOR TYPE	PROTECTORS	LIMITATIONS	NOT RECOMMENDED
IMPACT	Chipping, grinding, machining, masonry work, riveting, and sanding.	Flying fragments, objects, large chips, particles, sand, dirt, etc.	B,C,D, E,F,G, H,I,J, K,L,N	Spectacles, goggles faceshields SEE NOTES (1) (3) (5) (6) (10) For severe exposure add N	Protective devices do not provide unlimited protection. SEE NOTE (7)	Protectors that do not provide protection from side exposure. SEE NOTE (10) Filter or tinted lenses that restrict light transmittance, unless it is determined that a glare hazard exists. Refer to OPTICAL RADIATION.
HEAT	Furnace operations, pouring, casting, hot dipping, gas cutting, and welding.	Hot sparks	B,C,D, E,F,G, H,I,J, K,L,*N	Faceshields, goggles, spectacles *For severe exposure add N SEE NOTE (2) (3)	Spectacles, cup and cover type goggles do not provide unlimited facial protection. SEE NOTE (2)	Protectors that do not provide protection from side exposure.
		Splash from molten metals	*N	*Faceshields worn over goggles H,K SEE NOTE (2) (3)		
		High temperature exposure	N	Screen faceshields. Reflective faceshields. SEE NOTE (2) (3)	SEE NOTE (3)	
CHEMICAL	Acid and chemicals handling, degreasing, plating	Splash	G,H,K *N	Goggles, eyecup and cover types. *For severe exposure, add N	Ventilation should be adequate but well protected from splash entry	Spectacles, welding helmets, handshields
		Irritating mists	G	Special purpose goggles	SEE NOTE (3)	
DUST	Woodworking, buffing, general dusty conditions.	Nuisance dust	G,H,K	Goggles, eyecup and cover types	Atmospheric conditions and the restricted ventilation of the protector can cause lenses to fog. Frequent cleaning may be required.	
OPTICAL RADIATION	WELDING: Electric Arc		O,P,Q	TYPICAL FILTER LENS SHADE / PROTECTORS SEE NOTE (9) 10-14 Welding Helmets or Welding Shields	Protection from optical radiation is directly related to filter lens density. SEE NOTE (4). Select the darkest shade that allows adequate task performance.	Protectors that do not provide protection from optical radiation. SEE NOTE (4)
	WELDING: Gas		J,K,L, M,N,O, P,Q	SEE NOTE (9) 4-8 Welding Goggles or Welding Faceshield		
	CUTTING			3-6		
	TORCH BRAZING			3-4	SEE NOTE (3)	
	TORCH SOLDERING		B,C,D, E,F,N	1.5-3 Spectacles or Welding Faceshield		
	GLARE		A,B	Spectacle SEE NOTE (9) (10)	Shaded or Special Purpose lenses, as suitable. SEE NOTE (8)	

16

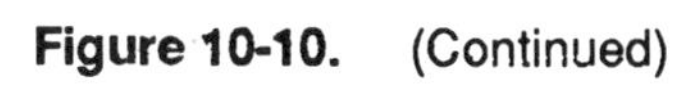

Figure 10-10. (Continued)

Table 10-4. OSHA permissible noise exposure

dB_A	Hours
90	8
92	6
95	4
97	3
100	2
102	1-1/2
105	1
110	1/2
115	1/4

Source: 29 CFR 1910.95.

Audiometric Testing

Because occupational hearing losses occur gradually over time, they may go unnoticed without audiometric testing. The OSHA standard recognizes this and requires that testing be made available to all employees whose exposures equal or exceed an 8-hour time-weighted average of 85 dB_A. The tests must be provided at no cost to employees and be performed by qualified personnel with properly calibrated equipment.

Baseline Audiograms

Within 6 months of an employee's first exposure at or above the action level (85 dB_A), the employer must establish a valid baseline audiogram against which subsequent audiograms can be compared.

Annual Audiograms

At least annually after obtaining the baseline audiogram, the employer must obtain a new audiogram for each employee who is exposed at or above the 85 dB_A action-level.

Audiogram Evaluation

Each employee's annual audiogram must be compared to the employee's baseline audiogram to determine if it is valid and if a standard threshold shift has occurred. A standard threshold shift (STS) is defined by the regulation as a change in hearing, relative to the baseline, of an average of 10 dB or more at 2,000, 3,000 and 4,000 Hz in either ear.

If a STS is noted, the employee may be retested within 30 days, and the results of the new test may used as the annual audiogram. However, if the STS persists, the employee must be informed in writing within 21 days.

Employees who have experienced a STS and who are not wearing ear protection must then be fitted with it, trained in its use and be required to wear it, unless a physician determines that the standard threshold shift is neither work related nor aggravated by occupational noise exposure. Employees who are already wearing ear protection must be refitted and retrained and, if necessary, provided with hearing protectors that afford more attenuation.

If additional testing is necessary, or if the employer suspects that a medical pathology of the ear is caused by or aggravated by ear protection, the employee must be referred for a clinical evaluation.

Hearing Protectors

A popular but false perception about ear protection is that it interferes with the ability to communicate. In reality, use of ear protection makes voice communication easier. As shown in Figure 10-11, ear protection attenuates high-frequency noise more than low-frequency noise. Since the high-frequency speech-interfering component of wide-band industrial noise is attenuated, it is actually easier to communicate while wearing ear protection.

The noise standard requires that ear protection be worn by employees who:

- Are exposed to sound levels exceeding those shown in Table 10-4
- Are exposed to an 8-hour time-weighted average of 85 dB_A and who have not received a baseline audiogram
- Have a standard threshold shift

Although the noise standard does not *legally* require ear protection to be worn except under the three situations listed above, it is prudent to wear it in any environment where noise levels exceed 85 dB_A *regardless of duration.*

Employee Training

A training program must be implemented for all employees exposed to noise at or above an 8-hour time-weighted average of 85 dB_A. The training program must be repeated annually and must include at least the following elements:

- An explanation of the effects of noise on hearing
- An explanation of the purpose of hearing protectors

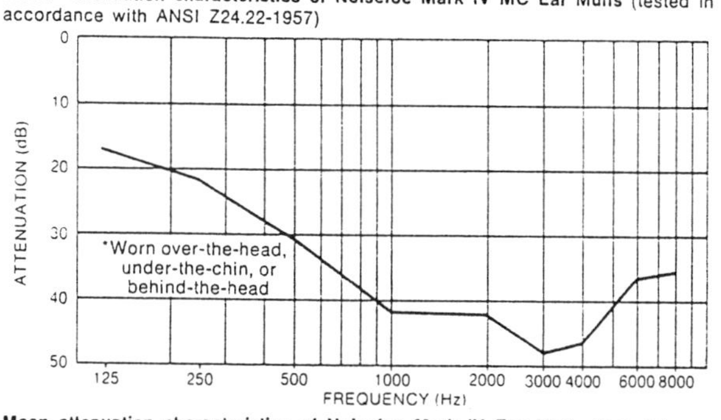

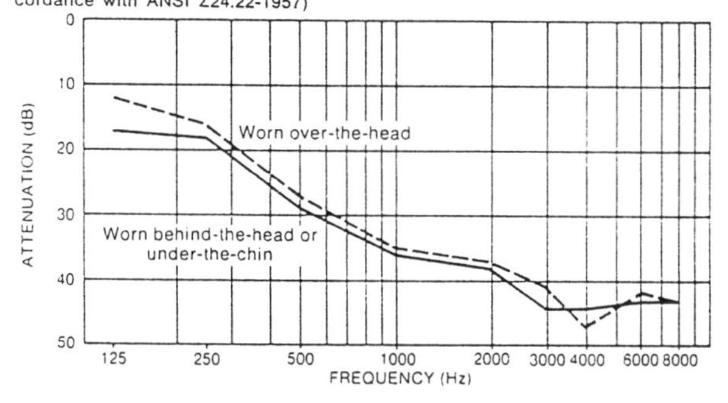

Figure 10-11. Attenuation characteristics for MSA Ear Defenders. (Courtesy MSA)

- The advantages, disadvantages and attenuation characteristics of various types of hearing protection
- Instructions on selection, fitting, use and care of hearing protection
- An explanation of the purpose of audiometric testing and description of the test procedures

Types of Hearing Protection

Hearing protection is generally available in three styles: ear-muffs, ear plugs and canal caps. It should also be noted that all devices do not provide the same level of protection and there are variations in protection even among the same style equipment. For example, noise reduction ratings from five different manufacturers of ear plugs range from 24–31 dB.

Ear-muffs

Ear-muffs (Figure 10-12), which are technically called *circumoral protectors,* consist of two acoustically insulated cups connected with a metal or plastic band. The cups are placed over the ears and the band

Figure 10-12. Ear muffs. (Cabot Safety)

is usually adjusted over the head. The band can also be placed behind the neck or under the chin, but different positions result in varying degrees of attenuation.

Earplugs

Earplugs are less expensive than muffs, but have much shorter service lives. They are available in three styles: molded, custom and formable.

Molded plugs (Figure 10-13) are manufactured from silicone rubber or plastic and are generally available in small, medium and large sizes. Because they are relatively firm, molded plugs tend to become uncomfortable when worn for more than a few hours.

Custom plugs are generally more comfortable to wear than molded plugs. They are prepared from a special pliable compound that is pushed gently into the ear canal by a trained technician. When the material cures, it retains the form of the individual's ear canal.

Formable plugs are made from resilient materials such as very fine fiber glass, expandable plastic foam or wax-impregnated cotton. The wearer manually compresses the plug material and inserts it the ear. After a few moments the compressed material expands, sealing off the ear canal.

Canal Caps

Unlike ear plugs which are inserted into the ear canal, canal caps (Figure 10-14) provide protection by

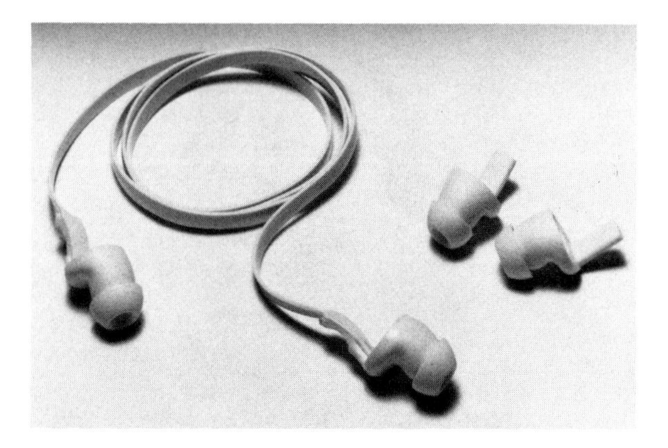

Figure 10-13. Molded ear plugs. (Cabot Safety)

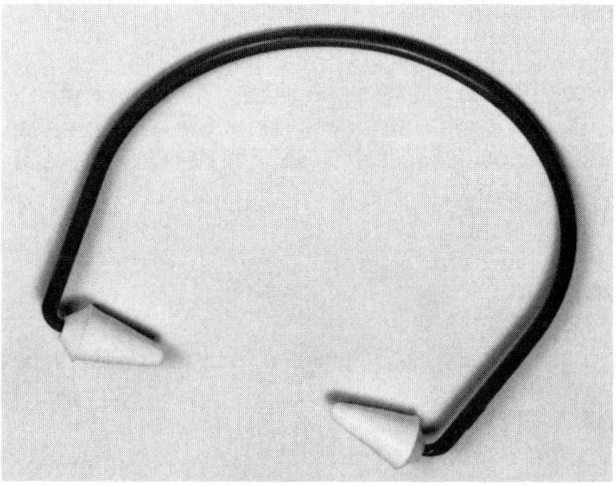

Figure 10-14. Canal caps. (Cabot Safety)

sealing off the opening in the outer ear. They consist of two small rubber caps connected together by a semicircular band. The band is fitted over the head, behind the neck or under the chin, and the caps are positioned over the canal opening. Like ear muffs, the level of attenuation depends on the position of the headband.

Determining the Level of Protection

The level of noise reduction afforded by a specific type of ear protector is indicated by the EPA Noise Reduction Rating (NRR). It should be noted, though, that NRRs are established on the basis of laboratory tests, and attenuation levels during actual use are less than those achieved in the laboratory.

OSHA recognizes that this reduction in protection exists and its *Technical Manual* offers guidelines for evaluating the effectiveness of ear protection used in the workplace. Using the OSHA method, 7 dB is first subtracted from the published NRR to compensate for spectral uncertainty. The result is then divided by 2 to provide a safety factor. As demonstrated below, a device with a published NRR of 27 would provide only a 10 dB attenuation in the workplace.

Correct for spectral uncertainty by subtracting 7 dB from NRR

$$27 \text{ dB} - 7 \text{ dB} = 20 \text{ dB}$$

Divide by 2 for safety factor

$$\frac{20 \text{ dB}}{2} = 10 \text{ dB}$$

FOOT PROTECTION

ANSI Standard Z-41 classifies safety-toe footwear into three categories depending on its ability to meet the compression and impact requirements specified in Table 10-5. Puncture resistant soles are available on some footwear styles, and the upper portion of the feet can be protected by an integral metatarsal guard. Some shoes and boots are also constructed with nonferrous components to reduce the possibility of fire or explosion in hazardous locations. Conductive soles are also available that allow static electricity to drain off more easily.

PROTECTION IN HOT ENVIRONMENTS

Above ground spaces such as chemical reactors, process vessels and storage tanks can become veritable ovens during the summer months, especially if they have dark external surfaces which absorb solar radiation.

Although ventilation can be employed to achieve some comfort cooling, it is not effective against radiant heat like that given off by tank walls. However, some protection against heat stress can be provided by personal cooling vests like those shown in Figures 10-15 and 10-16, or a tunic like that shown in Figure 10-17.

Table 10-5. ANSI classification of safety toe footwear

Class	Compression (Pounds)	Impact (Foot Pounds)
75	2,500	75
50	1,750	50
30	1,000	30

Adapted from ANSI Z41-1975.

The vest shown in Figure 10-15 is a passive device made of woven fabric. The front and back of the vest contain internal pockets that are sized to hold chemical gel-packs. The gel-packs are placed in a freezer until they solidify, and they are then slipped into the vest pockets.

Active devices circulate chilled water through a vest (Figure 10-16) of a tunic (Figure 10-17). The vest unit incorporates an integral pouch that houses a centrifugal pump, a battery, an on/off switch and a temperature regulation valve. The coolant bag is filled with water and crushed ice, ice cubes or frozen chemical gel-packs, but ice transfers heat more efficiently so it cools more quickly than gel-packs. The pump circulates chilled water through the vest, and the wearer uses the regulator valve to adjust the temperature to a comfortable level.

The vest is normally worn with the pouch in back, but it can be reversed to allow the use of an SCBA. At a minimum, a T-shirt should be worn under the vest, and if the wearer is still too cold after adjusting the regulator, he can control the temperature further by switching the pump on and off.

The tunic style is similar to the vest in operation; however, as seen in Figure 10-17, 10-18 and 10-19, the battery pack, circulating pump and coolant reservoir may be worn on the back or attached to a waist belt or SCBA harness.

Figure 10-15. Passive cooling vest. (Mainstream Engineering Corporation)

Figure 10-16. Active cooling vest style. (ILC Dover)

CHEMICAL PROTECTIVE CLOTHING

Chemical protective clothing (CPC) includes gloves, boots, pants, jackets, splash hoods and fully encapsulated suits. Although CPC may be required during some entries, it should be noted that no single material provides protection against all chemicals. Instead, protective equipment is available in a variety of materials which offer varying degrees of resistance to different substances.

Selection Process

Selection of CPC is largely influenced by three factors:

1. The nature of the hazard
2. The chemical resistance offered by the material
3. The physical conditions where the equipment will be used

Figure 10-17. Belt-mounted cooling tunic. (MSA)

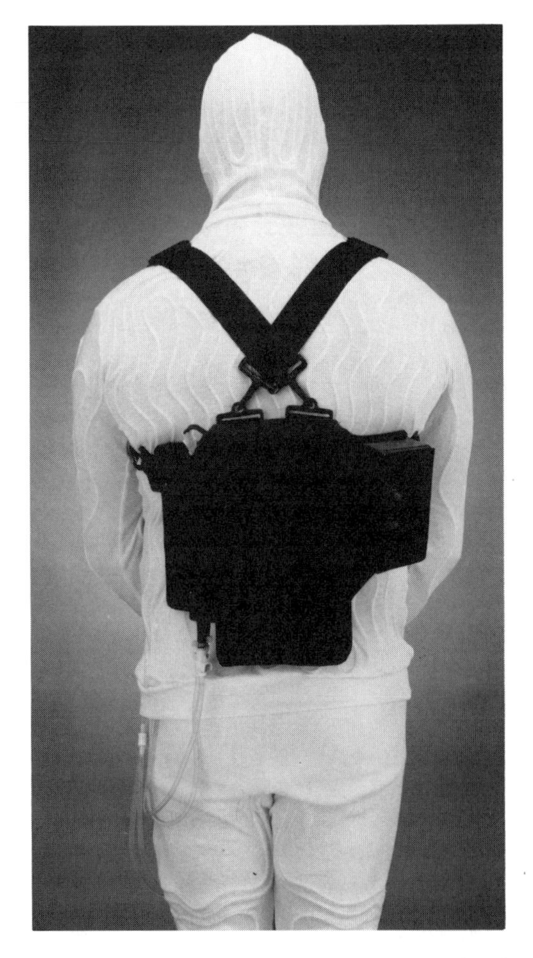

Figure 10-18. Cooling tunic worn on the back. (MSA)

Figure 10-19. Worn with SCBA. (MSA)

Since selection of appropriate CPCs can be very complicated, it should only be done by a qualified person.

Type of Hazard

Three questions must be asked when characterizing the nature of a chemical hazards:

1. What is the material's physical state—dust, liquid, vapor, mist or gas?
2. What is the nature of the contact—incidental contact, splash contact or complete immersion?
3. What part of the body is affected—face, head, torso, hands, arms, legs, feet or the entire body?

Chemical Resistance

As mentioned previously, different CPC materials offer varying degrees of resistance to different chemicals. The degree of chemical resistance offered by a specific material may be determined by consulting compatibility charts similar to Table 10-6. Rubber gloves, for instance, would be a good choice when

Table 10-6. Typical selection chart for chemical protective clothing

Substance	CPE	FRU	Viton	Butyl	Vinyl	Nitrile	Neoprene
Acetaldehyde	G	X	X	X	X	X	P
Acetic acid, glacial	E	X	P	G	P	P	X
Acetone	E	X	X	E	X	X	G
Allyl alcohol	E	P	E	E	E	E	E
Ammonia	E	P	E	E	E	E	E
Benzene	P	P	E	X	P	X	X
Butyl alcohol	E	X	E	G	G	E	E
Chlorobenzene	X	P	E	X	X	X	X
Cresol	E	X	E	X	P	X	P
Diesel oil	E	E	E	X	E	E	G
Dinitrotoluene	P	X	P	X	P	X	X
Ethyl alcohol	E	X	E	E	G	E	E
Ethyl benzene	P	X	E	X	G	X	X
Ferric sulfate	E	E	E	E	E	E	E
Formic acid	E	P	P	E	G	G	E
Gasoline	G	E	E	X	P	E	G
Hexane	E	E	E	X	G	E	G
Hydrogen cyanide	E	G	E	E	G	G	G
Hydrogen sulfide	E	E	X	E	E	X	E
Isopropyl alcohol	E	P	E	E	E	G	G
Lacquer thinner	X	X	X	P	X	X	X
Maleic anhydride	E	E	E	P	E	X	X
Methyl bromide	I	I	E	X	I	G	X
Methyl ethyl ketone	P	X	X	E	X	X	X
Nitric acid conc.	X	X	E	P	X	X	X
Perchloroethylene	P	G	E	X	X	P	X
Potassium hydroxide	E	G	G	E	E	G	G
Sodium hydroxide	E	X	G	E	G	G	E
Toluene	P	G	E	X	X	P	X
Trichloroethylene	P	G	E	X	X	P	X
Xylene	X	E	E	X	X	X	X

E = Excellent; G = Good; P = Poor; I = Insufficient information; X = Not recommended.

working with acids and alkalies, but a poor choice when working with solvents.

In addition to general selection guides like that shown in Table 10-6, many manufacturers of chemical protective clothing also offer much more detailed information on the performance characteristics of the materials from which they manufacture their products. These materials are usually characterized in terms of their resistance to penetration, degradation and permeation.

Penetration. Penetration is the passage of a liquid material through small openings in clothing. These include the gaps between zipper teeth, holes caused by stitched seams and minor surface imperfections like pores and pin holes.

Degradation. Degradation is a change in some physical property of a material that results in a reduction in the material's performance. These changes may be produced by chemical contact or by exposure to environmental conditions such heat, ambient ozone or UV radiation from sunlight. Some changes like swelling, wrinkling, hardening, softening and discoloration may be easy to detected by sight or touch. Others, such as the degree of change or reductions in physical strength that result in decreased resistance to snags, punctures or tears, are more difficult to detect.

Permeation. Permeation is the process by which a chemical passes through a piece of protective material that has no visible holes. It may occur with little or no visible effect on the material, and there may be no obvious indication of degradation. As shown in Figure 10-20, the permeation process takes place in three phases.

First, the chemical is absorbed by the exposed surface of the material. Next, the chemical diffuses through the material. Finally, it passes to the other side where it can contact the wearer. The time from initial contact to detection on the other side of the material is called the break-through time.

Break-through time is an important consideration when choosing between CPC materials that have similar degradation characteristics. However, as we'll see later, the detection limit of the instrument used to determining break-through must also be considered. Break-through time is affected by a number of factors including the:

- Type, thickness and solubility of CPC material
- Process used to manufacture the CPC
- Concentration of the challenge agent
- Ambient temperature and pressure
- The temperature of the challenge agent

Since permeation is inversely proportional to the material's thickness, doubling the thickness cuts the permeation rate in half. And because break-through time is directly proportional to the square of the material's thickness, doubling the thickness effectively quadruples the break-through time.

Batteries of challenge agents used by ASTM and NFPA are listed in Table 10-7.

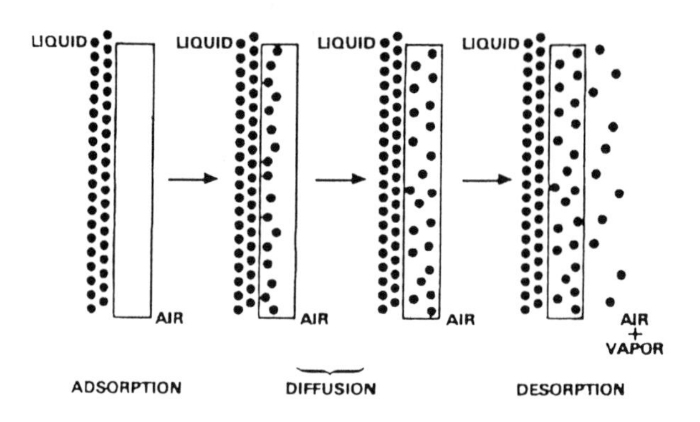

Figure 10-20. The permeation process. (TRI/Environmental)

Conditions of Use

The conditions under which CPC will be used must also be evaluated during the selection process. Among the factors to consider are whether the CPC will be used in environments where it might be subjected to open flames, cuts, tears, punctures and abrasion.

Use of Vendors' Data

Although many suppliers publish compatibility charts, the reported test results may be difficult to compare because tests may be conducted in different ways. Some vendors test materials themselves, others simply rely on data provided to them by the material's manufacturer.

The test most widely used to determine permeation is American Society for Testing and Materials (ASTM) Method F739, "Standard Method for Resistance of Protective Clothing Materials to Permeation By Liquids and Gasses." As shown in Figure 10-21, the ASTM test cell is divided into two compartments by the material to be tested. One side of the cell is then filled with the challenge agent and the instrumentation used to measure break-through is connected to the other side.

The ASTM method, however, does not specify the limit of detection for the analytical method, and different vendors may use instruments with different detection limits. One manufacturer, for example, might use an instrument with a detection limit of 0.001 ppm, while another uses an instrument with a detection limit of 0.1 ppm. In this situation, selecting

Table 10-7. Permeation test challenge agents

Chemical	ASTM F1001[1]	NFPA 1991[2]	NFPA 1992[3]
Acetone	o	o	o
Acetonitrile	o	o	
Ammonia	o	o	
1,3 Butadiene	o		
Carbon disulfide	o	o	
Chlorine	o	o	
Dichloromethane	o	o	
Dimethyl amine	o	o	o
Dimethylformamide	o	o	
Ethyl acetate	o	o	o
Ethyl oxide	o		
Hexane	o	o	o
Hydrogen chloride	o		
Methanol	o	o	
Methyl chloride	o		
Nitrobenzene	o	o	
Sodium hydroxide	o	o	o
Sulfuric acid	o	o	o
Tetrachloroethylene	o	o	
Tetrahydrofuran	o	o	o
Toluene	o	o	o

1 ASTM F1001 American Society for Testing Materials, *Standard Guide for Chemical Protective Clothing*.
2 NFPA 1991 National Fire Protection Association, Standard of Vapor-Protection Suits for Emergency Response.
3 NFPA 1992 National Fire Protection Association, Standard on Liquid Splash-Protective Suits for Emergency Response.

a material on the basis of break-though time alone would be a mistake because the analytical methods differ by a factor of 100.

If you're confused, consider a hypothetical situation involving two pairs of gloves. One manufacturer reports a break-through time of one hour, the other reports six hours. On the surface, it would appear that the six hour gloves are more protective because their break-through time is six times greater. However, if this six hour break-though was determined

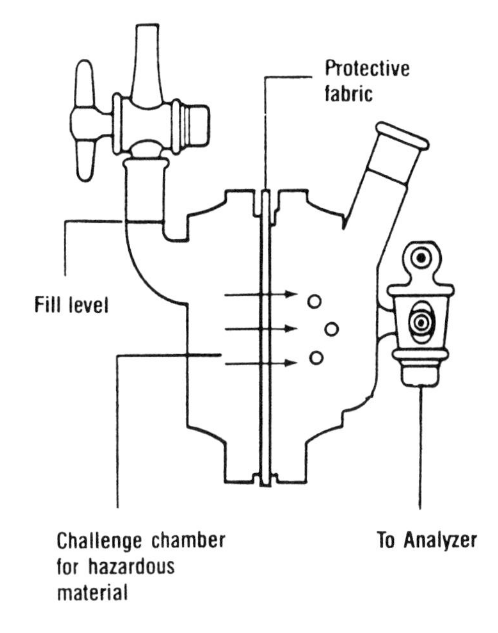

Figure 10-21. ASTM test cell. (DuPont)

using a instrument that was 100 times less sensitive than the one used to determine the one hour break-through, the one-hour rated glove could actually be more protective. Thus it is important to not only compare break-through times, but also the detection limit of the instrument used to establish them.

Fabrication Materials

Physical resistance characteristics of different types of materials used to fabricate chemical protective clothing are listed in Table 10-8, and chemical resistance characteristics are summarized below. More detailed chemical resistance information can be found in:

Quick Selection Guide to Chemical Protective Clothing, Mansdorf S.Z. and K. Forsber, Van Nostrand Reinhold, New York, NY (1988); *Guidelines to Selection of Chemical Protective Clothing,* American Industrial Hygiene Association, Fairfax, VA; and *Chemical Protective Clothing Permeation and Degradation Database,* Blotzer, M.J., Lewis Publishers, Chelsea, MI.

Butyl rubber, which is a member of the isobutylene/isoprene chemical family, is not resistant to grease and oil and is attacked by petroleum products, aromatic hydrocarbons and coal tars. However, butyl does affords good protection against other chemicals

Table 10-8. Physical characteristics of CPC materials

Material	Resistance to							Relative Cost
	Cuts	Abrasion	Tears	Puncture	Heat	Ozone	Flexibility	
Butyl rubber	G	F	G	G	E	E	G	High
Natural rubber	E	E	E	E	F	P	E	Medium
Neoprene	E	E	G	G	G	E	G	Medium
Neoprene/styrene-butadiene	G	G	G	G	G	G	G	Medium
Neoprene/natural Rubber	E	E	G	G	G	G	E	Medium
Nitrile rubber	E	E	G	E	G	F	E	Medium
Nitrile rubber/ polyvinyl chloride	G	G	G	G	F	E	G	Medium
Polyethylene	F	F	F	P	F	F	G	Low
Chlorinated polyethylene	G	E	G	G	G	E	G	Low
Polyurethane	G	E	G	G	G	G	E	High
Polyvinyl alcohol	F	F	G	F	G	E	P	Very High
Polyvinyl chloride	P	G	G	G	P	E	F	Low
Styrene-butadiene	G	E	F	F	G	F	G	Low
Viton	G	G	G	G	G	E	G	Very High

*Rating subject to variation depending on formulation thickness and whether material is supported by fabric.
E = Excellent G = Good F = Fair P = Poor
Adapted from *Guide to Selection of Chemical Protective Clothing,* ACGIH, Cincinnati, Ohio (1985) pg. 31.

such as ketones, esters, inorganic salts and most acids and alkalies.

Chlorinated polyethylene (CPE) provides good to excellent resistance against aliphatic hydrocarbons, phenols, ketones, esters, acids, bases and salts. It also exhibits excellent abrasion characteristics and offers good resistance to cuts, tears and punctures.

Natural rubber, or latex, offers good temperature resistance and tensile strength. It provides good resistance to acids, alkalies and alcohols, but is not recommended for aliphatic or aromatic compounds.

Neoprene is a synthetic material that offers good to excellent resistance to straight chain hydrocarbons, aliphatic hydroxy compounds, methanol, ethanol, ethylene glycol, animal and vegetable fats and oils, and fluorinated hydrocarbons like the Freons. It has excellent tensile strength and resists heat and ozone. It offers only moderate abrasion resistance but remains flexible at low temperatures.

Nitrile is virtually unaffected by saturated and unsaturated aliphatic hydrocarbons, alkali solutions, and saturated salt solutions. It is a good choice when working with oils, fats, acids, caustics and alcohols, but it is not generally recommended for use with strong oxidizing agents, ketones and acetates.

Polyvinyl chloride (PVC) is a synthetic thermoplastic that offers good to excellent resistance to oils, greases, acids, aromatic solvents and amines. However, it offers poor resistance to halogenated hydrocarbons, ketones and esters. One of PVC's most attractive features is its excellent wet grip and abrasion resistance.

Polyvinyl alcohol is a water soluble synthetic material. It is highly impermeable to gases and affords excellent chemical resistance to aromatic and chlorinated solvent. However, since it is adversely affected by water, it cannot be used in water or in aqueous solutions.

Viton® is a specialty fluoroelastomer that offers excellent resistance to petroleum products such as oils, fuels and lubricants, most mineral acids, hydraulic fluids, and aliphatic, aromatic and chlorinated solvents.

Clothing Types

Although an assortment of materials is used in the manufacture of chemical protective clothing, Table 10-9 indicates that specific items of protective clothing are not made from some materials.

Table 10-9. Commercially available types of CPC

Material	Gloves	Suits*	Boots
Butyl rubber	o	o	o
Natural rubber	o	o	o
Neoprene	o	o	o
Nitrile	o	o	
Polyethylene	o		
Polyvinyl chloride	o	o	o
Polyvinyl alcohol	o		
Polyurethane	o	o	
Viton	o		

* Separate jacket and pants.

Gloves

Gloves are perhaps the most frequently encountered type of CPC. Since there are dozens of makes, models and styles of gloves; choosing the right ones can a confusing task. The following guidelines may help to make the selection process easier.

1. Match the glove material to the application by using published permeation and degradation data.

2. Select the length and thickness required for the task.

   Wrist/forearm length (9–14 inches) protects the hand and wrist from a variety of hazards and provides good protection in most situations.
   Elbow length (14–18 inches) provides additional protection of the forearm.
   Shoulder length (30–36 inches) provides maximum protection and is especially well suited for situations that require the arms to be inserted into deep vats or tanks.
   Glove thickness is usually measured in thousandths of an inch or mills (1 mill = 0.001 inch). Thin gloves offer less overall protection than thick gloves, but they afford better tactile sensitivity, flexibility and dexterity.

3. Decide if the gloves should be lined.

   Unlined gloves offer the best sensitivity and dexterity.
   Flocked lining is a shredded fiber (usually cotton) applied to the inside surface of a glove. The flock helps to absorb moisture and makes the gloves easier to don and doff.
   Knit or jersey linings consist of cotton or synthetic material bonded to the inside of the surface of the glove. Jersey linings absorb perspiration and afford additional protection against temperature extremes.

4. Determine the level of grip required. Some gloves have a textured finish, others rely on the inherent gripping quality of the glove material.

5. Select the appropriate style of cuff.

   Rolled-edge cuffs provide additional strength as well as a barrier that keeps material from running into the glove and onto the hand.
   Knit wrists provide a snug fit that prevents material from entering the glove.
   Gauntlet cuffs provide extended length to protect the wrist and forearm.

Gloves should be sized for a proper fit by measuring the circumference of the hand around the palm area as shown in Figure 10-22. The measurement in inches indicates the glove size. For instance, an 8-inch circumference indicates a size 8 glove.

Glove sizes vary among manufacturers and are determined according to men's hands. A comparison between hands and generic glove sizes is provided in Table 10-10.

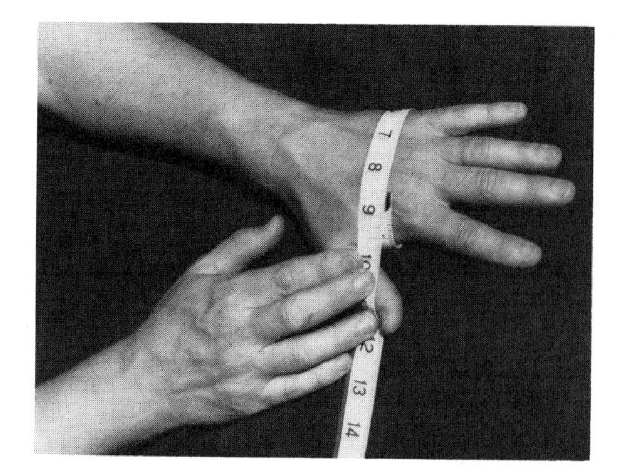

Figure 10-22. Measuring hand for proper glove fit. (John Rekus)

Table 10-10. Comparison of glove sizes

Glove Size	Hand Size
Extra small	6–7
Small	7–8
Medium	8–9
Large	9–10
Extra large	10–11

Figure 10-23. Chemical resistant overalls. (Standard Safety)

Jackets, Pants and Hoods

Chemical protective clothing that provides splash protection is available in jackets, overalls and pants, (Figure 10-23), hoods (Figure 10-24), suits (Figure 10-25), and aprons. Full-body protection against vapors and gases can also be provided by fully encapsulated suits like those shown in Figures 10-26 and 10-27.

Footwear

Chemical resistant footwear (Figure 10-28) is available in only a limited number of polymers that include natural rubber, PVC, neoprene and butyl rubber. Different tread styles offer varying degrees of slip resistance and some chemical protective footwear meets the ANSI requirements for safety-toed foot protection.

Protective Clothing Ensembles

EPA has grouped protective clothing ensembles into four categories. However, it should be noted that the EPA categories are based what on the ensemble looks like rather than on performance specifications.

Level A ensembles provide the highest level of protection and are used when a high degree of hazard to the skin, eyes, or respiratory system exists or is suspected of being present. They consist of a positive-pressure, self-contained breathing apparatus worn with a total encapsulating chemical protective suit. The breathing apparatus is worn inside the suit, where it is protected from damage that might result from chemical contact.

Level B ensembles maintain the same level of respiratory protection as Level A, but provide a lower level of skin protection by substituting chemical-resistant clothing consisting of a hooded jacket and pants in place of the totally encapsulated chemical suit. These suits may be reusable or disposable.

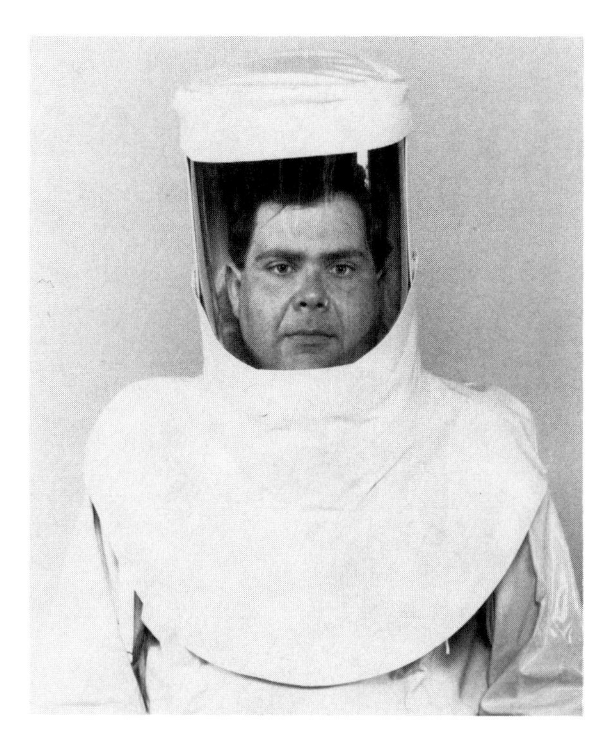

Figure 10-24. Chemical splash hood. (Standard Safety)

Figure 10-25. Chemical resistant splash-suits. (Standard Safety)

Level C protection consists of chemical resistant clothing and an air purifying respirator.

Level D does not require respiratory protection and consists of ordinary work clothing.

Decontaminating Chemical Protective Clothing

It is important to decontaminate used chemical protective clothing so that other people can handle it without being exposed to potentially toxic surface residue.

Decontamination is affected by variables that include: the physical and chemical characteristics of the contaminant, the nature of the CPC material and the duration of the chemical exposure. There are no published consensus standards which address the decontamination procedures, and in some situations decontamination may neither be cost effective nor appropriate. Your answers to the following three questions may help you decide whether to decontaminate or dispose of used CPC.

1. Will decontamination be effective? If not, then don't bother doing it.
2. Will decontamination damage the equipment? If it does, then it's not appropriate.
3. Is decontamination economically viable? If it costs more to decontaminate CPC than it does to properly dispose of it and purchase new gear, decontamination is not an economically attractive option.

Decontamination Methods

The three methods most widely used for decontamination are water washing, solvent flushing and aeration.

Water Washing. Water washing has two major limitations. First, it only removes surface contamination, not material that may have permeated into the CPC matrix. Second, it is only effective for water soluble contaminants.

Contaminated CPC should be washed for about 10 minutes with a commercial laundry detergent and water. Soft bristle brushes may be used to remove

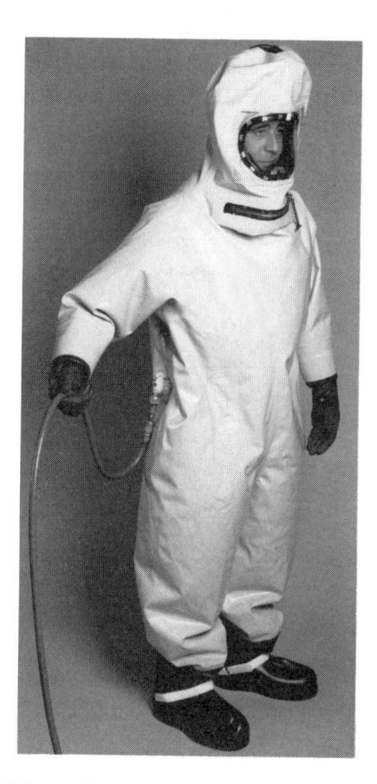

Figure 10-26. Gas- and vapor-tight fully encapsulated suit worn with airline respirator. (Standard Safety)

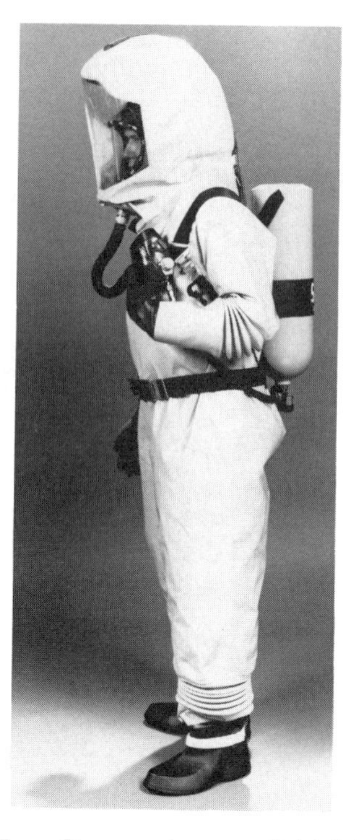

Figure 10-27. Gas- and vapor-tight fully encapsulated suit worn with SCBA. (Standard Safety)

stubborn visible contaminant deposits. The equipment should be rinsed, inspected and washed and rinsed again if necessary. It can then be air dried.

Water temperature for washing and rinsing should not exceed 180°F, and if hot air is used for drying, its temperature should not exceed 160°F. Local regulations and ordinances may require that the wash water be either collected and treated, or disposed of as a hazardous waste.

Solvent Washing. Another washing method employs solvents. However, like water washing, solvent washing is useful only for removing surface contamination.

The ideal solvent is one that:

- Provides effective decontamination
- Can be redistilled for reuse
- Has a low heat of vaporization
- Has low flammability
- Is not very toxic

While an ideal solvent does not exist, there are three which come close. They are Stoddard solvent, perchloroethylene and Freon TF, all of which have been used for years in the dry cleaning industry.

Figure 10-28. Chemical resistant boot for use with gas- and vapor-tight fully encapsulated suit. (Standard Safety)

The principal disadvantage of solvent washing is that it degrades some materials and strips the plasticizers from others. Rubber, for instance, is adversely affected by all three of the commonly used dry cleaning solvents; neoprene is affected by ing solvents; neoprene is affected by perchloethylene, and polyvinyl chloride is damaged by Freon.

Aeration. Since water- and solvent-washing are only effective in removing surface contaminants, a different method must be used to remove contaminants that have permeated the CPC material's matrix. These contaminants can be desorbed by hot air aeration. Aeration, however, is only effective for volatile materials and it must be done in a well-ventilated area because desorbed substances are released to the environment. Recommended aeration time-temperature parameters are listed in Table 10-11.

While aeration may be effective in desorbing contaminants it can cause stiffening, shrinkage and reduction of chemical resistance with no visible change in appearance.

Table 10-11. Recommended desorbtion parameters

Temperature (°F)	Time (Hours)
130	8
120	48
80	72

Proposed ASTM Standard

No standard methods or validated tests concerning the effectiveness of decontamination procedures are currently published, but ASTM is in the process of developing one. The proposed ASTM method requires the test garment to be placed in a container where it is heated for 24 hours. The container's air is then analyzed using a detector tube or direct reading instrument to determine if it meets "unacceptable" limits. If not, additional decontamination is performed until an acceptable level is achieved.

ESTABLISHING A PROGRAM

Protective clothing and equipment is of no practical value unless it is properly selected, correctly worn and adequately maintained. These goals can be achieved by establishing a personal protective management program that includes five key elements.

- Hazard assessment
- Equipment selection
- Employee training
- Workplace surveillance
- Management commitment

Hazard Assessment

The first step in setting up a personal protective equipment program is determining when and where it must be worn. Remember, protective equipment should be considered as the last alternative, and it's better to *eliminate* hazards at their source through engineering controls. Although hazard assessment is not a particularly complex task, it is highly subjective because of the wide variety of hazardous situations that may exist in a workplace. However, most hazards can be identified by obtaining information on the space, the reason for entry and the nature of the operations to be performed.

Equipment Selection

The marketplace offers a large number of makes, models and styles of equipment, and all of it does not provide the same level of protection. For example, the noise attenuating characteristics of ear protection varies widely depending on the make and model. In addition, even equipment which meets identical technical specifications can vary significantly in terms of style, appearance and comfort. These subjective factors should also be considered when selecting protective equipment because they play a decided role in gaining wearers' acceptance.

In other words, a pair of safety glasses that looks "classy" and feels comfortable is more likely to be worn than a pair that pinches the nose and ears and makes the wearer look like a raccoon.

Many employers have found that when workers participate in the equipment selection process, they are more likely to wear it. Allowing workers to choose from a variety of protective devices can also increase the frequency of use. For example, three types of ear protection might be made available so that employees can choose the type that they like best. While stocking a variety of devices may initially be more expensive, it is cost effective in the long term because employees are more likely to wear equipment that suits their personal preference. This increased use has two benefits. First, compensation claims resulting from injuries will decrease. Second, supervisors will spend less time filling out accident report forms, reminding employees to wear their equipment and disciplining those who don't.

Employee Training

Passing out equipment and ordering employees to wear it will be only marginally effective for getting employees to participate in the program. A far better approach is to spend some time explaining why protective equipment is necessary, and describing when, where and how it should be worn. Personal accounts from employees whose injuries could have been prevented through the use of protective equipment or case histories gleaned from the OSHA log of occupational injuries can be invaluable tools to reinforce the training. Employees should also be instructed on how to properly clean, inspect and maintain their assigned equipment and informed of the procedure for replacing worn or damaged devices.

Workplace Surveillance

All personal protective equipment must retain its original level of effectiveness and Table 10-12 identifies selected items that should be regularly inspected. The use of worn or defective equipment is not only a violation of OSHA standards, but it could also result in injury if it fails to perform its intended function.

Users should clean, maintain and inspect protective equipment in accordance with the training they received. Any damaged or broken equipment should be immediately removed from service.

Supervisors have an obligation to monitor and enforce the wearing of protective equipment among those whose work they direct. They must remain ever vigilant and assure that required equipment is not just worn, but worn properly.

Management Example

Since protective equipment can only be effective if it's worn, managers and supervisors must set a good example by wearing their equipment properly and in designated areas. Nothing will undermine a personal protective equipment program faster than a management official who fails to wear his equipment or who sets the example by wearing his safety glasses on top of his head or in his breast pocket. Actions do speak louder than words, and employees who see their bosses disregard the rules will perceive the protective equipment program as cheap talk and empty words.

Table 10-12. Visual inspection items

All equipment generally for:

- Punctures
- Tears
- Cracks
- Brittleness
- Tackiness
- Discoloration
- Abrasion
- Pin holes
- Nonuniform coating
- Flaking/peeling

Shoes and boots for:

- Sole separation
- Tread wear

Seams on jackets and pants for:

- Imperfect material alignment
- Lessening of coating at fold points

Zippers for:

- Tooth alignment
- Complete closures
- Separation from garment
- Trapped foreign materials
- Corrosion

SUMMARY

Workplace hazards can best be controlled through engineering controls or by substituting less hazardous materials, equipment or processes. In those situations where these approaches are not feasible, personal protective equipment should be used. However, for protective equipment to be effective, its use must be integrated into a comprehensive program that includes proper selection, user training and management oversight.

REFERENCES

A.D. Little Inc., *Development of Performance Criteria for Protective Clothing Used Against Carcinogenic Liquids*, Arlington, VA: National Technical Information Service, Publication 299-318, (1978).

American National Standard for Occupational and Educational Eye and Face Protection: Z87.1-1989, New York, NY: American National Standards Institute, (1989).

American National Standard for Protective Headwear for Industrial Workers: Z89.1-1986, New York, NY: American National Standards Institute, (1986).

American National Standard for Safety-Toe Footwear: Z41-1975, New York, NY: American National Standards Institute, (1975).

Beraninelli, S.P. and R. Hall, Site specific whole glove chemical permeation, *Am. Ind. Hyg. J.* 46:60–64 (1985).

Bowen, J.E., Chemical protective suits, *Am. Fire J.* 41: 36 (1989).

Carrol, T.R. and A.D. Schwope, Non-destructive Testing and Field Evaluation of Chemical Protective Clothing, Report to the US Fire Administration, Cambridge, MA: Arthur D. Little, (1990).

Coletta, G.C., A.D. Schwope, I.J. Arons, J.W. King, and A. Sivak, *Development of Performance Criteria for Protective Clothing Used Against Carcinogenic Liquids*, DHEW (NIOSH) Publication 79-106 Cincinnati, OH: National Institute for Occupational Safety and Health, (1978).

Eye and Face Protection, Code of Federal Regulations, Vol. 29, Part 1910.133.

Forsberg, K. and S.Z. Mansdorf, *Quick Selection Guide to Chemical Protective Clothing*, New York: NY, Van Nostrand Reinhold (1989).

Grey, G.L. and K.J. York, Chemical protective clothing: do we understand it, *Fire Eng.* 139(2):28–38.

McGary, R.A., Chemical resistant environmental suits: some questions to answer before you invest, *Fire Eng.* 139(5):47–50 (1986).

Merkitch, A., Chemical protective suits: in-house maintenance, *Fire Eng.* 139(11):36 (1986).

Mikatavage, M., S.S. Que Hee and A.A. Ayer, Permeation of chlorinated aromatic compounds through viton and nitrile glove materials, *Am. Ind. Hyg. J.* 45: 617–821 (1984).

Nelson, G.O. and J.S. Johnson, Glove permeation by organic solvents, *Am. Ind. Hyg. J.* 42:217–225 (1981).

Occupational Exposure to Noise, Code of Federal Regulations, Vol. 29, Part 1910.95.

Occupation Foot Protection, Code of Federal Regulations, Vol. 29, Part 1910.136.

Occupational Head Protection, Code of Federal Regulations, Vol. 29, Part 1910.135.

Oxy-Fuel Gas Welding and Cutting, Code of Federal Regulations, Vol. 29, Part 1910.253.

Perkins J.L. and J.O. Stull, Chemical Protective Clothing in Emergency Response, Special Technical Publication 1037, International Symposium on Protective Clothing, San Diego, CA, January 16–17, 1989, American Society for Testing and Materials Philadelphia, PA, (1989).

Personal Protective Equipment: General Requirements, Code of Federal Regulations, Vol. 29, Part 1910.133.

Sansone, E.B. and Y.B. Tewori, The permeability of laboratory gloves to selected solvents, *Am. Ind. Hyg. J.* 39:169–174 (1978).

Say, D.J., Chemical protective clothing: still a long way to go, *Fire Eng.* 114(8):86–88 (1991).

Schwope, A.D., *Guidelines for the Selection of Chemical Protective Clothing*, Cincinnati, OH: American Conference of Governmental Industrial Hygienists, (1983).

Silkowski, J.B., S.W. Horstman, and M.S. Morgan, Permeation of five commercially available glove materials by two pentachlorophenol formulations, *Am. Ind. Hyg. J.* 45:501–504 (1984).

Standard Test Method for Resistance of Protective Clothing Materials to Permeation by Hazardous Liquid Chemicals, F739-81, American Society for Testing and Materials, Philadelphia, PA (1981).

Stull, J.O., Selecting chemical protective clothing: Part 1, *Fire Eng.* 142(9):39–42 (1989).

Stull, J.O., Selecting chemical protective clothing: Part 2, *Fire Eng.* 142(9):45.

Stull, J.O., Considerations for the design and selection of chemical protective clothing, *J. Haz. Mat.* 14(2): 165–189 (1987).

Vapor-Protective Suits for Hazardous Chemical Emergencies, NFPA 1991-1990, Quincy, MA: National Fire Protection Association, (1990).

Welding, Cutting and Brazing: General Requirements, Code of Federal Regulations, Vol. 29, Part 1910.252.

Williams, J.R., Permeation of glove materials by physiologically harmful chemicals, *Am. Ind. Hyg. J.* 40: 877–882 (1979).

Williams, J.R., Evaluation of intact gloves and boots for chemical permeation, *Am. Ind. Hyg. J.* 41: 884–887 (1980).

CHAPTER 11

Respiratory Protection

INTRODUCTION

Chapter 8 explained that atmospheric hazards are best controlled by providing adequate ventilation. However, there are some situations where ventilation may not be feasible, practical or effective. For example, mechanical ventilation may only succeed in keeping spray paint vapors below 10% of the lower explosive level. Although this level is acceptable from a fire and explosion perspective, the atmosphere still presents a serious inhalation hazard and entrants would have to be protected with respirators.

As a rule, respirators may only by used in five situations:

1. When adequate engineering controls cannot be provided
2. While engineering controls are being installed
3. When feasible engineering controls are unable to reduce air contaminants to within acceptable levels
4. During short-term maintenance operations
5. In emergencies, such as escape from a hazardous atmosphere or entry into a confined space for rescue

There is much more to respiratory protection than simply passing out equipment and telling employees to wear it. In fact, OSHA requires employers to establish *written* programs that address a variety of concerns.

The seemingly infinite variety of makes, models and styles of respirators demands that equipment be selected by a qualified person who considers such factors as:

- The potential for oxygen deficiency
- The nature of the work activity
- The worker's health status
- The level of protection provided by the respirator
- The contaminant's concentration
- Occupational exposure limits, e.g., PELs, TLVs and RELs
- The chemical and physical properties of the contaminant including its odor threshold

Respirators must fit properly and users must be provided with an opportunity to wear them in a test atmosphere. If atmosphere-supplying equipment such as self-contained breathing apparatus or air-line devices is used, the breathing air must meet specific criteria for quality.

Training is a very important component of any respirator program, and users as well as their supervisors must be trained in basic respiratory protection practices. Finally, all equipment must be properly inspected, maintained and stored.

PERMISSIBLE PRACTICE

Respirators are considered the last line of defense in protecting employees from harmful dusts, mists, fumes, gases and vapors. In fact, respirator standards published by both OSHA and ANSI require that atmospheric hazards preferably be controlled by substituting less toxic materials or by providing engineering controls such as ventilation.

For example, OSHA would not consider respiratory protection acceptable in a situation where an entrant is over-exposed to welding fumes which could easily be controlled by a relatively inexpensive local exhaust system. However, OSHA would accept respiratory protection in situations where ventilation systems are infeasible, impractical or incapable of fully controlling the hazard.

Respirators may also be used in emergencies such as escape and rescue. In these cases, protection is best afforded by an atmosphere-supplying respirator such as a self-contained breathing apparatus that provides the wearer with an independent air supply. Regardless of what type of equipment is used, OSHA requires employers to establish a comprehensive, written respirator protection program.

RESPIRATOR STANDARDS

The full text of the OSHA respirator standard can be found in 29 CFR 1910.134 which is included in Appendix A at the end of this book. Like many OSHA regulations, the respirator standard was adopted in the early 1970s from an existing ANSI standard. Unfortunately, the 1969 edition of the ANSI standard adopted by OSHA does not discuss advances in respiratory protection that have been made over the last two decades. For example, paragraph (e)(3)(ii) of the standard indicates that hose masks with blowers can be used in atmospheres immediately dangerous to life and health. In actuality these devices are *no longer approved by NIOSH* for use in atmospheres that are immediately dangerous to life or health (IDLH). The standard also shows its age by citing the US Department of the Interior's Bureau of Mines as an authority on respiratory protection, even though the Bureau of Mines transferred all of its regulatory authority for respirators to NIOSH in 1977! Fortunately ANSI has kept pace with advances in the field of respiratory protection and last revised its standard in 1992.

In addition to being woefully outdated, the OSHA standard annoys, irritates and frustrates many readers because instead of being self-contained, it incorporates-by-reference the following five documents which are difficult for anyone but respirator aficionados to locate.

- *Practices for Respiratory Protection,* American National Standards Institute Standard ANSI Z-88.2-1969.
- *National Standard Method of Marking Portable Gas Containers to Identify the Material Contained,* ANSI Z-48.1-1954.
- *Commodity Specification,* Compressed Gas Association Standard CGA-G7.1 (1966) which has been superseded by CGA G7.1 (1989).
- *Air Compressed for Breathing Air Purposes,* Federal Specification BB-1034a, June 21,1868.
- *Breathing Apparatus, Self-contained,* Interim Federal Specification, GG-B-00675b, April 27, 1965.

Sources of Documents

Intrepid readers interested in reviewing these documents can obtain them from the sources listed below:

The two ANSI Standards may be obtained by writing the American National Standards Institute, 1430 Broadway, New York, NY, 10018 of by calling (212) 642-4900.

Commodity specification G-7.1 may be obtained from the Compressed Gas Association, 1235 Jefferson Davis Highway, Arlington, VA 22202, (703) 979-0900.

The two federal specifications may be obtained by contacting your nearest OSHA Assistant Regional Administrator for Technical Support. OSHA Regional Office address and telephone numbers are listed in Appendix B. Alternatively, the OSHA National Office of Health Standards may be contacted by writing USDOL-OSHA, Health Standards Office, 200 Constitution Ave., Washington, DC 20210, (202) 219-7065.

Technical Recommendation

In light of the technical problems associated with the existing standard, prudent readers are cautioned not to be lulled into a false sense of security by believing everything will be all right if they simply comply with OSHAs requirements. Like all OSHA standards, those governing respirators establish only minimal levels of protection, and following standards that are only minimally protective and outdated provides little assurance that employees will not be injured or killed. Consequently, responsible employers are encouraged to consult the latest edition of ANSI documents for the most current information on respiratory protection.

APPROVED EQUIPMENT

Whenever an OSHA or ANSI standard refers to an "approved" respirator, it means a device that has passed a series of laboratory performance tests specified in 30 CFR 11. While responsibility for respirator testing and certification is shared jointly by NIOSH and the Mine Safety and Health Administration (MSHA), the actual testing is performed at the NIOSH Appalachian Laboratory in Morgantown, WV.

All certified equipment is accompanied by labels similar to those shown in Figure 11-1, which describes the respirator's limitations, conditions of use and, in the case of air-purifying devices, the contaminants for which it provides protection. The label also includes a unique "TC" (tested and certified) number issued by NIOSH/MSHA which indicates that the respirator meets the specified performance tests. The first two digits of the "TC" number indicate the schedule under which the respirator was approved (Table 11-1) and the last three are the sequential order in which approvals were granted, beginning with number 001. Thus, a label saying TC-23-123 means that the respirator was the 123rd dust, mist and fume respirator approved under Schedule 23.

NIOSH periodically publishes a list of all certified equipment, the most recent of which was issued on September 30, 1993 (NIOSH Publication 92-101) and is available by writing NIOSH at 4676 Columbia Parkway, Cincinnati, OH 45226, or by calling (513) 533-8287.

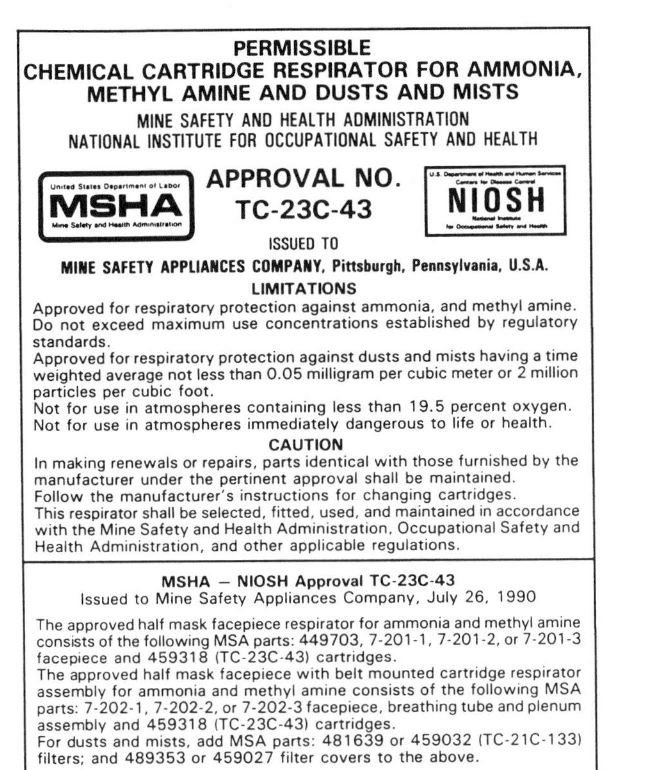

PERMISSIBLE
CHEMICAL CARTRIDGE RESPIRATOR FOR AMMONIA, METHYL AMINE AND DUSTS AND MISTS

MINE SAFETY AND HEALTH ADMINISTRATION
NATIONAL INSTITUTE FOR OCCUPATIONAL SAFETY AND HEALTH

United States Department of Labor
MSHA
Mine Safety and Health Administration

APPROVAL NO.
TC-23C-43

U.S. Department of Health and Human Services
Centers for Disease Control
NIOSH
National Institute for Occupational Safety and Health

ISSUED TO

MINE SAFETY APPLIANCES COMPANY, Pittsburgh, Pennsylvania, U.S.A.

LIMITATIONS

Approved for respiratory protection against ammonia, and methyl amine. Do not exceed maximum use concentrations established by regulatory standards.
Approved for respiratory protection against dusts and mists having a time weighted average not less than 0.05 milligram per cubic meter or 2 million particles per cubic foot.
Not for use in atmospheres containing less than 19.5 percent oxygen.
Not for use in atmospheres immediately dangerous to life or health.

CAUTION

In making renewals or repairs, parts identical with those furnished by the manufacturer under the pertinent approval shall be maintained.
Follow the manufacturer's instructions for changing cartridges.
This respirator shall be selected, fitted, used, and maintained in accordance with the Mine Safety and Health Administration, Occupational Safety and Health Administration, and other applicable regulations.

MSHA – NIOSH Approval TC-23C-43
Issued to Mine Safety Appliances Company, July 26, 1990

The approved half mask facepiece respirator for ammonia and methyl amine consists of the following MSA parts: 449703, 7-201-1, 7-201-2, or 7-201-3 facepiece and 459318 (TC-23C-43) cartridges.
The approved half mask facepiece with belt mounted cartridge respirator assembly for ammonia and methyl amine consists of the following MSA parts: 7-202-1, 7-202-2, or 7-202-3 facepiece, breathing tube and plenum assembly and 459318 (TC-23C-43) cartridges.
For dusts and mists, add MSA parts: 481639 or 459032 (TC-21C-133) filters; and 489353 or 459027 filter covers to the above.

Figure 11-1. Typical respirator approval plate. (MSA)

Table 11-1. NIOSH approval schedules

Type of Respirator	30 Part 11 Subpart	NIOSH/MSHA Schedule
Self-contained breathing apparatus	H	TC-13-F
Gas masks	I	TC-14-G
Supplied-air respirators	J	TC-19-C
Air purifying: dust, fume, mists	K	TC-21
Air purifying: gas and vapor	L	TC-23

In addition to initial testing, NIOSH conducts periodic audits to assure that manufacturers' quality control remains within acceptable limits by purchasing representative respirators in the open market and testing them to determine if they continue to meet the specifications in 30 CFR 11. Equipment that fails to perform can have its approval terminated.

RESPIRATOR NOMENCLATURE

Respirator Classification

Respirators may be broadly divided into two categories: air-purifying and air-supplied.

Air-purifying devices use mechanical filters, chemical cartridges or canisters to remove contaminants from the air that workers breathe. Filters remove particulates such as dusts, mists and fumes, while chemical cartridges and canisters remove gases and vapors.

Air-supplied respirators, on the other hand, provide the wearer with a continuous supply of clean, breathable air which is delivered to a facepiece, hood or helmet via a hose attached to a compressor, high-pressure air cylinder or ambient air pump.

Styles of Devices

Respirators are generally available in two styles: tight-fitting facepieces or loose-fitting hoods or helmets.

Tight-Fitting Respirators

Tight-fitting respirator facepieces are made of flexible, molded rubber, silicone, neoprene or other elastomeric materials. They are held on the face by four to six plastic or woven straps which are attached to the facepiece and are buckled behind, or slipped over, the head. Three types of tight-fitting facepieces are available.

Quarter-face (Figure 11-2) covers only the nose and mouth. These devices provide the least protection because the lower sealing surface merely rests on the chin and can be easily jarred loose, allowing contaminants to leak in.

Half-face (Figure 11-3) devices also cover only the nose and mouth, but they seal more readily than the quarter-mask because the lower sealing edge fits tightly under the chin.

Full-face devices (Figure 11-4) cover from about the top of the forehead to well below the chin. They generally provide the greatest level of protection because they seat most reliably.

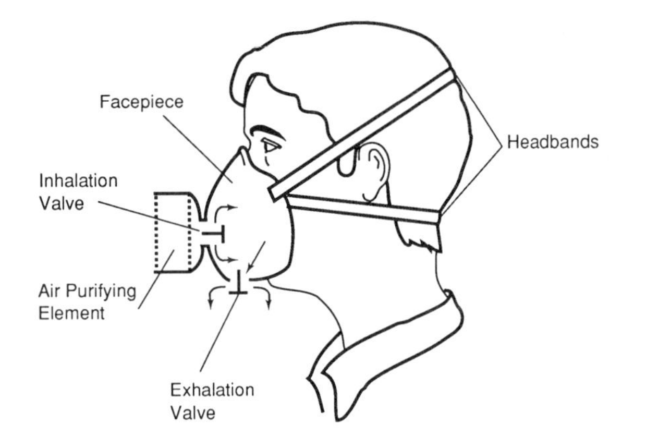

Figure 11-2. Quarter-facepiece respirator. (Paul Trattner)

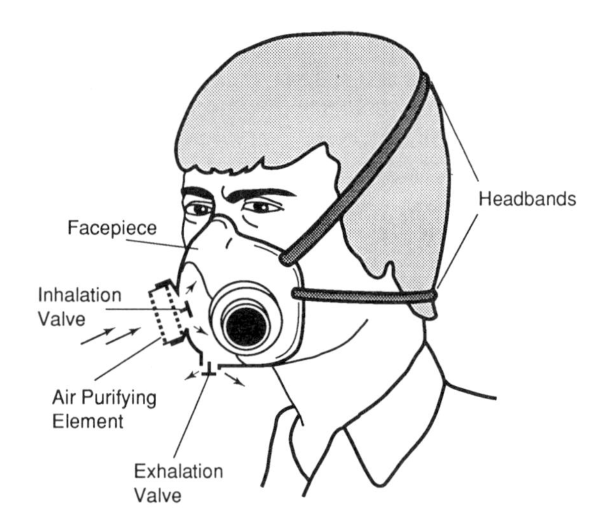

Figure 11-3. Half-facepiece respirator. (Paul Trattner)

Loose Fitting Respirators

Loose fitting devices exist as hoods and helmets. Hoods (Figure 11-5) are flexible coverings that enclose at least the head and neck; however, some hoods cover the head, neck and shoulders. Helmets are more rigid than hoods and also provide impact protection.

A special type of helmet (Figure 11-6) used for abrasive blasting also provides protection from rebounding shot. The unit is fitted with a heavy fabric cape that covers the worker's torso, and the viewing lens is protected from scratching and pitting by a fine mesh screen or disposable plastic cover shields.

One advantage that hoods and helmets have over tight-fitting facepieces is that they can be worn by people who ordinarily have difficulty getting a tight face-to-facepiece seal, such as individuals who wear beards or glasses, or who have loose or missing dentures.

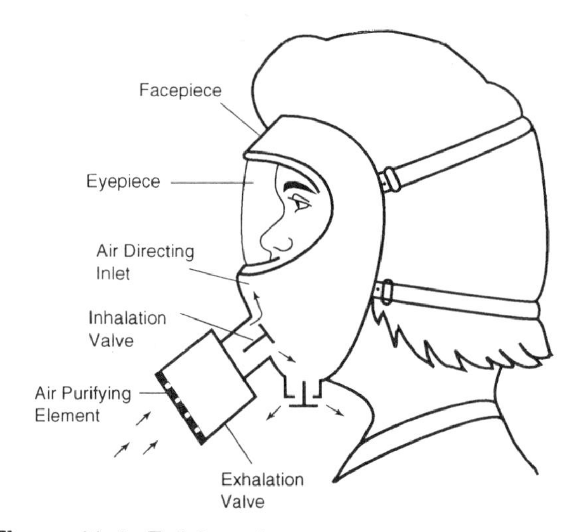

Figure 11-4. Full-facepiece respirator. (Paul Trattner)

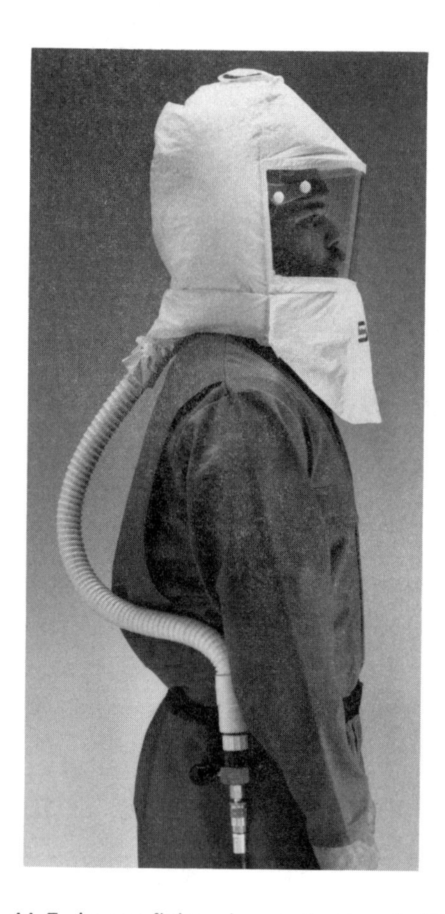

Figure 11-5. Loose-fitting air-supplied hood. (Bullard)

Figure 11-6. Loose-fitting air-supplied helmet. (Bullard)

AIR-PURIFYING RESPIRATORS

Air-purifying respirators are small, inexpensive and, compared to air-supplied devices, relatively easy to maintain. A wide variety of facepieces and purifying elements allows wearers to select a suitable respirator for many applications.

Negative-Pressure vs. Powered Devices

Air-purifying respirators may be broadly classified as negative-pressure devices or powered devices.

Negative-pressure devices are perhaps the most familiar type of respirator. The term "negative-pressure" derives from the fact that when the wearer inhales, a negative pressure (relative to atmospheric air) is created inside the facepiece. This negative pressure allows ambient air to flow through filters, cartridges or canisters which remove contaminants.

To keep the inhaled and exhaled air flowing through the respirator in the right directions, the facepiece is equipped with a set of 1-way flapper valves that is visible in Figures 11-2 through 11-4. When the wearer breathes in, the inhalation valve opens and the exhalation valve closes. When he breathes out, the inhalation valve closes and the exhalation valve opens.

Negative pressure respirators are the least protective types of devices because contaminants can easily leak in through the edges of poorly fitting facepieces.

Powered air-purifying respirators (PAPRs) use a small, battery-powered blower (Figure 11-7) rather than the wearer's lungs to draw contaminated air through the purifying elements. The purified air is then delivered to a facepiece, hood or helmet (Figures 11-8 and 11-9) either by a breathing tube or by air channels molded into the shell of a helmet. Since the facepiece, hood or helmet is under positive pressure, leakage is theoretically outward. However, it is possible to over-breathe a PAPR, that is, to inhale more air than is being provided by the blower. If this happens, say during strenuous work activities, then contaminants can be drawn inside the respirator just like they are with a negative-pressure device.

The blower motors and batteries of a PAPR may not be "approved" for use in hazardous (classified) locations as defined by Article 500 of the National Electrical Code.® Consequently, users who intend to use PAPRs in classified locations should check the unit's electrical approval before taking it into an area where it might ignite combustible gases, vapors or dusts.

Gas Masks

Air-purifying respirators that employ large air purifying elements, technically called canisters, are called gas masks. The principal advantage of gas masks is that they may be used in atmospheres containing higher contaminant concentrations than can be tolerated by cartridges. It should be noted, though, that gas masks may used for escape from, but not entry into, IDLH atmospheres.

Historically, gas masks have been negative pressure devices; however, one manufacturer now offers a family of PAPRs that are also approved as gas masks under NIOSH/MSHA Schedule TC-14G. Approvals issued as of December 31, 1991 include: organic vapors, hydrogen chloride, ammonia, chlorine, methyl amine and acid gases.

While gas masks offer an attractive alternative in situations where contaminant levels may preclude the use of cartridge-type respirators, they have certain limitations which will be discussed later.

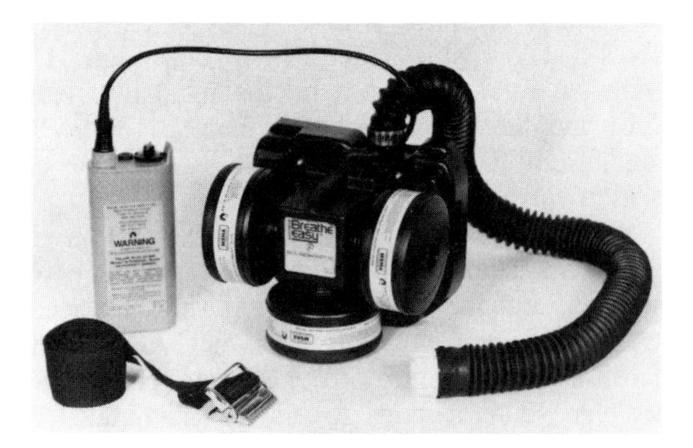

Figure 11-7. PAPR battery pack and blower assembly. (Racal)

Figure 11-8. PAPR helmet style. (Racal)

Particulate Filter Devices

Type of Filters

Replaceable dust and mist filters are designed for protection against dusts and mists having an exposure limit of not less than 0.05 milligrams per cubic meter (mg/m^3).

Replaceable fume filters protect against metal fumes having an exposure limit of not less than 0.05 mg/m^3.

Figure 11-9. PAPR facepiece style. (Racal)

Replaceable dust, mist and fume filters provide protection against dusts, mists and fumes with an exposure limit less than 0.05 mg/m^3.

Disposable respirators, often called single use, protect against pneumoconiosis and fibrosis producing dusts and mists. As shown in Figure 11-10, the filter media comprises the entire facepiece of these respirators.

Filter Efficiency

High efficiency filters are 99.97% effective against 0.3 micrometer diameter droplets of dioctylthalate. As such, they are certified for protection against dusts, fumes and mists having an exposure limit of less than 0.05 mg/m^3.

Low efficiency filters are approximately 99.5% effective against silica dust particles with a geometric mean diameter between 0.4–0.6 micrometers. They are suitable for use for materials with a PEL of 0.05 mg/m^3 or greater.

Gas and Vapor Removing Devices

Removal Mechanisms

Gases and vapors can be removed from the atmosphere by three mechanisms: adsorption, absorption and catalysis.

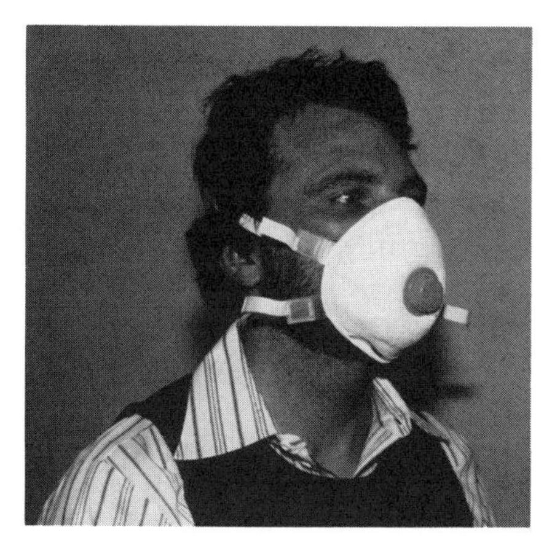

Figure 11-10. Disposable respirator. (John Rekus)

Adsorption is a process in which contaminant molecules are trapped on the surface of a sorbent granule by physical attraction, the degree of which varies depending on the contaminant and the nature of the sorbent.

All adsorbents possess a high surface area to mass ratio, and one gram of sorbent typically has a surface area of about 1,500 square meters. Activated charcoal is the most commonly employed adsorbent and is widely used to remove organic vapors. It can also be impregnated with certain materials to make it more selective against specific gases and vapors. For example, charcoal impregnated with metal salts and metal oxides is used to remove acids, gases and ammonia, and iodine-impregnated charcoal is used to remove mercury.

Absorption differs from adsorption in that the gas or vapor molecules penetrate deeply into the molecular spaces, where they are held chemically. Like adsorbents, absorbents are porous but their surface to mass ratio is not as great. Absorbers are used for protection against acid gases, and they commonly include sodium or potassium hydroxide mixed with lime and/or caustic silicates.

Catalysis is a process that employs catalysts which are substances which influence the rate of chemical reaction without themselves entering into the reaction. A catalyst used for respiratory protection is Hopcalite, which is a mixture of copper and manganese oxides that accelerates the reaction between carbon monoxide and oxygen to form carbon dioxide.

Cartridges and Canisters

The main difference between cartridges and canisters is the volume of sorbent that they hold. Cartridges are gas- and vapor-removing elements that are usually used in pairs on half- and full-facepieces respirators. Since cartridge sorbent volumes are on the order of about 50 to 200 cm^3, service life is relatively short, especially in situations where contaminant concentrations are high. Canisters, on the other hand, contain much larger sorbent volumes. They may be chin-, front-, or back-mounted and can be used in high concentrations including escape from IDLH atmospheres.

Chin style canisters (Figure 11-11) typically contain between 250 to 500 cm^3 of sorbent and are worn attached to a full-facepiece. Back- and chest-mounted canisters (Figures 11-12 and 11-13) contain between 1,000 to 2,000 cm^3 of sorbent. They are supported by a body harness and are connected to the respirator facepiece by means of a flexible corrugated breathing tube.

Limitations of Use

Ambient Oxygen

Air purifying devices cannot be worn in atmospheres containing less than 19.5% oxygen or in atmospheres immediately dangerous to life and health, except in the case of gas masks approved for emergency escape. In addition, air purifying devices cannot be used for protection against gases and vapors that have poor warning properties, except in those cases where the respirator is equipped with an end-of-service-life indicator like that on canister shown in Figure 11-14.

Adequate Warning Properties

According to its *Technical Manual,* OSHA considers substances with odor thresholds greater than ten times the permissible exposure level to have poor warning properties. A comprehensive listing of odor thresholds may be found in *Odor Thresholds for Chemicals with Established Occupational Health Standards* which may be obtained from the American Industrial Hygiene Association, 2700 Prosperity Ave., Suite 250, Fairfax, VA 22031. A typical page of this document is shown in Figure 11-15.

Maximum Use Concentrations

When NIOSH and the U.S. Bureau of Mines initiated their joint respirator certification program in 1972, they established Maximum Use Concentrations

Figure 11-11. Chin-mounted canister. (MSA)

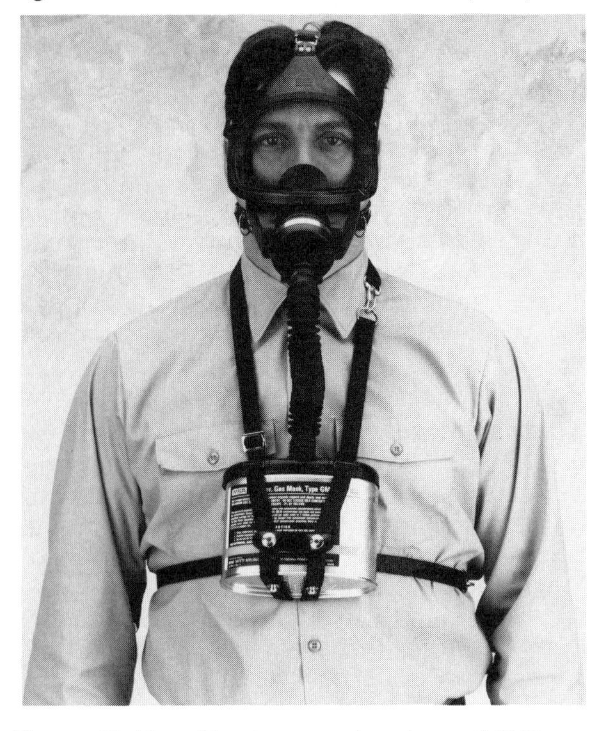

Figure 11-12. Front-mounted canister. (MSA)

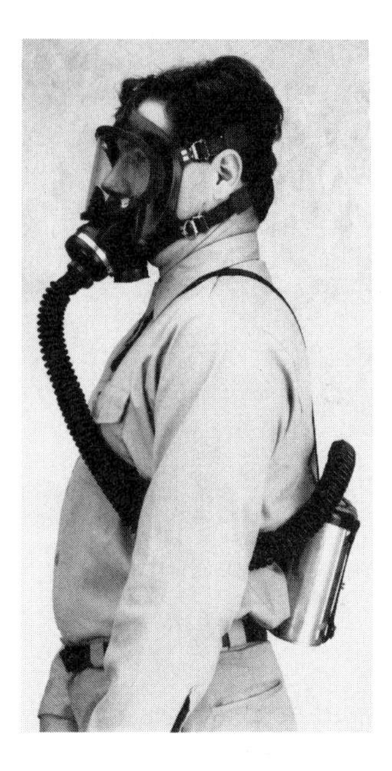

Figure 11-13. Back-mounted canister. (MSA)

Figure 11-14. A typical end-of-service is shown on the center of the canister. (MSA)

(MUCs) for chemical cartridges. The MUC for organic vapors was set at 1,000 ppm, and MUCs for six specific substances (Table 11-2) were calculated by multiplying the PEL by 10, the assigned protection factor for half-face respirators. Although not required by 30 Part 11, NIOSH had included MUCs on chemical cartridge approval labels. However, after

Compound Name Synonyms	Formula	Molecular Weight	TLV (ppm)	Geometric Mean Air Odor Threshold[A] (ppm)	Type of Threshold[B]	Odor Character	Range of Acceptable Values[C] (ppm)	Range of All Referenced Values (ppm)
Butane	C_4H_{10}	58.12	TWA = 800	none		natural gas	none	1262–5048
n-Butanol See: n-Butyl Alcohol								
2-Butanol See: sec-Butyl Alcohol								
Butanethiol See: Butyl Mercaptan								
2-Butanone See: Methyl Ketone								
2-Butoxyethanol Butyl Cellosolve; Ethylene Glycol Monobutyl Ether	$C_6H_{14}O_2$	118.17	TWA = 25 STEL = 75	0.10 0.35	d r	sweet/ ester/musty	* *	0.10–0.35
n-Butyl Acetate	$C_6H_{12}O_2$	116.16	TWA = 150 STEL = 200	0.31 0.68	d r	sweet/ banana	0.063–7.4 0.038–12	0.0063–368
Butyl Acrylate Butyl 2-Propenoate; 2-Propenoic Acid Butyl Ester; Acrylic Acid n-Butyl Ester	$C_7H_{12}O_2$	128	TWA = 10	none		sweet/ rancid/ plastic	none	0.00096–0.10
n-Butyl Alcohol n-Butanol	$C_4H_{10}O$	74.12	C = 50[D]	1.2 5.8	d r	sweet/ alcohol	0.12–11 1–20	0.05–990
sec-Butyl Alcohol 2-Butanol	$C_4H_{10}O$	74.12	TWA = 100 STEL = 150	3.2[F] 0.41	d r	sweet/alcohol	0.12–13.8 *	0.12–26
tert-Butyl Alcohol 2-Methyl-2-propanol	$C_4H_{10}O$	74.12	TWA = 100 STEL = 150	960	d	sweet/alcohol	*	3.3–957
n-Butylamine	$C_4H_{11}N$	73.14	C = 5[D] skin	0.080 1.8	d r	sour/ ammoniacal	* 0.24–13.9	0.08–13.9
Butyl Cellosolve See: 2-Butoxyethanol								
Butyl Mercaptan n-Butanethiol	$C_4H_{10}S$	90.19	TWA = 0.5	0.0010[F] 0.00073	d r	skunk	0.0073–0.001	0.00041–4.9
Butyl 2-Propenoate See: Butyl Acrylate								
Camphor	$C_{10}H_{16}O$	152.23	TWA = 2 STEL = 3	0.079	d	camphorous	*	0.0026–0.96
Carbolic Acid See: Phenol								
Carbon Disulfide	CS_2	76.14	TWA = 10 skin	none		vegetable sulfide/ medicinal	none	0.016–0.42
Carbon Tetrachloride Tetrachloromethane; Perchloromethane; Necatorina; Benzinoform	CCl_4	153.84	TWA = 5 skin	252[F] 250	d r	sweet/dry cleaner	140–584 *	1.6–706

Figure 11-15. Example page from *Odor Threshold for Chemicals with Established Health Standards.* (Used with permission, American Industrial Hygiene Association, Fairfax, VA.)

Table 11-2. NIOSH/BOM maximum use concentrations

Contaminant	MUC (ppm)
Ammonia	300
Chlorine	10
Hydrogen chloride	50
Methyl amine	100
Sulfur dioxide	50
Vinyl chloride	10
Organic vapors	1,000

OSHA revised the PELs in 1989, NIOSH reconsidered the value of labeling cartridges and issued a Respirator Information Notice which stated that MUCs would no longer be listed on approval labels.

The following June, OSHA Office of Technical Support issued a memorandum entitled "Guidelines on the Use of Chemical Cartridges for Protection Against Organic Vapors" which indicated that the use of organic vapor cartridge was limited by the lowest of the three concentrations—the IDLH, 1,000 ppm or 10 times the OSHA PEL.

The OSHA memorandum further stipulated that, with three exceptions, cartridges were acceptable only for organic vapors which had definite odor warning properties. This means that the air contaminant has to have a distinctive odor at concentrations at or below the PEL and that the odor is not obscured by olfactory fatigue. Selected materials which do not meet these criteria are listed in Table 11-3.

The three exceptions cited by the memorandum were:

1. Where cartridge use was permitted by a specific OSHA standard, e.g., acrylonitrile
2. When the cartridge is equipped with an approved end-of-service-life indicator
3. When cartridges were not used beyond the break-through time determined by laboratory testing

As a frame of reference for the third exception, published break-through times for selected materials are listed in Table 11-4.

Table 11-3. Selected materials with poor warning properties

Acrolein	Methyl chloride
Aniline	Isocyanates (MDI, TDI, etc.)
Arsine	Nickel carbonyl
Bromine	Nitro benzene
Carbon disulfide	Nitrogen oxides
Carbon monoxide	Nitroglycerine
Dimethyl aniline	Nitromethane
Dimethyl sulfate	Ozone
Hydrogen cyanide	Phosgene
Hydrogen fluoride	Phosphine
Hydrogen sulfide	Phosphorous trichloride
Hydrogen selenide	Stibine
Methanol	Vinyl chloride

Adapted from Odor Detection and Respirator Cartridge Replacements, P.C., F. Rex, *Am. Ind. Hyg. J.*, 563–566, 1977.

Changing Purifying Elements

Cartridges, canisters and filters must be periodically replaced because, with use, sorbents become depleted and filters clog up. While OSHA regulations do not stipulate a replacement schedule, the NIOSH *Guide to Industrial Respiratory Protection* recommends that filters be "...replaced at least daily or more often if breathing resistance becomes excessive...," or if the filter becomes damaged, for example, by tears, punctures or burn holes. Reusable filters should be cleaned in accordance with the manufacturer's instructions.

With respect to cartridges and canisters, NIOSH recommends replacement "...daily or after each use, or even more frequently if the wearer detects odor, taste or irritation." Because of the tendency of gases and vapors to desorb through the sorbent while respirators are stored overnight, NIOSH further recommends that cartridges be discarded at the end of the day even if the wearer does not taste or smell anything.

AIR-SUPPLIED RESPIRATORS

In contrast to air-purifying respirators which rely on filters or sorbents to remove contaminants, the facepieces, hoods and helmets of air-supplied respirators are connected directly to a source of clean, breathable air. Air-supplied equipment exists as air

Table 11-4. Break-through times for selected organic vapors

Solvent	Time to reach 1% breakthrough 10 ppm (min)	Solvent	Time to reach 1% breakthrough 10 ppm (min)
Acetone	37	Ethanol	28
Acrylonitrile	49	Ethyl acetate	67
Allyl alcohol	66	Hexane	52
Amyl acetate	73	Heptane	78
Amyl alcohol	102	Iso-amyl alcohol	97
Benzene	73	Iso-propanol	54
Butanol	115	Methanol	0.2
2-Butanone (MEK)	82	Methyl acetate	33
Butyl acetate	77	Pentane	61
Carbon tetrachloride	77	Toluene	94
Chlorobenzene	107	Vinyl chloride	3.8
Chloroform	33	*m*-Xylene	99

Adapted from Respirator Cartridge Efficiency Studies, G.O. Nelson and C.A. Harder, *Am. Ind. Hyg. J.* (35): 7491–510, 1974.

line units, self-contained breathing apparatus and combination devices.

Air-line respirators, also referred to as supplied-air respirators or SARs, consist of a facepiece, hood or helmet, connected to a compressor, high pressure air cylinder or ambient pump.

Self-contained breathing apparatus, or SCBAs, are devices in which the user carries a supply of breathing air—or in some cases oxygen—in a compressed gas cylinder worn in a backpack.

Combination devices, as their name suggests, are air-line devices that also function as air-purifying respirators or SCBAs. All three types of equipment will be discussed in more detail later.

Air-supplied respirators may operate in one of three modes: continuous-flow, demand-flow or pressure-demand.

Continuous-Flow

In the continuous-flow mode, air is constantly delivered to the respirator. Although loose-fitting hoods and helmets use this mode exclusively, tight-fitting facepieces may also be operated in the continuous-flow mode.

Demand-Flow

The demand-flow mode provides air to the respirator only when the wearer takes a breath. In other words, it supplies the wearer with breathing air "on demand." The air flow is regulated by a diaphragm-activated admission valve located in the air-line.

When the respirator wearer inhales, the negative pressure created inside the air-line sucks the regulator diaphragm down, opening the admission valve. When the valve opens, air flows through the line and into the facepiece. The slight positive pressure created during exhalation closes the admission valve and forces expired air through exhaust valves attached to the facepiece.

Since the facepiece is under negative-pressure during inhalation, contaminants can leak in through the facepiece-sealing edges. For this reason, demand-flow respirators do not provide any more protection than their air-purifying counterparts.

Pressure-Demand

Pressure-demand respirators overcome this short coming by fitting the diaphragm with a small spring that applies a slight pressure on the admission valve. The spring keeps the admission valve from closing all the way, allowing air to trickle into the facepiece. To prevent air from escaping through the exhalation valve, it is fitted with a spring that maintains a pressure of 1.5 inches of water inside the facepiece. This combination of springs assures that the pressure inside the facepiece remains positive with respect to the air outside. On inhalation, the admission valve opens all the way to provide full air flow "on demand," explaining the name pressure-demand. This positive-pressure feature affords a much higher level of employee protection than the demand mode.

AIR-LINE RESPIRATORS

Historical Perspective

The earliest air-line respirators consisted of a facepiece attached to a rigid hose, the end of which was placed in fresh, uncontaminated air. When the wearer inhaled, air was drawn into the hose and delivered to the facepiece. As time progressed, this "hose mask," as it was called, was improved through the addition of a hand-cranked air mover which made breathing easier by blowing air into the facepiece. The hand crank was later replaced by an electric motor. Air-line respirators have continued to evolve and modern design improvements have resulted in products that are more comfortable to wear and more protective.

It should be noted that some textbooks, including a few with recent copyright dates, include selection-logic decision trees which erroneously indicate that hose masks may be used in atmospheres immediately dangerous to life and health. This is not true since these devices are no longer approved by NIOSH for use in such atmospheres. Today, hose masks are more an historical artifact than a protective device likely to be used during confined space entry. In fact, the *Certified Equipment List* indicates that only one manufacturer still has a NIOSH certification for hose masks.

Air Flow Requirements

As shown schematically in Figures 11-16 and 11-17, air-line respirators are available as either loose-fitting or tight-fitting styles. Loose-fitting devices are available either as disposable hoods or reusable rigid helmets. Disposable hoods are used widely for spray painting and other operations where the equipment may be quickly rendered unusable because of extensive surface contamination.

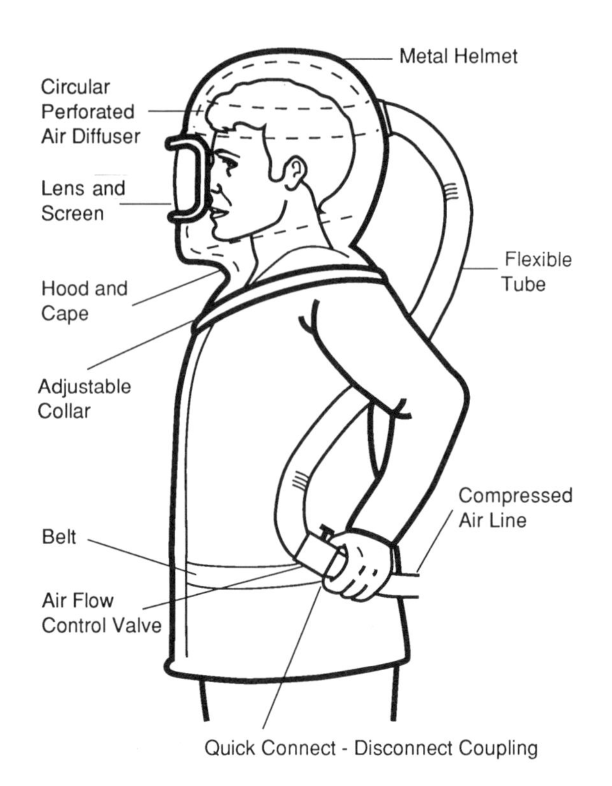

Figure 11-16. Loose fitting air-line respirator. (Paul Trattner)

Although rigid helmets may be used in many situations, one application where they are used almost exclusively is abrasive blasting. Abrasive blasting helmets, also known as type "CE" respirators, have integral neck closures and bibs that provide the wearer with protection from rebounding shot. In addition, the facepiece is covered with a fine mesh screen that keeps the lens from being scratched by abrasive particles.

NIOSH certification requirements dictate that air-lines cannot be longer than 300 feet and that the pressure at the air supply attachment point cannot exceed 125 psig. The maximum air flow to an air-line respirator cannot exceed 15 CFM (425 Lpm) when connected to the shortest hose and operating at the highest pressure specified by the manufacturer. Conversely, when connected to the longest hose and

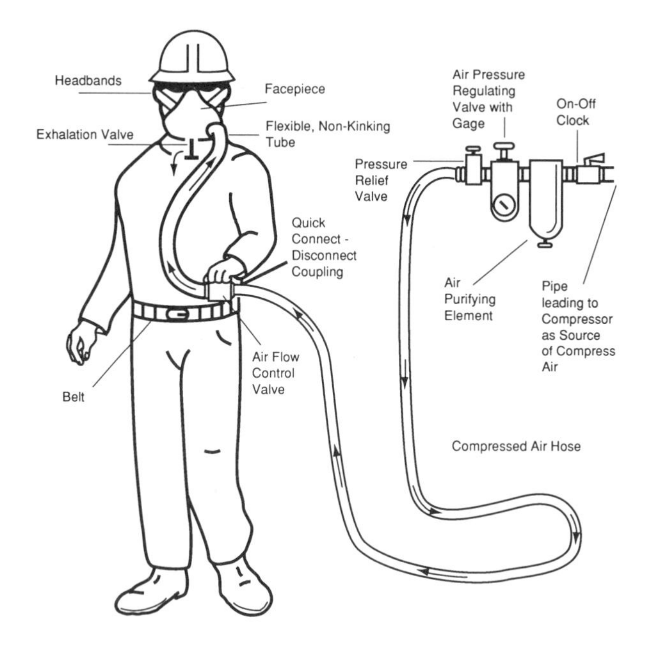

Figure 11-17. Tight-fitting air-line respirator. (Paul Trattner)

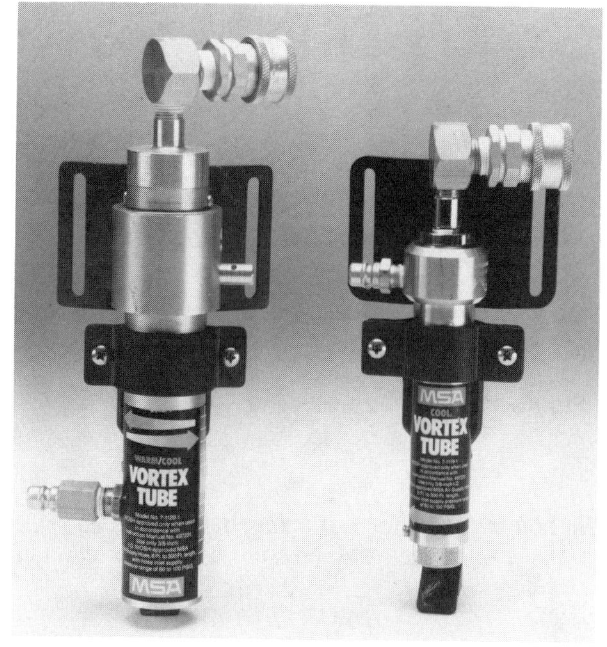

Figure 11-18. Vortex tube. (MSA)

lowest specified operating pressure, air flow cannot be less than 6 CFM (170 Lpm) to hoods and helmets, or 4 CFM (115 Lpm) to tight-fitting facepieces.

Purity specifications and sources of breathing air will be discussed later.

Heating and Cooling

Both hoods and helmets may be attached to a device called a vortex tube (Figure 11-18) which can temper ambient air by as much as 50–60°F. Amazingly, the vortex tube performs this feat without any moving parts.

Figure 11-19 illustrates how the vortex tube works. High-pressure compressed air (90–100 psi) enters the tube through a side-mounted nozzle. Since the air enters tangentially, it spins through the tube cyclonically, eventually reaching speeds of over a million revolutions per second. As it moves through the tube, air close to the surface heats up due to friction. A portion of this hot air is then exhausted through a control valve located at the end opposite the inlet nozzle. The air that does not escape through the valve flows back up the center of the tube at a slower speed and gives up its heat to the outer, faster-moving airstream. When the slower inner airstream exits the tube, it has been cooled to a very low temperature.

Vortex tube heating and cooling comes at a cost. The tubes consume air at a rate of about 25 CFM and usually operate at pressures of about 90 to 100 psig. This means that the air supply must be rated to operate at the required air volume and pressure. Approximate temperature changes produced by vortex tubes are shown in Table 11-5.

SELF-CONTAINED BREATHING APPARATUS

Self-contained breathing apparatus (SCBA) are a form of air-supplied respirator in which users carry their air supply with them. Breathing air is usually supplied by a compressed air or oxygen cylinder carried on a backpack, but a few units generate oxygen through a reaction between expired carbon dioxide and potassium superoxide. SCBAs are available

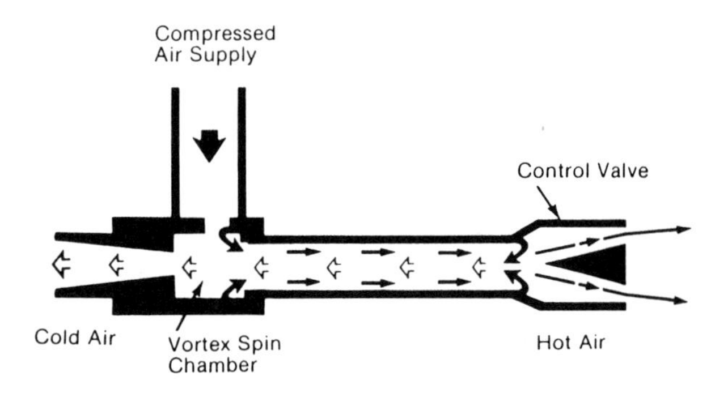

Figure 11-19. Vortex tube. (Exair)

in both closed-circuit or open-circuit styles and may operate in the demand, pressure-demand or continuous flow mode.

Closed-Circuit SCBAs

Closed-circuit devices (Figure 11-20) are also known as "rebreathers" because they recirculate expired air. Carbon dioxide in the exhaled airstream is first removed by a chemical sorbent, supplemental oxygen is added and the resulting gas mixture is recirculated back to the facepiece where the cycle begins again. Closed circuit devices exist as self-generating or compressed oxygen type devices.

Compressed oxygen devices are by far the most common type of rebreathers. As shown in Figure 11-21, supplemental oxygen is provided by small compressed oxygen cylinders inside the units, which have rated service lives of 30, 45, 60 or 240 minutes.

Oxygen-generating rebreathers (Figure 11-22) were once very popular, but like their cousin the hose mask, they are now an artifact of a by-gone era and are seldom employed in industrial applications. In fact, virtually the only situation for which they are still used is fire fighting on board U.S. Navy warships.

Open-Circuit SCBAs

Open circuit devices (Figure 11-23) do not recirculate expired air, but instead vent it to the atmosphere through an exhalation valve mounted in the facepiece. The breathing air supply is provided by a high pressure (2,000–4,500 psig) compressed air cylinder which is connected to a regulator mounted on the facepiece or worn on the belt. (Figure 11-24 and 11-25).

Open circuit devices may operate in the demand, pressure-demand or continuous flow mode.

While rebreathers generally tend to be more complicated in design, they have the advantage of providing greater service life, up to 4 hours vs. 30–45 minutes for open-circuit devices. The bulk and weight of most SCBAs make them unsuitable for strenuous work, and their limited service life makes them unsuitable for use for long continuous periods.

SCBA Approvals

SCBAs are approved by NIOSH/MSHA for use under either of two conditions: (1) entry and/or escape from hazardous atmospheres or (2) escape only.

Entry and Egress SCBAs

To be approved for entry under 30 Part 11, a SCBA must provide at least 30 minutes service time when subjected to a NIOSH breathing machine test and be equipped with the following safety features:

- Gauges which indicate to the wearer the quantity of gas remaining in the cylinder
- A remaining service-life indicator that provides an audible warning when the remaining service time or service volume reaches 20–25%
- A bypass valve that allows the user to manually control airflow in the event of regulator failure

Escape-Only SCBAs

Some SCBAs, like that shown in Figure 11-26, are certified for emergency escape use only. The units consist of an easily donned facepiece or clear plastic hood attached to a small, lightweight compressed air cylinder rated for 3, 5 or 10 minutes. The cylinder is usually hip-mounted or slung over the shoulder in a carrying bag with the air valve positioned so that it is readily accessible for immediate activation.

COMBINATION AIR-LINE RESPIRATORS

Combination Supplied Air/SCBA

The principal advantage of air-line respirators is that they provide wearers with a virtually unlimited

Table 11-5. Approximate temperature changes produced by a 25 CFM vortex tube

Air Inlet Pressure (psig)	Cooling: Approximate Valve Turns Open						Heating: Approximate Valve Turns Open					
	Full	2	1-1/2	1	1/2	0	Full	2	1-1/2	1	1/2	0
60	84	73	59	44	28	2	80	104	132	168	236	—
80	92	80	65	49	31	3	86	113	143	181	249	—
100	99	86	70	53	33	4	91	119	151	192	257	—
120	104	90	74	55	34	5	94	123	156	195	256	—
140	109	94	76	56	35	6	96	124	156	193	250	—

Source: MSA.

Figure 11-20. Closed circuit SCBA. (Biomarine)

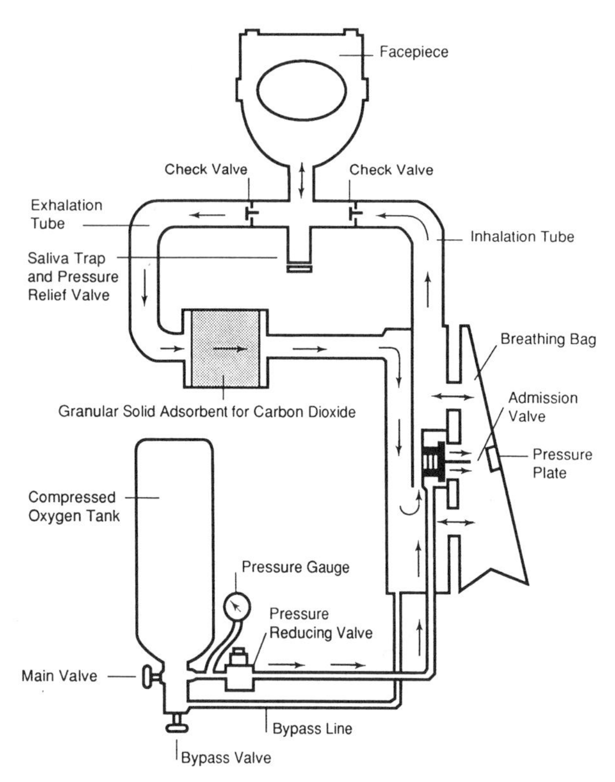

Figure 11-21. Compressed oxygen closed-circuit SCBA. (Paul Trattner)

supply of breathing air. However, if the primary air supply fails, the user must be able to escape from the area uninjured. This is usually accomplished by providing an auxiliary air supply in a small compressed gas cylinder worn on the belt or carried in a pouch over the shoulder (Figure 11-27). The self-contained portion of the devices is used only when the air line fails and the wearer must escape, or when the wearer must disconnect from the primary air supply to change locations.

Escape supply cylinders provide 3, 5, 10 or 15 minutes of service time. Because of this short service life, these units are generally used only for emergency entry or escape from IDLH atmospheres. A combination air-line SCBA may be used for emergency entry into a hazardous atmosphere to connect to an air line if the SCBA is rated for 15 minutes or more and no more than 20% of the air supply is used during entry. The importance of providing a readily accessible

Figure 11-22. SCBA rebreather self-generating. (John Rekus)

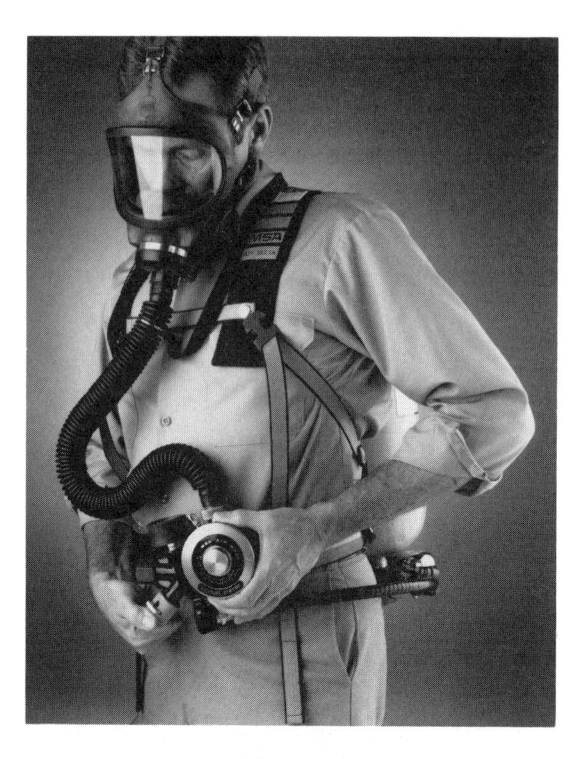

Figure 11-24. Belt-mounted SCBA regulator. (MSA)

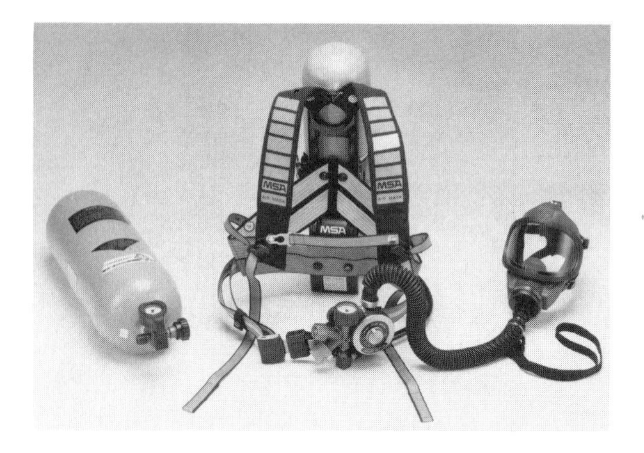

Figure 11-23. SCBA open circuit. (MSA)

reserve air supply is underscored by the following case history.

Case History No. 11-1: Air Line Not Equipped With Escape Supply

A maintenance worker was assigned the task of removing some residue from a toxaphene stripper tank. The tank was about 11 feet deep and 10 feet in diameter. It had been cleaned with carbon tetrachloride and then rinsed with water.

The worker wore a full-face airline respirator and entered the vessel through an 18-inch manway. The ladder he used to climb to the bottom of the tank was removed to permit a bucket employed in the cleaning process to be passed up and down.

After two buckets of residue were removed, the worker requested that a short ladder be placed in the vessel so that he could climb up with the last bucket. As he climbed up, his air line separated at a quick-connect fitting. There was no emergency escape cylinder and as he tried to reconnect the air line, the ladder slipped, rotated 180 degrees, and pinned him against the vessel wall. As he attempted to self-rescue, he fell to the bottom of the tank.

Another ladder was placed in the vessel and an assistant operator climbed down but had to evacuate because of the vapors. A maintenance mechanic entered and he too was overcome. A foreman then entered, but he was able to self-rescue before being overcome. After ventilating the space with a blower, rescue squad members entered and removed the two unconscious men who were later pronounced dead on arrival at the hospital. The assistant operator and foreman were hospitalized, treated and released.

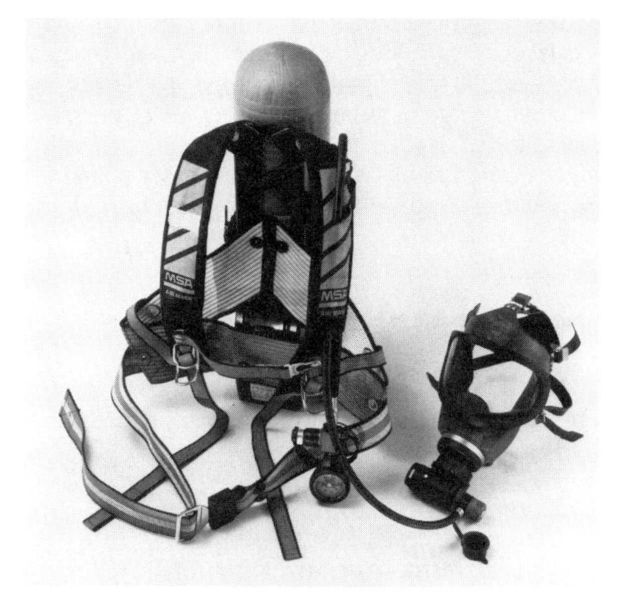

Figure 11-25. Facepiece-mounted SCBA regulator. (MSA)

Figure 11-26. Egress SCBA. (Scott)

Combination Supplied-Air/Air-Purifying

As shown in Figure 11-27, these devices consist of a supplied-air facepiece fitted with air-purifying cartridges or canisters. The supplied air portion may be either pressure-demand or continuous flow. The principal advantage of these units is that the air-purifying elements weigh considerably less than the SCBA escape cylinder.

These devices are affected by all of the limitations of the air-purifying elements and can only be used under the same conditions specified by the element's approval certificate. Some approval plates may specify that the air purifying element can only be used:

- To enter an area prior to connecting to the air supply
- For egress after disconnecting or loss of air
- To move from one air supply to another
- To escape in the event of loss of air

BREATHING AIR SPECIFICATIONS

OSHA respirator standard requires that air-supplied respirators be provided with at least Grade "D" breathing air. However, rather than explaining what Grade "D" air is, the standard refers to the 1969

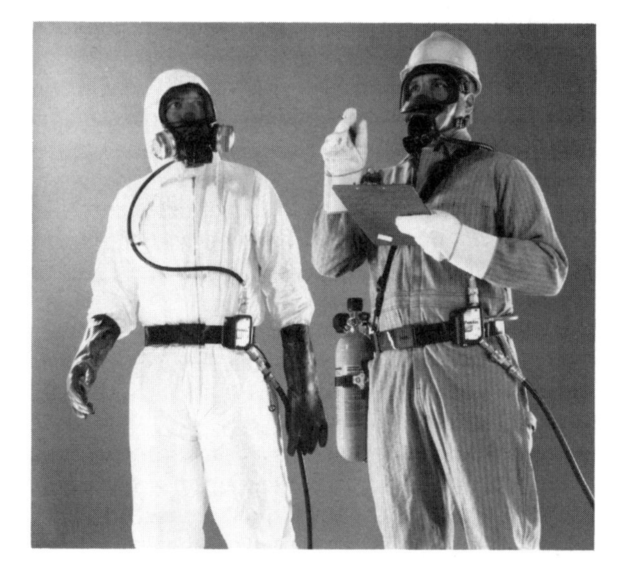

Figure 11-27. Combination air-line/air-purifying respirator (left) and combination air-line/SCBA (right). (MSA)

edition of the Compressed Gas Association (CGA) publication *Commodity Specification G-7.1*.

Recognizing the adverse consequences that could result from the presence of carbon monoxide in breathing air, CGA revised its standard in 1989 and reduced the acceptable level of carbon monoxide to 10 ppm. It's important to note, though, that since NIOSH bases its approvals on the 1989 CGA standard, it is OSHA policy to enforce the 10 ppm carbon monoxide level. Criteria for selected grades of breathing air are given in Table 11-6.

The following case history demonstrates the importance of assuring that breathing air meets at least Grade "D" specifications.

Case History No. 11-2: Use of Non-Grade "D" Breathing Air

A painting company contracted to sandblast and paint both the interior and exterior of a water tank located on the roof of a pharmaceutical plant. The tank measured 20 feet long, 11 feet wide and 7 feet high. On top were three 23½-inch diameter manholes.

One employee was assigned the task of tending the sand pot outside the tank, while a second served as a spotter for the blaster inside. The blaster entered the vessel and instructed the spotter to remain topside to finish scraping paint from the side of the tank. After five or ten minutes the blaster had not signaled for the sandblast air to be turned on, so the pot tender looked down the manhole to see what was going on.

He saw the blaster lying at the bottom of the tank wearing his air supplied hood. Apparently recognizing that something was wrong, the spotter entered the tank and made an unsuccessful attempt to revive the blaster. He then called for rescue services who later pronounced the blaster dead at the scene. The air line supplying the blasting hood was later tested and found to contain only 4% oxygen.

Case History 11-3: Oxygen Used Instead of Breathing Air

An employee wore a full-face air-line respirator as he performed oxy-acetylene cutting in a sewer line. The respirator was fed from a compressed gas cylinder on the surface. As he started cutting, the edges of the facepiece began to burn. He responded by immediately pulling off the mask, but it was burning so intensely that he burned his hands.

Investigators discovered that although the cylinder was labeled "breathing air," it was filled with oxygen. Apparently a spark ignited the upper portion of the mask which began to burn intensely under the influence of the oxygen escaping through the sealing edges of the facepiece. When the worker pulled the facepiece off, the sudden rush of oxygen caused it to burst into flames.

SOURCES OF BREATHING AIR

Breathing air may be provided by compressors, ambient air pumps or high-pressure gas cylinders.

Air Compressors

Air compressors can provide a virtually unlimited supply of air. They have the added advantage of being readily available, portable and available for rent. OSHA standard 29 CFR 1910.134 (d)(2)(ii) requires that a "breathing grade compressor" be used, but it is unclear what constitutes a breathing air compressor since no governmental agency or nationally recognized testing laboratory certifies compressors for breathing air.

Ambient Air Pumps

Ambient air pumps (Figure 11-28) consist of a small electric or compressed air-driven motor attached to a rotating vane impeller similar to a water pump on a car. A flexible tube attached to the inlet side of the pump is placed in "clean air." And the outlet side is connected to the air line that feeds the respirator. Ambient air pumps are relatively inexpensive, highly portable and may even be fitted with explosion-proof motors so they can be used in hazardous locations.

Some manufacturers' product literature suggests that these devices don't need to be equipped with sorbent purification beds or carbon monoxide alarms because they are not oil-lubricated. This information is misleading because it makes two assumptions which are not always true.

The first assumption is that ambient air meets CGA Grade D specifications. Although ambient air may initially meet Grade D standards, there is no guarantee that the air quality won't change. Consider, for example, an air inlet line stretched through a plant area where the air initially meets Grade "D"

Table 11-6. CGA specification for breathing air

Limiting Characteristics	Grades of Breathing Air					
	D	E	F	G	H	I
% Oxygen (vol/vol) Balance nitrogen	←———19.5 – 23.5%———→					
Hydrocarbon, condensed mg/M3 of gas	5	5				
Hydrocarbons, gaseous as CH4			25	15	10	0.5
Carbon monoxide (ppm)	10					
Carbon dioxide (ppm)	1,000	500	500	500	0.5	
Nitrogen dioxide (ppm)				2.5	0.5	0.1
Nitrous oxide						0.1
Sulfur dioxide (ppm)				2.5	1.0	0.1
Halogenated solvents (ppm)				10	1	0.01
Acetylene (ppm)						0.05
No pronounced odors (ppm)	√	√	√	√	√	√

Adapted from CGA 7.1–1989.

criteria. The quality of the air delivered to the respirator will deteriorate rapidly when welding fumes or spray paint vapors formed by adjacent operations are sucked into the air inlet.

The second assumption—that breathing air can only be contaminated with carbon monoxide when the compressor's lube oil burns—is flawed because breathing air may be contaminated with carbon monoxide from many other external sources.

Compressed Air Cylinders

Compressed gas cylinders provide the third alternative for supplying breathing air. A typical air cylinder is 9 inches in diameter, 51 inches tall and is pressurized at about 2,200 psig. The air volume at 70°F is 224 cubic feet, which means that a single cylinder will provide about 60 minutes of service to a tight-fitting facepiece, but service time can be extended through a multiple cylinder cascade system like that shown in Figure 11-29.

BREATHING AIR SYSTEM REQUIREMENTS

Specified Air Pressure

Air-supplied respirators must be provided with an adequate supply of air if they are to provide their rated level of protection. At least 6 CFM are required for hoods and helmets and 4 CFM is required for tight-fitting facepieces. Since too much air can dry out a wearers eyes and respiratory tract, the maximum flow is limited to 15 CFM.

The easiest way to assure delivery of the proper air flow is to consult the manufacturer's instructions to determine what pressure must be provided at the air line connection point. It should be noted, though, that inlet pressures vary not only depending on the respirator make and model, but that the pressure required for a particular device can also vary depending on diameter and length of the air line, and even the types of quick-connect fittings used.

As shown in Table 11-7, an air-supplied device may require one pressure when attached to short lengths of hose and a different pressure when connected to long lengths. Different operating pressures may also be required for identical lengths of hose, depending on whether the quick-connect couplings are Schrader or Hansen-type fittings so its important to read and follow the manufacturer's instructions.

Compressor Safety Features

In general, three precautions must be taken to assure that breathing air delivered from compressors meets Grade "D" specifications.

Figure 11-28. Ambient air pumps. (Bullard)

Figure 11-29. Typical cascade system. (Air Systems)

Position of Air Intake

The compressor's air in-take must be situated in a location reasonably free from contamination. For example, it should not be located near exhaust stacks for spray booths, furnaces or fuel-fired space heaters.

Sorbent-Beds and Filters

OSHA standard 1910.134 (d)(2)(ii) specifically requires that "...suitable in-line sorbent beds and filters [be] installed to further assure breathing air quality." These beds and filters typically remove odors, condensed oil mist and water vapor.

CO Monitoring or Alarm

OSHA standard 29 CFR 1910,134 requires that oil-lubricated compressors used to provide breathing be equipped high-temperature alarms, carbon monoxide alarms or both. However, if the compressor is equipped only with a high-temperature alarm, the air must be tested regularly for carbon monoxide.

In theory, the high-temperature alarm is supposed to provide warning of conditions which would cause cylinder lubricating oil to thermally decompose forming carbon monoxide. However, a study performed by researchers at Lawrence Livermore National Laboratory suggests that high-temperature alarms may not provide sufficient warning of excessive carbon monoxide.

The study simulated overheating of a compressor by wrapping it with a 576 watt heat tape covered with a 4-inch thick blanket of foil-backed fiberglass building insulation. Crankcase oil and cylinder-head temperatures were monitored with thermocouples, while air in the receiver was tested for carbon monoxide.

Experimental results indicated that cylinder head temperatures of 175°C were sufficient to produce CO levels which exceeded the 20 ppm limit set by OSHA for breathing air. However, crankcase oil temperatures never exceeded 81°C, which was within the safe operating range. Since high-temperature sensors typically use crankcase oil temperature as the indicator of overheating, they will not provide sufficient warning of carbon monoxide and should not be considered as a substitute for CO monitoring.

If a carbon monoxide alarm (Figure 11-30) is used, the set points should be adjusted so the alarm sounds at 10 ppm. In addition, the calibration of the carbon monoxide detector should be checked by a qualified person on a regular basis—perhaps daily for mobile compressors that are moved from site to site—to assure that it is functioning within acceptable limits. Calibration is usually performed by introducing a test gas of known CO concentration into the sensor and observing the instrument's indicator to verify that it reads the same as the concentration of the test gas.

The following two case histories graphically demonstrate the importance of monitoring for carbon monoxide.

Table 11-7. Pressure requirements for selected air-supplied respirators

Make and Model	Approval Number	Airline Length (feet)	Pressure Range (psig)	
Bullard CC20	TC-19-154	10 only	8–17	
		25–100	12–17	
		120–200	20–35	
		225–300	24–45	
		100 only	5–13	
		200 only	8–20	
		300 only	11–22	
MSA 463277	TC-19-113	8–50	30–35	
MSA 468697	TC-19-129	8–300	35–40	
US Safety 802-SL 802-WL 802-WW 802-WW	TC-19-193	Without Control Valve		
		25–50	15–30	
		75–125	25–45	
		150–200	30–60	
		225–300	40–70	
		With Control Valve		
		25–50	25–30	
		75–125	30–40	
		150–200	35–50	
		225–300	40–60	
North 85210	TC-19-204	3/8-inch Diameter Hose		
		12.5–50	5–15	
		75–150	9–24	
		175–300	15–35	
		1/2-inch Diameter Hose		
		12.5–50	3–18	
		75–150	5–22	
		175–300	8–27	
			1/4 NPT-OBAC	Schrader-Foster
Survive Air 9810-02 9810-04	TC-19-231	25	2.5–10	3.5–13
		50	4–15	5–18
		75	6–20	7–23
		100	7–24	8–27
		125	8–25	9–28
		150	9.5–28	10.5–31
		175	10–32	11–35
		200	11–33	12–36
		225	11.5–35	12.5–38
		250	12–38	13–41
		300	14–41	15–44

Adapted from NIOSH Certified Equipment List.

Figure 11-30. CO meter and alarm. (Enmet)

Case History No. 11-3: CO Poisoning From Air Compressor Exhaust

A contractor was hired to spray the inside of an underground sewer line with a tar-based protective coating. The job foreman recognized the potential for atmospheric contamination and instructed the spray gun operator to wear a full-face air-line respirator.

Breathing air was provided by a compressor which had been rented from a local construction equipment supplier. Because the unit was intended primarily for use with pneumatic tools like jack hammers, it was not equipped with a carbon monoxide monitor or the purification devices needed to supply breathing grade air.

The operator was about 100 feet into the sewer line when another worker who had been paying out the air supply line noticed that it was no longer being pulled into the manhole. When he climbed down to determine what the problem was, he found that the spray operator was unresponsive.

Firefighters were able to rescue the unconscious sprayer but he was pronounced dead upon arrival at the hospital. The cause of death was determined to be carbon monoxide poisoning.

Investigators who later examined the compressor determined that exhaust from its gasoline engine had entered the air intake manifold and was subsequently fed to the respirator's facepiece.

Case History No. 11-4: CO Poisoning From Compressor Fire

Four sandblasters who were working in a compartment of a gas turbine tanker were sent to the hospital after being suddenly overcome. Two were dead on arrival. The other two were in critical condition. High levels of carbon monoxide were found in their blood.

The air-lines they used were individually connected to an on-deck manifold which, in turn, was connected to a bank of four compressors located on the pier. The output of the compressor bank was fed to an air scrubber, dryer, cooler and then distributed through pipes and hoses. The system provided all air for breathing, ventilation and hand tools.

One of the oil-lubricated compressors had overheated about 15 minutes before the blasters passed out, so it had been shut down. After the incident it was determined that this oil-lubricated compressor had caught fire internally and that the fire had spread to the air receiver. Although the unit was equipped with a high-temperature shut-off valve, it was mounted in such a way that it jammed. Subsequent test of the air in the receiver showed carbon monoxide levels of 3,000 ppm.

Air-Line Couplings

Couplings for connections to breathing air must be incompatible with other in-plant piping systems to prevent air lines from being inadvertently misconnected to other gases such as nitrogen, argon or fuel gases. Safety can be further enhanced by labeling and color coding all pipes to indicate their contents. The consequences of an accident where these precautions were not taken is summarized below.

Case History No. 11-5: Lack of Incompatible Couplings

A foreman who was in charge of a crew that cleaned chemical tank trucks and rail cars prepared a tank truck for cleaning. The tank truck, which previously contained toluene diamine, was dried out by blowing compressed air into it. A manifold for service air, nitrogen, water and steam, which was being used for the first time, was supposed to have been fitted with a quick connect "crow-foot" fitting for air and a threaded connection for nitrogen, but the connections were reversed during installation.

On the morning of the incident, the foreman placed what he thought was an air line down into the tank. Before lunch, the foreman used a full-face air line attached to a compressed air cylinder. After lunch he changed the cylinders and switched to a half-face air-line respirator. He and another tank cleaner were working and ran out of rags. The other tank cleaner went to get some more. He returned 5 minutes later and found the foreman unconscious.

The foreman was pulled out with ropes attached to his wrists and administered CPR but he remained unresponsive. The cause of death was determined to be asphyxiation by nitrogen. Investigation showed that the respirator was attached to the nitrogen line rather than the breathing air.

Case History No. 11-6: Incompatible Couplings

A sandblaster entered a water tank to clean the interior surfaces before repainting. About 5 or 10 minutes went by and he still had not signaled for the air to be turned on, so the pot tender looked in to see what was causing the delay. He saw the blaster collapsed at the bottom of the tank wearing his air-supplied helmet. Recognizing that something was wrong, the pot tender entered the tank and attempted to revive the blaster.

When his efforts proved unsuccessful, he called for rescue services. The arriving rescuers pronounced the blaster dead at the scene. Subsequent investigation showed that the blaster was asphyxiated when he inadvertently connected his air hose to an in-plant nitrogen line.

Air Receiver and Failure Alarm

OSHA standard 29 CFR 1910.134(d)(2)(ii) requires that compressors be equipped with failure alarms and..."a receiver of sufficient capacity to enable [a] respirator wearer to escape from a contaminated atmosphere in [the] event of a compressor failure..." In other words, the compressor must be equipped with a device that lets workers know that the compressor has failed and supply tank (the receiver) that provides sufficient breathing air for them to escape from a contaminated area.

RESPIRATOR USE UNDER SPECIAL CONDITIONS

Facial Hair and Deformities

Facial deformities such as prominent cheek bones, deep skin creases, scars, severe acne and lack of teeth or dentures can prevent respirators from sealing properly. Facial hair between the wearer's skin and the sealing surface of a respirator can also be a problem because it prevents a good seal. Consequently, tight-fitting facepieces should not be worn by employees with sideburns, beards and moustaches which could interfere with respirators' sealing edges.

Even a few days worth of stubble can permit contaminants to leak into negative pressure facepieces. Or in the case of positive pressure devices, outward leakage of breathing air will reduce the service time of the air supply. This is a particularly serious concern when small compressed air cylinders are employed for escape.

Glasses and Contact Lenses

Ordinary glasses cannot be worn with full-facepiece respirators because temple bars that protrude through the sides of the facepiece will break the seal. Respirator manufacturers, however, offer special corrective lenses like those shown in Figure 11-31 which can be mounted inside the facepiece. To assure good vision, comfort and a proper seal, the lenses should be mounted by an individual designated and qualified by the respirator manufacturer.

ANSI Z88.2-1992 *Practices for Respiratory Protection* does not prohibit contact lenses from being worn with a respirator, but OSHA respirator standard does. However, it is the agency's policy not to issue citations for contact lens violations. This decision is based on research conducted by Lawrence Livermore National Laboratories (LLNR).

Lawrence Livermore researchers surveyed 9,100 firefighters in the U.S. and Canada to determine if contact lens wearers had experienced any serious problems while using SCBAs. Of the 1,405 firefighters who completed the survey 403 indicated that they wore contact lenses. Of these, only six indicated that they had encountered problems which caused them to remove their facepieces. In contrast, the number of incidents related to wearing insert glasses were proportionately higher.

These results lead LLNR researchers to conclude that contact lenses were not significantly more hazardous than insert-glasses, and that their use should not be prohibited. ANSI adopted a similar stand when it revised its respirator standards in 1992.

Temperature Extremes

Facepiece fogging, which results when warm, moist, exhaled air contacts cold facepiece lenses, can

Figure 11-31. Facepiece insert-glasses. (MSA)

be prevented in two ways. First, the interior lens surface can be coated with antifogging agents available from the respirator's manufacturer, which are effective to about 32°F. Second, the facepieces can be fitted with nose cups which direct exhaled air through the exhalation valve without touching the lens. Nosecups are effective to temperatures of -30°F.

At subzero temperatures, high pressure connections on SCBAs can leak. Caution must be taken not to overtighten these connections because they could break on expansion when the SCBA returns to room temperature.

Elastomeric components such as facepieces, gaskets, valves and diaphragms can become stiff at low temperatures, and emergency respirators stored in cold environments may require special elastomeric components that will retain their elasticity. Another factor to consider is that cold facepieces may become so distorted that an adequate face-to-facepiece seal cannot be achieved. At high temperatures, elastomeric components will deteriorate at an accelerated rate and exposure to intense heat can cause facepieces to become permanently distorted.

Communication Difficulty

Communication is difficult when wearing any type of respirator and loud ambient noises can further obscure the voices of respirator wearers. Special speaking diaphragms in full-face devices can make communication easier, but voices are still distorted.

Some manufacturers have overcome these problems by fitting their facepieces with miniature microphones that can be plugged into two-way radios or small portable amplifiers like that shown in Figure 11-32. Remember, though, that if intrinsically safe microphones, radios and amplifiers are used in "classified" locations, the devices must be "approved" *as an integrated unit.*

Air Cylinder Ignition Hazard

Aluminum, like brass and copper, is normally considered to be a non-sparking metal. However, there is one anecdotal report that suggests that under some conditions, aluminum SCBA cylinders could become ignition sources.

An eyewitness told me a story about being part of a team that wore SCBAs fitted with aluminum cylinders while inspecting a heavily rusted pipe. According to the witness, bright white sparks were emitted every time someone's cylinder banged against the pipe.

This description suggests a thermite reaction similar to that which I had seen demonstrated at the United Kingdom's Health and Safety Executive laboratory in Sheffield, England. In that demonstration, a heavily rusted steel block was covered with aluminum foil food wrap and struck with a hammer. The impact was sufficient to initiate a reaction between the rust and aluminum producing sparks and intense heat.

ELEMENTS OF A RESPIRATOR PROGRAM

Both OSHA and ANSI standards require employers to develop written respirator programs regardless of whether protection is provided by simple disposable air-purifying respirators or complex self-contained breathing apparatus. At a minimum, the program should include the following provisions:

1. Administration by a qualified person who possesses the requisite technical skills and the authority to assure that the program is carried out properly
2. Written standard operating procedures governing respirator selection, use, cleaning, maintenance and storage
3. Medical evaluation of users to assure that they are physically capable of wearing respirators
4. Use of only NIOSH/MSHA certified devices that are selected on the basis of the hazard to which users are exposed
5. Training employees on respirator use, limitations, fitting, care and maintenance

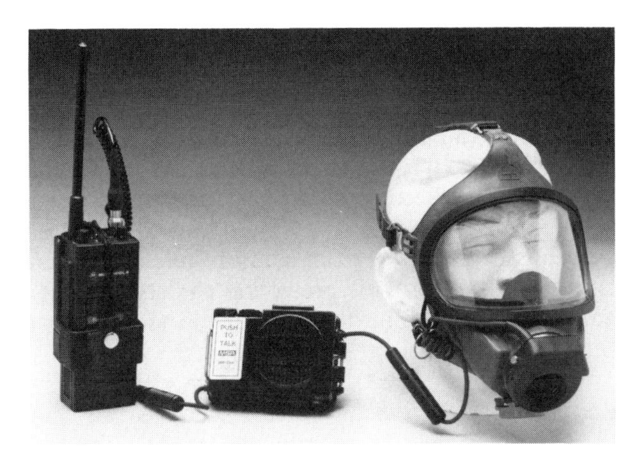

Figure 11-32. Facepiece amplifier. (MSA)

6. Fit-testing respirator users to assure that they are provided with adequate protection
7. Frequent inspection of respirators for worn, missing, unapproved or defective parts
8. Adequate maintenance including regular cleaning and disinfection and storage of equipment in a convenient, sanitary location
9. On-going workplace surveillance for changes in process, materials or employee work practices that might affect respirator usage
10. Annual review of the entire program to assure its continued effectiveness

PROGRAM ADMINISTRATIVE ASPECTS

A single person should be given the responsibility and commensurate authority necessary to administer the respirator protection program. The designated person should be an industrial hygienist, safety specialist or, in the case of a small employer, an upper level manager, senior supervisor or other qualified individual. The administrator must have sufficient technical knowledge and experience to be able to manage the program and to select appropriate protective equipment for specific applications.

STANDARD OPERATION PROCEDURES

Written standard operation procedures should be established for the following five items:

1. Respirator selection process
2. User medical evaluations
3. Respirator cleaning and maintenance
4. Inspection of respirators
5. Respirator storage

Suggestions covering each of these procedural areas will be discussed later in this chapter.

MEDICAL EVALUATION

Since filters and chemical cartridges used by negative pressure air-purifying respirators increase breathing resistance, respiratory conditions like emphysema and bronchitis can greatly reduce a worker's ability to wear a respirator. The weight added by a self-contained breathing apparatus may increase the risk of cardiac arrest for some workers, especially when combined with the physical and emotional stress associated with an emergency or rescue operation. Other risk factors include high blood pressure, excessive body weight and chronic smoking. Recognizing the potential for these kinds of problems, the OSHA standard recommends that a physician determine if employees are indeed physically able to wear respirators. The standard does not describe the content of this medical determination, but instead relies on the attending physician's judgment. Fortunately, ANSI comes to the rescue again by offering guidance in its standard Z88.6-1984, *Respirator Use—Physical Qualifications for Personnel*. This standard offers physicians specific suggestions upon which they may base their opinions for qualifying an employee's ability to wear a respirator. The standards also identifies the conditions listed below as reasons for prohibiting respirator use.

- Facial deformities
- Facial hair
- Respiratory disease
- Cardiovascular disease
- Endocrinal disorders
- Neurological disability
- Past or current medication
- Psychological conditions such as claustrophobia

Information Provided to Physician

The program administrator, or other qualified person, should provide the physician with the following information.

1. A description of respiratory protection the examinee will use, including any special conditions such as the extra weight exhibited by SCBAs, or the fact that respirators may be used in hot environments.

2. A description of the level of work the examinee is typically required to perform, e.g., light, moderate, heavy or strenuous.
3. The anticipated extent of respirator use, for example, daily, weekly, monthly or rarely The process can be streamlined by using a form similar to that shown in Figure 11-33.

Physician Action

The physician should obtain a medical history, paying special attention to:

- Previously diagnosed diseases, particularly those of a cardiovascular or respiratory nature
- Problems associated with breathing during normal work activities, or any past problems with respirator use
- Psychological problems such as claustrophobia
- Known physical deformities or abnormalities, including those which may interfere with respirators use
- Tolerance to tachycardia (rapid heart beat) produced by inhalation of heated air

The physician may also want to consider performing special tests such as spirometry that measures the lung's expiratory volume and vital capacity, or exercise stress tests to evaluate employees who might use SCBAs. Any individual who has apparent ischemic disease who cannot perform well on the tread mill because of musculoskeletal, respiratory or other physical problems should not be assigned to an emergency team or other position that might require the use of such equipment.

Physicians Report

After the examination, the physician must certify the examinee into one of the following three classes.

Class 1: Unrestricted respirator use
Class 2: Some specific restrictions
Class 3: No respirator use under any circumstances

Exam Frequency

ANSI recommends that respirator medical qualification examinations be provided every five years for users up to 35 years old, every two years for those up to 45 and annually thereafter.

RESPIRATOR SELECTION

With the dozens of makes, models and styles of respirators available in the marketplace, choosing the right one for a particular application can be a real nightmare. In general, though, respirators must be selected on the basis of the hazard to which an employee is exposed. A paint spray respirator, for example, will not provide protection against chlorine, and an air purifying device will provide no protection in an oxygen-deficient atmosphere.

Some of the factors that influence respirator selection are discussed below. However, because the selection process is so complicated, readers who are responsible for selecting respirators are referred to the NIOSH Respirator Selection Decision Logic (NIOSH Publication 87-108).

Oxygen Deficiency

NIOSH approval labels indicate that air-purifying respirators cannot be used in atmospheres containing less than 19.5% oxygen by volume. However, recall from Chapter 2 that oxygen percentage by volume is not the only consideration in evaluating oxygen deficiency. Because of the reduced partial pressure of oxygen at high altitudes, workers who are not acclimatized may experience more than normal difficulty breathing when using air-purifying devices.

Contaminant Properties

The chemical and physical properties (Table 11-8) of a contaminant influence the selection of air-purifying respirators. Different air purifying elements are required depending on whether the contaminant is gaseous or particulate. Chemical properties such as molecular structure, acidity and alkalinity must also be considered when selecting sorbent-based cartridges and canisters.

Airborne Concentration

Since respirators provide varying degrees of protection, it is necessary to know the approximate level of contamination in order to select an appropriate respirator. The contaminant concentration can be determined by personal air sampling, but if exposures are unknown, it is prudent to provide entrants with supplied-air respirators until air-sampling demonstrates that a lower level of protection is sufficient.

Respirator User Qualification

Supervisor completes sections between bold lines only

Employee Name:	ID Number:
Supervisor:	Department:

Type of respirator to be used (Check all that apply):

Negative-pressure air purifying: ❒ Half-face ❒ Full-face

Powered air-purifying: ❒ Half-face ❒ Full-face ❒ Hood ❒ Helmet

Continuous flow air-line: ❒ Half-face ❒ Full-face ❒ Hood ❒ Helmet

Demand-flow air-line: ❒ Half-face ❒ Full-face ❒ Hood ❒ Helmet

Combination air-line/SCBA: ❒

Combination air-line/air-purifying: ❒

SCBA: ❒ Open circuit ❒ Closed circuit

Level of work effort (check one): ❒ Light ❒ Moderate ❒ Heavy ❒ Strenuous

Anticipated usage: Duration of use:___ Hours Frequency: ❒ Daily ❒ Once a week

❒ Occasionally, but more than once a week ❒ Rarely, or likely to be used only in emergency

List any special work considerations (e.g. high altitudes, use of chemical protective clothing, hot environments, work in confined spaces, etc.):

Physician's Evaluation:

User classification: ❒ No restriction ❒ Restrictions indicated below ❒ No respirator use permitted

Restrictions:____

Physician's Signature:_______________________________ Date:____________

Figure 11-33. Respirator medical qualification form.

Table 11-8. Classification of respiratory hazards according to their properties which influence respirator selection

Gases and Vapors	Particulates
Inert: Substances that do not react with other substances under most conditions, but create a respirator hazard by displacing oxygen producing and oxygen deficiency (for example, helium, argon, neon, nitrogen).	Particles are produced by mechanical means through disintegration processes such as grinding, crushing, drilling, blasting and spraying; or by physiochemical reactions such as combustion, vaporization, distillation, sublimation, calcination and condensation. Particulates are defined as follows:
Acidic: Substances that are acids or that react with water to produce an acid. In water they produce positively charged hydrogen ions (H^+) and a pH of less than 7. They taste sour and many are corrosive to tissues (for example, hydrogen chloride, sulfur oxide, fluorine, acetic acid, nitrogen dioxide, carbon dioxide, hydrogen sulfide and hydrogen cyanide).	**Dust**: A solid, mechanically produced particle, with sizes varying from submicroscopic to visible or macroscopic.
	Spray: A liquid, mechanically produced particle with sizes generally in the visible or macroscopic range.
Alkaline: Substances that are like alkalis or that react with water to produce alkali. In water they may result in the production of negatively charged hydroxyl ions ($-OH^-$) and a pH greater than 7. They taste bitter and may be corrosive to tissues (for example, ammonia, amines, phosphine, arsine and stibine).	**Fume**: A solid condensation particle of extremely small particle size, generally less than one micrometer in diameter.
	Mist: A liquid condensation particle with sizes ranging from submicroscopic to visible or macroscopic.
	Fog: A mist of sufficient concentration to perceptibly obscure vision.
Organic: The compounds of carbon. Examples are saturated hydrocarbons (methane, ethane, butane) unsaturated hydrocarbons (ethylene, acetylene), alcohols (methanol, ethanol, propanol), aldehydes (formaldehyde), ketones (methyl ethyl ketone), organic acids (formic acid, acetic acid), halides (chloroform, carbon tetrachloride), amides, (formamide, acetamide), nitriles (acetonitrile), isocyanates (Toluene disocyanate), amines (methyl amine), epoxies (epoxyethane, propylene oxide) and aromatics (benzene, toluene, xylene).	**Smoke**: A system which includes the products of combustion, pyrolysis of chemical reaction of a substance in the form of visible and invisible solid and liquid particles and gaseous products in air. Smoke is usually of sufficient concentration to perceptibly obscure vision.
Organometallics: Compounds in which metals are chemically bonded to organo groups (for example, ethyl silicate, tetraethyl lead and organic phosphate).	
Hydrides: Compounds in which hydrogen is chemically bonded to a metal and certain other elements (for example, diborane and tetraborane).	

Adapted from American National Standard, Practice for Respiratory Protection, ANSI Z88.2-1980, pg. 15.

Remember too, that air-purifying devices cannot be worn in atmospheres Immediately Dangerous to Life or Health (IDLH). A list of IDLH concentrations can be found in the NIOSH Pocket Guide to Chemical Hazards (NIOSH Publication 90-117).

Odor Threshold

The odor threshold of some solvents is much greater than the PEL, thus precluding the use of air-purifying devices since the presence of an odor inside the facepiece signals that it's time to change the cartridge. If the wearer can't smell a chemical until it's concentration is greater than the PEL, he could be unwittingly overexposed.

Protection Factor

The protection factor is a measure of a respirator's effectiveness and is defined as the ratio of the contaminant concentration outside the facepiece compared to the concentration inside the facepiece. For example, a protection factor of 10 means that the contaminant level inside the respirator facepiece is 1/10 of the contaminant level outside. Looking at it another way, if the air outside the respirator contained 500 parts per million (ppm) of a toluene vapor, the concentration inside a properly fitted respirator would at most be 1/10 of 500 ppm, or about 50 ppm.

Assigned protection factor is the minimum protection provided by a properly functioning respirator, or class of respirators, to a given percentage of properly fitted and trained users.

Workplace protection factor is a measure of the protection provided in the workplace by a properly functioning respirator when correctly worn and used.

Simulated workplace protection factor is a surrogate measure of the workplace protection provided by a respirator.

There may be a big difference between a respirator's assigned protection factor and the level of protection actually achieved in the workplace. In fact, two NIOSH studies found that the level of protection afforded by some powered air-purifying devices was substantially lower than the protection factor specified in OSHA regulations.

One study that NIOSH conducted at a primary lead smelter showed the workplace protection factor for tight-fitting PAPRs to be only 376 as opposed to a 1,000 assigned by the OSHA lead standard. (The protection factor was reported as a geometric mean of 376 with a standard deviation of 2.64.) Of the protection factors observed, 95% were greater than 77, and 84% were less than 1,000.

In another study involving helmet-type PAPRs used at a secondary lead smelter, NIOSH found that 98% of the observed protection factors were below the OSHA assigned level of 1,000, 95% of the observations were greater than 33, and the geometric mean and standard deviation were 182 and 3.2, respectively. In addition to the discrepancies between the protection factor assigned by OSHA and those measured by NIOSH, there are inconsistencies among OSHA assigned protection factors. For example, the level of protection ascribed to a powered air-purifying respirator is 200, 250 or 1,000, depending on whether it is being used for asbestos, lead or coke-oven emissions. (Table 11-9). The 1992 ANSI assigned protection factors are summarized in Table 11-10.

The need for respirator selection by a qualified person is further supported by the following case history.

Case History No. 11-7: Improper Respirator Selection

A plumber entered a pit measuring 27 feet long, 18½ feet wide and 11 feet deep located under a degreasing tank that previously contained 1,1,1-trichloroethane.

Another plumber who arrived to help, brought a full-face canister-type respirator which he wore for about fifteen minutes. After completing his task, he took off his respirator and left the area.

The first plumber, believing that the work had not been done correctly, decided to reenter the pit to correct the problem. He put on the respirator that the other plumber had left behind, reentered the pit and removed a cap that allowed 1,1,1-trichloroethane to enter through an attached pipe.

He was later discovered unconscious at the bottom of the pit by a maintenance helper. The helper notified plant security who subsequently called the fire department. During the five minutes it took the fire department to arrive, six other employees tried to rescue the injured plumber. All were unsuccessful because of the overwhelming concentration of the 1,1,1-trichloroethane vapors.

Table 11-9. OSHA assigned protection factors for PAPRs

Substance	Assigned Protection Factor
Asbestos	100
Arsenic	1,000
Cadmium	250
Coke oven emissions	Unlimited
Cotton dust	100
Lead: General industry	1,000
Construction	25

Table 11-10. ANSI assigned protection factors

Type of Respirator	Half-Face	Full-Face	Hood or Helmet	Loose-Fitting Facepiece
Negative-pressure Air purifying	10[1]	100	—	—
Powered air-purifying	50	1,000[2]	1,000[2]	25
Air-line (demand)	10	100	—	—
Air-line (pressure-demand)	50	1,000	—	—
Air-line (continuous flow)	50	1,000	1,000	25
SCBA (demand)[3]	10	100	—	—
SCBA (pressure-demand) Self-contained	—	—[4]	—	—

1 Includes 1/4 mask, disposable half-masks and half-masks with elastomeric facepieces.
2 Protection factors listed are for high-efficiency filters and sorbents. With dust filters, an assigned protection factor of 100 should be used due to limitations of the filter.
3 Demand SCBA shall not be used for emergency situations such as fire fighting.
4 Although positive-pressure respirators are currently regarded as providing the highest level of protection, a limited number of recent simulated workplace studies concluded that all users may not achieve protection factors of 10,000. Because of this data a definitive protection factor cannot be assigned. For emergency planning purposes where concentrations can be estimated, a protection factor of no more than 10,000 should be used.

Firefighters wearing SCBAs recovered the plumber's body and discovered that he was still wearing the respirator. The medical examiner later attributed the cause of death to extensive inhalation of 1,1,1-trichloroethane vapors. The six would-be rescuers were transported to the hospital, treated and released. Investigators found that the respirator the plumber wore had been incorrectly selected and was a type approved for use in atmospheres containing ammonia—not organic vapors.

Employee Training

Users are not the only people who must be trained in respiratory protection. Individuals who issue respirators and supervisors who monitor the activities of respirator wearers must also receive basic respirator training.

Respirator Issuers Training

The individual responsible for assigning respirators must be trained to ensure that the correct respirator is issued for each application in accordance with the program's written standard operating procedures.

Supervisors Training

Supervisors and foremen who oversee the work of employees who wear respirators must be trained so that they can determine if respirators are being used properly. ANSI recommends that basic supervisors' training should consist of at least the following elements:

- Introduction to the nature and extent of the respiratory hazards to which employees may be exposed
- The principles of respirator selection
- Respirator inspection techniques
- Procedures for maintenance and storage

Users' Training

Contrary to what some employers may believe, users' training involves a lot more than merely telling workers to read the instructions on the respirator's box. As a minimum, user training should cover the following items:

- The reasons for respiratory protection
- The nature, extent and effects of respiratory hazards to which the employee may be exposed

- An explanation of the reasons why a particular respirator is selected for a specific respiratory hazard
- An explanation of the respirator use and limitations
- Instruction on adjusting the fit, including procedures for both positive- and negative-pressure field checks
- An explanation of how respirator should be maintained, cleaned and stored
- How to obtain replacement parts
- To whom problems should be reported
- An overview of respirator regulations
- An explanation of how to recognize and cope with emergency situations

Employees must also be informed that they are permitted to leave the work area for any respirator-related reason such as malfunction, detection of contaminant leakage, increase in breathing resistance and illness, including dizziness, coughing, nausea, sneezing or vomiting.

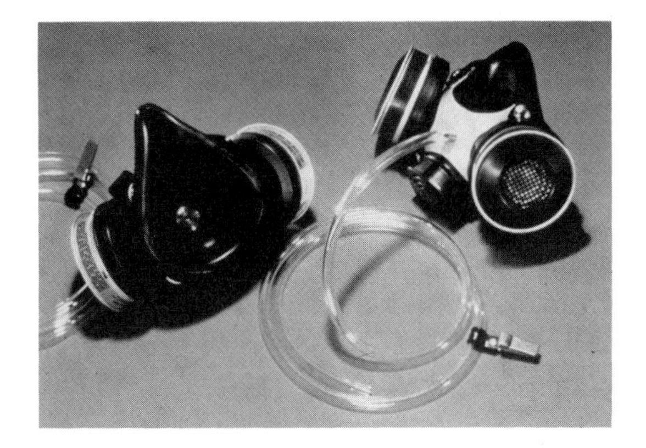

Figure 11-34. Probed facepieces. (MSA)

RESPIRATOR FIT-TESTING

A respirator that doesn't fit properly won't provide adequate protection. Hoods and helmets can be fitted easily because they cover the entire head. Tight-fitting respirators, on the other hand, are a greater challenge since the facepiece and the straps must be adjusted properly. For a proper fit, straps should be positioned so that one is behind the neck and the other over the back of the head. When both straps are worn behind the neck in the style preferred by many workers, the respirator becomes deformed, allowing contaminants to leak in around the facepiece sealing edges. Other factors which can influence the fit of a respirator include beards, loose or missing dentures and deep facial scars all of which were discussed previously.

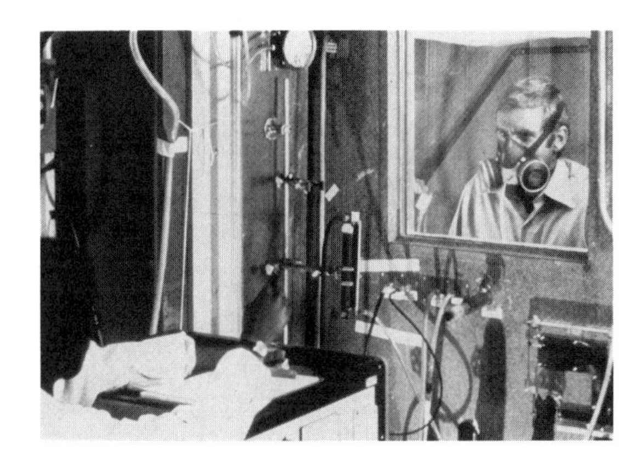

Figure 11-35. Fit test chamber. (MSA)

Fit-Testing Methods

Quantitative fit-testing is a method which measures the actual concentration of a contaminant that reaches the inside of the respirator. The test subject, equipped with a specially probed facepiece (Figure 11-34), enters a chamber similar to that shown in Figure 11-35. Once inside the chamber, the subject is exposed to a known concentration of an aerosol such as sodium chloride or corn oil mist. Instrumentation measures and compares the contaminant concentrations inside and outside the facepiece.

Quantitative fit-testing methods indicate the level of protection afforded under controlled conditions, and actual workplace conditions may result in a lower level of protection.

At one time, quantitative fit-testing was an onerous task, requiring substantial operator skill and training. However, recent advances in technology have now brought quantitative fit-testing within the grasp of anyone who can afford to spend $6,000 for the equipment.

One manufacturer offers an industrial version of a portable quantitative fit-tester (Figure 11-36) originally developed to allow soldiers in the field to check the fit of their respirators for protection against chemical and biological warfare agents. The device works on the principle of measuring the light scattered by atmospheric dust particles.

Qualitative fit-testing involves a simple "go/no-go" test based on one of three criteria: the ability to smell "banana oil" (isoamyl acetate), taste a saccharin mist or cough in the presence of an irritant smoke.

Prior to the adoption of the OSHA lead standard, qualitative fit testing was performed using rather primitive techniques. Two of the most popular consisted of testing the facepiece seal by either waving a stencil brush filled isoamyl acetate around it, or using a squeeze bulb attached to a ventilation smoke tube to blow irritant "fume" into the subject's face. If the wearer smelled the "banana oil," or coughed from exposure to the smoke, he failed the test.

The OSHA lead standard (29 CFR 1910.1025) superseded these "quick and dirty" approaches by establishing very rigorous fit-testing protocols that must be followed exactly as specified. The fit-test procedures are summarized below. Readers who are interested in the full text of the protocols can find them in Appendix "D" of the OSHA lead standard, 29 CFR 1910.1025.

Qualitative Fit-Testing Protocols

Employees are eligible for fit testing only after they have been trained in the use and limitations of respiratory protective equipment. Prior to the selection process described below, the test subjects should also receive a refresher demonstration on how to properly don, position and adjust the respirator for a snug, comfortable fit.

Facepiece Selection

When the irritant fume and banana oil test methods are used, selection should be performed in a room separate from the testing chamber to prevent olfactory fatigue. The test subject is allowed to select the most comfortable respirator from an array of facepieces that must consist of at least three sizes from two different manufacturers. A mirror must be available to assist the subjects in evaluating the fit and positioning of the respirator. Once the subject chooses the facepiece that appears most suitable, he must wear it for at least ten minutes before proceeding with the test.

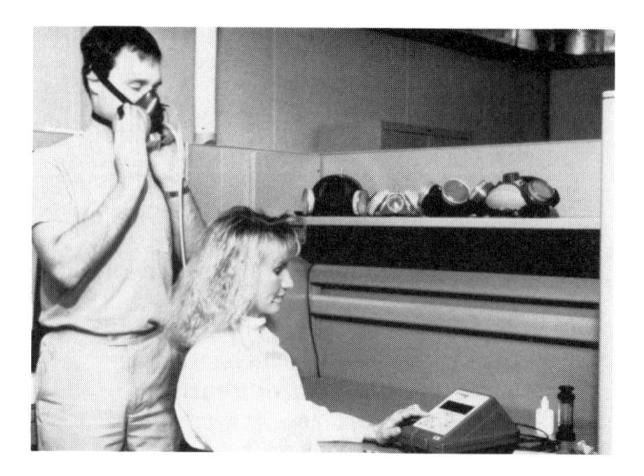

Figure 11-36. Portacount. (TSI)

Test Exercises

Under the direction of the test administrator, the wearer performs a series of test exercises that simulate work movements. These include normal breathing, deep breathing, nodding up and down, and moving the head from side to side. Care must be taken not to bump the respirator during these movements. The subject should talk loudly and slowly for several minutes or, alternatively, he can read the "rainbow passage" below. The rainbow passage is a specially designed text that elicits a wide variety of facial movements.

The Rainbow Passage

> When sunlight strikes raindrops in the air, they act like a prism and form a rainbow. The rainbow is a division of white light into many beautiful colors these take the shape of a long round arch, with its path high above, and its two ends apparently beyond the horizon. There is, according to legend, a boiling pot of gold at one end. People look but no one finds it. When a man looks for something beyond reach his friends say he is looking for the pot of gold at the end of the rainbow.

Test Atmospheres

Fit-testing employs one of three test atmospheres: isoamyl alcohol, sodium saccharin and oxystantic chloride.

Isoamyl Acetate Protocol. This test is based on the ability to smell isoamyl alcohol (IAA), which is an organic liquid that smells like bananas. A stock solution is prepared by adding 1 mL of IAA to a quart jar containing 800 mL of odor-free water and shaking it for about 30 seconds. From this stock solution, 4 mL is then added to a second jar containing 500 mL of water. A third jar, which serves as a bank sample, is filled with 500 mL of water. The last two jars should be marked 1 and 2 for identification purposes, and if the lids are marked instead of the glass, they can be interchanged so that people won't think that the same jar always contains the IAA.

A screening test to determine if the subject can smell a low concentration of banana oil is performed in a room separate from the test chamber. The following instructions should be printed on a card and left on a table with the two test jars.

> "The purpose of this test is to determine if you can smell banana oil at low concentration. The two bottles in front of you contain water. One of the bottles also contains a small amount of banana oil. Be sure the covers are on tight and shake each bottle for two seconds. Unscrew the lid of each bottle one at a time and sniff the mouth of the bottles. Indicate to the test conductor which bottle contains banana oil."

A test chamber is prepared by using a two foot diameter piece of wood or cardboard to support a plastic 55 gallon drum liner. A length of stiff wire is extended through the center of the support and formed into a hook on both ends. Organic vapor cartridges are installed on the facepiece and the subject is provided with a 6-inch × 5-inch piece of paper towel which has been folded in half and wetted with 0.75 mL of pure IAA. Upon entering the chamber, the subject hangs the towel on the hook provided. About two minutes is then allowed for the vapor to reach the desired concentration. The waiting time can be spent on reexplaining the test or demonstrating some of the exercises.

The subject then performs the test exercises discussed above. If the subject detects a banana-like odor at any time, he should so indicate, exit the test chamber and leave the area to prevent olfactory fatigue. The test can be repeated after the subject checks and repositions the facepiece for a better fit. Satisfactory fit is demonstrated when the wearer can perform all of the test exercises without smelling the banana oil.

Saccharin Mist Protocol. The saccharin mist test allows fit-testing of disposable dust, fume and mist respirators in which the filter media itself comprises the entire facepiece. The test requires the use of two Devilbis Model 40 Inhalation Medication Nebulizers, marked so that they can be differentiated, and a hood. Test subjects cannot eat, chew gum or drink anything except plain water for fifteen minutes prior to the test and must be instructed to breathe through an open mouth with tongue extended when the test is conducted.

A test solution is prepared by adding 83 grams U.S. Pharmocopia Grade sodium saccharin to 100 mL of warm water, and a threshold-screening solution is prepared by adding 1 mL of the test solution to 100 mL of water.

After being informed of the test protocol, the subject dons the hood, and the first nebulizer filled with the threshold screen solution is inserted through the hole in the front hood. The test administrator then squeezes the nebulizer bulb ten times and asks if subject tastes the saccharin. If not, the administrator squeezes the bulb ten more times and inquires again. If the subject cannot taste the saccharin after 30 squeezes the saccharin fit test may not be performed.

Subject who pass the threshold screening may select an appropriate respirator from the array and wear it for ten minutes as previously described. The second nebulizer is then filled with the test solution, the subject dons the hood again and the test solution is sprayed using the same technique employed during threshold screening. After the test aerosol is generated, the subject performs the test exercises. The nebulizer bulb should be squeezed every 30 seconds during the test to maintain the aerosol concentration inside the hood. If subject tastes saccharin, the fit is deemed to be unsatisfactory and a different respirator should be tried.

Users of sodium saccharin should know that the International Agency For Research on Cancer (IARC) has identified sodium saccharin as a class 2B carcinogen, i.e., an agent that is possibly carcinogenic to humans. The National Toxicology Program (NTP) has also indicated that there was sufficient evidence to conclude that sodium saccharin was carcinogenic in experimental animals. As a result of these reports, NIOSH recommends that sodium saccharin not be used as a respirator fit-testing agent. It should be noted that while NIOSH certifies respirators, it does not have the authority to prescribe fit-testing procedures related to OSHA standards. That means that even though there is scientific evidence of the carcinogenicity of sodium saccharin, its use in respirator fit-testing is not a violation of OSHA standards. Nevertheless, the consequences of exposure to sodium saccharin should be carefully considered before it is selected as a fit-testing agent.

Irritant Fume Protocol. One problem with both the IAA and saccharin mist tests is that subjects can cheat. In other words, they can deny that they smell banana oil or taste saccharin. This problem can be largely eliminated by using an irritant fume, since only a very small number of people are unaffected by it. The area chosen for the test should have adequate exhaust ventilation, and respirators should be fitted with a magenta-colored High Efficiency Particulate Aerosol (HEPA) filters.

To conduct the test, a stannic oxychloride smoke similar to a MSA part number 56454 is attached to a piece of plastic tubing that is connected to the exhaust side of either an aquarium pump or personal air-sampling pump adjusted to deliver about 200 mL/min. It should be noted that all ventilation smoke tubes are not alike and the one used for this test must be the stannic oxychloride type.

Test subjects are first allowed to smell a weak concentration of smoke to familiarize themselves with the odor. They then don their respirators and wear them for about 10 minutes. When they are ready to be tested, they close their eyes to prevent irritation. The tester, who is also wearing a HEPA filtered respirator, directs a stream of smoke towards the facepiece from a distance of about twelve inches. The tester gradually moves the tube to within an inch of the facepiece and directs smoke around the entire perimeter. If a subject coughs, he fails the test. If he passes, the tester then conducts a sensitivity check by having the wearer remove the facepiece to determine if the subject coughs in the presence of the smoke.

Record Keeping

Individuals should be allowed to use only those makes and models of respirator for which they could demonstrate an adequate fit. ANSI also recommends that records be kept of the following:

- Test date and type of test
- Names of test subject and person conducting the test
- Make and model of respirator for which fit was achieved
- Test results: successes or failure for qualitative test, and measured fit factor for quantitative test

Field Fit-Checks

In addition to qualitative and quantitative fit-testing, OSHA standard 1910.134(e)(5)(i) requires that workers conduct a *fit-check* every time they put a respirator on. This check can be performed using two techniques, referred to as the positive- and negative-pressure checks.

Positive-Pressure Checks

A positive-pressure check is performed by covering the exhalation valve and breathing out gently as shown in Figure 11-37. If the wearer feels the respirator puff up slightly, it is properly adjusted.

Negative-Pressure Checks

A negative-pressure test is performed by covering the cartridges—or the end of the breathing tube for respirators which are so equipped—with the hands, or a piece of paper, or polyethylene film, as shown Figure 11-38. The wearer then inhales deeply and holds his breath for about 10 seconds. If the facepiece collapses slightly and stays collapsed, the respirator is properly adjusted.

Figure 11-37. Positive-pressure check. (John Rekus)

Figure 11-38. Negative-pressure test. (John Rekus)

If either field check indicates leakage, then the facepiece should be repositioned and the head straps adjusted until an acceptable fit can be achieved.

RESPIRATOR CLEANING

Respirators should be cleaned and sanitized after each use. Although alcohol is widely used as a disinfectant, it is not recommended for disinfecting respirators since it can adversely affect rubber components, causing premature drying and cracking of some parts. In addition, wash water temperature should not exceed 110°F to prevent facepiece distortion. Most respirator manufacturers sell cleaner-sanitizers that are mixtures of a detergent and disinfectant. The contents of the package are mixed with water to form a solution that both cleans and disinfects the respirator. Alternatively, respirators can be washed in warm soapy water and sanitized by dipping them in a solution consisting of a cap full of household bleach added to a gallon of water. Cleaned respirators should then be allowed to air dry. Hoods, helmets, air cylinders and harness webbing can all be scrubbed clean with a mild soap and water.

Air-line helmets can be cleaned with a detergent and damp cloth, and facepieces can be scrubbed with mild soapy water. However, the water temperature should not exceed 140°F to prevent damage to rubber and plastic facepiece parts. After washing, facepieces should be immersed in a disinfectant rinse. A safe and effective rinse solution can be prepared by adding about a capful of laundry bleach to a quart a water. Immersion for about two minutes provides sufficient time for the disinfectant to do its job.

EQUIPMENT STORAGE

Respirators that have been cleaned and sanitized must be properly stored to prevent contamination. Self-contained breathing apparatus and gas masks may be stored in carrying cases. Half- and full-face air-purifying respirators can be conveniently stored in heavy duty Ziploc® freezer bags.

EQUIPMENT INSPECTION

Like any other personal protective equipment, respirators must be regularly inspected. The inspection should include the facepiece, headband, eyepiece lenses and all connecting hoses and tubing. Components to inspect and inspection criteria are described below and summarized in Table 11-11.

General Inspection Items

Facepieces should be examined for excessive dirt and contamination as well as cracks, tears, holes or distortion that may have been caused by improper storage (Figure 11-39). Elastomeric parts should be pliable and there should be no foreign material, such as detergent residue or dust particles trapped in the exhalation valve. Air-purifying element holders should be inspected for cracks, breaks and worn threads.

Cartridges and canisters should be examined for punctures, dented threads, missing, worn or broken gaskets and attachment fittings.

Headbands should be inspected for broken straps, malfunctioning buckles and attachments, and excessively worn serrations which might permit slippage.

Lenses on helmets and full-facepieces should be checked for correct mounting, broken or missing mounting hardware and scratches that could impair vision.

Table 11-11. Respirator inspection criteria

Facepieces for:

- Excessive dirt or contamination
- Cracks, tears, holes or distortion resulting from improper storage
- Inflexibility of elastomeric parts
- Cracked or badly scratched lenses in full-facepiece
- Incorrectly mounted full-facepiece lenses, broken or missing mounting clips
- Cracked or broken air-purifying element holders, badly worn threads or missing gaskets (if required)

Headstraps or Head Harness for:

- Broken straps or headbands
- Loss of elasticity
- Broken or malfunctioning buckles and attachments
- Excessively worn serrations on full-facepiece head harness which might permit slippage

Facepiece Interior for:

- Foreign material, such as detergent residue, dust particles or hair under the valve seat
- Cracks, tears or distortion in valve material
- Cracks, breaks or chips in the valve body, particularly at the sealing surfaces
- Missing or defective valve covers
- Improper installation of valve bodies

Air Purifying Elements for:

- Incorrect cartridges, canister or filter for the hazard
- Incorrect installation, loose connections, missing or worn gaskets (if required)
- Cross-threading in cartridge holder
- Expired cartridge or canister shelf-life date
- Cracks or dents in outside case of filter, cartridge or canister
- Evidence of prior use of sorbent cartridge or canister as evidenced by absence of sealing, material, tape or foil over element inlet

Breathing Tube (if present) for:

- Broken or missing gaskets or O-rings
- Missing or loose hose clamps
- Deterioration found by stretching hose and looking for cracks and tears

Air Supply Systems for:

- Compressor or air pump located in area not subject to ambient contamination
- Compressors and air pumps are equipped with accurate reading pressure gages
- Air is delivered to a specific length hose at the pressure stated on the respirator approval plate
- Grade "D" breathing air is provided
- The breathing-air system is equipped with a carbon monoxide monitor or self-indicating CO removing canisters

Valve assemblies should be examined for proper installation as well as cracks, breaks or chips, particularly at the sealing surfaces.

Breathing tubes, if present, should be stretched to their full length to inspect for cracks and tears (Figure 11-40). Breathing tubes should also be examined for broken or missing gaskets, o-rings and hose clamps. Leaks can be identified by plugging the ends of the tube with a cork and compressing it while holding underwater (Figure 11-41). Any pin holes will show up as air bubbles.

Figure 11-39. Respirator facepieces should be thoroughly inspected. (MSA)

Figure 11-40. Breathing tubes should be inspected by stretching them out their entire length. (MSA)

Hoods and helmets should be inspected for general mechanical condition paying attention to cracks and missing parts such as a worn or missing suspension. If the unit is equipped with a cape, it should be examined for cuts, rips and tears. Any seams should also be checked for integrity.

Compressed air cylinders on SCBA should be checked to assure that they are pressurized to the manufacturers recommended pressure.

Harness and backpack webbing should be examined for wear, cuts, nicks or abrasions.

Special Requirements for SCBAs

Emergency Use

OSHA standard 29 CFR 1910.134(f)(2)(ii) specifies that SCBA's designated for emergency use be inspected monthly. As a practical matter, SCBAs that are intended only for emergency confined space entry can be inspected as part of the pre-entry procedure.

The standard requires that the inspection determine that air cylinders are fully charged and that the regulator and the end-of-service-life warning alarm are functioning properly. While the pressure can be checked by looking at the gauge, the SCBA maintenance manual must be consulted for instructions on testing the regulator and alarm.

Hydrostatic Testing

OSHA standard 29 CFR 1910.134(d)(2)(i) requires that compressed air cylinders be "...tested and maintained as prescribed in the Shipping Container Specification Regulations of the Department of Transportation (49 CFR 173)." The important thing to know about 49 CFR is that paragraph 173.34 includes a provision for periodic hydrostatic testing.

The date of the last hydrostatic test can be determined by looking for a date-stamp. On steel cylinders the date will be marked with an imprint as shown in Figure 11-42, composite cylinders, on the other hand, will be marked with a label similar to that shown in Figure 11-43. For the cylinder to be in compliance, the date must be within the retest periods shown in Table 11-12.

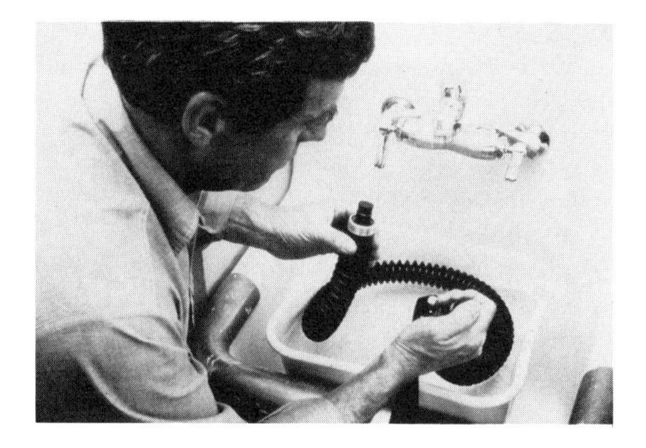

Figure 11-41. Checking breathing tube for leaks. (MSA)

Figure 11-42. Hydrostatic test date stamped on the body of a steel cylinder. (ITS)

Figure 11-43. Hydrostatic test label in a composite cylinder. (MSA)

Flow Testing

Respirator manufacturers recommend periodic flow-testing to assure that all SCBA components are working together properly. While the recommended frequency for flow-testing varies from one to two years depending on the manufacturer, the procedure is essentially the same. The facepiece is placed on a special mannequin head and instrumentation measures the SCBA performance under specified test conditions.

Since the NIOSH certified equipment list stipulates that SCBA must be maintained in accordance with the manufacturer's recommendations, failure to perform flow-testing essentially voids the NIOSH approval.

Table 11-12. Hydrostatic test schedule for SCBA cylinders

Construction Material	DOT Cylinder Classification	Retest Period (Years)
Steel	Type 3AA	5
Aluminum	Type E-6498	5
Composite	Type E-7235	3
Composite	Type E-7277	3
Composite	Type E-8059	3

WORK AREA SURVEILLANCE

An on-going surveillance program is important to assure that employees continue to wear their respirators properly. When left to their own devices, employees often demonstrate innovative ways of modifying their equipment in a fashion never imagined by the respirator designers.

Some employees have discovered that breathing resistance on negative-pressure air-purifying devices can be reduced by removing the exhalation valve. Unfortunately, this modification also eliminates all protection!

Other employees have overcome design incompatibilities that prevent one manufacturer's cartridges from being used on another's facepiece by attaching the cartridge to the facepiece with electrician's tape. Some have improperly substituted welding oxygen for breathing air. Still others have bored out fittings on air-supplied devices to increase air flow to hoods and helmets.

While these examples may provide an amusing insight to workers' imagination and creative spirit, they are all *strictly prohibited and potentially fatal!*

ANNUAL PROGRAM REVIEW

At least once a year the program administrator should make a thorough audit of the program to determine if there are any weaknesses, problems or changes in procedures, processes or equipment that could affect respirator selection and use. Any deficiencies that are noted should be promptly corrected.

SUMMARY

Whenever possible, atmospheric hazards in confined spaces should be eliminated through ventilation or by substituting less hazardous materials. If these controls are not feasible, then respirators may be used as the last line of defense for protecting workers from the harmful effects of dusts, mists, fumes, gases and vapors. However, if respirators are employed, their use must be governed by a comprehensive written program that includes provisions for selection, worker training, fit testing, inspection, maintenance and storage.

A qualified person should be designated to manage the respirator program and only NIOSH/MSHA approved devices should be used. Respirator users should receive a comprehensive medical examination to determine if they are physically capable of wearing a respirator. The attending physician should be informed of the nature of the worker's duties and the types of respirators he is likely to wear.

Users and their supervisors must be trained in the basics of respiratory protection, and the supervisor must provide on-going surveillance to assure that their employees continue to wear respirators properly. Users must also receive a fit-test that verifies that their respirators provide an adequate face-to-facepiece seal.

If air-supplied respirators are used, several special precautions must be taken. At a minimum, breathing air must conform to Compressed Gas Associations Grade "D" specifications. Air must be provided to the facepiece, hood or helmet only through lengths of hose and at operating pressure specified by the manufacturer. Respirator air-line coupling must be incompatible with other compressed gas fittings so that they can't be interchanged. The air supply must be continuously monitored for carbon monoxide, and compressed air receivers must be sized to provide an adequate reserve supply in case of compressor failure.

All respirators should be routinely inspected for defects that might affect their performance. They should also be regularly cleaned, disinfected and stored in a sanitary place when not in use. Finally, the entire respirator program should be audited annually to assure the detection and correction of problems or deficiencies.

REFERENCES

Air Contaminants, *Code of Federal Regulations*, Title 29 Part 1910.1000.

Amoore, J.E. and E. Hautala, Odor as an aid to chemical safety: odor thresholds compared with threshold limit values and volatilities for 214 industrial chemicals in air and water dilution, *J. Appl. Toxicol.* 3(6):272–290 (1983).

American National Standard for Respiratory Protection—Respirator Use—Physical Qualifications For Personnel, ANSI Z88.6-1984, New York, NY: American Nation Standards Institute, (1984).

American National Standards Institute, American National Standard for Respiratory Protection, ANSI Z88.2-1992, New York, NY: American National Standards Institute.

Asbestos, *Code of Federal Regulations*, Title 29 Part 1910.1001.

Bollinger, N.J. and R.H. Schultz, *A Guide to Industrial Respiratory Protection*, DHEW (NIOSH) Pub. No 87-116, Cincinnati, OH: U.S. Department of Health and Human Service, Centers for Disease Control, National Institute for Occupational Safety and Health, (1987).

Blackwell, D.S. and G.S. Rajhans, *Practical Guide to Respirator Usage in Industry*, Boston, MA: Butterworths, (1985).

Clark, R.A., memo dated October 28, 1992, to OSHA Regional Administrators, breathing air requirements.

Commodity Specification for Air CGA G 7.1-1989, Arlington, VA: Compressed Gas Association, (1989).

Coke Oven Emissions, *Code of Federal Regulations*, Title 29 Part 1910.1029.

daRoza, R.A. and C. Weaver, Is it Safe to Wear Contact Lenses With a Full Facepiece Respirator, Lawrence Livermore National Laboratory, Manuscript UCRL-536553, (1985).

Dixon S.W., T.J. Nelson and J.E. Wright, Workplace protection factors for negative pressure half-mask facepiece respirators, *J. Int. Soc. Respir. Prot.* 2(4): 347–361.

Douglas, D.D., W. Revoir, J.A. Pritchard, A.L. Hack, L.A. Geoffrion, T.O. Davis, P.L. Lowry, C.P. Richards, L.D. Wheat, J.M. Bustos, and P.R. Hesch, Respirator Studies for the National Institute for Occupational Safety and Health, July 1, 1974 through June 30, 1975, Los Alamos Scientific Laboratories Report, LA 6386-PR, (1976)

Hack, A.L., E.C. Hyatt, B. Held, T. More, C. Richards, and J. McConville, Selection of Respirator Test Panel Representatives of U.S. Adult Facial Sizes, Los Alamos Scientific Laboratory Report No. LA-5488, 1974.

Hyatt, E.C., J.A. Pritchard, C.P. Richards, and L.A. Geoffrion, Effects of facial hair on respirator performance, *Am. Ind. Hyg. J.* 34(4):135–142.

Inorganic Arsenic, *Code of Federal Regulations*, Title 29 Part 1910.1018.

Lead, *Code of Federal Regulations*, Title 29 Part 1910.1025.

Lenhart, S.W. and D.L. Campbell, Assigned protection factors for two respirator types based on workplace performance testing, *Ann. Occup. Hyg.* 28(2): 173–182, 1984.

Meyers, W.R., M.J. Peach, K. Cutright, and W. Iskander, Workplace protection factor measurements on powered air purifying respirators at a secondary lead smelter: results and discussion, *Am. Ind. Hyg. J.* 45(10):681–688.

Meyers, W.R., M.J. Peach, and K. Cutright, Field test of powered air purifying respirators at a battery manufacturing facility, *J. Int Soc. Resp. Prot.* 4(1):62–89 (1985).

Millar, J.D., Letter dated January 21, 1992 to Darell Bevis, use of sodium saccharin as a respirator fit testing agent.

Moore, D.E. and T.J. Smith, Measurement of protection factors of chemical cartridge, half-mask respirators under working conditions in a copper mine, *Am. Ind. Hyg. J.* 1976(8):453–458.

Nelson, G.O. and C.A. Harder, Respirator cartridge efficiency studies, *Am. Ind. Hyg. Assn. J.* 35(7): 491–510 (1974).

NIOSH Certified Equipment List as of December 31, 1989, Cincinnati, OH: U.S. Department of Health and Human Services, Centers for Disease Control, National Institute for Occupational Safety and Health, (1989).

Occupational Safety and Health Administration, memorandum from Thomas Scepich, to regional administrator, sodium saccharin as a fit testing agent, July 22, 1992.

Odor Thresholds for Chemicals with Established Occupational Health Standards, Fairfax, VA: American Industrial Hygiene Association, (1989).

Raven, B., A.T. Dodson and T.O Davis, The physiological consequences of wearing industrial respirators: a review, *Am. Ind. Hyg. Assoc. J.* 40(6): 517–534, 1979.

Respiratory Protection, *Code of Federal Regulations,* Title 29 Part 1910.134.

Respiratory Protective Devices; Tests for Permissibility; Fees, *Code of Federal Regulations,* Title 30 Part 11.

Shipping Contained Specifications, *Code of Federal Regulations,* Title 49 Part 178.

CHAPTER 12

Emergency Planning and Rescue

INTRODUCTION

The preceding ten chapters described a variety of precautions that should be taken prior to entering a confined space. These precautions included controlling mechanical hazards, disconnecting fluid lines, providing ventilation, using protective equipment and testing the space's atmosphere. Each of these preventive measures may be viewed as a link in a chain that protects entrants from injury. The more links there are in the chain, the greater the margin of safety. But a chain is only as strong as its weakest link, and sometimes, even the strongest chain breaks.

An effective confined space entry program attempts to identify and control all potential hazards to which entrants are likely to be exposed. If appropriate precautions are taken, work can be performed safely. But we live in an imperfect world, and things do go wrong. A hazard goes unnoticed, someone ignores or forgets to follow a procedure, a piece of mechanical equipment fails—and an emergency results. But there are varying degrees of confined space emergencies, and the response to some is more critical than the response to others.

In establishing a frame of reference for this chapter, we need to distinguish between two very different types of confined space emergencies. The first type are those where confined space hazards are recognized and precautions have been taken, but the control program has broken down. In these cases, emergency planning efforts provide a safety net to catch the human errors, equipment failures and unforeseen circumstances that slip through holes in the program.

The second type of emergencies are those exemplified by many of the case histories discussed earlier in this book. They are characterized by situations where confined hazards were not recognized, and consequently, few, if any, precautions were taken prior to entry. In the most severe situations, entrants were overcome by a hazardous atmosphere and incident management was provided by the local fire department or other emergency services personnel. Since these responders are often unfamiliar with the hazards in the space, their entry may be delayed while they size up the situation and develop an appropriate plan of action.

While it is important for rescue services to develop operating procedures to address the second type of emergency, the focus of this book is confined space accident *prevention*. As a result, this chapter is *not intended to be a guide for emergency services* and therefore it will not discuss tactical considerations such as incident command systems, rescue rigging and emergency medical practices. Instead, the focus will be on emergency planning as yet another element of a comprehensive entry program.

CLASSIFICATIONS OF INCIDENTS

Mention the words "confined space emergency" and many people immediately conjure up the image of workers unconscious at the bottom of a space. While this situation is indeed an emergency, it is not the only one that can arise during entry. Other, less life-threatening accidents, such as cuts, fractures and chemical burns can also constitute emergencies. In general, emergencies might be described as sudden unplanned events that demand immediate action. Examples of emergency situations that may arise in confined spaces include fires, injuries to personnel and sudden changes in physical or atmospheric conditions.

Response to some confined space emergencies will be more involved than response to others, and all will not necessarily require rescue efforts. An effective response plan recognizes that there are different degrees of emergencies and provides a measured response appropriate to the magnitude of the incident. For example, consider a situation where an entrant slips and sprains an ankle. The space has be isolated and ventilated and there are no atmospheric or physical hazards which threaten the entrants safety. Responders who enter the space merely to splint the entrant's ankle would not have to wear breathing apparatus and full protective clothing since there are no hazards threatening them. On the other hand, protective equipment would be appropriate in situations where responders enter hazardous atmospheres or are faced with chemical or physical hazards.

A framework for emergency planning can be established by identifying and characterizing the types of emergencies that are likely to occur in a confined space. While there are any number of ways of classifying confined space emergencies, the following scheme provides one approach that divides incidents into five broad categories. It should be noted that while this scheme identifies a variety of incident categories, all of them are not necessarily relevant to every space, facility or situation.

Evacuation Situations

Evacuation, or self-rescue as it is sometimes called, is perhaps the lowest level of emergency response. Conditions that precipitate evacuation often arise in a manner that allows entrants to escape without extraordinary effort. Some examples of situations necessitating evacuation are summarized below:

- An attendant observes a potential problem that can affect the entrants, such as failure of a ventilation blower or a solvent spill near the blower inlet
- Activation of an automatic alarm that signals a hazardous change in atmospheric conditions, such as a decrease in oxygen or an increase in "toxic" or flammable gases or vapors
- Failure of the main air supply when air-line respirators are used
- Entrants believe they are in danger because they hear unusual or threatening noises, experience signs or symptoms that suggest chemical exposure or smell unfamiliar or characteristic chemical odors

Specific self-rescue methods will vary with the space. In an above ground storage tank, the entrants might simply walk over to a side wall manway and crawl out. In an underground vault, they might climb up a ladder. However, in all cases the entrants are uninjured, in full command of their faculties and able to escape the space without any assistance from those outside.

Incidents with Moderate Injury

The second category of emergency incidents are those where an entrant has received moderate injury. Moderate injury for our purposes is that which requires medical attention beyond first aid. Typical injuries could include severe contusions, sprains, deep wounds and minor fractures. The victim is still conscious and the degree of injury is such that he can still evacuate the space unaided. However, once outside the space, he will require assistance from others.

Entry to Provide Initial Treatment

These are situations where there is severe trauma or other problem that must be attended to before the victim can be extricated from the space, for instance, broken bones that must be immobilized before the victim can be moved. The victim may or may not be conscious, but is not threatened by other acute hazards such as flooding or a hazardous atmosphere. These incidents differ from the two proceeding types in that responders must enter the space to begin initial treatment. The responders, however, are not threatened by any hazards that would require special precautions such as the use of SCBAs, chemical protective clothing, etc.

Assisted Rescue from Outside

Assisted rescue from outside usually involves a single victim who may or may not be conscious. The entrant wears a full-body harness and is tethered to a lifeline connected to retrieval equipment located outside the space. In an emergency, the attendant operates the equipment which pulls the entrant from the space without the need for entry by additional personnel.

Rescue Requiring Emergency Entry

These are the most critical types of incidents. The size, shape or configuration of the space frequently prevents rescue from outside. Multiple victims may be involved, and all or some of them may be unconscious. If the victims are not breathing, time is critical because permanent brain damage will occur within four to six minutes.

Rescuers themselves may be threatened by hazards in the space such as high concentrations of air contaminants or a deficiency in oxygen. They may also encounter great physical and psychological stress as they race against time to extract entrants from the space. Of the five categories of emergencies, this one is the most severe. Not only is time of the essence, but additional personnel are also placed at risk when they enter the space.

ROLE OF THE ATTENDANT

A major problem with confined space entry is that people working outside the space can't normally see an incapacitated entrant. As a result, the incapacitated employee could die before anyone even realized that something was wrong. In fact a similarity that exits among many confined space accidents is that incapacitated employees were not found until it was too late for rescue.

Clearly then, one of the fundamental precepts of emergency planning is to assure that an attendant is stationed outside the space. The attendant, who is sometimes called a stand-by man or hole watcher, serves as the keystone of an effective emergency response plan. The attendant's duties and responsibilities include:

- Keeping track of the identity and number of entrants in the space
- Monitoring activities in and around the space and being alert to potential hazards that might compromise entrants' safety
- Maintaining effective and continuous contact with those inside the space
- Initiating the emergency action plan if a problem arises

Requiring an attendant outside of a permit space is a widely accepted method for assuring entrants' safety. Consequently, the OSHA standard requires an attendant to be on duty for the duration of the entry. Attendants must be able at all times to monitor conditions in and around the space, track the activities of entrants and summon rescue services in an emergency. However, attendants are not precluded from performing other tasks provided the tasks do not interfere with their primary duties. For example, they can pay out air-lines, welding cables or cutting torch hoses as well as pass tools and equipment into and out of the space.

Tracking Authorized Entrants

It is not sufficient for attendants to simply know how many entrants are inside the space. Instead, they must be able to track and *identify* the authorized entrants entering and exiting the space. The reason for having to track the identity of entrants is twofold.

First, a tracking system assures that all entrants have exited the space at the conclusion of the job. Second, it provides rescuers with a positive means of identifying entrants.

This is critical during rescue operations, because with all the excitement, it becomes easy to loose track of exactly who exited the space. A system which only keeps count of the number of authorized entrants is not acceptable since it cannot ensure that all entrants have been rescued. Without a systematic approach for tracking employees, there is no way to account for entrants who failed to inform the attendant of their successful self-rescue from the space. Rescue personnel would then be exposed unnecessarily when they entered the space in an attempt to rescue entrants—who, unknown to emergency responders—had already evacuated.

Another problem with the simple count method is that it does not address situations where unauthorized personnel may have entered the space. These people—who may have even caused the emergency in the first place—could easily be counted as "entrants" as they emerged from the space. The resulting miscount could lead the attendant to believe that everyone had evacuated when, in fact, some authorized entrants were still inside. These employees might then suffer further injury or death as a result of the mix-up.

Clearly, it is essential during rescue operations to be able to determine whether or not all authorized entrants have evacuated. Any system for monitoring employee activity (badge-boards, sign-in/sign-out sheets, etc.) is acceptable as long as it identifies entrants inside the space.

Ordering An Evacuation

The attendant must be vested with the authority to warn curious or unauthorized people away from the space and to order entrants out of the space under any of the following situations:

1. If he observes processes or procedures not allowed by the permit, for example, solvent use, hot work or unauthorized work practices
2. If he notices behavioral changes such as euphoria or giddiness that might result from oxygen deficiency or from exposure to some anesthetic gases or solvent vapors
3. If he detects situations outside the space that could endanger those inside, such as a vehicle idling with its tailpipe next to the fresh air inlet of a ventilator
4. If he sees uncontrolled hazards such as exposed energized conductors or leaking fluid lines
5. If he is monitoring more than one space, perhaps two side by side, and has to divert his attention exclusively to one because of operational requirements or rescue efforts
6. If he can no longer perform his duties as an attendant

Attendants' Emergency Duties

Since the attendants' chief rescue responsibility is to summon emergency services, it's their ability to determine that entrants need help, not their proximity to the space, that is critically important. Acknowledging that electronic surveillance and communication equipment can assist attendants in carrying out their duties and responsibilities, the OSHA standard allows attendants to monitor more than one space. The regulation also permits attendants to be stationed anywhere outside the space, provided they can effectively carry out their assigned duties. For example, attendants could be in control rooms that allow them to monitor entrants remotely using electronic tracking devices, closed-circuit television or public address systems.

EMERGENCY RESPONSE OPTIONS

An effective confined space entry program attempts to identify and control all of the hazards to which entrants may be exposed. If everything goes as expected, work proceeds safely and nobody gets hurt. But we live in an imperfect world, things sometimes go wrong and emergencies arise. Clearly then, the time to plan for an emergency is before it happens, not when it happens.

Rescue may be necessary either because of extraordinary circumstances that arise suddenly and without warning or because of some deficiency in the permit program. The confined space standard *does not* require on-scene rescue capability. Instead, it stipulates that employers develop and implement *procedures.*

These procedures must address:

- Summoning rescue and emergency services
- Rescuing entrants from permit spaces
- Providing necessary emergency services to rescued employees
- Preventing unauthorized personnel from attempting rescue

Some employers may prefer to provide on-site emergency responders because off-site response may be too slow, inadequate or ineffective. Other employers may prefer to rely on off-site rescue services perhaps because they believe they do not have the resources to train employees to perform rescue or because of the ready availability of highly trained, quick-responding, off-site services makes on-site capability unnecessary.

Non-Entry Retrieval

The OSHA standard requires that a non-entry retrieval system consisting of a rescue harness and lifeline be employed whenever an authorized entrant goes into a permit space. An exception is made for situations where the retrieval equipment itself would not contribute to the rescue effort or would increase the overall risk of the entry.

One end of the retrieval line must be attached to the harness either above the head, or at the center of the back near the shoulders. The other end of the line must be attached to a mechanical device or fixed point outside the space to facilitate prompt response.

Wristlets may be used in lieu of a harness if the employer can demonstrate that the harness use is infeasible, or creates a greater hazard, and that wristlets provide the safest and most effective alternative. If it is necessary to retrieve entrants from vertical permit spaces deeper than 5 feet, a *mechanical* lifting device must be available. Various types of rescue and retrieval equipment will be discussed in more detail later in this chapter.

Rescue Services

Rescue services may be provided by either on-site or off-site responders. On-site rescue teams have the advantage of being immediately available and intimately familiar with the facility. Team members may also have existing relationships with other craftsmen such as electricians, pipefitters and mechanics who may be called on in an emergency. Since response time is critical in any emergency, an on-site team reduces response time to a minimum. The principle disadvantage of a on-site teams, unless contract services are employed, is the time and expense required for initial and annual training of team members.

The expense for off-site municipal rescue services, on the other hand, is borne by the community. But not every community has the resources to train and equip a confined space rescue team. Assuming that a properly equipped municipal rescue squad does exist, the time spent in dispatch and travel can delay its arrival. Even with the best pre-emergency site familiarization and planning, response will not be as fast a that of an on-site team.

On-Site Rescue Service

If an employer chooses to use the on-site approach, then the rescue team members must be properly trained. At a minimum, the standard requires that they receive the same level of training required of authorized entrants. They must also be trained in the proper use of personal protective and rescue equipment.

Each member of the team must be trained in basic first-aid and in cardiopulmonary resuscitation (CPR), and at least one member must be *currently certified* in both.

Rescue team members must also participate in an annual hands-on drill. The drill should simulate rescue operations in which dummies, mannequins, or actual persons are removed from confined spaces. Either actual or simulated spaces may be used, provided the opening size, configuration and degree of accessibility approximate the spaces from which rescue may be performed.

Rather than incur the expenses for equipping and training their own in-house teams, some employers have opted for contract rescue services. For a fee, arrangements can be made with commercial enterprises, state fire-training academies or local volunteer fire departments to provide fully trained and equipped *on-site* responders for the duration of the entry.

Off-Site Rescue Service

Off-site emergency services can be employed provided that two conditions are met. First, rescue service must be informed of the hazards that may be encountered when responding to a confined space incident. Second, responders must be provided with access to all permit spaces from which rescue may be necessary so that they can develop appropriate rescue plans and conduct practice operations.

OSHA enforcement policy allows employers to rely on 911 emergency services provided the two conditions above have been met. As a practical matter though, it is essential to verify that 911 service can be provided *reliably*. In other words, you better be sure that the rescue services that are supposed to respond to an incident at your facility are not fighting a fire in the next county when you need them!

EMERGENCY PLANNING CONSIDERATIONS

A recurring theme throughout this book is that there is a wide diversity not only among confined spaces but also the reasons for entry. An effective emergency plan must consider both of these factors if it is going to clearly identify and adequately address potential problems that may arise.

Since the operations and processes performed in the space have a decided influence on what types of emergencies may occur, they should be among the first things considered. For example, if hot work such as welding, cutting and brazing is being performed, it is reasonable to anticipate the outbreak of unintended fires. Similarly, a splash to the eyes may be anticipated if a vessel is being manually cleaned with a corrosive liquid. In some cases, the interior walls of water towers are inspected from rubber rafts that are lowered as water is drained from the bottom of the tank. Since a fall overboard can be anticipated, life jackets should be worn.

GENERAL CONSIDERATIONS

In planning for emergencies it's wise to remember Murphy's Law, which states that if anything can go wrong, it will. Two important corollaries to Murphy's Law are (1) if there is a chance of several things going wrong, the one that goes wrong is the one that does the most damage and (2) when left to themselves, things always go from bad to worse. Clearly then, the cardinal rule for emergency planning is to start planning before an incident occurs, not when it occurs.

Accident case studies are a valuable tool in emergency planning because they graphically illustrate things that have gone wrong in the past and provide us with an opportunity to learn from others' mistakes. Using the knowledge gained from past experiences, we can take appropriate steps to prevent similar occurrences in the future. An analysis of dozens of case histories reported by OSHA and NIOSH suggests that most ill-fated rescues could have been prevented by following one or more of the do's and don'ts listed below.

1. Don't enter a space unless absolutely necessary. There is no need to place additional people at risk if a rescue can be effected from outside. Numerous devices which allow entrants to be extracted from the space by attendants or rescuers outside are commercially available. Some of these devices will be described in detail later.
2. Do wait for help before effecting rescue. If it is necessary to enter the space for rescue purposes, rescuers should enter the space only after sufficient help has arrived to assure that the response can be performed without further incident.
3. Do assume that the atmosphere is immediately dangerous to life or health. Unless air monitoring indicates otherwise, it is wise to presume that the atmosphere is immediately dangerous to life or health whenever entrants are unconscious. As a result, rescuers must be equipped with SCBAs or air-line respirators with escape bottles.

4. Do limit the number of entrants. If entry is required for rescue, the number of rescuers should be limited to the minimum necessary to perform the job safely.
5. Don't use entrants' breathing air. If air-line respirators will be used during the rescue effort, they should be fed from a supply independent of that used by entrants.
6. Don't count on local emergency responders. Capabilities of fire departments and rescue squads vary from jurisdiction to jurisdiction. Their effectiveness in responding to a confined space emergency cannot be assured without extensive evaluation on a local basis.

In addition to these general guidelines, there are a number of specific considerations that will influence the scope and depth of the emergency plan. Some of the more important items are discussed below.

MEDICAL RESPONSE CONSIDERATIONS

The geographic location of the site greatly influences the extent of planning that must be done relative to emergency medical response, transportation and treatment.

Emergency Response

Emergency medical response planning is a little less complicated in metropolitan areas than in rural areas. Establishments located in metropolitan areas have ready access to municipal ambulance services, hospitals and good roads. Job sites are also identified by street addresses which facilitates easy identification by emergency responders.

Some rural areas may also be served by paid emergency service units, but most are supported by volunteer fire departments. While volunteer departments may be highly trained and well-equipped for fighting dwelling fires and attending to vehicle accidents, they may not be prepared to respond to a confined space emergency. The service area of rural departments is often larger than that of their urban counterparts, so response times may be more variable. Travel distances to hospitals may also be longer, roads may not be as good and without pre-planning, responders may have difficulty in locating sites that have no street address.

Emergency Transportation

Once a victim has been extricated from a space, he may need to be transported to a medical facility. In metropolitan areas this is not usually a major problem since timely transport can often be arranged through a municipal ambulance service. The situation is different in remote locations where ambulance response and travel times may be substantial.

Depending on the nature of the emergency, it may be more appropriate to transport a victim using an on-site vehicle rather than waiting for an ambulance. This decision must be made on a case-by-case basis and will be predicated on the time that local emergency services estimate it will take to respond and transport the victim to the nearest medical facility. In an emergency, this time delay must be weighed against the time it would take to transport the victim directly from the site to the hospital. Another factor that must be considered is that even if the direct transport option is quicker, emergency treatment can be provided while an ambulance is en route to the hospital.

If potential responders are not familiar with the site location, it will be necessary to provide them with directions. Transportation pre-planning should also assure that employees at the site know how to get to the nearest medical facility or hospital. Travel routes should be preplanned and documented to minimize delays in an emergency.

Emergency Treatment

The site location will influence the level of medical preparedness that will be required during the entry. In situations where professional medical resources are readily available, an adequate initial response may be afforded by employees trained in first aid and cardiovascular resuscitation. However, in other situations, particularly those where entry is performed at remote locations or where the nature of the work presents some unusual risks, it may be prudent to have a qualified Emergency Medical Technician (EMT) on-site.

If an injured entrant is exposed to a substance for which there is written information such as a material safety data sheet, a procedure must be developed that assures that the information is provided to the medical facility treating the exposed employee.

Bloodborne Pathogens

Any responders who are occupationally exposed to blood, such as *designated first aid providers* and EMTs, are covered by the OSHA Bloodborne Pathogen Standard. This regulation presumes that all human blood may contain microorganisms which can cause disease and, consequently, demands that precautions be taken to protect potentially exposed employees. The two pathogens that are of most concern are hepatitis B (HBV) and the human immunodeficiency virus (HIV) that causes AIDS. The specific requirements of the standard are detailed in 29 CFR 1910.1030 and selected provisions are summarized below.

Exposure Control

You must develop an exposure control plan that documents (1) the job categories where *all* employees are occupationally exposed to blood and (2) the job categories where *some* employees are occupationally exposed to blood. For those job categories where only some employees are exposed, you must also identify the tasks that may result in exposure. The control plan must also explain how incidents that result in exposure will be investigated so that future occurrences can be prevented.

Employee Training

Employees who may be potentially exposed to blood must be informed of the:

- Nature of the risks posed by hepatitis B and the human immunodeficiency virus
- Precautions they can take to prevent or minimize exposure, such as the use of work practices and protective equipment
- Signs and symptoms of bloodborne diseases
- Modes by which bloodborne diseases may be transmitted
- Safety, efficacy, benefits and availability of the hepatitis B vaccine
- Procedures to follow and person to contact relative to an emergency involving contact with blood

Universal Precautions

Employees must be provided with and required to use personal protective equipment such as gloves, eye protection and resuscitator masks that prevent blood from contacting the body.

Hand Washing

Employees must wash their hands as soon as feasible after removing gloves or other protective equipment. If running water is not available, an antiseptic cleaner and hand towels may be provided for immediate use, but employees are still required to wash with soap and water as soon as practical.

Hepatitis Vaccination

Hepatitis vaccination must be made available to all potentially exposed employees. The vaccine must be administered at reasonable times and places at no cost to the employee. An employee who declines the vaccination must sign a waiver, the wording of which is specified by the standard. If an employee initially declines the vaccination, but reconsiders at a later date, it must be provided at that time.

Post Exposure Follow-up

Employees must be provided with a confidential medical evaluation following any exposure incident involving blood contact with the eyes, mouth or mucous membrane. This evaluation must include provisions for serological testing of the source individual's blood, when possible, and counseling for the exposed employee.

PHYSICAL CHARACTERISTICS OF THE SPACE

The physical characteristics of a space play a key role in helping to identify various emergency response options. Planning will be greatly simplified if all the spaces are generally alike because a single rescue procedure can be developed and implemented for all the spaces. On the other hand, a variety of methods and techniques will have to be developed to address situations where individual spaces are configured differently.

Some of the more important considerations that should be evaluated during the planning process are listed below.

Geometry of the Space

The size, shape and configuration of the space will influence the overall rescue strategy. Large spaces such as petroleum storage tanks afford rescuers more room to maneuver than small spaces such as valve pits or box beams. The shape of the space also dictates the type of extraction equipment that can be used, how it can be set up and the relative ease by which a victim can be removed from a space. For example, retrieval from open-topped spaces such as degreasing, paint dipping and electro-plating tanks could be accomplished by rigging a lifeline through pulleys attached to overhead structural beams. In an emergency, the victim could be easily hoisted to safety through the top of tank. On the other hand, a fully enclosed process vessel that is entered through a 24-inch side-wall manhole will require a much more complicated approach and equipment will have to be set up so that a victim can be pulled toward the manway. If the victim is unconscious, rescuers may also have to work both inside and outside the vessel to get him out.

Portal Characteristics

The ease by which emergency responders can both enter a space and get a victim out is affected by the size, location and orientation of the portals.

Portal Size. Large openings obviously pose fewer problems than small openings because they offer less resistance to entry and egress (Figure 12-1). In order to fit through small openings, responders sometimes temporarily remove their SCBA air cylinders and pass them through the opening to assistants standing outside. This practice not only delays the rescue, but also voids the SCBA's NIOSH certification.

The problem posed by restrictive openings can be solved in some instances by retrofitting spaces with larger portals. Orders for new tanks, vessels and reactors, etc., should specify that, when possible, equipment be supplied with larger access openings.

Portal Orientation. The portal orientation influences the type of equipment that can be used for extraction. Commercially available rescue tripods with self-contained hoists work well in situations that permit an unobstructed vertical lift through a horizontal opening, but they are virtually useless for rescues involving lateral movement through vertical openings in vessel side-walls. In these cases, lifelines must be rigged so that an entrant can be pulled toward the portal where rescuers can provide the assistance necessary to get a victim out of the space.

Portal Elevation. The elevation of the portal also affects response strategy. Vertical at-grade portals are relatively easy to deal with because rescuers can enter them without the need to climb stairs or ladders. But many spaces are not accessible from grade level. For example, portals for some chemical process equipment are located dozens of feet above ground. Since these portals are often reached via ladders and stairways, the rescuers' ability to maneuver is severely limited. In addition, once an injured entrant is removed from the space, specialized high angle rescue techniques like those employed in mountaineering, may be necessary to lower the victim to the ground Figure 12-2).

Obstructions in the Space

The number and nature of obstructions inside a space will determine whether or not it is possible to wear a lifeline at all times. In some spaces, the location of baffle plates, pipes and valves may preclude the use of lifelines. This point will be discussed more thoroughly later.

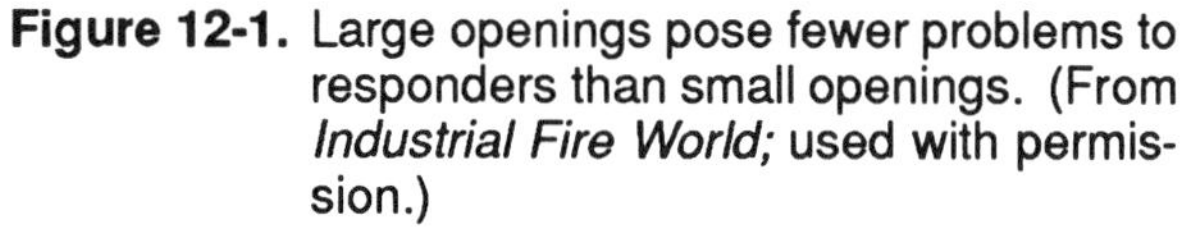

Figure 12-1. Large openings pose fewer problems to responders than small openings. (From *Industrial Fire World;* used with permission.)

The size and shape of the obstructions also influences whether or not SCBA can be worn. In some cases, openings through obstructions may be so small that airline respirators with escape supplies are the only realistic alternative.

Interior Lighting

The interior of the space may be dark. Even if lighting had been provided, it may have been rendered inoperative by the emergency. If portable 110 volt lighting is used during rescue, it should be either a low-voltage isolated system or protected by a ground-fault circuit interrupter, as described in Chapter 4. Remember too that only "approved" electrical equipment can be used in "classified" locations.

The Space's Contents

Both physical and health hazards may be posed by the contents of the space. If the space has not been drained and flushed, it may contain product residue which could produce hazardous air contaminants, be absorbed through the skin or cause surfaces to become slick or slippery. Entering responders must be suitably protected from these hazards.

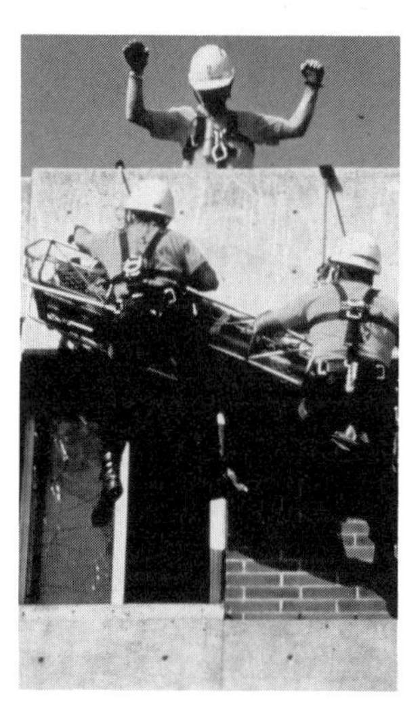

Figure 12-2. High-angle rescue techniques like those shown here may be needed to recover confined space accident victims from process vessels (a) or towers (b). (Roco, Inc. Don Allen (a) and Robert L. Davis (b))

COMMUNICATIONS CONSIDERATIONS

Two emergency communication concerns must be addressed. First, provision must be made for rescuers who enter the space to communicate with attendants who remain outside. Second, consideration must be given to how emergency medical assistance will be summoned to the work area in the event of an incident.

Factors influencing communications between rescue personnel inside and outside the space are essentially the same as those affecting entrants and attendants that were discussed in Chapter 6. In many cases, ordinary voice communication will be sufficient. In other cases, two-way radios may be required. Wired systems are not recommended for emergency response because the time spent in preventing wires from getting caught and tangled diverts attention from more critical rescue efforts.

Methods that can be used to summon emergency services to a site include two-way radios, telephones, intercoms and in-plant alarm systems. Although cellular telephones offer an attractive method of summoning help to remote work sites, there are certain locations within all cellular service areas where communication is marginal or nonexistent. So cellular phones should be field-tested for reliability before they are needed in an emergency.

If non-voice systems such as air horns will be used to sound an alarm for summoning additional help, the meaning of various signals and the expected response must be established ahead of time as part of the pre-planning.

EVACUATION CONSIDERATIONS

It must be clear to all entrants when and how evacuation will be performed. As mentioned previously, examples of situations that may prompt an evacuation include orders from the attendant, changes in atmospheric conditions or any unusual conditions perceived by the entrants. In all cases, planning must assure that evacuations are calm and orderly. If multiple entrants are involved, a rallying point should be designated for entrants to assemble after leaving the space.

RETRIEVAL CONSIDERATIONS

Ideally, every entrant should wear a lifeline attached to extraction equipment that allows speedy retrieval from outside the space. This approach is certainly appropriate when entry is made into spaces such as utility manholes, open-topped tanks and silos used to store grain, cement and plastic pellets. Since these spaces are generally free of internal obstructions that could entangle a lifeline or impede retrieval, it makes sense for an entrant to remain tethered at all times. But the world is not an ideal place, and in some situations it may actually be more hazardous to wear a lifeline at all times.

If a space contains pipes, valves, baffle plates or other equipment that could snare a lifeline, the benefit of wearing it would be lost as soon as it got caught or tangled. Lifelines can also become fouled in situations where multiple entrants are involved, especially if the portal is relatively small.

Professional judgment and good sense will ultimately dictate whether or not entrants should continuously wear lifelines, and in situations where their use is determined to be more hazardous, an alternative method of employee protection must be developed.

One approach would be to require entrants to wear a full body harnesses and to have *both* an attendant and a properly equipped rescuer present at the portal. The entrant would remain tethered as long as practical, and would disconnect the lifeline only when absolutely necessary. The free end of the line should be kept readily accessible, and if a problem developed the entrant would reconnect it to facilitate extraction. As demonstrated in Figure 12-3 a short length of rope tied to the harness D-ring and draped over a shoulder makes it easier for entrants to reconnect their lifelines because they don't have to reach behind their backs and fumble with the D-rings.

If the entrant is overcome, or if he needs help in getting out of the space, the already equipped rescuer could quickly respond to provide the necessary assistance. The rescuer could also guide the lifeline around obstacles that could cut or snag it, while the attendant operated retrieval equipment outside the space.

Figure 12-3. A length of rope attached to the D-ring makes reattachment of life lines easier. (John Rekus)

TRAINING CONSIDERATIONS

Although classroom activities are an important element of training, they are not sufficient by themselves to equip emergency personnel with the skills they need to respond in an emergency. It is essential that responders also acquire practical, "hands-on" experience with equipment and techniques they are likely to use. In an emergency, time is critical and responders cannot be wasting precious moments trying to figure out how to operate their equipment. This is not to imply that they should act impulsively. There will naturally be a delay as responders size up the situation and determine the best course of action, but they must be familiar enough with their equipment that its use is virtually second nature.

It is important for training to address even the most basic rescue scenarios, such as those where an attendant simply turns the crank on a winch to retrieve an entrant. The winch may incorporate some special feature like a locking pin or button that engages the lifting mechanism. If attendants are unaware of these features, they may not be able to operate the equipment when the need arises.

The amount of training required is dictated by the level of skill that rescuers are expected to possess. An hour or two may be sufficient to train an attendant how to operate a rescue winch, on the other hand, days, or even weeks, of intensive training may be necessary to equip responders with the skills needed to safely perform high angle rescue, advanced first aid or emergency medical procedures.

Classroom Training

Initial familiarization with emergency equipment such as SCBAs, rescue harnesses and extraction equipment can be performed in the classroom. The classroom setting is also appropriate for conducting some performance oriented training such as first aid and CPR. Table top exercises can be used to introduce rescue concepts and to provide participants an opportunity to practice their problem solving skills by discussing how they might respond to hypothetical emergency scenarios.

Practical Exercises

Since the goal of emergency response training is for rescuers to develop proficiency in the skills needed to respond to an incident, they must have an opportunity to apply classroom knowledge during practical field exercises. Field exercises also offer the advantage of being able to uncover hidden flaws that may exist in even the most well thought out response plan. These previously unforeseen problems can then be corrected before they jeopardize safety in an actual incident.

Field exercises must provide responders with an opportunity to use all the equipment they would employ in an emergency. The hypothetical scenarios that were discussed in the classroom can also be played out in environments similar to those likely to be encountered during an actual incident. If the response plan requires rescuers to enter a space to retrieve a victim, practice scenarios should incorporate volunteer "victims" or suitably weighted mannequins.

Actual spaces similar to those which would be entered in an emergency are ideal training aids since they introduce an element of realism. But if actual spaces are going to be used for training purposes, all of the normal precautions such as atmospheric testing, ventilation, isolation, etc., must be taken prior to entry.

Simulated spaces can be fabricated from a variety of building materials. For example, simulated tanks, vessels and manholes can be constructed from sections of steel or concrete pipe (Figure 12-4). Plywood sheeting with holes cut out can be arranged to simulate either horizontal or vertical portals (Figure 12-5). Rescue scenarios involving changes in elevation can be simulated by placing the plywood portals on scaffold frames.

Figure 12-4. A piece of corrugated steel pipe can serve as a confined space simulator. (Montgomery County, MD)

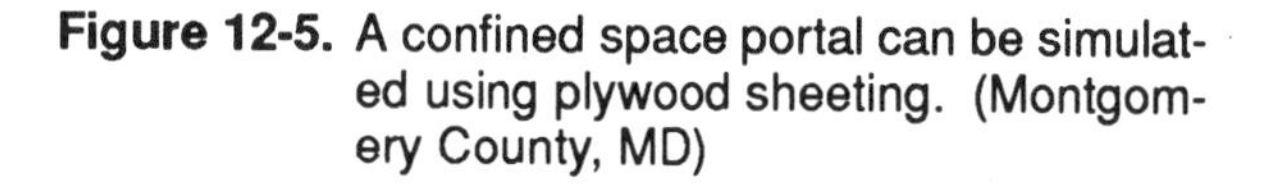

Figure 12-5. A confined space portal can be simulated using plywood sheeting. (Montgomery County, MD)

FAMILIARITY WITH THE SITE

Different approaches to emergency planning will be required depending on how familiar workers are with the site. A captive work force will possess certain innate knowledge of the plant environment such as site layout, personnel qualifications and the location of emergency equipment. This existing body of knowledge provides the foundation for building a response plan.

On the other hand, a mobile work force that travels from site to site will have little, if any, knowledge of its environment. Consequently, much more time and effort will have to be invested in identifying available emergency options and resources.

EMERGENCY EQUIPMENT

The specific equipment required for emergency response depends on the types of incidents expected. In general, though, the inventory is likely to include air-supplied respirators, first aid supplies, portable fire extinguishers, communications equipment and retrieval devices. Organizations that plan to enter spaces on a regular basis may want to dedicate specific equipment exclusively for confined space use. Necessary equipment can then be assembled in a job box, trailer or cart so that it is readily available and easy to transport.

Other organizations may choose to consolidate equipment from other plant areas when it is needed for an entry. While there is certainly nothing wrong with this approach, care must be taken to assure that equipment is not removed from other areas where it may be needed. For example, a self-contained breathing apparatus designated for emergency shutoff of leaking chlorine cylinders should not be commandeered for confined space entry. Likewise, a fire extinguisher positioned to protect a flammable storage room should not be appropriated from its assigned station.

Regardless of the method chosen, all emergency equipment should be inspected before use to determine that it is free from defects and complete and in good operating condition. A checklist that identifies specific conditions to examine is a handy inspection tool, and if equipment inspections are kept simple and easy, there's a good chance they will be performed.

Respiratory Protection

If entry into an atmosphere immediately dangerous to life or health is anticipated, then either self-contained breathing apparatus or pressure-demand airlines respirators with escape bottles should be on hand. The small size of airline respirator escape cylinders (Figure 12-6) makes them more maneuverable than full-size, back-mounted SCBAs. This feature is particularly advantageous if an entry must be made through a small opening. However, the reserve air cylinder must be sized for a capacity that provides sufficient time to escape in an emergency—a 3-minute escape supply will be of little value if it takes 5 minutes to get out of the space.

Communication Equipment

As mentioned previously, ordinary voice communication will be sufficient in many cases. If portable two-way radios are required, they should be operated on a channel separate from other in-plant communications to eliminate extraneous radio chatter. All equipment, including commercial telephones and in-plant intercoms, should also be checked prior to entry to assure that it is in proper working order and that personnel know how to operate it.

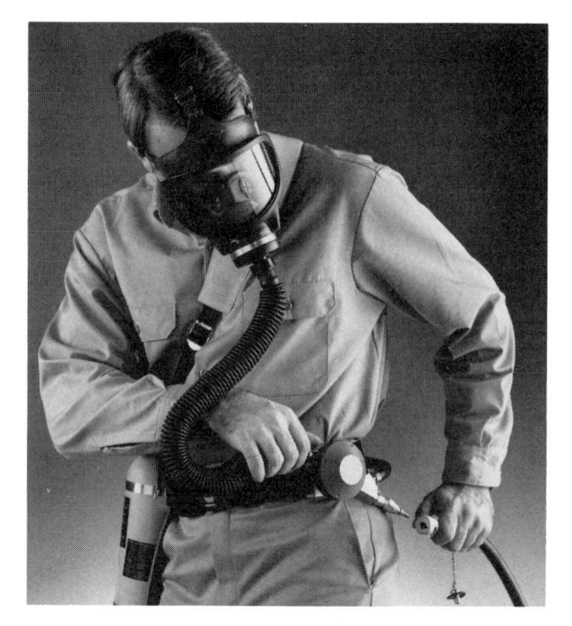

Figure 12-6. Air-line respirator with escape cylinder. (MSA)

First Aid Supplies

Selection of appropriate first aid supplies depends on the site characteristics, the responders' level of training and the nature of the anticipated injuries. An industrial grade first aid kit may be adequate for sites with ready access to professional medical services, but more elaborate provisions may be required when entry will be made at remote sites where access to medical facilities is more variable. In either situation, supplies must be provided that are consistent with the type of care expected to be rendered, and first aid providers and emergency medical technicians should consult with supervising medical authorities to determine what specific supplies are needed.

If employees may be exposed to injurious corrosive liquids, such as acidic or alkali solutions, quick drenching facilities for the eyes and body must be provided. In situations where running water is not available, portable self-contained devices like that shown in Figure 12-7 can be used. The unit is equipped with a detachable spray nozzle fed by a 30-gallon pressurized water tank.

Fire Extinguishers

Portable fire extinguishers should be on hand whenever hot work is performed. The number and size of extinguishers will be dictated by the extent of

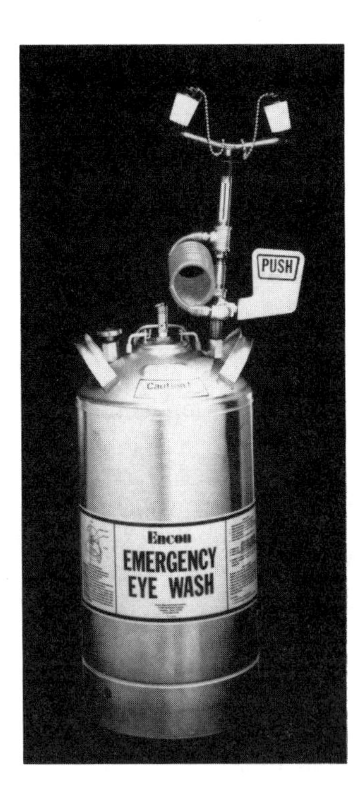

Figure 12-7. Self-contained eyewash drench shower. (Encon)

reasonably anticipated fire hazards. OSHA standard 1910.157 also requires that employees expected to use extinguishers be trained in their use. As a result, any employee expected to use an extinguisher should have an opportunity to use it on a live fire. Live fire training can usually be inexpensively arranged through a local fire department or through the contract service company that performs extinguisher maintenance.

Extinguishers should be inspected to verify that they are fully charged, that the tamper seal is in place and that all components such as gauges, hoses and discharge nozzles are in good repair.

Victim Transport

One device particularly well-suited for getting an injured victim out of a confined space is the Sked® stretcher. The Sked is fabricated from pliable heavy-duty plastic sheeting. Strategically placed grommets and openings provide tie points that permit an injured victim to be semi-rigidly secured much like a mummy (Figure 12-8). Once secured, the slippery surface of the Sked allows the victim to be slipped through narrow portals much like thread through a needle (Figures 12-9 and 12-10).

RETRIEVAL EQUIPMENT

The much ballyhooed myth that a single attendant outside the space can lift an entrant who has a rope

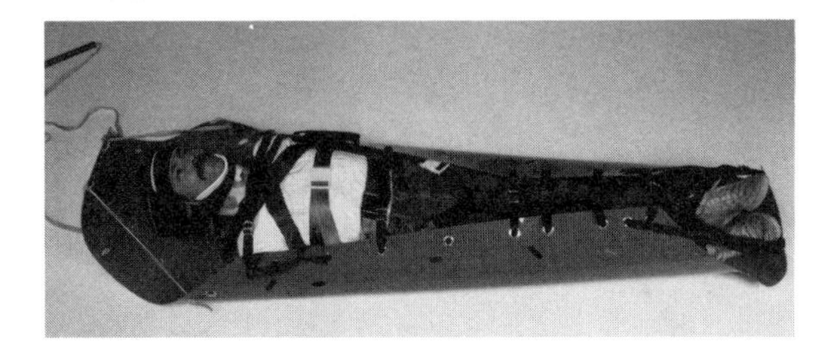

Figure 12-8. A victim can be secured to the Sked much like a mummy. (Skedco)

Figure 12-9. Sked stretchers can be used to move victims through horizontal openings. (Skedco)

Figure 12-10. A Sked stretcher being used to move a victim through a vertical opening. (Skedco)

tied around his waist can be dispelled with a simple demonstration. Have a volunteer loop a lifeline around his waist and lie on the floor. Have a second volunteer stand on a stool or other elevated platform and attempt to raise the "victim." He won't be able to, and even if he could, it would very difficult to get an unconscious worker out of the space because he would be in a horizontal position once he reached the portal.

To be effective, rescue equipment must incorporate devices that provide a mechanical advantage that allows the victim to be retrieved by a single person outside the space. Mechanical advantage is defined as the amount of force needed to move a load and is generally expressed as the ratio between a load and the force required to move it. For example, a mechanical advantage of 4:1 means that a 25 pound force can move a 100 pound load. Similarly, a 5:1 advantage means that a 125 pound load could be moved with the same 25 force.

From a practical perspective, high mechanical advantage means that attendants will have to exert less effort to pull a victim out of a space. At a minimum, retrieval systems should provide a mechanical advantage of 4:1. This would allow a 90 pound attendant to lift a 360 pound "victim." While a 4:1 mechanical advantage is acceptable, higher ratios are preferred because they make the retrieval easier and faster.

A manual retrieval system is adequate for depths up to about 50 feet, but powered devices should be considered for greater depths to assure a speedy ascent within the first few critical moments. Some manufacturers offer an air-driven motor accessory that attaches to the rescue winch. Compressed air can be provided either by cylinders or an in-plant piping system. Powered devices should only be used in spaces free from internal obstructions, including ladders or climbing rungs which could catch a victim's limbs and cause additional injury. They should also be fitted with torque limiters of approximately 450 pounds to avoid injuring an incapacitated victim.

As mentioned previously, all confined spaces are not entered vertically. Some spaces such as boiler mud drums and above ground storage tanks may be entered through side openings that require horizontal extraction. In these cases, interior surface finish must be considered and less resistance to sliding will be offered by smooth surfaces like glass linings than rough surfaces like concrete.

Retrieval equipment can be broadly divided into two types: preassembled and field-assembled systems.

Preassembled Systems

Preassembled systems consisted of a support structure such as a tripod (Figure 12-11) or a boom (Figure 12-12) which can be positioned over an opening to allow vertical lift. Tripod devices are free standing and typically have telescoping legs that can be adjusted to varying angles and heights to accommodate different field conditions. Some boom units like those shown in Figures 12-13 and 12-14 are free standing, others may be mounted on supports such as an existing structure (Figures 12-15 through 12-17), a truck trailer hitch (Figure 12-18) or a tank portal opening (Figure 12-19). Some preassembled units like those shown in Figures 12-20 and 12-21 are configured to allow horizontal extraction.

Most units are equipped with a winch containing a length of approximately 1/4-inch diameter galvanized wire rope, but some manufacturers offer both

Figure 12-11. A tripod-type vertical retrieval system. (DBI/SALA)

Figure 12-12. A boom-type retrieval system. (DBI/SALA)

Figure 12-13. A free-standing boom retrieval system. (Uni-hoist)

larger diameter and stainless steel ropes as options. These devices are relatively easy to set up and require very little skill to operate.

Field-Assembled Systems

Field-assembled systems offer more flexibility than preassembled systems but are more complicated to erect and require a greater degree of user sophistication. The basic device consists of a length of fiber rope that is reeved between pulleys to provide a mechanical advantage (Figure 12-22). One system that enjoys wide popularity in the fire and rescue service is the "Z-rig" shown in Figure 12-23.

Rescue rope as well as single-sheave (Figure 12-24) and double-sheave (Figure 12-25) pulleys may be purchased separately, but some manufacturers also offer kits which include all equipment necessary to assemble a complete system. As demonstrated by Figures 12-26 through 12-29, the mechanical advantage of field-assembled systems can increased by using configurations which incorporate additional pulleys.

LIFELINES

Lifelines used with retrieval equipment may be made of either fiber or wire rope. Wire rope is em-

Figure 12-14. Free-standing boom retrieval system adapted to a curved surface. (Uni-hoist)

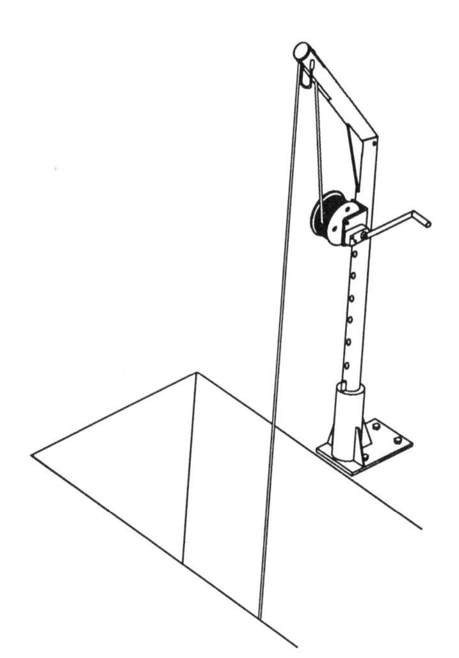

Figure 12-15. Boom-type retrieval system mounted on platform. (Uni-hoist)

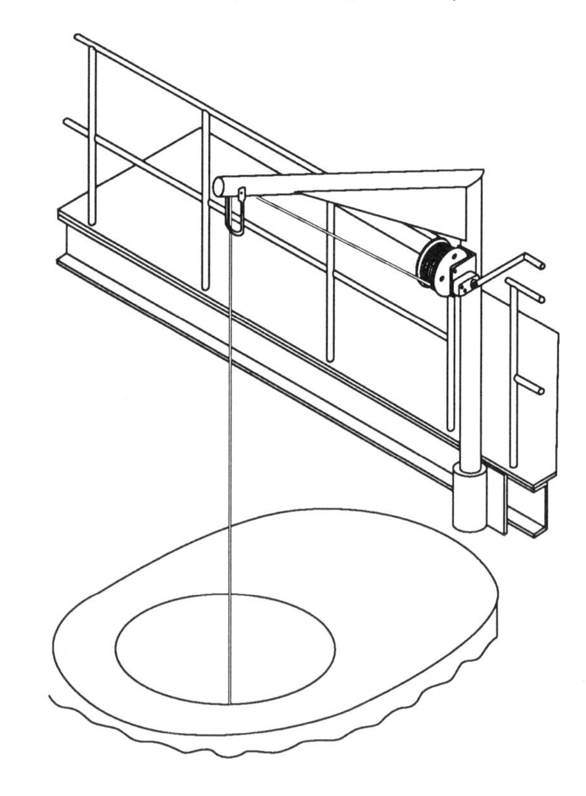

Figure 12-16. A boom-type retrieval system mounted on wall. (Uni-hoist)

ployed on winching devices provided with preassembled systems, while fiber rope is used with field-assembled systems. Rope designated for use as a lifeline should be used only for that purpose. Due to the critical nature of rescue, it is not enough for a rope to be strong; it must also be of a quality high enough to assure life safety. Ordinary commercial grade ropes are inferior to rescue ropes and are unsuitable for rescue operations.

Fiber ropes may be constructed using either a laid or braided design.

Laid rope (Figure 12-30) is perhaps the oldest and most familiar type. It is made by twisting lengths of individual fibers into larger bundles called strands. Multiple strands are then twisted together to form the rope. Laid rope tends to untwist when loaded, resulting in undesirable spinning and kinking. It is also easily weakened by abrasion, because all of the load-bearing strands are exposed and subject to wear.

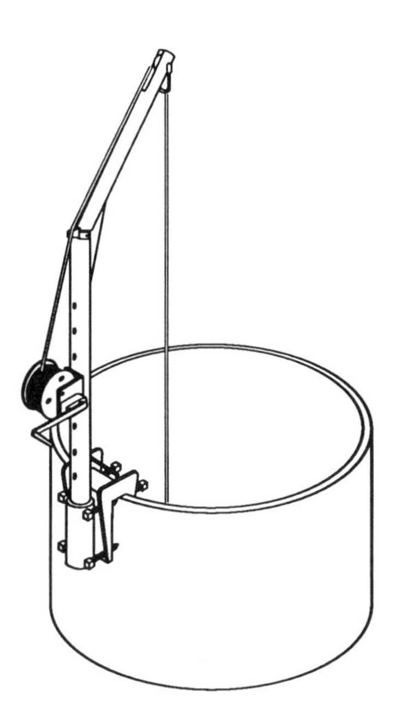

Figure 12-17. Boom-type retrieval system mounted on edge of vessel. (Uni-hoist)

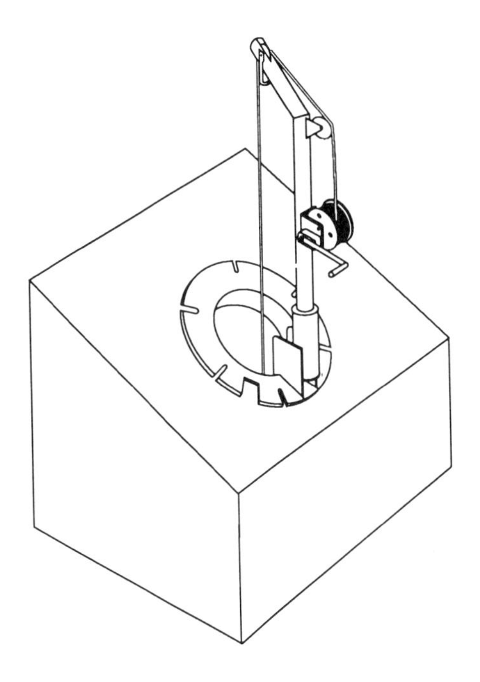

Figure 12-19. Tank collar for mounting boom-type retrieval system. (Uni-hoist)

Figure 12-18. Boom-type retrieval system mounted on trailer hitch. (Uni-hoist)

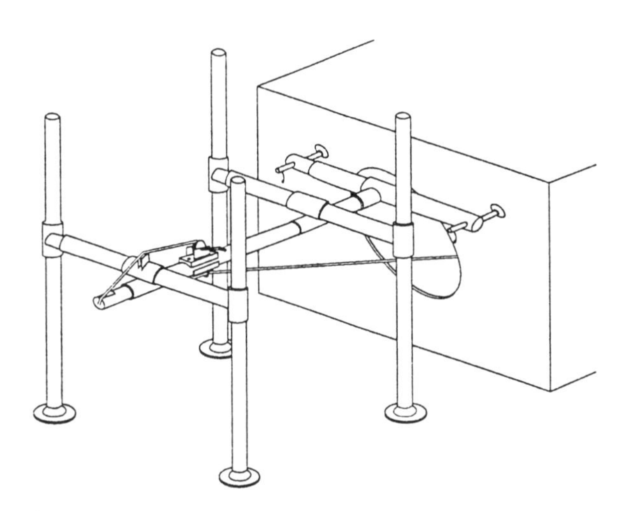

Figure 12-20. Horizontal fixed-position retrieval system. (Uni-hoist)

Braided rope is available in several designs, but the most suitable for rescue is the kernmantle style. Kernmantle rope consists of a central fiber core, the kern, covered by a braided sheath or mantle (Figure

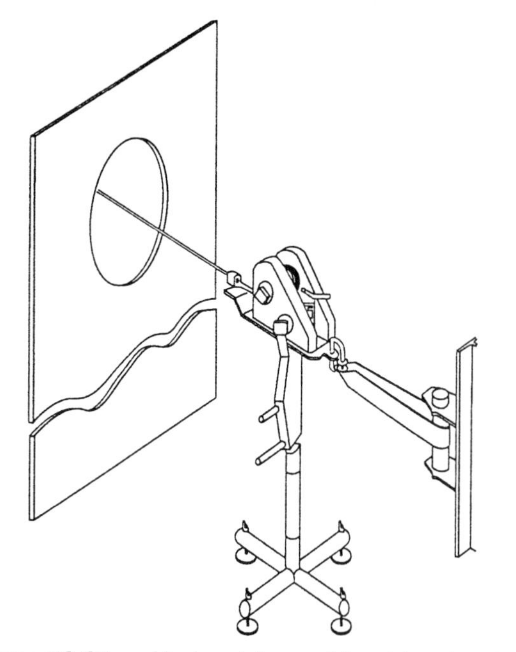

Figure 12-21. Horizontal movable retrieval system. (Uni-hoist)

Figure 12-22. Field assembled system with rope and pulleys. (Miller Equipment)

12-31). To minimize stretch and spin, the load carrying core is made up of bundles of fibers that run parallel and unbroken through the length of the rope. The tightly woven sheath surrounding the core protects it from dirt, abrasions and cuts. Kernmantle can generally withstand up to four times the abrasion of the same diameter of laid rope before it fails.

Materials of Construction

Natural Fibers

A variety of natural fibers such as manila, hemp, sisal, jute and cotton are used to manufacture rope. Of these, only manila is strong enough for life safety use. Although manila rope has been used traditionally for rescue operations, it is easily damaged by water, chemicals and petroleum products. It also tends to deteriorate over time, and even with the best of care, it can loose 10% of its strength per year due to aging. For these reasons, *manila rope is no longer deemed to be acceptable for life safety applications.*

Synthetic Fibers

Synthetic materials not only afford superior physical properties to natural fibers, but highly controlled manufacturing processes also produce products that are more consistent than those found in nature. Rope is commercially manufactured from four synthetic fibers. Table 12-1 compares the characteristics of these materials to manila and cotton.

Nylon is considerably stronger than manila having over 2½ times its breaking strength. This increased strength permits the use of smaller diameter lines which are lighter and easier to handle. In addition to greater strength, nylon also offers greater elasticity and shock loading resistance than manila. Nylon also has good abrasion resistance characteristics and is not prone to damage from mildew and dry rot. While it resists alkalis, petroleum products and many solvents, it is adversely affected by acids, natural oils, phenols and phenolic compounds. It is also affected by prolonged exposure to heat and UV rays from the sun.

Polyester has good abrasion resistance, stretches about half as much as nylon, resists mildew and rot, and is unaffected by sunlight. While it also offers increased resistance to acids, it is more easily damaged by alkalis than nylon.

Polyolefin (polypropylene and polyethylene) is not as strong or wear resistant as either nylon or polyester, but it does offer some other advantages. It is the lightest of rope materials and exhibits excellent resistance to acids, alkalis and mildew. Because of its low

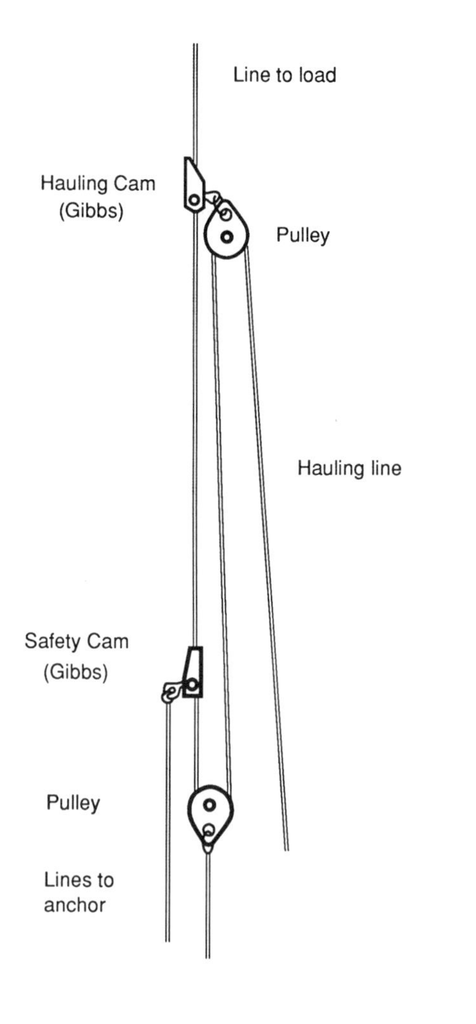

Figure 12-23. "Z"-rig retrieval system. (Paul Trattner)

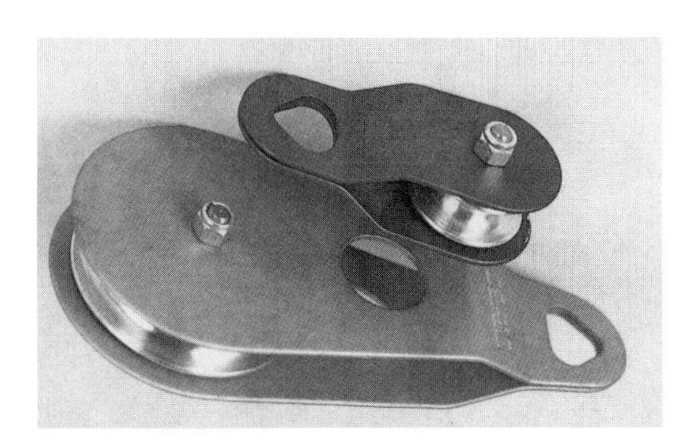

Figure 12-24. Single-sheave pulleys. (Rescue Systems)

Figure 12-25. Double-sheave pulleys. (Rescue Systems)

Figure 12-26. Rescue rigging that provided a mechanical advantage of 2:1. (Rescue Systems)

density, it will not sink even after prolonged exposure to water. However, polyolefin fibers degrade rapidly in sunlight. Since it also has low abrasion characteristics, low strength and low life expectancy, it is not generally recommended for rescue operations.

Rope Care

The useful life of both wire and fiber rope can be extended by proper care and storage. Tension caused by kinking, sharp bending and by bending over inflexible surfaces can stress individual wires and fibers, resulting in a decrease in a rope's service life. Dirt can also cause ropes to wear out prematurely, so

Figure 12-27. Rescue rigging that provided a mechanical advantage of 3:1. (Rescue Systems)

Figure 12-28. Rescue rigging that provided a mechanical advantage of 4:1. (Rescue Systems)

they should be cleaned prior to storage. Wire rope can be wiped off with a rag and lubricated with a light coat of oil. Fiber ropes may be cleaned with mild soap and water but care must be taken not to force abrasive particles into the fibers. Cleaning solvents should never be used, since they can damage and degrade the fiber. Wet rope should be air dried and stored where it will not be exposed to heat, chemicals, moisture, direct sunlight or rodents.

Figure 12-29. Rescue rigging that provided a mechanical advantage of 5:1. (Rescue Systems)

Figure 12-30. Laid rope. (John Rekus)

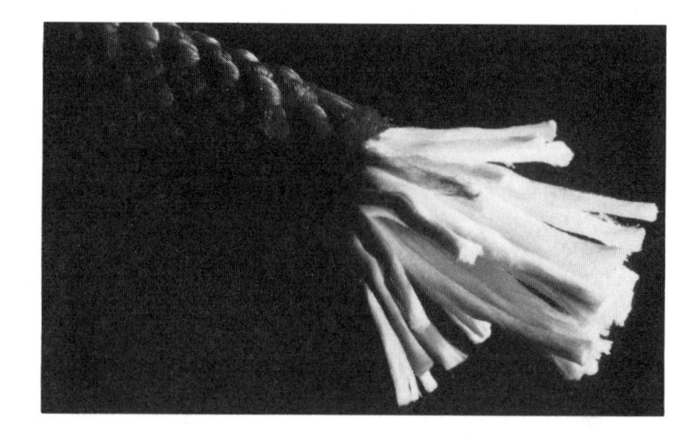

Figure 12-31. Kernmantle rope. (Rescue Systems)

Rope Inspection

It is not possible to predict how long a rope will last because service life is affected by variables such as how the rope is stored, its frequency of use, the environment it is used in and the type of loading it is subjected to. However, a routine inspection program can identify defects before they reach a point where a rope's performance will be critically affected. Both wire and fiber ropes should be inspected over their entire length before and after each use for irregularities, deformation, nicks, degradation, wear, abrasion, cut or broken fibers, rotting and chemical damage. Defective rope should be removed from service.

RESCUE HARNESSES

Even the most carefully maintained lifeline will be useless if it is simply tied around an entrant's waist. Instead, lifelines must be attached to a harness that both distributes the wearer's load and facilitates retrieval of an unconscious victim. However, to be effective, a harness must be selected properly, sized for a correct fit and inspected to assure that it retains its original effectiveness.

At one time, safety harnesses were constructed exclusively of leather, but today they are made of nylon and polyester webbing, which provide superior wear-resistance over natural materials.

Classes of Devices

Four classes of harnesses are described in American National Standards Institute (ANSI) standard A 10.14 (1975) *Requirements for Safety Belts, Harnesses, Lanyards, Lifelines and Drop Lines for Construction and Industrial Use*. Although ANSI withdrew this standard a few years ago, the nomenclature it employed continues to be used to identify different types of devices.

Body Support Devices

Body belts (ANSI Class I devices) consist of a simple or compound strap that secures around the wearer's waist and includes a means for attaching a lanyard (Figure 12-32). Body belts, frequently called "safety belts," are intended to retain a person in a hazardous work position. Although they can reduce the probability of falls, they are neither fall protection nor retrieval devices.

Chest Harnesses

Chest harnesses (ANSI Class II devices) consist of two sets of straps. One set secures around the rib cage while the other fits over the shoulders (Figure 12-33). The back of the harness is equipped with a ring for attaching a retrieval line. Chest harnesses may be used where there is a limited vertical free fall hazard and for retrieval purposes.

Full-Body Harnesses

A full-body harness (ANSI Class III device) resembles a parachute harness. It is secured around the wearer in a manner that distributes the load over the thighs, buttocks, chest and shoulders (Figure 12-34). A back-mounted D-ring is provided at about shoulder level for attaching a lifeline. As shown in Figure 12-35, one manufacturer has developed a pair of coveralls that has an integral Class III harness built in.

A full body harness can be used for retrieval, and with proper equipment such as a shock-absorbing lanyard, it can be used for fall arrest.

Independent Support Devices

A bosuns' chair (Figure 12-36) is a good example of independent work support (ANSI Class IV device) that can be used to raise or lower a worker in a confined space. Typical applications for these devices include inspecting, testing and painting the interior walls of vessels and silos. The occupant of a bosuns' chair must also be provided with an independent lifeline and fall arrest equipment for protection in the event of failure of the hoist line.

Equipment Sizing

Regardless of the type of harness used, it must be individually sized to the wearer. In general, all straps should be snug but not so tight as to be uncomfortable. Excessively long free ends can be tucked under or secured with tape.

Equipment Inspection

Users should inspect harnesses for damage prior to each use. At a minimum, the following items should be checked:

Table 12-1. Fiber rope characteristics

	Nylon	Polyester	Polypropylene	Polyethylene	Manila	Cotton
Dry strength	1	2	3	4	5	6
Wet strength (% of dry)	85	100	100	100	115	115
Shock load ability	1	3	2	4	5	6
Water flotation	Sinks	Sinks	Floats	Floats	Sinks	Sinks
Elongation at breaking	20–34%	15–20%	15–20%	10–15%	10–15%	5–10%
Water absorption	6%	1%	0	0	100%	100%
Melting point (°F)	480	500	330	275	Chars* at 350	Chars* at 300
Abrasion resistance	2	1	4	5	3	6
Resistance to sunlight	Good	Excellent	Poor	Fair	Good	Good
Resistance to rot	Excellent	Excellent	Excellent	Excellent	Poor	Poor
Resistance to acids	Poor	Good	Good	Good	Poor	Poor
Resistance to alkali	Good	Poor	Good	Good	Poor	Poor
Resistance to gasoline/oil	Good	Good	Good	Good	Poor	Poor
Electrical conductivity	Poor	Good	Good	Good	Poor	Poor
Flexing endurance	1	2	3	6	4	5
Specific gravity	1.14	1.38	0.90	0.95	1.38	1.54
Storage requirements	Wet or Dry	Wet or Dry	Wet or Dry	Wet or Dry	Dry only	Dry only

Rating scale: 1 = best, 6 = worst; * Does not melt

Webbing and its associated stitching should be examined for frayed or broken fibers and strands. Special attention should be paid to the belt tongues because of the extra wear they receive due to repeated buckling and unbuckling. Bending webbing over a round surface such as a piece of pipe will help to make cuts or breaks more visible. Broken strands will generally appear as tufts on the webbing surface and swelling, discoloration, cracks and charring are signs of possible chemical or heat damage.

Rivets should be tight and not able to be moved with the fingers. Rivet faces should lie flat against the belt surfaces and not be bent so that the edges are cutting into the webbing.

D-rings and their attachments should be checked for general condition, wear, pitting, cracks and any discoloration which may indicate corrosion.

Buckles should be free of distortion, strain, cracks or sharp edges that can affect the webbing.

RESCUE CASE STUDIES

The following four case studies of fatal confined space accidents were selected to emphasize two important points.

First, if a confined space entry program had been established the event which triggered the incident

Figure 12-32. ANSI Class I device. (MSA)

Figure 12-33. ANSI Class II device. (John Rekus)

would probably never have occurred. Second, the absence of an emergency plan resulted in the incidents rapidly escalating into full blown catastrophes.

It is interesting to note that in all four cases something as basic as testing the atmosphere could have prevented the death of entrants and would-be rescuers. In one case, the entrant was wearing a lifeline that permitted an attendant outside the space to retrieve him, but the attendant did not use the equipment provided.

It is hoped that some readers will use these case studies to stimulate group discussion in confined space training programs.

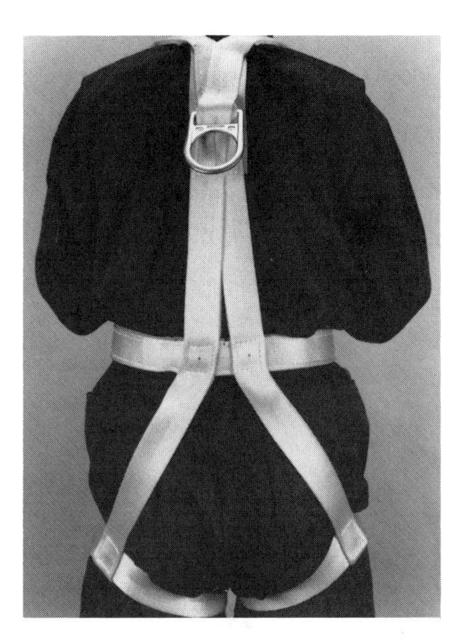

Figure 12-34. ANSI Class III device. (MSA)

Figure 12-35. ANSI Class III device built into coveralls. (MSA)

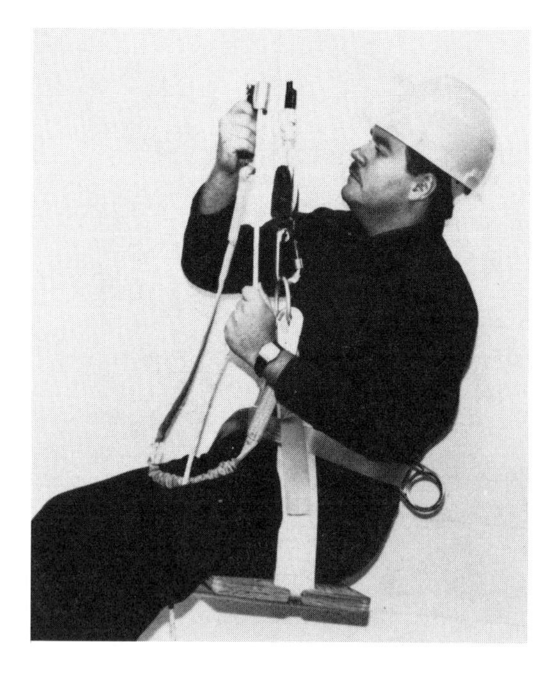

Figure 12-36. ANSI Class IV device. (Miller)

Case History No. 12-1: Barge Conversion

Synopsis of Events

A barge was purchased by a marine towing company and was in the process of being converted into a floating warehouse. One of the tasks involved removal of hatch covers leading to each of four cargo holds which had been sealed for about 10 years.

The foreman removed a bolt in the center of one of the man-way covers and tested for "bad air" by sniffing for odors. Not smelling anything, he proceeded to burn off the bolts holding the hatch in place.

He finished this job about 9 a.m., and, as was customary, left the site to go for coffee and doughnuts for his crew. On the way back, he heard a radio news report that three men had been overcome on the barge. Arriving on the scene, he met the Fire Chief, and, being a volunteer firefighter himself, grabbed a self-contained breathing apparatus and entered the hold. Upon entry, he observed the three employees floating face down in waist deep water. He recovered the bodies and was able to get them out of the hold with the assistance of on-deck firefighters.

Reconstruction of the accident suggested that the following chain of events occurred. The first employee entered the hold for reasons that remained undetermined and was apparently overcome due to a deficiency in oxygen. The owner's son saw him floating and called to another worker who entered to affect rescue. He too was overcome and a *third* employee tried to rescue him. A carpenter who was nearby building shelves also entered. He was overcome, but his arm caught in a ladder rung near the top of the hold. He was rescued by a fifth employee who entered and was almost overcome. Fortunately, he was able to evacuate in time.

Of the workers retrieved, the second entrant was declared dead on the scene, the third was alive but died while en route to the hospital. Surprisingly, the first entrant survived even though he was face down in the water for more than fifteen minutes. He remained in a coma for weeks but eventually recovered consciousness—fortunately without diminished mental capacity.

The medical examiner and principal investigator speculated that survival may have been attributed to the "divers' reflex." The "reflex" shunts oxygen to vital organs like the brain and has been observed in drowning victims in icy waters.

One of the tragic ironies of this case study was that company officials were aware of barge-related confined space hazards. In fact, they learned of them through a local newspaper article that described an almost identical incident. The company owner clipped the article and required all employees to read it as part of their "safety" training.

The proximate cause of the accident was determined to be oxygen deficiency, which may have been created in one of three ways.

First, it was possible that the steel in the hold may have oxidized to form rust. Second was the possibility that the oxygen was consumed by bacteria. The third possibility was that the oxygen could have been displaced by carbon dioxide formed by the decomposition of organic matter.

While the exact cause will never be known, one thing is certain—two people died and three more narrowly escaped!

Analysis and Commentary

Perhaps one of the most interesting aspects of this case is that even though the employer had read a news account about a virtually identical situation on another barge, he failed to incorporate protective measures into his own operation. Similarly, the employees who had read the newspaper article did not alter their behavior during the emergency. Instead, they acted instinctively, providing us with a classic example of the "domino effect" where several workers are killed as each tries to rescue those who had gone before.

Rather than sniffing the space for bad air, employees should have tested it, and testing would probably have revealed that the atmosphere was deficient in oxygen. This condition could easily have been corrected by mechanical ventilation.

The fact that the foreman had to borrow a selfcontained breathing apparatus from the fire department suggests that no rescue planning had been performed. Given the physical layout of the barge, it would have been a simple task to set up a rescue tripod on the deck and provide the entrant with a full-body harness and lifeline.

Case History No. 12-2: Painting Interior of Water Tank

Synopsis of Events

At approximately 4:30 p.m. one April afternoon, two workers encountered a problem while spray-painting the interior of a 500,000 gallon water storage tank. They signaled a coworker stationed outside, who then summoned additional help from several passersby before he too climbed into the tank through a 3-foot diameter hatch.

Emergency services personnel eventually arrived, but rescue efforts were hampered by high solvent vapor concentrations in the tank. Another complication was that entry into the tank was limited to a single, unstable, rope ladder. This combination of problems delayed rescue efforts for hours!

Finally, at about 7:00 p.m., one of the workers regained consciousness, but was killed when he fell to the bottom of the tank through a 110 foot riser pipe. By about 10:30 p.m. a police officer was able to secure one of the workers to a stretcher and pulled him to safety. This worker was hospitalized and later recovered. The third worker remained in the tank until 1:30 a.m. the following morning. He died in the hospital the next day. The cause of death was hypothermia resulting from the delayed rescue.

Analysis and Commentary

The facts of this case suggest that not even a rudimentary hazard evaluation had been performed. If it had, the hazards of the paint surely would have been known, and it would have been reasonable to expect that solvent vapors would accumulate inside the space if no ventilation was provided. The solvent vapor hazard could have been easily managed by ventilation or respiratory protection, but there is no evidence that either was employed.

In addition, it is clear that no rescue planning had been performed. Even though responders arrived in a timely fashion, they couldn't enter the space because of the high concentration of solvent vapors. If the unconscious workers could have been removed more expeditiously, there probably would have been no fatalities.

CASE HISTORY NO. 12-3: TOLUENE TANK

One firefighter was killed and sixteen more were injured while trying to rescue a worker who was overcome in a toluene storage tank. The tank was a vertical cone-roof design about 20 feet high and 9½ feet in diameter. It had a 16-inch diameter roof manway and a bottom outlet drain about four inches from the bottom. Residual toluene was present up to the level of the bottom outlet drain.

In an effort to ventilate the tank, the hatch cover and a 2-inch threaded bottom outlet cap were removed and a 2½-inch pressure-vacuum vent on the roof was tied open. The two workers who were assigned the task of cleaning the tank tied the middle of a piece of ¼-inch nylon rope around the pressure-vacuum relief vent and dropped one end into the tank. One worker removed his tennis shoes and proceeded to climb down the rope hand-under-hand but when he reached the bottom of the tank he was quickly overcome. The worker on the roof lowered a SCBA that had been rented that afternoon and nudged his coworker with it. When he got no response, he called for help.

Arriving fire officials were advised that the worker was unconscious and believed to be on his back with his face out of the liquid. Firefighters immediately attempted to force air into the tank by discharging extra SCBA air cylinders into the bottom outlet drain. After considering a number of options they decided the best course of action was to cut through the tank wall with a portable gasoline powered abrasive wheel saw.

Before beginning the cut, they ran an inch and a half hose line to the top of the tank and directed a water spray in through the manway. The spray helped to ventilate the tank by drafting fresh air in through the manway and also cooled the sparks produced by cutting through the wall. A second hose stream was used to cool the saw blade from outside, but it was shut off several times because the water interfered with the cutting.

The first cut was vertical and 19 inches long. A second vertical cut was made 18 inches to the left of the first, and the third and final cut was to be made horizontally connecting the first two together.

A Deputy Chief who became concerned about the flammable nature of the product inside the tank ordered the roof top firefighters to shut off the water spray and come down from the manway. About this same time a small fire was noted about 30 feet from the cutting operation, and since the other hose line was not being used to cool the cut, it was momentarily diverted to extinguish the blaze.

Just then, the vapor inside the tank ignited. A flame was observed flashing out of the open roof-hatch and the partially cut section of the tank was blown out and down in about a 75 degree angle. Although the ignition lasted less than a second, the resulting internal pressure was enough to distort the tank's bottom, rounding it out about 5 to 9 inches. The cone shape of the roof was also rounded out and wrinkled.

The force of the explosion through the opening was so severe that the firefighter who was operating the saw was thrown against a dike wall and the saw he was using flew through the air and hit the side of one of the fire engines. The firefighter who was backing up the saw operator was also thrown against the dike wall and received massive head and internal injuries—he was killed instantly. Several other firefighters were thrown about ten feet and had their helmets blown off and their protective clothing torn open. The heat was so intense that the protective hoods of the men nearest to the blast were damaged.

After the explosion, efforts were redirected to medical treatment and evacuation of the injured firefighters. When it appeared that there was no further danger of ignition, the body of the man inside the tank was recovered. The autopsy showed that he had only received minor burns to one arm. The cause of death was determined to be asphyxiation and inhalation of toluene vapors. The time of death was estimated to have been within ten minutes of entry, suggesting that he may have been dead about the time the fire department arrived, but certainly before the cutting began.

Analysis and Commentary

The facts in this case suggest that the tank's owner had some familiarity with confined space hazards. He apparently expected an atmospheric hazard because he had rented a SCBA and attempted to naturally ventilate the tank. However, he failed to test the tank's atmosphere to determine if the ventilation was adequate.

But even if the tank had been tested and found suitable for entry, climbing up and down a ¼-inch rope tied to a vent pipe is not the best way of getting in and out. Clearly this fall hazard could have been eliminated by using a rescue tripod and full-body harness connected to a lifeline independent of the wire rope to raise and lower the entrant. Another alternative would have been to fit the tank with a larger ground level manway which would have eliminated the need to enter from the roof.

The particular fire department involved is this case is one that has a national reputation for being knowledgeable about hazardous materials. The responding fire fighters were aware of toluene's hazards and they developed a fairly well-thought plan of action. Recognizing that time was of the essence, they chose to cut through the tank rather than attempt a more time consuming—and perhaps even more dangerous-rescue through the manway. The application of cooling water sprays seems like a reasonable method for controlling the potential fire hazard posed by the toluene vapor, but the Deputy Chief's concern about the safety of fire fighters manning the roof top hose line was also justifiable.

While there is no way of knowing whether or not the explosion would have occurred had the roof top spray been maintained, it is unfortunate that the hose was not secured so it could function unattended. It is also interesting to note that firefighters injuries would probably have been more severe had they not been wearing state of the art protective clothing. Clearly though, the whole catastrophe could have been prevented had the tank's owner developed an appropriate entry program.

CASE HISTORY NO. 12-4: GRAIN SILO

A grain processor hired a contractor to resurface the inside of a silo and to install some ventilation equipment. The silo was about 170 feet high and about 20 feet in diameter. Before the work began, an industrial hygienist from the OSHA Consultation Program met with the contractor to explain the requirements of the State's recently adopted confined space entry regulation. The work crew was experienced and had previously entered numerous silos without incident.

The bottom hopper was opened and the remaining grain was drained out, but when the silo was inspected, a grain bridge was noted at about mid-level. A bosuns' chair was rigged and a worker was lowered to break up the bridge with a piece of pipe. The worker wore an independent lifeline, but the chair was not equipped with a seat belt.

After the entrant was lowered about 60 feet, he called out to the top man to stop. He gave no explanation for his request and quickly ordered that lowering resume. A short time later the standby man noticed that the entrant was about 12 feet above the bridge and swinging his arm wildly in a circle. He then slipped off the seat, and when he hit the grain the portable light he was carrying went out. The standby called out to him, and when he got no response, he ran down 17 flights of stairs and explained what had happened to the foreman. The two of them then climbed back up the 17 flights of stairs. When they arrived at the portal, they raised the chair so the foreman could be lowered into the silo, but, unlike the entrant, the foreman did not wear an independent lifeline.

When the foreman got close enough to see that the entrant was still wearing his lifeline, he told the top man to raise the bosuns' chair. As he was about to climb through the portal, he slipped, and would have fallen if the attendant had not grabbed his collar in time. After getting out of the chair, he and the attendant hoisted up the victim, who was dead when he reached the portal.

State investigators who conducted air monitoring the next day found an oxygen level of 21% just below the portal. However, the percentage of oxygen decreased with depth and a concentration of only 2% was noted at the level of the grain bridge. Subsequent laboratory simulations using samples of the grain suggested that it had fermented and formed carbon dioxide which, being heavier than air, displaced the oxygen.

When the investigators asked the standby man why he didn't use the lifeline to immediately retrieve the victim, he explained that he had recalled from a high school health class not to move a person who had fallen because broken bones could puncture the lungs.

Analysis and Commentary

This case study provides a good example of the old maxim that a little knowledge is a dangerous thing. Unfortunately, the attendant's "little knowledge" concerning first aid proved fatal to the entrant. But like so many accidents, this one has no single cause. Instead, it resulted from a series of errors, the correction of any one of which may have prevented the fatality.

As in the previously discussed cases involving the barge and toluene tank, if nothing else other than testing the atmosphere had been done, the hazardous conditions in the space may have been identified and entry delayed until adequate ventilation or respiratory protection could have been provided.

The attendant's apparent failure to communicate effectively with the entrant constitutes another error. If the attendant had determined why the entrant initially gave the order to stop lowering, he may have become suspicious and raised the entrant when he did not get a good explanation. But even if the attendant had acted promptly, precious moments would have been lost during the 60 foot ascent to the manway. A powered take up should have been considered due to the silo's depth and could have easily been used because there were no internal obstructions.

The absence of restraining straps on the bosuns' chair represents another error. Had the chair been equipped with a harness or belt, the entrant may not have fallen off the seat, even if he lapsed into unconsciousness.

Obviously, there was no rescue plan and the attendant's knowledge of first aid was woefully lacking, but the real irony of this case is that it occurred in a state that had recently promulgated a confined space regulation. The driving force for the regulation was that three confined space incidents resulted in six fatalities over a three week period. The State's regulation required atmospheric testing, ventilation and a preparation of written emergency response plan. None had been performed.

SUMMARY

Emergency planning is an essential element of any confined space program because even the best programs can be affected by human failings which can lead to an emergency incident. However, the time to plan for an emergency is not when it happens, but *before* it happens.

Since confined space emergencies can differ in severity, an effective plan must provide a measured response that addresses the needs of a variety of incidents. In some instances, the entrants will be able to self-rescue or evacuate, in others they will be able to be retrieved by an attendant working outside. In still other cases, properly equipped rescuers will have to enter the space to assist entrants who are injured or unconscious.

The emergency planning process involves a host of considerations which must be evaluated to identify the most appropriate level of response. These considerations include the following:

- *Medical procedures* such as the means and methods for providing emergency response, transportation and treatment of injured personnel
- *Physical characteristics* of the space such as its size, configuration and location of portals which will determine various rescue options as well as the type of rescue equipment that will be required
- *Communications methods* both for summoning emergency assistance and for maintaining communication between rescuers inside and outside the space
- *Specialized equipment* needs such as SCBAs, retrieval devices, rescue harnesses, first aid supplies and appropriate portable fire extinguishers
- *Equipment inspection* procedures to verify that all emergency equipment is in good repair and proper working order
- *Training* that provides rescue personnel with hands-on experience with equipment and techniques that they are likely to employ in an emergency incident
- *Providing an attendant* who maintains effective and continuous communication with the entrants and monitors conditions in and around the space that may affect the entrants' safety; attendant also activates the response plan when an emergency arises

A comprehensive response plan that incorporates these provisions can make the difference between life and death in an emergency.

REFERENCES

Brown, M.G., Keeping your cool in a confined space, *Fire House,* June 1989, pgs. 69, 71–73.

Brown, M.G., Controlling fear and claustrophobia, *Response,* January/February 1988, pgs. 13–15.

Cloe, W.W., *Selected Occupational Fatalities and Asphyxiating Atmospheres in Confined Spaces as Found in Reports of OSHA Fatality/Catastrophe Investigations,* Washington, DC, U.S. Department of Labor, Occupational Safety and Health Administration, (1985).

Donahue, M.L., Confined space rescue: a neglected area of training, *Fire Engineering,* November 1983, pgs. 16, 18, 23.

Donahue, M.L., How to develop a standard operating procedure for confined space emergencies," *International Fire Chief,* October 1982, pgs. 22–25.

Downey, R., Confined space: rescuer or victim, Part 1, *Fire Engineering,* April 1989, pgs. 17–19.

Downey, R., Confined space: rescuer or victim, Part 2, *Fire Engineering,* May 1989, pgs. 16–19.

Eye and Face Protection, Code of Federal Regulations, Vol. 29, Part 1910.133.

Fire Department Occupational Safety and Health Programs NFPA-1500s, Quincy, MA: National Fire Protection Association, (1987).

Fire Service Life Safety Rope NFPA-1983, Quincy, MA: National Fire Protection Association, (1990).

Emergency Action Plans, Code of Federal Regulations, Vol. 29, Part 1910.38.

Henry, M.F., *Investigation Report: Flammable Liquid Tank Explosion, Phoenix Arizona, November 15, 1984, Fire Fighter Fatality,* Quincy, MA: National Fire Protection Association.

Kelly, T.E., Confined space rescue and entry, *Industrial Fire World,* June 1987 pgs. 6–7.

Mitera, J.C., Grain storage structure special rescue tactics, *Emergency Medical Services,* Vol. 16. No 8 September 1989 pgs. 61–64.

Moore, J., Recognizing and working safely in confined spaces, *International Society of Fire Service Instructors Voice,* February 1989, pgs. 28–29.

Occupational Foot Protection, Code of Federal Regulations, Vol. 29, Part 1910.136.

Occupational Head Protection, Code of Federal Regulations, Vol. 29, Part 1910.135.

Parsley, B.L., The twists and turns of rope, *Firehouse,* November 1987, pgs. 85–87.

Permit-Required Confined Spaces for General Industry, Code of Federal Regulations, Vol. 29, Part 1910.146.

Personal Protective Equipment: General Requirements, Code of Federal Regulations, Vol. 29, Part 1910.133.

Pettit, T.A., P.M. Gussey and R.S. Simmons, *Criteria for a Recommended Standard: Working in Confined Spaces,* Washington, DC, National Institute for Occupational Safety and Health, DHEW/NIOSH Pub. 80-106, (1980).

Portable Fire Extinguishers, Code of Federal Regulations, Vol. 29, Part 1910.157.

Respiratory Protection, Code of Federal Regulations, Vol. 29, Part 1910.134.

Rubin, D. and W. Luebberman, Confined Space Rescue, *Fire House,* May 1984, pgs. 57–59.

Sargent, C. and M. Brown, Confined space rescue: tragedy aboard ship kills firefighter and civilian, *Fire House,* March 1989, pgs. 39–47.

Schroll, R.C., Confined space rescue, *Fire Chief,* November 1988, pgs. 43–45.

Smith, D.L., The dangers of confined space emergency operations, will you be a rescuer or a fatality? *Fire Engineering*, July 1989, pgs. 44–48.

Vines, T. and S. Hudson, *High Angle Rescue Techniques*, Dubuque, IA: National Association for Search and Rescue.

Workmens' Compensation Board of British Columbia, *Industrial First Aid: A Training Manual*, New York, NY: Van Nostrand Reinhold, (1991).

CHAPTER 13

Entry and Hot Work Permits

INTRODUCTION

Entry permits are documents which certify that specific precautions such as isolation, atmospheric testing and ventilation have been taken before workers enter a confined space. Since a qualified person must complete a permit before each entry, the permitting process provides a systematic method for verifying that all elements of the entry program are in effect. The completed permit may then be used to inform entrants of potential confined space hazards, either by posting it outside the entry portal, or by having workers review it before they enter the space.

Both the form and content of a permit are important. A permit must be sufficiently detailed to address all the hazards associated with the entry, yet not so complicated that it is burdensome to use. But simply having a well-designed permit is not enough. The person who completes it must be able to evaluate potential hazards and document that each hazard is controlled before allowing the entry to proceed.

Using these broad concepts as a springboard, this chapter has the following four goals:

- To get you thinking about some organizational variables which influence permit design
- To identify specific core elements that must be included on all permits
- To offer some practical advice on designing effective permits
- To describe special permits and procedures related to hot work

GENERAL CONSIDERATIONS

An effective permit is one that addresses the needs of the organization, the people who have to complete it and the entrants. While all permits have certain things in common, their form and appearance may vary widely as a result of:

- Corporate philosophy and organizational structure
- The nature of the entry, including the characteristics of spaces and types of tasks to be performed
- The experience level of people who complete them

Organizational Structure

Organizational structure and management philosophy will generally dictate who is going to fill out the permit. Some facilities may choose to have one person complete the entire permit. Others may prefer to have individual departments address those permit elements for which they are responsible.

For example, electricians who lock-out motors would complete one section, pipefitters who isolate lines would complete another and industrial hygienists who evaluate the confined space atmosphere would complete yet another.

These two different approaches would require two different styles of permits. Otherwise, mass confusion would result when a half a dozen people tried to fill out a permit that was designed to be used by a single individual. If a number of people are going to complete the permit, then it had better be designed in a way that allows each person to document the tasks that he or she performs.

Nature of the Entry

To appreciate how the nature of the entry influences permit design, consider the following three scenarios.

- Company A is a distributor of motor fuels. It operates a fleet of ten identical tank trucks that delivers gasoline to service stations. The interior of the tank trailers must be visually inspected monthly.
- Company B is a municipal wastewater treatment plant that has confined spaces such as grease traps, wet wells, dry wells, lift stations, aerators, clarifiers and digesters that must be periodically entered for a variety of reasons.
- Company C is a contractor who cleans, repairs, inspects and installs equipment in confined spaces. No job is too large or small, too simple or complex for this company. For a price, they'll do any entry, anywhere, any time, for any reason.

A permit designed to respond to company A's needs would obviously be less complicated than that required for company B or C. To begin with, all of company A's spaces are identical, they all contain the same material (gasoline) and the reason for entry is always the same. Since there are no attached lines or internal mechanical equipment, isolation and lockout are not relevant. However, precautions would have to be taken to prevent the tank trailers from being moved unexpectedly.

Since the characteristics of the spaces and the nature of the entries are defined so exactly, it would be possible to develop a very detailed checklist style of permit that addressed every aspect of the entry. For example, the permit could specify the exact type of protective clothing, the protocol for atmospheric testing and the procedures for ventilating the space. After initial training, employees could use the permit as a guide to assure that each entry is conducted in exactly the same fashion.

Company B, the wastewater plant, presents a more complicated situation requiring a different kind of permit. While many of the wastewater plant's spaces present similar hazards such as flowing liquids, slippery surfaces, methane and hydrogen sulfide, the spaces differ with respect to function, size and shape. The reason for entry also varies. Some spaces may be entered for inspection, others for cleaning and still others for repair. Different entries may require different methods of isolation, ventilation and atmospheric testing.

Even though this variability exists, personnel, processes and equipment at wastewater plants remain pretty much the same day after day. As a result, standard operating procedures could be developed for specific confined space entry tasks such as cleaning digesters, inspecting bar screens, replacing ejector pumps, etc. These procedures could incorporate detailed information relating to isolation, ventilation, rescue equipment and so forth. The permit would then be used as a checklist to document that the specific procedures were followed.

The "guns-for-hire" scenario posed by Company C presents a real challenge. Unlike company A and B where the hazards were finite, Company C employees may be exposed to a virtually limitless number of hazards as they perform a variety of tasks involving thousands of chemicals in spaces of all shapes and sizes.

A further complication is that clients may contract for different levels of service. For example, some clients may drain, flush and isolate the space, others may simply turn it over to the contractor. The OSHA standard requires the host employer to *apprise* the contractor of any hazards associated with the space and to inform the contractor of any precautions or procedures taken to control those hazards. But the host employer is not obligated to prepare the space, and may instead defer to the contractor to take whatever steps are necessary to control the hazards.

The owner of a storage tank, for example, is obligated to inform a contractor of the previous contents of the tank, provide a material safety data sheet for the product and share any other information he or she has concerning hazards associated with the tank. But the contract agreement may stipulate that the contractor is responsible for performing all pre-entry work such a draining, disconnecting lines and ventilating.

Permit User's Experience

The user's experience and level of sophistication relating to occupational health and safety will significantly influence the permit's form and content. In general, the less-experienced employees are, the more specific the permit will have to be. For example, a permit used by employees who have extensive training and experience in selecting chemical protective clothing might simply provide a check-off box to indicate that appropriate clothing had been selected. On the other hand, a permit used by less experienced workers should provide space where a person knowledgeable in the selection process can specify the exact type of clothing to be used. Similarly, employees who possess extensive knowledge of atmospheric testing methods, ventilation principles and lockout/tagout procedures could be provided with a much less complicated permit than those with limited experience in these areas.

CORE CONTENT

The amount of information that can be squeezed onto a permit is limited only by its size; however, specific core elements are mandated by the OSHA standard (Table 13-1). Although all of these elements are discussed in detail below, it should be noted that each element need not necessarily be included on every permit. For example, if a space doesn't have any attached lines or sources of hazardous energy, then the permit need not address isolation or lockout/tagout.

Both checklist and open-ended formats may be used on an entry permit, but each format has its strengths and weaknesses. The checklist format may be used successfully in situations where there are only a limited number of choices, or where workers

Table 13-1. Permit elements required by the OSHA Standard

- The identification of the space to be entered
- The purpose of the entry
- The date and authorized duration of the entry
- Description of the hazards of the space
- Measures taken to isolate the space and manage the hazards
- The acceptable entry conditions
- Results of initial and periodic tests, including the name or initials of the tester and indication of when tests were made
- Communications procedures
- Special equipment required
- Identity of the authorized entrants and attendant
- Any additional permits such as those required for hot work
- The rescue and emergency services that can be summoned and the means for summoning them
- Any other relevant information
- The entry supervisor's signature or initials

are highly trained and experienced. Fill-in-the blank formats are usually more appropriate in situations where workers are less experienced or where a variety of methods, procedures or equipment may be employed.

In some cases, a combination of open-ended and checklist formats may be appropriate. Consider, for example, a batch process chemical plant where employees could enter dozens of spaces that could contain any of a thousand different chemicals. The entrants may be well-trained in safety procedures, but they may not know what contaminants to look for. An industrial hygienist could prescribe specific detector tubes on the permit, and the entrants could use their existing knowledge of exposure standards and their skill in making measurements with detector tubes to test space and evaluate the results.

Descriptive Information

Permits should contain a section that generally describes the purpose, nature and scope of the entry. Specific items that must be included are the identity of the space, the date of entry, a brief explanation of the operations or processes to be performed, and the approximate duration of the entry.

Hazard Identification

Actual and potential hazards that exist in or around the space should be identified. The hazard assessment must consider both the characteristics of the space and the nature of the work to be performed. Bear in mind that the hazard identification process should consider not only existing conditions, but also potential hazards that may arise from entry related operations and processes. As noted previously, the decision whether to use a checklist or open-ended format depends on organizational and site-specific variables. Readers interested in developing a checklist may obtain some ideas from the partial list of hazards presented in Table 13-2.

Preparation and Isolation

Spaces such as storage tanks, reactors and electroplating tanks may need to be drained, flushed and decontaminated before entry. Other spaces may need to be isolated or prepared in some special way. For example, the protective shutter on radioactive source level gauges must be closed to prevent irradiating the entrants.

If a single, well-established method is used to prepare and isolate the space, the permit may only need a check-off block. But if the procedures for preparation and isolation are complicated, or if more than one method is employed, an open-ended format would be more appropriate.

Purging and Ventilating

If tanks are purged with steam, the permit should identify the vacuum relief precautions that should be taken to prevent the shell from collapsing as it cools. If flammable atmospheres are purged with inert gases such as nitrogen or carbon dioxide, the permit should indicate that they, too, must be purged with fresh air before entry.

If ventilation procedures are well-established and only one type of blower is used, a check-off box may be sufficient. However, if there is a variety of ventilation equipment available, or if different spaces require different equipment configurations, a fill-in-the-blank format may be more appropriate. Any special precautions such as electrical bonding, venturi eductors or use of explosion-proof equipment should also be noted on the permit.

Table 13-2. Potential confined space hazards

Mechanical Hazards
- Agitators
- Blenders
- Stirrers
- Conveyors

Environmental Hazards
- Heat stress
- Wind chill (attendants)
- Wet, damp, humid environments
- Insects
- Snakes and vermin
- Slippery surfaces

Radiation
- Lasers
- Welding flash
- RF and microwaves
- Radioactive sources

Electrical Hazard
- Lines and cables
- Transformers
- Capacitors
- Relays
- Switch gear
- Exposed terminals

Noise
- Ambient noise levels
- From fans and blowers
- From operations in the space

Atmospheric Hazards
- Oxygen deficiency
- Oxygen enrichment
- Combustible materials
- Toxic air contaminants

Pressurized Fluids
- Acids
- Corrosives
- Hydraulic fluid

Engulfing Materials
- Plastics and chemicals
- Agricultural products
- Coal and coal product
- Wood chips

Ignition Hazards
- Open flames
- Heat sources
- Frictional sparks
- Unapproved electrical equipment
- Welding and cutting
- Hot riveting
- Hot forging
- Salamanders
- Internal combustion engines
- Portable electric tools
- Grinding
- Chipping
- Sandblasting

Traffic Hazards
- Pedestrian
- Public traffic
- Forklifts

Chemical Contact Hazards
- Acids
- Alkalis
- Coal tar products
- Sensitizers
- Skin irritants

Atmospheric Testing

The permit must include the results of atmospheric measurements. Since measurements must be made in a specific sequence—oxygen first, then combustibles, then toxic materials—the fields for entering test results should be arranged in the same order that the measurements are made. Specific values, or the range of values that are deemed acceptable for entry, must also be indicated on the permit. Conditions that are acceptable for entry, from a *regulatory perspective,* are provided in Table 13-3, but remember these values provide only the minimum level of protection. You may want to consider establishing more rigorous criteria such as 20% oxygen and *no detectable levels* of flammables, combustibles or toxics.

Table 13-3. Acceptable atmospheric conditions

Oxygen	19.5–23.5%
Combustibles	< 10% of LEL
Toxic materials	< PEL or TLV

Recall from Chapter 8 that confined spaces should be evaluated *before and after* they are ventilated. If a fill-in-the-blank format is used to record these measurements, be sure to provide an adequate number of blanks for the anticipated number of measurements. In other words, don't give people a 1-inch space to record six dozen readings. If certain contaminants will be evaluated on a regular basis, you may want to list them on the permit as shown in Figure 13-1.

The permit should provide space for recording the results of the "periodic" tests required by the standard. Provision should also be made for recording results of remonitoring that is performed after the space has been vacated for an extended period of time such as during lunch, coffee breaks and shift changes. As an alternative, if the instrument can print out electronically logged data, the hard copy can simply be attached to the permit.

When trying to decide on the best format for recording atmospheric measurements, think about your sampling strategy and aim to develop a format that complements that strategy. For example, if measurements are required at the top, middle and bottom of the space, you might want to provide a separate field for each measurement. The permit must also provide space for recording the testers name or initials, as well as the time the tests were made.

I know of one facility where oxygen measurements were made by the Fuel Department, combustibles by the Fire Department and toxics by Industrial Hygiene. Such a bureaucratic approach might seem pretty silly, but if it's required because of company "politics," the permit had better have a provision that allows individual testers to identify which measurements they made. Otherwise, there will be a lot of finger pointing when something goes wrong.

Other information that might be appropriate in the atmospheric monitoring section includes:

- The make, model and serial number of the instrument or, alternatively, an internal control or inventory number
- An indication that the instrument's response was field-checked before and after use
- The date of the last factory calibration

Test	Date/Time		Date/Time		Date/Time		Date/Time		Date/Time		Date/Time	
	/		/		/		/		/		/	
	Results	Initial	Results	Initial	Results	Initial	Results	Initial	Results	Initial	Results	Initial
Oxygen %												
% LEL												
CO ppm												
H_2S ppm												
Benzene ppm												
Hexane ppm												
Other:												
Other:												
Other:												

Figure 13-1. This format for recording air sampling results provides space for the tester's initials.

Some example permit formats for recording atmospheric test data are shown in Figures 13-1 through 13-3.

Personal Protective Equipment

Any personal protective equipment required for entry should be noted on the permit. Again, the choice of whether to use a checklist or open-ended format depends on organizational and site-specific variables previously described. Sample checklist formats are shown in Figures 13-4 and 13-5.

Another factor to consider in selecting protective equipment is that different levels of protective equipment may be required for different tasks. Level B protection, for example, may be required while removing caustic sludge. But after the sludge is removed, Level D protection may be suitable for other tasks such as thickness gauging and visual inspection.

Communications Procedures

The permit must indicate the preferred method by which entrants and attendants will communicate. It should also identify any special communications equipment that is required. Acceptable communications methods include direct voice communication, radios, wired headsets or even tugs on a rope. Each of these methods offers certain advantages and limitations, and the choice of the most appropriate method varies depending on the nature of the space and of the work being performed. For example, wired headsets may not be the best choice in situations where interconnecting cables could be cut, tangled or damaged by heat or chemical contact.

If you plan to use radios, in-plant intercoms or telephones to summon emergency response personnel, you might want to provide a check-off box on the permit to indicate that equipment was tested before entry and found to be operational. Remember too that any electrical equipment used in a hazardous location must be of an "approved" type.

Emergency Response and Rescue

Any special retrieval or rescue equipment should be specified on the permit. Typical equipment may include hoisting devices, first aid supplies, self-contained breathing apparatus and portable fire extinguishers. You might also want to consider providing a checklist that identifies critical components on emergency equipment that should be tested or inspected. For example, ropes and rescue harnesses should be inspected for cuts, burns or other damage. Similarly, SCBAs should be inspected to assure that their air tanks are full and that their low-air warning alarms function properly.

The permit must also provide space for summarizing the procedures to be followed in the event of an emergency. At a minimum, evacuation procedures and the preferred method for summoning rescue services should be noted. If your emergency plan utilizes any special in-plant or off-site telephone numbers, they should be listed on the permit.

Written directions and accompanying photocopies of local street maps should be provided whenever onsite personnel may be required to transport injured workers to a medical facility. This is especially crucial in situations where the people involved in transporting the victim may not be familiar with the local geography.

Oxygen 21-22%		Combustibles <10% LEL		CO <25 ppm		H_2S <10 ppm		Other:_______		Other:_______	
Time	Results	Time	Results	Time	Results	Time	Results	Time	Results	Time	Results

Figure 13-2. This is a vertical format that could be used for recording air sampling results.

Time	___	___	___	___	___	___	___	___	___	___
Oxygen	___	___	___	___	___	___	___	___	___	___
%LEL	___	___	___	___	___	___	___	___	___	___
H_2S	___	___	___	___	___	___	___	___	___	___
SO_2	___	___	___	___	___	___	___	___	___	___
Cl_2	___	___	___	___	___	___	___	___	___	___
Acetic acid	___	___	___	___	___	___	___	___	___	___
Benzene	___	___	___	___	___	___	___	___	___	___
Ethanol	___	___	___	___	___	___	___	___	___	___
Hexane	___	___	___	___	___	___	___	___	___	___
Tolune	___	___	___	___	___	___	___	___	___	___
Xylene	___	___	___	___	___	___	___	___	___	___
_________	___	___	___	___	___	___	___	___	___	___
_________	___	___	___	___	___	___	___	___	___	___
_________	___	___	___	___	___	___	___	___	___	___
_________	___	___	___	___	___	___	___	___	___	___
_________	___	___	___	___	___	___	___	___	___	___
_________	___	___	___	___	___	___	___	___	___	___
_________	___	___	___	___	___	___	___	___	___	___

Figure 13-3. This is a horizontal format that could be used for recording air sampling results.

Employee Designation

The permit must list the name of the current attendant and entry supervisor. It must also indicate the means by which attendants will keep track of authorized entrants. The primary reason for tracking the identity of employees inside the space is to quickly and accurately determine if all the entrants have been rescued in an emergency. An accurate accounting of entrants is also necessary to assure that no employees remain in the space at the end of the job.

The easiest way of keeping track of entrants is simply to write their names on the permit. But this approach isn't always practical or efficient. For ex-

Check Off All Required Personal Protective Equipment

Eye protection
❑ Chemical splash goggles
❑ Face shield
❑ Safety glasses

Chemical Protective Clothing (*specify type*)

❑ Gloves ( )	❑ Suit ( )
❑ Boots ( )	❑ Level A
❑ Jacket ( )	❑ Level B
❑ Pants ( )	❑ Level C
❑ Hoods ( )	❑ Level D

Respiratory Protection
❑ Self-contained breathing apparatus
❑ Positive-pressure air-line; if IDLH specify:
❑ 5 minute escape bottle
❑ 10 minute escape bottle
❑ 15 minute escape bottle
❑ Continuous flow hood or helmet
❑ Half-face ❑ Full-face air purifying
Specify filter/cartridge
❑ HEPA ❑ DMF ❑ AG ❑ OV
❑ NH_3 ❑ CL_2 ❑ AG/DMF

❑ Hearing protection
❑ Hard hat
Other
(specify):____________________
(specify):____________________
(specify):____________________

Figure 13-4. This is a checklist format that could be used for identifying personal protective equipment.

ample, consider the situation where multiple entrants continually enter and exit the space. Rather than having the attendant constantly add and delete names on the permit, alternative methods such as tagboards, entry badges, sign-in/sign-out sheets or electronic tracking systems could be used. In fact, any system for monitoring employee activity is acceptable as long it both accurately tells who is in the space at any given moment and is immediately accessible to the attendant.

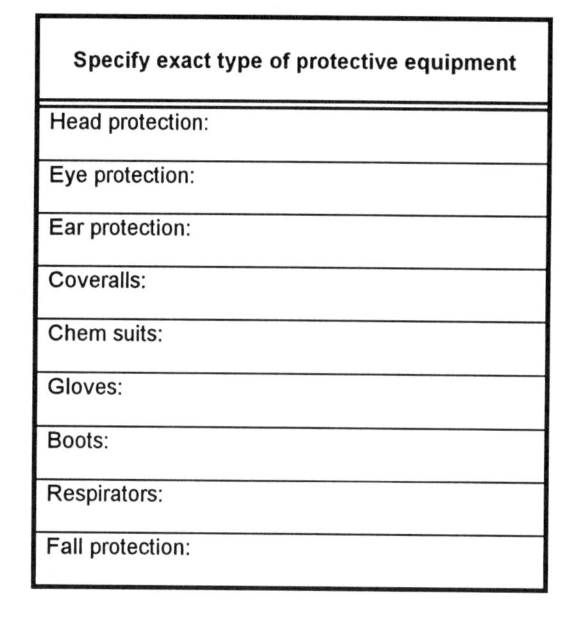

Specify exact type of protective equipment
Head protection:
Eye protection:
Ear protection:
Coveralls:
Chem suits:
Gloves:
Boots:
Respirators:
Fall protection:

Figure 13-5. This is a fill-in-the-blank format that could be used for identifying personal protective equipment.

Specialized Equipment

The permit should provide space for listing any specialized tools or equipment that may be required for entry. This may include such things as fall protection gear, portable lights, pedestrian barriers, manhole shields, personal radiation monitors, traffic control devices, etc.

Miscellaneous Information

No matter how well-designed a permit is, it won't be able to address every conceivable situation that may arise. Consequently, you should provide some blank space for additional comments or miscellaneous information. The blank space could also be used to reference other permits for safe work procedures, hot work or line breaking like those shown in Figures 13-6 through 13-9.

Authorizing Initials or Signatures

The permit must have space for the signature or initials of the individual who authorizes the entry. An additional tier of protection can be achieved by

Contractors' Safe Work Permit
Return This Permit Upon
Leaving The Designated Area

Date:__________ Time In:______ Time out: ______
Work area:__________
Work crew ID:______ Number of employees:______

All contractor workers received pre-job safety briefing. ☐
All contractor workers know the emergency evacuation signal. ☐
Contractor workers know the location of the emergency assembly area. ☐
Plant personnel know the location of this contract work crew. ☐

Tools and equipment workers are authorized to use

Hand tools ☐	Winch ☐	Air compressor ☐
Electric tools ☐	Hilti-gun ☐	Oxy torch ☐
Pneumatic tools ☐	Fork lift ☐	Welder ☐

Other (specify):__________

Additional permits required

Hot work ☐	Rail roadright-of way ☐	Waste disposal ☐
Line opening ☐	Road blockage ☐	Vessel entry ☐

If excavation is required:
Competent person is designated ☐ Miss utility contacted ☐
Underground reference drawings have been reviewed ☐
Soil classified ☐ Protective method selected ☐

Critical lift summary has been completed for all lifts over power lines structures or process equipment. ☐

Signature of person authorizing this safe work permit

Signature of contractor accepting safe work permit

Form SWP (Rev.3) ©1993 John F. Rekus

Figure 13-6. A typical contractors' safe work permit.

Hot Work Permit

Jobs cannot be authorized until all items are completed

Date	From AM PM	To AM PM	Location

Job description: __________

Site Preparation
Equipment intended for work has been ☐ washed ☐ steamed ☐ purged
Mechanical equipment is ☐ isolated ☐ locked or tagged out of service
Equipment has been checked for linings ☐
Equipment checked for flammable, corrosive, or toxic residue ☐
Combustibles in the area are protected or have been moved. ☐
Hazards posed by conductive heat transfer have been evaluated. ☐

Safety Equipment Required
☐ Eye protection ☐ Gloves ☐ Supplied-air respirator
Other (specify):__________

Work Practices
Fire inspection has been performed. ☐
Fire watch is present. ☐
Special procedures:__________

Department or group assigned to job	
Signature of employees assigned to job	
Signature of person authorizing the permit	

Form HWP (Rev.1) ©1993 John F. Rekus

Figure 13-7. A typical hot work permit.

requiring two authorizing signatures. One should be that of the person who is in direct charge of the work (the entry supervisor), the other should be that of an independent third party such as a supervisor from a different department, or a representative from the safety or industrial hygiene unit. This double-check approach often removes the temptation for supervisors to short-cut safety procedures to meet production goals.

DESIGNING THE PERMIT

Since entry permits are communications tools, people who design them should be familiar with some basic graphic communications principles. The appropriate application of these principles not only improves a permit's appearance, but, more importantly, it increases the effectiveness with which critical, perhaps even life-saving, information is communicated.

The permit design process involves collecting information from a number of sources and organizing it a clear, logical and easy-to-use format. A well-designed permit should be virtually self-explanatory and require minimal training. This might sound simple, but trying to strike the right balance between a permit that is both simple to use and adequate for the need can be quite challenging.

Line Breaking Permit

Jobs cannot be authorized until all items are completed

Date	From AM PM	To AM PM	Line Location

Job description: ____________________

Material safety data sheet has been reviewed. ☐

Material in line is: ☐ flammable ☐ toxic ☐ corrosive ☐ shock sensitive

Line has been: ☐ flushed ☐ steamed ☐ cleaned ☐ inerted

Operators have been notified of line opening. ☐

Spill and leak protection equipment is present ☐

Workers know the location of nearest emergency shower ☐
Workers know the location of the nearest eye wash station ☐
Workers know the location of the nearest plant telephone ☐

Specify protective equipment required:

Department or group assigned to job	
Signature of employees assigned to job	
Signature of person authorizing the permit	

Form LBP (Rev.2) ©1993 John F. Rekus

Figure 13-8. A typical line opening permit.

Bear in mind that a permit should be a tool that makes the job easier. If it doesn't, or if workers see it more as an obstacle than a benefit, they'll probably find a shortcut around it—perhaps even by making false entries. This last statement might startle some readers. "Oh no!" they'll say, "No one would deliberately falsify an official document." More pragmatic readers will agree that established safety procedures are circumvented every day. It's simply human nature: make your permit helpful and easy to use and people will use it. Make it stupid, pointless and a pain in the neck and they'll find a way around it. You choose.

CONFINED SPACE ATTENDANT'S PERMIT
Entry cannot be authorized until all items are completed

Date	From AM PM	To AM PM	Location

Special instructions: ____________________

There is a valid entry permit on the job site ☐
Potential hazards in and around the space have been identified ☐
Access to the space is restricted ☐
Communication method has been confirmed with entrants ☐
Emergency communications system is functioning ☐
Emergency procedures have been discussed and understood ☐
Rescue harness has been inspected ☐
Retrieval equipment functioning properly ☐
All personnel have been briefed on emergency plan ☐
Work coordinated with other crafts working near by ☐

Describe emergency plan:

Attendant's Signature

Form AP (Rev.5) ©1993 John F. Rekus

Figure 13-9. An attendant's permit.

Sizing Data Entry Fields

Who hasn't wanted to clobber the form designer who gives you a thumbnail-sized space to write a paragraph-sized answer? Remember that frustration when you're sizing data entry fields on your permit. Some entries will require more space than others, so consider the size of the anticipated response and provide a corresponding amount of room for it. The space required for recording air sampling results, for example, might be smaller than that needed to list required protective equipment. A good rule of thumb is to approximate how much space you would need if you were filling out the permit and provide at least twice as much room.

Ambiguous Wording

Everything we read is colored by our point of view. For example, if I were to ask you to visualize a farm house on a hill, what would you see? Do you see the same farm house I see? My farm house is a new ranch style house constructed of brick and aluminum siding, and I'm looking down at it from a vantage point high on another hill. Is that what you envisioned? I'd guess not. Perhaps you imagined a weathered wooden house viewed from the bottom of the hill. Regardless of what you imagined, I think you will agree that the five words "farm house on a hill" creates different images for different people. Clearly then, even simple words, phrases, and expressions may take on many dimensions of meaning. As a result, wording on a permit which seems simple, clear and logical to you, may not be simple, clear and logical to everyone else.

For example, consider a permit that has a data field captioned "Personal Protective Equipment?" Some people may interpret this field as a question to which they can answer "yes" or "no." But is that the response you were looking for? If instead you wanted people to *identify* the specific equipment needed for the job, it would be more appropriate to say "List required protective equipment." Now people know exactly what your expectation is, and they can respond accordingly.

You can identify text which is fuzzy, ambiguous or unclear by circulating a draft permit to a cross-section of people who will have to use it. Have them discuss it as a group to determine if each of them understands everything the same way. Chances are that different people will have different perspectives on some points, but by working as a group they will be able to come up with clearer less, ambiguous phrasing.

Another thing to consider when designing a permit is that an affirmative answer to a "yes-no" question ought to indicate a safe rather than hazardous condition. In this light, consider the following checklist items.

	Yes	No
Sources of ignition present?	[ ]	[ ]
Lines disconnected or blanked?	[ ]	[ ]
Hazardous fluids present?	[ ]	[ ]
Ventilation provided?	[ ]	[ ]
Hazardous atmosphere present?	[ ]	[ ]
Rescue equipment provided?	[ ]	[ ]

Did you notice anything peculiar about these questions? An affirmative answer to the first, third and fifth question indicates the presence of a hazardous condition! But that's not all. As indicated below, even appropriately marked responses remain *visually ambiguous* since it not obvious which is the appropriate answer.

	Yes	No
Sources of ignition present?	[ ]	[x]
Lines disconnected or blanked?	[x]	[ ]
Hazardous fluids present?	[ ]	[x]
Ventilation provided?	[x]	[ ]
Hazardous atmosphere present?	[ ]	[x]
Rescue equipment provided?	[x]	[ ]

But contrast these questions with the phrasing below.

	Yes	No
No sources of ignition present?	[x]	[ ]
Lines disconnected or blanked?	[x]	[ ]
No hazardous fluids present?	[x]	[ ]
Ventilation provided?	[x]	[ ]
Atmosphere acceptable?	[x]	[ ]
Rescue equipment provided?	[x]	[ ]

Now a "no" answer not only indicates a hazardous condition, but it also stands out *visually* and flags attention to the problem.

But this leads us to another point, and that's the value of even offering "yes-no" choices. Think about it—who in their right mind is going to check a "no" box indicating the existence of a hazard and then allow the entry to proceed? Another problem with "yes-no" questions is that they are not relevant in some situations. For example, not all spaces are connected to pipes containing hazardous materials. Providing a question about blanking, blinding or isolation that demands a "yes" or "no" answer is not appropriate since *there are no lines* to be blanked, blinded or isolated! As a practical matter, the only two logical choices to many questions are "yes", or an indication that the question is not applicable.

Format and Layout

Remember that a well-designed permit should make the user's job easier, not harder. So, if you want people to use your permit, you better make it easy for them. To that end the permit's layout should be simple, logical and uncluttered. Different sections such as hazard identification, atmospheric testing and rescue procedures should be clearly identified. You might also want to consider separating different sections by blank space or bold lines.

Avoid using footnotes. Instead, use parentheses to integrate secondary information into the main text. Text should be horizontal and flow from left to right, which is natural for readers of English. Avoid frequent variations in line spacing, especially when trying to crowd a lot of information into a limited space, and remember to allow adequate space for the anticipated response. In other words, don't allow a 1/2-inch space for a 3-inch answer!

Decimally numbered headings and subheadings such 1.1.1, 1.1.2.1 and 1.1.2.2 tend to confuse many people, so avoid using them. If you feel that you must number headings, don't use more than a *single* subheading and employ the clearer dashed format (1-1, 1-2, 1-3, 2-1, 2-2, 2-3, etc.). Pictographs and symbols are wonderful for quickly conveying a lot of information in a limited space, but if you use them on your permit make sure everyone recognizes them and understands their meaning.

Type Styles and Size

Printed type should be clear and legible. If possible, 9 to 12 point type should be used. Eight point type is pushing the limit of readability, and anything less than six point will be virtually indecipherable without a magnifying glass (Figure 13-10). Ornate lettering styles should not be used since they are harder to read. Instead, use clean, simple type faces such as Helvetica or Times Roman (Figure 13-11). Also avoid using a variety of type styles on the same permit. Choose one typeface and stick with it.

If you want to add emphasis or draw attention to specific items, use heavier weight (bold) letters in the same type face. Reversal effects such as white letters on a black background are visually dramatic but less readable, especially if large blocks of text are used. Long strings of capital letters are also harder to read than a mix of upper and lower case letters. When numbering, use Arabic numbers (1, 2, 3) which are more familiar to most readers than Roman numbers (I, II, III).

Do not depend on color to highlight important text or information. Some users may be colorblind, and if the permit is photocopied, the highlighting color may either be lost or it may obliterate the underlying text.

Physical Construction

Permits intended for use in damp, wet, corrosive or other hostile environments should be designed to withstand the conditions to which they may be exposed. Paper and cardboard permits can be laminated, placed in polyethylene envelopes or covered with plastic sheet protectors that are available from most stationery stores.

Multiple copies of the permit may be required in some situations. For example, as previously explained, some facilities use different crafts to perform different pre-entry tasks such as isolation, ventilation and testing. A representative of each craft involved in the pre-entry process should complete the relevant portion of the permit. If carbonless paper is used, each representative can retain a record copy of the permit. The last page or bottom copy is the final version that is communicated to entrants and posted at the space.

Permits printed on red, orange, or virtually any color of day-glow paper will be highly visible. Unfortunately, these colors don't provide very much contrast for black letters, making them difficult to read, especially at night or under low light levels. If you want to add visual impact to your permit use a bright colored border instead of colored stock. A 1/4- to 1/2-inch wide stripe around the perimeter of the permit is sufficient to make it stand out.

Rigid permits such as those that are laminated or made of card stock should be fitted with a grommet so they can be more easily attached to pipes, bolt holes and valves with string, wire and cable ties.

Make a mock-up of your permit using a piece of graph paper and hand lettering. Run off a few dozen copies and let people try them out to see how they work. Talk to users and find out what works and what doesn't. Make the necessary changes before you order 10,000 from a printer.

Figure 13-12 provides an example of a permit which combines the regulatory requirements of the standard with principles and practices of good graphic design.

This is 6 point Helvetica

This is 7 point Helvetica

This is 8 point Helvetica

This is 9 point Helvetica

This is 10 point Helvetica

This is 11 point Helvetica

This is 12 point Helvetica

This is 13 point Helvetica

This is 14 point Helvetica

This is 6 point Times Roman

This is 7 point Times Roman

This is 8 point Times Roman

This is 9 point Times Roman

This is 10 point Times Roman

This is 11 Point Times Roman

This is 12 point Times Roman

This is 13 Point Times Roman

This is 14 point Times Roman

Figure 13-10. Examples of type sizes.

Revising the Permit

The usefulness and applicability of the permit should be audited at least annually. Reorganizations, departmental expansions or contractions, or changes in procedures may require the permit to be revised.

ENTRY PERMIT PROCEDURES

Written procedures must be established for preparing, using, issuing and canceling permits. Since a permit is valid only as long as specified conditions are maintained, the permit procedures permit must assure that entry supervisors are aware of their ongoing responsibility to monitor work activities. Supervisors must also know that they are to revoke the permit whenever they detect conditions or work activities that introduce hazards not addressed by the existing document.

abcdefghijklmnopqrstuvwxyz

abcdefghijklmnopqrstuvwxyz

abcdefghijklmnopqrstuvwxyz

abcdefghijklmnopqrstuvwxyz

abcdefghijklmnopqrstuvwxyz

abcdefghijklmnopqrstuvwxyz

Figure 13-11. Fancy type faces like this should not be used on a permit because some people may have difficulty reading them.

Posting the Permit

Completed permits should be posted at the portal or made available to entrants so that they can personally verify that pre-entry precautions have been taken. Permits may not be signed until all actions required for safe entry have been completed, nor may they be removed until the last entrant leaves the space.

Canceling the Permit

Before canceling the permit the entry supervisor should verify that all entrants are out of the space and that appropriate measures have been taken to prevent unauthorized entry. After determining that all work has been completed, the space can be returned to service, and the permit can be canceled.

Annual Permit Review

Canceled permits should be retained since the standard requires that they be reviewed annually.

Confined Space Permit

Space ID:	Date:	Duration:

Purpose of entry:

Identify hazard associated with entry:

Preparation:	Isolation
Drained ☐ Flushed ☐ Inerted ☐ Purged ☐ Ventilated ☐ Barricaded ☐	Lines: Disconnected ☐ Blanked ☐ Equipment: Locked-out ☐

Communications procedures: Voice ☐ Radio ☐ Intercom ☐ Phone-set ☐ Rope signals ☐
Emergency response communications system functioning ☐

List emergency services and means for summoning them (use other side if necessary):

List any special equipment required (use other side if necessary):

Identify type of protective equipment required:
Head protection:__________ Eye protection:__________ Ear protection:__________
Respirators:__________
Chemical protective clothing:__________

Air Monitoring: acceptable limits Oxygen 20-21%, Flammable <10% LEL Toxics< PEL/TLV/REL

Time										
O_2%										
LEL%										
H_2S_{ppm}										
CO_{ppm}										
SO_{2ppm}										
Other contaminants (specify)										

Instrument ID and S/N:	Last calibrated:

Tester's signature:

Authorized entrants:

Attendant:	Entry supervisor:

Form CSP (Rev2)

Figure 13-12. 10 or 12 point type sizes should be used on confined space permits.

The purpose of this review is to identify and correct any procedural deficiencies in the permitting system.

SPECIAL PERMITS FOR HOT WORK

Hot work is a term used to describe heat-producing operations such as welding, flame-cutting and grinding. Hot work presents two significant hazards. First, open flames, flying sparks and hot surfaces can ignite flammable gases and vapors. Second, hot work may produce toxic fumes and gases. The fact that hot work is going to be performed in a confined space must be noted prominently either on the entry permit or on a special "hot work permit" attached to the entry permit. Some of the more important precautions and work practices that should be considered when permitting hot work are summarized below.

Preservative Coatings

Some preservative coatings may be flammable, others may produce toxic air contaminants when heated. Consequently, it is important to evaluate the potential for fire and health hazards posed by surface coatings before beginning hot work. Soft, greasy coatings generally present a greater fire hazard than hard coatings since they tend to off-gas more easily and ignite more quickly. They also tend to be slippery, making it more difficult to walk, stand and maneuver. This increases the chances of slipping, tripping or falling, perhaps resulting in someone dropping a lighted torch.

Work performed on preservative coatings may also produce toxic fumes, gases and vapors. For example, welding or cutting on surfaces that are galvanized, electroplated or protected with lead-based paints can produce fumes containing heavy metals such as zinc, cadmium, chromium, nickel and lead. Welding on phosphate coated steel can produce phosphine gas, which is irritating to the lungs.

Coating Removal

Coatings that may present a hazard upon heating should be stripped for a distance sufficient to prevent heat transferred by conduction from igniting or decomposing unstripped material. The distance will vary depending on the characteristics of the material and the nature of the hot work. As a general rule, surfaces should be stripped back at least six inches from the furthest edge of the hot work. However, if off-gassing, charring or blistering is observed in the unstripped area, hot work should be suspended and the area should be stripped back farther.

Preservative coatings may be removed by grinding, abrasive blasting and chemical stripping. If grinding or abrasive blasting are employed, the equipment should be provided with local exhaust systems that remove contaminants at their source. Chemical strippers are effective for cleaning painted surfaces and are available in solvent-based and caustic-based formulations. Solvent-based formulations are normally a gel-like material that is applied by brush. As the paint layers dissolve in the gel, it is scrapped from the substrate with a sharp knife.

Since the solvent's vapors are frequently flammable and toxic, additional ventilation or respiratory protection may be required.

Caustic strippers present a more manageable hazard in confined spaces because they are nonflammable and nonvolatile. However, their corrosive nature makes them skin and eye irritants.

The area to be stripped is first coated with a proprietary chemical paste which is then covered with a piece of polyethylene sheeting and allowed to stand for a prescribed time. When the polyethylene is removed, it lifts the underlying layers of paint with it. Any remaining paint traces are removed by wiping the surface with a dilute acid which also neutralizes the caustic residue. Finally, the surface is wiped down with a damp rag. In many situations it is possible to reach base metal with one application, but heavily painted surfaces may require multiple treatments.

When Removal is not Possible

It may not always be possible or practical to remove protective coatings before doing hot work. In those situations a fire watcher equipped with portable extinguishers should be provided to assure immediate suppression of incipient-stage fires. Hazardous air contaminants created by hot work should be controlled with local exhaust ventilation. If ventilation is not possible, employees should be provided with pressure-demand air-supplied respirators with escape bottles unless air-monitoring results from similar operations suggest that a lower lever of protection will be adequate.

Scale and Blisters

Flammable residue may be trapped in blisters, scale and rust deposits that form on the walls of some spaces. Scale in some crude oil tanks may also contain deposits of iron polysulfide, a pyrophoric material which ignites spontaneously when exposed to air. These conditions present an unusual hazard because flammable residues may persist even after the space has been steamed, rinsed and ventilated. In evaluating the potential fire hazard posed by these conditions, consider the scale's depth and porosity, the amount of surface area it covers and the flammable material's flashpoint and autoignition temperature. Deposits deemed to be hazardous should be removed before performing any hot work.

Pipes, Tubes and Coils

Some spaces contain pipes, tubes, coils and lines that may be part of hydraulic, pneumatic, heating or cooling systems. These systems should be depressurized, and if they contain materials which could pose a hazard when heated, they should be drained and flushed. All loose combustible materials should be removed, and any reside such as spilled hydraulic fluid should be cleaned up before hot work begins. Pipes, fittings valves and other system components that could be adversely affected by flames, hot slag, or sparks should also be protected with heat resistant coverings.

SUMMARY

The process of completing an entry permit provides a systematic method for verifying that all pre-entry precautions have been taken. The completed permit can then be used to inform entrants that potential confined space hazards have been controlled. Although all entry permits serve the same purpose, they may differ in appearance and format due to differences in organizational culture, the nature of the entries, and the experience level of permit users.

Specific core elements that should be considered for all permits include a general description of the space, an explanation of the reason for entry, identification of potential hazards, identification of the measures taken to control those hazards such as isolation, lockout/tagout and ventilation, provision for atmospheric testing, identification of special protective, communications and rescue equipment and a brief description of the emergency plan.

Permits should be clear, logical and easy to use. Data entry fields should be large enough to record the desired information and the lettering should be easy to read. If hot work is to be performed, that fact must be noted either on the entry permit or on a special hot work permit that prescribes specific fire safety precautions.

REFERENCES

Control of Hazardous Energy, Code of Federal Regulations, Vol. 29, Part 1910.147.

Oxy-Fuel Gas Welding and Cutting, Code of Federal Regulations, Vol. 29, Part 1910.253.

Permit-Required Confined Spaces for General Industry, Code of Federal Regulations, Vol. 29, Part 1910.146.

Personal Protective Equipment: General Requirements, Code of Federal Regulations, Vol. 29, Part 1910.133.

Pettit, T.A., P.M. Gussey and R.S. Simmons, 1980, *Criteria for a Recommended Standard: Working in Confined Spaces,* National Institute for Occupational Safety and Health, (DHEW/NIOSH Pub. 80-106) Government Printing Office, Washington, DC.

Portable Fire Extinguishers, Code of Federal Regulations, Vol. 29, Part 1910.157.

Welding, Cutting and Brazing: General Requirements, Code of Federal Regulations, Vol. 29, Part 1910.252.

CHAPTER 14

Employee Training

INTRODUCTION

Virtually every aspect of confined space entry requires employee training. Individuals who supervise entries and complete entry permits must be trained in pre-entry requirements such as isolation, ventilation and atmospheric testing. Entrants must be trained in hazard recognition, safe work practices, and protective equipment. Attendants must be trained in hazard surveillance methods and emergency procedures. Finally, rescuers must be trained in emergency response techniques.

Because of the wide variety of confined space hazards, there is no such thing as a one-size-fits-all confined space training program. A program designed to meet the needs of a municipal wastewater treatment plant, for example, would probably not be adequate for an itinerant contractor who cleans chemical storage tanks and process vessels. Similarly, a program designed to meet the complex needs of refinery workers would probably overwhelm public works employees who only enter storm drains. But regardless of the characteristics of the space, or the size and type of the facility, there are some universal educational principles that can be applied to all confined space training.

This chapter addresses those universal principles in three ways. First, it identifies the goals that confined space training must achieve. Second, it explains how you can reach those goals through the practical application of adult learning theory. Third, it outlines a step-by-step approach you can use to develop and tailor a confined space training program that fits your organization's needs.

CONFINED SPACE TRAINING NEEDS

Every confined space training program must address specific training requirements governing four categories of employees.

- Those who authorize the entry
- Those who enter the spaces
- Those who serve as attendants
- Those who provide emergency and/or rescue services

While the training requirements for each of these categories will be discussed separately, it should be noted that a single employee may perform more than one job. For example, the person who completes the entry permit may also serve as an entrant or attendant. Similarly, an attendant who is trained in emergency procedures could serve as a member of a rescue team.

It is important to understand that all four jobs are interwoven and that confined spaces can only be entered safely when everyone works together as a team. With this in mind, let's look at the specific training requirements for each team member.

TRAINING FOR THOSE AUTHORIZING ENTRY

The standard refers to the person who authorizes the entry as the "entry supervisor." This title does not imply that individuals in charge of the entry are necessarily management level employees, but rather that they have been vested by their employer with specific duties, responsibilities and authority with respect to verifying that the permit has been completed properly before authorizing the entry.

Entry Supervisors' Duties and Responsibilities

The entry supervisor has overall responsibility for checking the accuracy of permit and for evaluating conditions in and around the space to determine that all necessary precautions have been taken to protect the entrants. To accomplish this, entry supervisors must have intimate knowledge of confined space entry procedures, practices and methods. They must be aware of the hazards entrants may face during entry including signs and symptoms of exposure, as well as the possible consequences of those exposures.

Supervisors' Duties

Specific duties that the supervisor must perform before signing the permit and allowing the entry to proceed include:

- Verifying that all tests and measurements required by the permit have been made
- Verifying that all procedures and equipment listed on the permit are in place
- Verifying that rescue services are available and that the means for summoning them are operable
- Determining that acceptable entry conditions exist initially and that conditions remain acceptable throughout the duration of the entry

Supervisors' Responsibilities

Entry supervisors are responsible for enforcing work practices necessary to assure continued employee safety. To that end they must have the authority to cancel the permit and terminate the entry whenever they detect a condition not allowed by the permit. They must also be empowered to remove—*by force if necessary*—unauthorized individuals who enter, or attempt to enter, a permit spaces during entry operations. Finally, when responsibility for the entry is transferred from one supervisor to another, for example, during shift changes, the incoming entry supervisor must confirm that acceptable entry conditions continue to exist before allowing workers to reenter the space.

Training Strategy

Entry supervisors' training must be consistent with the duties and responsibilities with which they are charged. Consequently, the training program must develop skills which enable supervisors to verify that the permit has been completed properly, that all specified tests have been conducted and that all necessary procedures and equipment are in place.

The methods used to determine if the required precautions have been taken will vary depending on the facility's organizational structure. In many situations, entry supervisors will complete the permit themselves. But this is not always the case. In some facilities, responsibility for pre-entry hazard control—and consequently responsibility for completing the permit—may be divided among different crafts. Laborers, for example, may drain and flush the space while pipefitters isolate or disconnect lines and electricians disable motors. In other facilities, the entrants themselves may be responsible for performing all pre-entry tasks.

Clearly, these organizational differences will influence the nature and type of training required by the entry supervisor. In organizations where the entry supervisor is responsible for performing specific tasks, such as lockout/tagout or atmospheric testing, the training program must be designed to develop the skills necessary to perform those tasks. However, a different training approach is required in situations where responsibility for various pre-entry tasks is divided among different crafts. In these situations, the entry supervisor does not necessarily have to be able to *perform the tasks,* but instead must be capable of determining *if they have been done.*

While all entry supervisors must possess the skills needed to verify that the permit is properly completed, the exact nature and extent of their training depends on what they are expected to do. For example, an entry supervisor who does not personally lock or tag a piece of equipment must be able to determine if all designated energy isolation points are indeed locked or tagged.

He could do this in any number of ways. For instance, he could watch the "authorized employee" attach the locks or tags. Or, using a checklist that identified the isolation points, he could audit the "authorized employee's" work to verify that locks and tags are properly placed. Or he could simply consult with the "authorized employee" and confirm verbally that lockout/tagout has been completed. Each of these methods has certain advantages and disadvantages, and individual organizations must determine which methods and procedures best suit their needs.

Once an organization has identified the specific duties it expects entry supervisors to perform, it can begin to design a training curriculum that addresses those expectations. It is important to note, though, that methods, procedures and techniques that work well in one facility may not work in another that has different management philosophy, corporate culture or organizational structure. The performance-based nature of the standard accommodates these organizational differences by allowing flexible approaches to training. Virtually any approach can be used provided that entry supervisors can meet the four previously described performance goals.

ENTRANTS' TRAINING

Confined space entrants must be proficient in four areas:

1. Recognizing hazards in and around the space
2. Using any special equipment required for entry
3. Communicating with the attendant
4. Self-rescuing under emergency situations

Hazard Recognition

Hazard recognition is perhaps the most important aspect of entrants' training because entrants who understand the nature of workplace hazards are more likely to protect themselves than those who don't. For example, employees who recognize the hazards posed by flying chips or chemically contaminated surfaces are more likely to wear eye protection and gloves without being told to by their supervisor.

The nature of hazards to which entrants may be exposed depends on the tasks they perform and the characteristics of the space. For example, laborers who squeegee acid sludge out of an open-surface pickling tank could be exposed to air contaminants, acid splash and slipping hazards. Similarly, maintenance workers who apply protective coatings to the interior walls of storage tanks could be exposed to irritating, toxic or combustible atmospheres. Although the hazards in these two cases are different, the approach used to recognize them is the same.

The training method used to develop hazard recognition skills depends largely on how dynamic the work environment is. Municipal wastewater treatment plants, for instance, are good examples of relatively stable work environments. Admittedly, the constituents of influent can change from moment to moment, but, in general, the facilities, the employees, the processes and the characteristics of the spaces remain pretty much the same day after day. In this situation, it would be possible to survey each space and prepare an inventory of all the hazards to which employees might be exposed. After first determining that no changes to the space have been made, the entry supervisor could use the previous inventory as a pre-entry checklist to assure that all hazards are addressed.

But contrast the relatively stable environment of the wastewater plant with the ever-changing environment faced by employees of a contract tank cleaning service. In a given week, itinerant workers could be in a tank truck, a barge, a chemical reactor, a grain silo and an underground storage tank. Clearly, it would be impossible to anticipate, list and describe every conceivable hazard to which these employees might be exposed. Instead, it is much more practical to train workers in a systematic approach they can use to identify hazards on a case-by-case basis.

While it is not an exhaustive list, some of the most important aspects of entrants' training are discussed below.

Chemical Information

The OSHA Hazard Communication Standard requires that employees be informed of the nature of chemical hazards to which they may be exposed, and the precautions they should take to protect themselves from those hazards. Although this information is generally conveyed to employees through material safety data sheets, the quality of data sheets varies widely. Some are so technical that you have to be a Certified Industrial Hygienist to understand them, while others are so vague that they are virtually useless.

For example, some data sheets indicate that gloves should be worn when working with the product, but they don't specify what kind of gloves. Recall from Chapter 10 that different glove materials afford different levels of protection for different substances, and a glove that is suitable for one chemical may not be appropriate for another. Still other data sheets recommend wearing respirators, but as explained in Chapter 11, respirators are considered to be the last line of defense against atmospheric hazards and should only be used if engineering controls are inadequate or infeasible.

For these and other reasons, it might be more prudent to also train entrants in the use of specific reference texts, chemical hazard guide books or computer data bases which provide more meaningful information on chemical hazards.

Air Contaminants

Entrants who are likely to be exposed to air contaminants should be informed of the signs and symptoms of exposure. For example, employees who work in areas where carbon monoxide or hydrogen sulfide might be present should be informed that the respective early warning signs of exposure include headache and an odor reminiscent of rotten eggs. Likewise, entrants who work with solvents should know that warning signs and symptoms of exposure include dizziness, giddiness, light-headedness and loss of coordination.

Slips and Falls

Employees working in confined spaces may be faced with a variety of fall hazards. These include slips and trips when working on slick, uneven or unstable surfaces, or falls from an elevation when climbing up and down ladders, working on scaffolds, or entering through portals where there is a change in grade such as street manholes. If a mechanical winching device is used to lower an entrant into the space, he or she must be equipped with a full-body harness and an independent lifeline to arrest a fall if the line used for lowering fails.

Electrical Hazards

As a general rule, entrants should be trained to inspect electrical cords, plugs, receptacles and power tools for defects such as exposed conductors or worn and frayed insulation. Workers should also understand why equipment grounding or ground-fault circuit interrupters are required. Finally, they must know that only "approved" electrical equipment can be used in hazardous locations.

Specialized Equipment

Entrants must be trained in the proper use of any specialized equipment such as instruments, ventilators, radiation monitors, lockout locks and traffic control devices they may employ during entry.

Personal Protective Equipment

Entrants should be instructed on the care, use and limitations of any personal protective equipment that they are expected to wear. This includes eye and head protection, respirators, chemical protective clothing and fall protection devices. They should also be trained how to inspect their equipment for defects and instructed what to do if any defects are found.

Instrumentation

Entrants who are provided with gas monitors should generally understand the operating theory and limitations of their assigned instruments. They should understand the results presented by any meter displays and be able to interpret the meaning of alarm signals. Entrants should also know the importance of keeping an instrument's protective grill work free from obstructions that could block the flow of ambient air into the sensor elements.

Ventilation Equipment

Entrants should understand the purpose of any mechanical systems used to ventilate the space. Those who use local exhaust systems should understand that the hood should be as close as possible to the point where contaminants are being generated. Any entrants who are expected to set up ventilation equipment should be instructed on the methods and techniques necessary to properly configure the system.

Lockout/Tagout Devices

Training on lockout/tagout procedures is required for entrants who are "authorized employees" or "affected employees" as defined by 29 CFR 1910.147. Authorized employees must be equipped with the skills they need to recognize hazardous energy sources and must know means and methods by which hazardous energy will be isolated or controlled. Affected employees must generally be instructed on the purpose and use of the lockout/tagout procedures and be informed of the effects that those procedures have in relation to the work they do inside the space.

Barriers

Entrants should be familiar with any physical barriers or barricades used to provide protection from external hazards such as electrical or mechanical equipment, hot surfaces, highway traffic or pedestrians.

Communications Procedures

Entrants should be instructed on the methods that will be used to maintain contact with the attendant. No special training is required when ordinary voice communication is used. However, if more sophisticated communications methods, such as radios or sound-powered headsets are used, employees must be trained in the proper operation of the equipment.

Entrants should be instructed to alert the attendant whenever they notice a prohibited activity, or whenever they detect warning signs or symptoms of exposure to dangerous conditions.

Entrants must also know to alert the attendant prior to evacuation so that the attendant can be prepared to render assistance and, if necessary, notify emergency services.

Self-Rescue

Entrants need to know that they should evacuate a space under any of four conditions:

1. When they are ordered out by an attendant or entry supervisor
2. If they detect any signs or symptoms that suggest the existence of a hazardous situation
3. If they detect any prohibited activities
4. If an evacuation alarm is activated

Evacuation training should include information on any special evacuation methods, such as the use of retrieval lines and winching systems, and a discussion on the need for a rallying point when multiple entrants are involved.

ATTENDANTS' TRAINING

Attendants' training should reflect the responsibilities with which they are charged. These responsibilities include:

- Monitoring activities both inside and outside the space, being alert to conditions that might pose a potential hazard to those in the space
- Keeping track of the workers authorized to be in the space
- Control of unauthorized access by directing unauthorized intruders or bystanders away from the space
- Assuring continuous and effective contact with those in the space
- Properly using emergency equipment and practices to effect rescue without entering the space.

Hazard Recognition

Attendants must be trained to monitor activities inside and outside the space and be able to recognize conditions that might pose a hazard to the entrants. They must also know the mode, signs, symptoms and consequences of exposure to these hazards. Typical hazards attendants might be alert to include:

- Electric tools used without ground-fault circuit interrupters
- Nonexplosion-proof electrical equipment used in hazardous locations
- Unapproved materials or processes used in the space
- Tangling of hoses, cords, lifelines
- Hazardous liquids which leak or seep into the space
- Exposed energized conductors

Tracking Authorized Entrants

Attendants must maintain a *continuous* accounting of all authorized entrants that in an emergency they can quickly and accurately determine who is in the space. It is essential that they be trained the *specific methods* used for tracking the entrants. If a small number of workers is involved—perhaps five or less—and, if the entrants come and go infrequently, the attendant may be able to keep track them right on the permit. However, if a large number of entrants is involved, or if there is a small number who enter and exit frequently, some type of formal check-in/check-out system will be required. Systems such as tag boards, entry badges, sign-in sheets and electronic tracking devices are all acceptable, but attendants must know how to use them.

Controlling Access

Attendants must be ever-vigilant for the presence of unauthorized people in the work area. They must also control access to the space and allow only authorized employees to enter. They should be instructed to warn unauthorized people away from the space. If an unauthorized person does enter, the attendant should order the intruder out and inform both the authorized entrants and the entry supervisor.

Communications Procedures

Communications training for attendants should address two concerns. First, attendants should be trained in the use of any special communications equipment such as radios, field telephones or sound-powered headsets used to communicate with the entrants. Second, they should be trained in the use of any communications equipment such as in-plant intercoms or cellular telephones that could be used to summon emergency services.

Attendants must also be trained to order an evacuation under any of the following situations:

1. If they observe any conditions not allowed by the permit, for example, unauthorized hot work, abrasive blasting or solvent usage

2. If they detect behavioral changes like euphoria or giddiness which might result from oxygen deficiency or excessive exposure to some gases and vapors
3. If they detect situations outside the space which could endanger those inside; examples include a pool of spilled solvent or a vehicle idling near the ventilation system intake
4. If they notice an uncontrolled hazard like an exposed energized circuit or a leaking fluid line in the space
5. If they cannot effectively and safely perform their duties as an attendant

Emergency Response

Attendants must understand that their emergency response actions are generally limited to two duties:

1. To summon emergency assistance such as an ambulance or a rescue team
2. To operate any retrieval equipment that may be provided

As discussed in Chapter 12, it is essential that attendants develop hands-on skill with any rescue equipment they are expected to use. To accomplish this goal, they must have an opportunity to practice using the equipment to retrieve simulated victims.

As a general rule, attendants should be trained in first aid and cardiopulmonary resuscitation so that they will be able to render assistance until emergency services arrive. Remember, though, that employees who are expected to administer first aid as part of their normal job duties are covered by OSHA's bloodborne pathogen standard. The training requirements associated with that standard are summarized in Table 14-1.

It is critical that attendants understand that they are not to enter a space to provide assistance until after additional help arrives, and then only if they are "qualified" emergency responders.

RESCUERS AND EMERGENCY PERSONNEL

Recall from Chapter 6 that a permit-required entry program must include provisions for rescue and emergency response. In fact, that standard requires employers to develop and implement procedures for:

- Summoning rescue and emergency services
- Rescuing entrants from permit spaces
- Providing necessary emergency services to rescued employee

Note carefully that the standard stipulates that *procedures* be established. It does not require that emergency responders necessarily be on-site. This does not relieve employers of the responsibility of evaluating emergency response times, and under some conditions it may be necessary to have on-scene responders equipped to immediately enter the space, for example, during entries where workers might get trapped in atmospheres that are immediately dangerous to life or health.

Obviously, response time in an emergency can be reduced to an absolute minimum by having on-scene rescuers who are immediately available and intimately familiar with the facility. On-site emergency responders may also have existing working relationships with craftsmen such as pipefitters, electricians, crane operators, etc. who may assist in an emergency response. The principal disadvantages of electing an in-house response are the time and expense required for initial and annual training of emergency responders.

The expense for municipal services, on the other hand, is borne by the community, but not every community has the financial resources to equip and train a confined space rescue team. Assuming that a properly equipped municipal rescue squad does exist, its arrival will be delayed by dispatching and travel to the site. Even with the best pre-emergency site familiarization and planning, municipal services will not be able to respond as fast as an on-scene personnel.

If an employer decides that his or her employees will enter permit spaces to perform rescue, then those employees must be properly trained. The training requirements of the confined space standard also apply to municipal responders in states that operate their own OSHA approved state-plan program (Table 14-2).

On-Site Rescue Services

If an employer chooses to use the on-site approach, then responders must be properly trained. At a minimum, they must receive basic entrants' training, including techniques for recognizing hazards in the space, communications methods and self-rescue procedures. They must also be trained in the proper use of protective equipment, including rescue devices and respirators. Training on self-contained breathing apparatus is dictated by the potential for emergency

Table 14-1. Training requirements for exposure to bloodborne pathogens

- A general explanation of the content of 29 CFR 1910.1030
- A general explanation of the epidemiology, modes of transmission and symptoms of bloodborne pathogens
- An explanation of the exposure control plan and means by which it may be obtained
- An explanation of the methods of recognizing tasks and activities that may involve exposure to potentially infectious material
- An explanation of the use and limitations of the methods used to prevent or reduce exposure such as engineering controls, protective equipment and work practices
- Information on the types, selection criteria, use, location, removal, handling, decontamination and disposal of personal protective equipment
- Information on the hepatitis B vaccination including information on its efficacy, safety, method of administration, benefits of being vaccinated and that the vaccination is offered free of charge
- Information on appropriate action to take and persons to contact in the event of an emergency involving blood or other potentially infectious materials
- An explanation of the procedures to follow if an exposure incident occurs, including the method of reporting the incident
- Information on the post-exposure follow-up evaluation following an exposure incident

entry into atmospheres immediately dangerous to life and health.

Other training elements include first aid, cardiopulmonary resuscitation and the hazards posed by bloodborne pathogens. Each member of the team must be trained in basic first-aid and in cardiopulmonary resuscitation (CPR), and at least one member must be *currently certified* in both. As noted in Chapter 12, the physical characteristics of some spaces will require that responders be provided with highly specialized training on topics such as emergency rigging and high-angle rope rescue.

Rescue team members must also participate in an annual hands-on drill. The drill should simulate operations in which dummies, mannequins or actual persons are rescued. Either actual or simulated spaces may be used, provided the opening size, configuration, and degree of accessibility approximate the spaces from which rescue may be performed.

Table 14-2. Coverage of emergency services in state plan states

	Emergency Services	
State	**Paid**	**Volunteer**
Alaska	Yes	Yes
Arizona	Yes	Yes
California	Yes	No
Connecticut*	Yes	Yes
Hawaii	Yes	Yes
Indiana	Yes	Yes
Iowa	Yes	Yes
Kentucky	Yes	No
Maryland	Yes	No
Michigan	Yes	Yes
Minnesota	Yes	Yes
Nevada	Yes	No
New Mexico	Yes	Yes
New York*	Yes	Yes
North Carolina	Yes	No
Oregon	Yes	No
Puerto Rico	Yes	No
South Carolina	Yes	Yes
Tennessee	Yes	Yes
Utah	Yes	No
Vermont	Yes	No
Virginia	Yes	No
Virgin Islands	Yes	N/A
Washington	Yes	Yes
Wyoming	Yes	No

* Plan covers only state and local government.

Outside Rescue Service

Employers who arrange for another employer's employees to perform rescue operations still have certain training responsibilities. Although they are not required to be involved in the mechanics of training, they have an obligation to provide potential responders with access to all permit spaces from which rescue may be necessary. Without access to these spaces, responders would not be able to pre-plan their response, develop emergency action plans or conduct practice sessions.

ADULT LEARNING THEORY

Why, you might ask, should you bother reading anything about learning theory? You went to school, you know how to stand in front of people and talk, and if you didn't know anything about confined spaces before reading this book—you sure as heck should know something about them by now! So what's the big deal about learning theory? All you have to do is stand in front of a bunch of people and talk. Right? Wrong! All good instructors know that technical knowledge and good expressive skills are not enough. For training to be effective, instructional materials must be logically organized and clearly presented. Success can be further enhanced by designing a program that incorporates adult learning principles.

Training vs. Education

Many industrial training programs fail because instructors don't understand the difference between training and education.

Education is a process through which learners gain new understanding, acquire new skills or change their attitudes or behaviors. The educational process is very complex, and learning usually occurs on many levels. An educational program can be successful even if learners can't do anything new or different at the end of the program. For example, if the program's goal was simply to inform the participants of certain facts, it would be considered a success if the participants could demonstrate that they knew those facts at the end of the session.

Training, on the other hand, is a specialized form of education that focuses on developing skills. While training incorporates educational theories, principles and practices, it is performance-oriented. The goal of training is for learners to be able to do *something* new or better than before.

Anyone developing a confined space training program must understand the difference between training and education because it influences the teaching strategies and methods that are used to communicate with the learners. Since the goal of training is not merely for learners to know things, but to be able to do things, the program's effectiveness is measured by determining whether or not the learners can demonstrate the desired skills.

For example, at the end of a session on detector tubes, an employee may *know* that the instruction sheet contains correction factors that must be used to compensate for differences in temperature and humidity, but this doesn't mean that he has been trained! For *training* to have occurred, the employee must not only *know* that the correction factors must be applied, he must be able to *use them* to determine the actual concentration of the contaminant.

How Adults Learn

Even though adults do not learn the same way as children, many instructors continue to teach adults as if they were children. Since adults carry with them a lifetime of relevant personal and work experiences, they learn best in an atmosphere that permits and encourages them to use their existing knowledge to question, debate and discuss the relevance of any material presented in class.

As a result, instructors of adult learners need to move away from the traditional role of the teacher as the "giver of all knowledge" and adopt the role of a facilitator who guides participants, coordinates their activities and encourages things to happen. Unlike traditional teaching, facilitating involves much more than just transmitting information. The facilitator must help the participants to become self-directed. This can be accomplished in part by fostering interactions that encourage learners to make their own discoveries. The facilitative process must also provide learners with an opportunity to share their knowledge and learn from each other by presenting, examining and discussing their personal ideas, beliefs and experiences.

Teaching in the traditional sense tends to be content-oriented, and teachers often feel compelled to present a certain amount of information in the allotted time. Facilitating, on the other hand, is more participant-oriented, with learners playing an active part in the learning process. This is not to say that content is not important and should be ignored. Of course, content is important, but unlike traditional teaching, facilitating does not view content as an end in itself. Instead, it looks at content as the pathway by which participants develop specifically desired skills.

Some of the important differences between traditional and facilitative teaching are summarized in Table 14-3.

Table 14-3. Comparison between facilitative and traditional learning

Parameter	Traditional	Facilitative
Philosophy	Teacher is curator of knowledge or expertise that will be transferred to the students	Students need to develop the capacity to evaluate information and make their own value judgements
Objectives	Stated as the teacher's expectations, usually only cognitive results	Stated in cognitive, affective and psychomotor behavior that students will be able to demonstrate
Focal Point	Focus is on the teacher	Focus is on the students
Setting	Formal, frequently in organized rows	Informal, or more relaxed horseshoe seating pattern
Interaction	Teacher-dominated and teacher-directed	Students contribute through active participation
Method	Usually consists exclusively of lecture	Problem solving with a variety of teaching methods that encourage student participation
Evaluation	Factual, objective, knowledge recall	Application of information with an explanation

Adult Learning Atmosphere

Based on the discussion above, it stands to reason that program participants will learn best in an atmosphere which allows them to act as adults. The facilitative approach not only allows this kind of behavior, but, more importantly, it creates an atmosphere which encourages it.

Remember, the goal of training is not merely to *know* something new, but to be able to do something better or differently than before. Since "knowing" and "doing" are not the same, teaching in the traditional sense will have little effect in developing or improving skills. For learners to develop or improve their skills, they must be *actively involved* in the learning process.

Adult learning principles play a critical role in technical training because the goal of training is for participants to perform at a higher level of proficiency. To reach this goal, the learning event must be planned and implemented in such a way that the learners engage in some meaningful activity beyond lecture, reading and test taking. Classroom activities must also provide learners with an opportunity to abstract something useful that they can use in their jobs. This objective can be achieved by recognizing that three elements of learning must be incorporated into each training event. These elements are transmitting information, developing behavioral skills and changing attitudes.

Transmitting Information

Of the three aspects of learning, transmitting information is perhaps the easiest to accomplish. It involves presenting ideas, concepts and facts that provide a common foundation for the activities that will follow. Participants must be able to comprehend what is being transmitted orally or in writing, and they must have an opportunity to ask questions and to obtain clarification on points they find confusing. They must also be able to apply their knowledge by recalling appropriate ideas, theories or principles, and use these abstractions to respond to particular problems or situations.

Behavioral Skills

As noted previously, being aware of something is not the same as being able to do it. A worker may "know" that he should switch to his emergency escape bottle if the air supply feeding his air-line respirator fails. But if he has *never practiced this behavior,* he might panic in an emergency and pull his or her facepiece off instead.

Consequently, learning events must be designed to give participants an opportunity to put their knowledge into action by practicing what they have learned. Practice allows participants to "learn by doing," and repetition helps sharpen their skills. It is also important that participants receive feedback on whether or not they are succeeding at the tasks they are attempting, and a correct demonstration of the desired behavior is the best indication that learning has occurred.

Attitudinal Change

An employee may know that he is supposed to follow certain procedures. He may even be able to perform those procedures with a high degree of proficiency, but if he *believes* the procedures are unnecessary or inconvenient, he is not likely to follow them when no one's looking. Which leads us to the third, and perhaps most difficult level of learning, effecting changes in attitude.

An attitudinal change occurs when people alter their previously held beliefs, values or thoughts. The reasons why these changes occur are difficult to explain because they take place on an internal and deeply personal level. Attitudes and behaviors are closely linked and people generally resist doing things if they don't see how it benefits them. For this reason facilitators should focus their efforts on identifying benefits rather than features.

For example, many hand-held gas monitors incorporate features such as alarms, field replaceable sensors, multi-gas visual displays, etc. If the participants don't see how these features benefit them, their lack of interest may turn to apathy which, in turn, could lead to improper use of the instruments in the field. The facilitator's job is to skillfully guide the participants in identifying for themselves how they benefit from these features. When participants acknowledge the benefit of doing something a specific way and change their behavior as a result, an attitudinal change has occurred.

However, if the desired attitudes and behaviors are to persist, participants must continue to receive peer support even after the training event has ended. Without peer support, a crew that gets to the job site only to discover that its gas detector is back at the shop might decide to take a chance "just this one time," rather than go back and get the instrument. That one time could be the one that results in a fatality—a fatality that could have been prevented by a different attitude.

FIVE STEPS OF PROGRAM PLANNING

Developing a training program is a lot like going on a journey. First, you have to know where you are; next you have to know where you're going; then you have to decide how you're going to get there; you go; and finally, you need to know when you've arrived. In educational parlance these steps could be described as conducting a needs assessment, setting goals and objectives, selecting a training method, doing the training, and performing an evaluation.

STEP 1: KNOWING WHERE YOU ARE

Conducting a Needs Assessment

The first step in planning any program is to determine what participants need to know and what they already know. Educators call this a "needs assessment." While the reason for performing a needs assessment may seem fairly obvious, many inexperienced trainers ignore it and instead present what they want to talk about rather than what participants really need.

If the employees slated for confined space training are already proficient on topics such as protective equipment and lockout/tagout, then there is no need to focus attention on those topics. Instead, training efforts should be directed toward new or less understood topics such as atmospheric testing and ventilation. Selected areas of training that should be considered during the needs assessment phase are listed in Table 14-4.

STEP 2: DETERMINING WHERE YOU WANT TO GO

Setting Educational Goals and Objectives

Once the areas of training are identified, the next step is to determine what participants will be expected to do at the end of the program. Educators call this establishing goals and objectives.

Goals are broad statements that describe the desired end result of the learning experience. Typical goals of a confined space program might include being able to:

- Conduct a hazard assessment
- Properly ventilate a space
- Evaluate a space for atmospheric hazards
- Rescue a simulated unconscious entrant
- Select appropriate protective clothing and equipment

Objectives are specific elements that must be accomplished to reach the goal. Learning objectives

Table 14-4. Needs assessment considerations

- Recognition of confined space hazards
- Control of external hazards
- Traffic control methods
- Methods of draining and flushing the space
- Isolation methods including lockout/tagout
- Atmospheric testing
- Ventilation
- Communications methods and equipment
- Protective clothing and equipment
- Respiratory protection
- Special work practices
- Use of fire extinguishers
- Rescue methods and techniques
- First aid and CPR

Table 14-5. Behavioral objectives

	Words to Avoid	
Understand	Appreciate	Grasp the idea of
Know	Become familiar with	Recognize
	Words to Use	
Demonstrate	Evaluate	Apply
List	Analyze	Perform
Identify	Plot	Design
Indicate	Plan	Create
Select	Examine	Explain
Summarize	Describe	Compare
Organize	Contract	Express

may be based on knowledge or behavior, but they must be logical, relevant, feasible, observable and measurable. Since training is skills-oriented, learning objectives should be framed in a way that allows the instructor to determine if participants possess the desired skills. In other words, they should be based on behavior.

Formulating Behavioral Objectives

Behavioral objectives are statements that describe what participants should be able to do at the conclusion of the training program. The process of preparing learning objectives helps you to clearly identify your expectations. This is very important because if you don't know what you expect participants to be able to do, how will you be able to determine whether or not they can do it? The more focused the objectives are, the easier it is to determine if participants have met them.

A difficulty that many people have in writing behavioral objectives is that they chose verbs like "to know," "to appreciate," or "to understand." These are examples of cognitive processes, not behaviors, and there is no way of observing or measuring them without asking the question, "What would participants have to do to show that they "know," "appreciate" or "understand" something?" When you answer this question, you have identified a behavioral objective. Table 14-5 lists a variety of verbs that should be used and avoided in developing performance-based learning objectives.

STEP 3: GETTING THERE

Selecting the Method

A training program can incorporate a variety of teaching methods. Each method has advantages and disadvantages, and the method of choice depends on the lesson's stated goals and objectives.

Lecture Method

The lecture method is perhaps the most frequently used and least effective method of training. Communication is largely one way, from the instructor to the learners, and the only indication the instructor has that students understand is non-verbal.

In spite of these drawbacks, the lecture method is useful for introducing concepts, presenting factual information, summarizing important points and reviewing material presented previously. Some good applications of the lecture method would be

- Describing the use and theory of operation of air sampling instruments
- Introducing different types of respiratory protection
- Identifying reference resources such as the OSHA permissible exposure limits or chemical protective clothing selection guides
- Explaining the legal requirements of the OSHA lockout/tagout standard
- Describing the format of an entry permit

Although the lecture method can be used to transmit knowledge, it is ineffective for developing behavioral skills or altering attitudes.

Discussion Method

The discussion method allows a two way exchange of thoughts, information and ideas between the instructor and participants. Discussions are dynamic learning events that stimulate thinking by actively involving individual participants. The instructor, however, must avoid becoming the focal point of the discussion and instead encourage participants to interact with each other.

Participants must have some knowledge of the topic since they cannot discuss things they know nothing about. This knowledge can come from many sources including personal experiences, readings or information presented as part of classroom learning activity. Through group interaction, participants gain new insights and perspectives by processing ideas, concepts and approaches to problems that they may not have thought of before.

Some typical applications of the discussion method include:

- Debating the relative merits of various approaches to ventilating a space
- Identifying what type of personal protective equipment would be necessary in different situations
- Comparing the advantages and disadvantages of specific air monitoring instruments

To be effective, discussions must be informal and done in small groups. Large classes should be divided into smaller break-out groups to create a less threatening atmosphere which will promote the free and open exchange of ideas between group members. Working in small groups, participants can discuss a topic among themselves. Then, after allowing ample time for discussion, the facilitator can have a single representative present the group's findings. The finding can then be critiqued, challenged or discussed by the entire class.

Ventilation provides an example of a topic that lends itself nicely to the discussion method. As explained in Chapter 8, there are many factors which influence how a particular space should be ventilated. An instructor can provide participants with a dimensional sketch of a hypothetical space that shows the size and position of openings. Armed with this data and information about the space's physical characteristics and the reason for entry, participants could work together to identify the best approach to ventilation.

The discussion method can also be used very effectively as a means of learning from others' mistakes. Groups can be provided with a description of a confined space accident (this book is full of them) and asked to analyze the incident to determine what went wrong. They can then identify what steps could have been taken to have prevented the incident from happening.

A major advantage of discussions is that everyone is involved in the learning process. Group problem-solving sessions also provide participants with an opportunity to learn from each other so that everyone benefits from everyone else's experience. Best of all, the problem-solving skills developed through classroom exercises can be readily translated into "real world" applications.

Demonstration Method

The demonstration method works well whenever equipment is involved. But simply demonstrating equipment will not assure that participants will be able to use it. The second component of any demonstration is a successful redemonstration by the participants.

Inspection and cleaning of self-contained breathing apparatus provides a good example of effective use of the demonstration method. Using a SCBA, the instructor can point out the items to inspect. He can then describe the inspection criteria while showing participants what to look for. For example, he might say something like "Inspect the pressure gauge on the air cylinder to assure that it's tight, that the lens is not cracked, and that the indicator needle is attached and reading 2,216 psig." While handling the SCBA facepiece, he would indicate that the straps should be in place, that the serrations are not worn and that the lens is not cracked, scratched or broken.

Participants can then be provided with an opportunity to practice the demonstrated techniques either individually or in small groups. When they are confident that they can perform the tasks, they redemonstrate them to the instructor who verifies that they are being performed correctly.

Because of the show-and-tell aspect of this method, all equipment should be checked out well ahead of time to make sure it works. Nothing will undermine a demonstration faster than a broken piece of equipment or one that doesn't work the way it's supposed to, and the effectiveness of any demonstration is quickly lost if you have to describe how something is *supposed* to work rather than actually showing it.

Laboratory Method

The laboratory method can be used effectively to help participants develop skills in such areas as instrument calibration, atmospheric testing, ventilation, lockout/tagout and rescue. The laboratory approach is also very useful in situations where resources are limited and it is not possible to provide every participant with an individual piece of equipment.

Laboratory "learning stations"—each with a different piece of equipment—can be set up and participants can move from station to station on a prearranged schedule. One station may have testing instruments, another a SCBA, another retrieval gear and so forth.

It is often helpful to provide an activity worksheet that describes the nature of the laboratory exercises with space for the participants to record their observations and findings. Sufficient time should be allowed for participants to complete each exercise, after which they rotate in sequence to the next station. At the conclusion of the session, participants should be provided with an opportunity to discuss their experiences.

STEP 4: GETTING THERE

Delivering the Program: Lesson Planning

A lesson plan is a document that describes the steps you will follow to reach your goals and objectives. As a practical matter, the results indicated by the needs assessment must be weighed against organizational priorities to determine whether the training program should be presented all at once, or spread over a few days, weeks or even months.

Continuity on long-term training programs can be maintained by using a modular approach that organizes the program into training elements with a scope narrow enough to fit the available time slot. For example, individual modules could be developed for hazard recognition, lockout/tagout, ventilation, etc. Some topics may require more than one instructional module. Instrumentation, for example, may incorporate separate modules for selection, theory of operation, calibration, use and pre-entry evaluation.

Your lesson plan should generally include the following four elements.

- *A scope statement* which establishes the focus of the lesson and defines its boundaries
- *The lessons goals* which describes what participants should expect to achieve through the lesson
- *Learning objectives* which identify what participants should be able to do at the conclusion of the lesson
- *An outline* which shows the organizational structure of the lesson and provides a guide for where participants are going, how they are going to get there and how much time it should take

Program Preparation

Preparation is an important component of any training activity. It includes everything that is done to assure that lessons are presented in a way that encourages learning. While there is no such thing as being too prepared, it is possible to be over-prepared. Instructors who are over-prepared are so obsessed with form and structure that they become inflexible and allow no diversions or digressions from their planned lesson. Unfortunately, this approach tends to suppress the spontaneity which so often adds spice, flavor and excitement to the learning experience. So while it's smart to be prepared, it's wise to be flexible.

You should have a general idea of the points you want to make and can mentally practice what you are going to say. While you don't have to have a script memorized, you should have a clear idea of how you are going to start and where you are going to go. You should also identify questions that are likely to be asked, and formulate answers ahead of time.

Be sure that you are familiar with any audio-visual equipment you intend to use. While you don't have to be an electronics technician, you ought to know enough about your equipment to make minor repairs such as changing burned out bulbs, freeing jammed slide carousels and reconnecting VCRs whose cables become detached from their monitors. And if you're using slides or overhead transparencies, check them out ahead of time. *There is no simply no excuse for projecting materials upside-down or backwards!*

Selecting the Facility

The facility should be quiet, well-lit and well-ventilated. It should provide a reasonably comfortable environment that is neither too hot nor too cold. Allow time to arrange the room the way you want it. Furniture should be adjusted so participants can see and interact with each other. Circular or U-shaped seating (Figure 14-1) is preferable to having participants sit in rows.

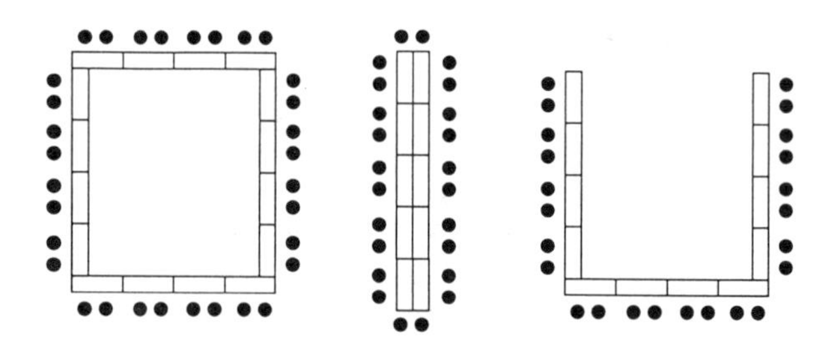

Figure 14-1. Circular or U-shaped seating arrangements allow participants to interact with each other. (Paul Trattner)

Printed Materials

Printed materials can be used to reinforce things that are presented or discussed in class, but any materials used must be relevant to the topic. An outline, or copies of slides and overheads if they are used, can also help to minimize note taking. If you use an outline, leave plenty of space for participants to make notes, and try to stick to it because digressions tend to confuse students who are trying to follow along.

Any handouts ought to be clean and readable, not copied so many times that they are faded, speckled and virtually impossible to decipher. Do not keep talking while distributing handouts. If you plan to refer to the handouts, pass them out before you begin. If you are using handouts simply to reinforce certain points, pass them out at the end so they won't divert participants' attention from the lesson materials. Be sure to have enough copies for everyone, and it's always smart to have a few extras.

Social Amenities

Little things like having refreshments available can make a big difference in establishing a mood that will help produce an effective training experience. Hot coffee, tea and doughnuts are always welcomed in the morning, and if a program is slated to last all day, cold drinks are nice to have for an afternoon break. While it might seem insignificant, these amenities influence participants' attitudes and can help maintain a positive atmosphere that is conducive to learning.

It's important to start on time and to keep on schedule. Be sure to allow adequate time for breaks, and no lecture or discussion should last for more than 45 minutes without a break. Also allow enough time for lunch so participants won't feel rushed. Or better yet, if your budget allows it, provide a catered lunch. It doesn't have to be anything fancy; cold cuts, potato salad and cole slaw are acceptable to almost everyone, and the goodwill generated from receiving a free lunch contributes immeasurably to the learning experience.

Delivery

Speak informally and face the audience whenever possible. Stand to the side when using flip charts, chalkboards or projection screens, and pause momentarily when you face a surface to write on it. Print large and clearly and use colored markers. They provide visual variety and also allow you to accent important points with contrasting colors.

Some participants may have difficulty in expressing themselves, others may feel awkward about asking questions in public. One of your jobs as a facilitator is to help these participants articulate their thoughts. In doing this, you might make a statement like "What I think you're trying to say is..." Alternatively, you could phrase your assistance in the form of a question like "Are you trying to say...?" With positive reinforcement and knowledge that you won't chastise or embarrass them, shy students will feel comfortable and be more likely to speak up.

Simple statements like "I'm really glad that you asked that question!" or "That's an excellent question!" also help to create a friendly, supportive atmosphere that encourages participants to ask questions.

Measuring Your Success

You will know you are succeeding if participants appear to be relaxed and involved in the program. Everyone seems to be awake and there is enthusiasm in the air. Tangible indicators of success also include good eye contact with the participants, no hesitation to ask questions and smiling faces with periodic nods of agreement.

On the other hand, nervous, stiff behavior, little social contact between you and the participants during breaks, a serious, somber atmosphere and the absence of any questions are signs that things are not going well. Novice instructors should take heart; everyone has to start somewhere and no one is perfect the first time. Keep trying and don't give up. You will become more successful with experience.

STEP 5: KNOWING WHEN YOU'VE ARRIVED

Evaluating Program Effectiveness

The primary purpose of an evaluation is to measure the program's effectiveness. Remember, the goal of training is for participants to be able to do something new or better than before. If they can't, then the training has not been effective. But evaluations also serve as a means of collecting information that can be used to improve the program content, format and delivery.

Process evaluations consider the participants' feelings, attitudes, reactions and opinions. As such, they typically include questions about what the participants thought of speakers, facilities or quality of the audiovisual materials. These evaluations are used primarily to obtain feedback that can be used for program improvement.

Product evaluations attempt to measure "how well" or "how much" people learned as the result of the training. In other words, they compare the outcome of a program against its stated goals and objectives.

Written Tests

Written tests are suitable for evaluating factual knowledge, but useless for measuring behavioral skills. They can be used to determine if participants can recall certain facts, such as the minimum level of oxygen that must be present for safe entry, where to obtain protective equipment, who to contact in an emergency, etc. It is important to review the results of written tests with participants so that they can learn from their mistakes.

Skills Demonstration

Performance-based evaluations provide the best way of determining if participants have developed the desired skills. They are relatively easy to conduct since the evaluator need only consult the learning objectives to determine the specific things participants should be able to do. The evaluation consists of the participant demonstrating the skills specified by the learning objectives while the evaluator observes their performance to determine if it meets specific criteria for proficiency.

One application of skills-based evaluation would be to determine if participants could perform a pre-entry evaluation. Typical criteria used to define acceptable performance might include a determination that the participant:

- Selects the proper instrumentation given information on the characteristics of the space, its previous contents and the operations to be performed
- Correctly performs specified functional checks such as instrument integrity, battery-check, calibration-check, etc.
- Takes an appropriate number of measurements that reflect the nature, size and configuration of the space
- Properly interprets measurement results and prescribes a course of action based on those results

TRAINING AIDS

Figures 14-2 and 14-3 suggest that much of what we learn, we learn though doing and seeing. Training aids take advantage of this learning principle by adding an exciting, visual dimension to any classroom activity. Blackboards, flip charts and demonstration equipment can be used to graphically illustrate abstract ideas and visually simplify complex concepts. Audio-visual materials such as films, videotapes and 35 mm color slides can be employed to show participants equipment, operations and processes that otherwise could not be duplicated in a classroom setting.

But training aids accomplish more than just making learning more fun and interesting. As shown in Figure 14-4, their use also helps participants to remember more of what they learn. Although there is a wide assortment of aids from which to choose, each has advantages and limitations. It is important for

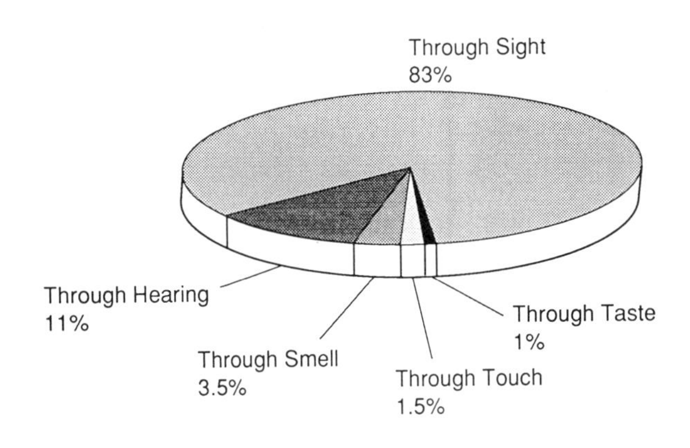

Figure 14-2. Modes of learning. (Paul Trattner)

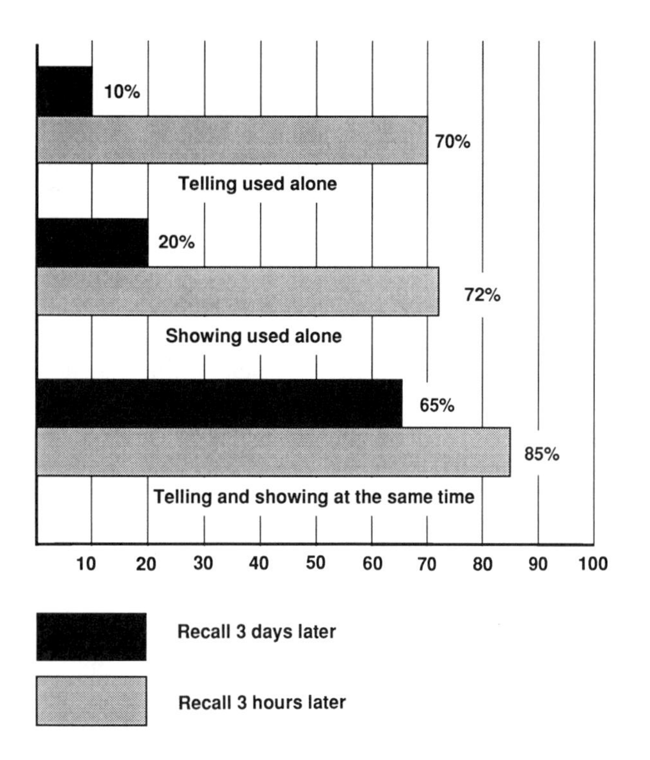

Figure 14-3. How we learn. (Paul Trattner)

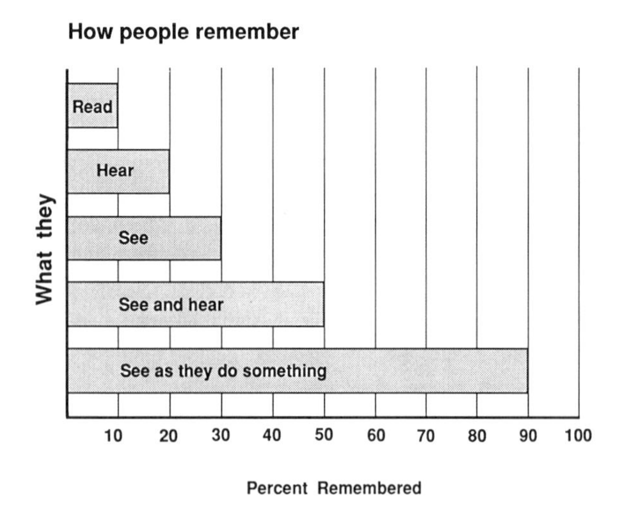

Figure 14-4. How we remember. (Paul Trattner)

instructors to carefully consider the learning objectives they want to achieve and to choose the training aids which are most appropriate for reaching those objectives. Table 14-6 offers some assistance in this area by providing time-proven guidelines for matching training aids and learning objectives.

Non-Projected Materials

Non-projected training aids include demonstration equipment, simulators, flip-charts and chalk boards.

Chalk Boards and White Boards

Chalk boards and, their modern counterpart, the hard plastic white board provide the simplest and least expensive way of graphically presenting complex verbal information. No other teaching tool is so readily available, easy to use and adaptable to so many learning situations.

While use of the board requires no special skill, talent or training, there are some techniques that can be employed to enhance its effectiveness. For example, an instructor can use the board to promote group interaction by asking a question and then writing participants' responses on the board. The list of answers can then be used as the foundation for a group discussion.

Different colored chalk or markers can be used for variety or emphasis, but some colors do not always show up clearly. For example, it is sometimes difficult to tell the difference between blue, black, purple and brown. So determine in advance which colors work and which don't.

Some boards can be adapted for use with magnetic materials. Pliable magnetic sheeting can be purchased from graphics supply houses and cut into shapes which can be stuck on the board. Color can be added by painting the cut-outs or covering them with colored construction paper. If text is required, professional looking lettering can be added by using self-adhesive vinyl die-cut letters which are available from most stationery stores.

In addition to being easy to use, boards can be erased and reused. They can also be used in conjunction with other training materials such as slides, overhead transparencies and demonstration equipment. The two principal limitations of boards are that material cannot be retrieved once it is erased and that the detail and complexity of graphics is limited by the user's artistic ability.

Flip Charts

Flip charts consist of a large pad of paper which is either attached to a wall or supported by an easel. The flip chart is used in much the same fashion as a board, and like the board, its use requires no special skill, talent or training. The principal disadvantage of

Table 14-6: Comparison of the effectiveness of various training aids

	Learning Objectives				
Instructional Material	**Learning Facts and Theories**	**Learning Visual Identifications**	**Comprehending and Applying Facts, Principles, Concepts**	**Performing Perceptual Motorskills**	**Influencing Attitudes, Opinions, Motivations**
Non-Projected					
Drawings and Illustrations	Medium	High	High	Medium	Medium
Photographs	Low	High	Medium	Low	Low
Blackboard	Medium	High	Medium	Low	Low
Models/mock-ups	Low	High	Medium	High	Low
Simulators	Medium	High	High	High	Low
Real objects	Low	High	High	High	Low
Printed material	High	Low	High	Low	Medium
Projected					
Color slides	Medium	High	Medium	Medium	Low
Overhead viewgraphs	Medium	High	Medium	Low	Low
Slide-tape	Medium	High	Medium	Medium	Low
Motion pictures	Medium	Medium	High	High	High

flip charts is that they are smaller than boards, making them a little harder to see. It is also more difficult to correct mistakes on flip charts since errors cannot be easily erased.

On the other hand, flip charts are more portable than boards and can be used to prepare material ahead of time. In addition, notes and drawings made on a flip chart can be saved for reuse at a later time.

Sheets of flip chart paper can also be used by break-out groups to summarize findings or conclusions arising from a group activity. Individual sheets from each group can then be taped to the wall to provide a focal point for a discussion by the entire class.

Demonstration Equipment

Demonstration equipment such as respirators, lockout/tagout devices and air monitoring instruments are indispensable for skills-based training. Since the goal of training is for participants to be able to do something new or better than before, it is essential that participants receive "hands-on" training on equipment they are expected to use.

Mock-ups of electrical disconnects, cut-off valves and flanged pipe sections can be assembled and used to demonstrate isolation methods and techniques. Following the demonstration, participants can practice on the mock-ups until they achieve the desired degree of proficiency.

A 30-gallon trash can or a clean, deheaded, 55-gallon drum can be converted into an environmental chamber for demonstrating direct reading instruments. A few drops of ammonia or bleach is first placed in the bottom of the trash container, which is then sealed by stretching plastic sheeting over the end. The sheeting is held in place by taping it to the side of the container. Participants can practice measuring toxic gases by inserting detector tubes through slits in the plastic and drawing samples of the test atmosphere. Training on combustible gas meters can be achieved in much the same way except that flammable liquids such as lacquer, thinner, gasoline or lighter fluid are substituted for ammonia or bleach. Similarly, compressed inert gases such as nitrogen, argon or carbon dioxide can be metered into the container to create oxygen deficient atmospheres for training participants in the use of oxygen meters.

Training Simulators

Short sections of corrugated steel tubing or concrete drain pipe can be used to simulate confined spaces. More elaborate simulators can be fabricated from lumber or surplus process vessels (Figure 14-5). Realism can be achieved by using spaces such as street manholes, rail cars, tank trucks and process vessels for practicing skills learned in the classroom. Remember, though, if actual spaces are used for training purposes, all of the pre-entry precautions must be taken before they are entered.

Simulators are particularly useful for teaching and practicing rescue techniques because they provide a realistic yet non-threatening environment for participants to gain experience in responding to emergency situations. Figure 14-6 is an example of a simulator fabricated from 4 × 8 foot sheets of plywood. One wall is fitted with a window which permits students to observe each other's performance so it can be critiqued at the end of the exercise.

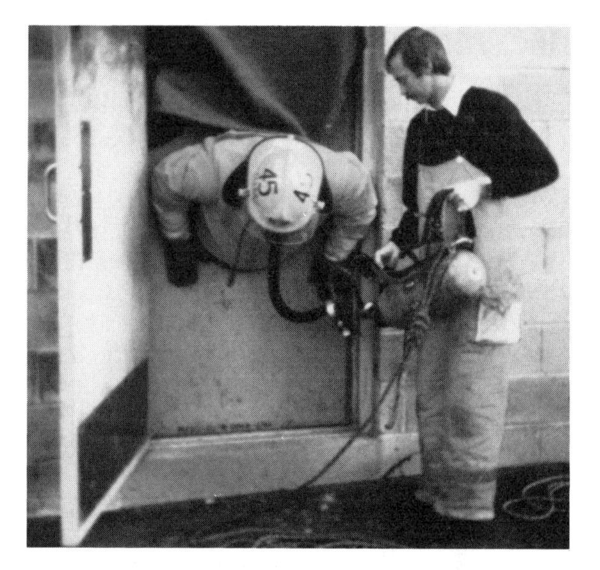

Figure 14-5. Portal simulated with plywood and doorway. (Montgomery County)

Figure 14-6. Confined space simulator. (General Motors)

Projected Materials

The marketplace is filled with projectable visuals that can be incorporated into confined space training programs, but be sure you preview any materials *before* you show them. If you don't, you might see your training efforts undermined by technical faux pas and unsafe work practices that are shown in some commercial products. A few of the errors that I have observed include users of negative-pressure respirators who are sporting full beards, air-monitoring instruments used incorrectly, and failure to perform lockout in accordance with OSHA requirements.

Make sure you know how to operate any equipment you plan to use. Set it all up ahead of time and check it out thoroughly. Bear in mind that you'll be jeopardizing your credibility as a "safety" instructor if your equipment poses a hazard to you or your students, so be sure everything is secure and in good repair. Electric cords should be arranged so that they don't create a tripping hazard, and if you use duct or masking tape to secure cords to floor, cover the entire length of the cord, not just a few widely spaced sections that can catch people's toes.

Position the equipment so that the projected image fills the entire screen, not just a little one foot by two foot rectangle. If you have slides in both vertical and horizontal formats, adjust the projector so that the image does not extend beyond the edge of the screen. And always have a spare projector lamp on hand in case the primary bulb burns out.

Overhead Transparencies

Overhead transparencies are inexpensive, reusable and relatively easy to prepare. Transparencies have historically been prepared by means of a heat transfer process that requires infrared-sensitive film and a special processor. A transparency sheet is placed over a photocopy of the desired image and fed into the processor which transfers the image to the film. While the process is not particularly complicated, the film is relatively expensive.

While the infrared process is still widely used, materials are now available which allow transparencies to be made on office photocopiers and laser printers. Special clear plastic sheets are loaded into the unit's paper carrier and the equipment is operated in the normal way, but images are transferred to the projectable plastic sheets rather than plain paper.

If you don't want to bother making overheads yourself, most commercial copy centers can make them for you. In fact, some copy centers are equipped with color copiers that can make color transparencies from color slides or prints.

Color 35 mm Slides

Color slides offer the same benefits as transparencies but are easier to handle and store, especially if a large number of visuals is employed. Slides also offer the advantage of being able to present large, life-like images which can add an element of realism to any presentation.

Computer-based software packages can be used to develop drawings and text which can be converted into a slide by commercial service bureaus. Since there is no much graphics software available, you'll have to first check with the service bureau you intend to use to determine what software it can process. You then use that software to prepare a graphics file in the normal fashion. The finished computer file is provided to the service bureau on disk, or transmitted via modem. The bureau then converts the computer file to a 35 mm color slide which it returns to you.

Preparation of slide presentations requires a 35 mm camera and some basic photographic skill. Substantial time must also be budgeted for editing and rehearsal since the instructor's remarks must coincide with the slides.

Films and Videotapes

The principal advantage of films and videotapes is that they depict live action. But no matter how exciting the action is, viewers will learn little from it unless they understand why they are seeing it. So don't—repeat don't—simply show a film. Instead, introduce it, explain why you are showing it and discuss how you think it relates to the lesson.

When using films or videotapes, you might want to consider providing participants with a worksheet containing questions they can answer as they watch. Alternatively, participants can use a worksheet at the conclusion of the film or video to summarize their observations, list key points or comment on what they saw. The worksheets can serve as a springboard for a group discussion. You could also initiate post-viewing discussion by asking questions like "What did you think about...?", "How do you feel this is relevant to...?" or "Did you notice that ...?", or in some cases, "What did you notice that was *wrong?*"

Don't be afraid to use old films! Some of the best demonstrations of static electricity and dust explosions I've ever seen are depicted in the U.S. Bureau of Mines films that were made in the 1950s. Admittedly, many vintage films might be considered corny and unsophisticated by today's standards, but that doesn't have to detract from their educational value.

If you use a "classic" film, tell participants right up front that in spite of film's age, it still contains some excellent information. Be as specific as possible by pointing out exactly why the film is still relevant by saying things like "Watch for the part where..." or "Notice how they do..."

If you follow this advice, you'll probably find that participants will not be distracted by the time warp produced by such incongruities as milk trucks making home deliveries and thigh-high skirts. On the other hand, if you fail to provide adequate warning of the films "quaint" character, you're likely to find that your audience tunes out as soon as they hear the goofy theme music or see cars with running boards or tail fins.

ANSWERING QUESTIONS

Questions from participants can be a double-edged sword. On the positive side, questions can spark interest that produces a dynamic group discussion which provides an excellent indication that learning is taking place. On the downside, questions not relevant to the topic tend to be distracting and divert time and attention away from more important items.

In general, you should attempt to answer relevant questions and then move on. If you plan to cover the topic later, say so. If the question is beyond the scope of the program, answer it briefly and offer to discuss it more thoroughly during a break.

If a questioner is confused about a particular point, back up and clarify it. If the issue is resolved, move on, but if it appears that more interaction is necessary, arrange to talk more at the break, unless the question seems to have sparked group interest. If the questions seems relevant to the group, or if many people seem to express an acute interest in the question, seize the opportunity for a group discussion.

Perhaps the most important thing to remember about answering questions is, *don't bluff!* If you don't know the answer, say so. You don't have to be the source from which all knowledge flows; besides, someone else in the class may know the answer, so why not direct the question to the group. Participants can usually sense when answers are not authentic, and phony answers can undermine your credibility as a facilitator. Another risk associated with providing phony answers is that they can often trigger a combative questioner who knows the "real" answer and has been waiting for the opportunity to pounce on you with both feet.

Some questions cannot be answered in class and may require that you consult other documents, procedures or policies. For instance, answers to questions like "What's the TLV for isopropyl alcohol?" and "Where is the disconnect switch for the A-32 reactor?" may not be on the tip of your tongue.

Answers to some questions are more critical than others. For example, questions like "When are we going to get new sensors for the toxic gas monitor?" or "When are the blowers going to be repaired?" are certainly more pressing than a question like "Why does the calibration gas come in 52 ppm rather than 50 ppm?" or "How come sampling lines come in only lengths of 15, 25 and 50 feet when our spaces are 20 feet deep?"

Your attitude in responding to questions is very important. Do not embarrass or belittle the questioner. You may have to bite your tongue because short-tempered or sarcastic answers may discourage someone else from asking a question whose answer might prevent a fatality.

Try to anticipate questions that might be asked and formulate answers ahead of time. Make it a point to avoid responses such as "That's what the regulation says," "That's just the way it is," and "We've always done it that way." Remember, this is an adult audience and adults want practical answers to their questions.

Loaded or "Curve Ball" Questions

These are the ones that are asked to embarrass you or put you on the spot. They do not usually have a simple answer and the people who ask them are usually trying to attract attention or recognition. In many cases, the questioner already has some idea of what your answer will be and is prepared to counter it with another question.

The best approach to these unanswerable questions is to try to disarm the questioner by reversing roles and saying something like, "That's an excellent question, what do you think?" or "That's a good question, let's ask the others what they think." Someone else may have an interesting insight which answers the question and diffuses a potentially explosive situation. If the questioner persists, indicate that in the interest of the group you would like to move on and suggest that you talk further during the break.

The Hypothetical Question

Hypothetical, or "what if," questions present facilitators with an interesting challenge because while the question might have learning value, it can also divert attention from more important matters. While many hypotheticals may be interesting, an instructor must try and determine if the question has some practical relevance or is being asked merely out of some obscure academic curiosity. While academic curiosity should not be discouraged, substantial time spent on answering an obscure question can divert time from other areas where it would be better spent.

Hypothetical questions can be generally divided into two categories: the relevant and the arcane.

Relevant hypotheticals are those that have some practical bearing on the matter at hand. They may be based on a participant's past experience or result from a synthesis of material that creates a question. Relevant hypothetical questions make excellent teaching tools because they allow facilitators to tap into adult learning processes.

Arcane hypotheticals, on the other hand, are of interest only to the person who asks them. They frequently involve some esoteric or highly unlikely situation that, while related to the topic in some way, are so obscure that they only succeed in being a distraction. While they may be mentally stimulating and provide grist for interesting philosophical discussions, they usually divert time away from more important matters and are better discussed during breaks or social hour after class.

The Combative Questioner

Combative questioners can be a real nuisance because they are often disruptive in a way that can destroy the group harmony that the facilitator has worked to build. The questioner is frequently motivated by unfocused anger and resentment and is often seeking attention or recognition. Combative questioners wait for you to make a minor mistake which they then delight in calling to your attention.

The best way to deal with the combative questioner is to disarm them. Don't hide, don't run away and don't fight with them. Instead, admit that you may have indeed made a mistake. You can say something like, "Oh I'm so sorry, I must have made a mistake." Or "Oh, you're right, I didn't think about that... consider that..." etc. Then ask the person if they would like to clarify things or elaborate on their point by providing the "correct" information.

The Long-Winded Question

This is the questioner who takes up a lot of time droning on and on about some irrelevant issue or concern. In many cases the questioner is not really asking a question, but instead is expressing a particularly strong opinion or viewpoint. By the time the questioner gets to the end, everyone already knows what the answer is. The best approach in handling long-winded questioners is to interrupt and tell them how interesting their question is, so interesting in fact, that you'd love to chat at length about it during the break or after class.

SUMMARY

Virtually every aspect of confined space entry requires employee training. Individuals who complete entry permits must be trained in pre-entry requirements such as isolation, ventilation and atmospheric testing. Entrants must be trained in hazard recognition, safe work practices and protective equipment. Attendants must be trained in hazard surveillance methods and emergency procedures. Rescuers must be trained in emergency response techniques.

Although the terms "training" and "education" are often used interchangeably, there is a significant difference between them. Education is a process through which learners gain new understanding, acquire new skills or change their attitudes or behaviors. Training, on the other hand, is a specialized form of education which focuses on developing skills. While training incorporates educational theories, principles and practices, its goal is for learners to be able do something new or better than before.

Adults carry with them a lifetime of relevant personal and work experiences, and they learn best in an atmosphere that permits and encourages them to use their existing knowledge to question, debate and discuss the relevance of any material presented in class. As a result, instructors of adult learners must become facilitators. Facilitating involves much more than transmitting information. Facilitators must also help participants to become self-directed. This can be accomplished in part by developing training programs that provide participants with an opportunity to learn by doing.

Effective training programs can be planned using a 5-step process. The first step is to conduct a needs assessment that determines what participants already know and what they need to know. The next step is to set goals and objectives. Goals are broad statements that describe the desired end result of the learning experience, and objectives are specific elements that must be accomplished to reach the goal. Learning objectives should be framed in a way that allows the instructor to determine if participants possess the desired skills.

The third step is to select a training method appropriate for the lesson's goals and objectives. A training program may incorporate a variety of methods such as lectures, discussions, demonstrations and laboratory sessions, but each method has certain advantages and limitations.

The fourth step is to develop and present the lesson. This task can be made easier by preparing a lesson plan that describes how the goal and objectives will be achieved. Since we learn and remember mostly through what we see, effective use of audiovisual materials can speed up learning and increase retention.

Finally, the training program's effectiveness is measured by an evaluation. Since the goal of training is for participants to be able to do something new or better than before, performance-based evaluations provide the best indication of whether or not learning has occurred.

REFERENCES

Alexander, L.T., R.H. Davis and S.L. Yelton, *Learning System Design: An Approach to the Improvement of Instruction*, New York, NY: McGraw Hill, (1974).

Axford, R., *Adult Education: The Open Door to Lifelong Learning*, Indiana, PA: A.G. Hallin, (1980).

Briggs, L.J., K.L. Gustafson and M.H. Tillman, *Instructional Design Principles and Applications*, 2nd Ed., Englewood Cliff, NJ: Educational Technology Publishing, (1991).

Carey, L. and D. Walter, *The Systematic Design of Instruction, 3rd Ed.*, Glenview, IL: Foresman Scott, (1990).

Davies, I.K., *Instructional Techniques*, New York, NY: McGraw-Hill, (1981).

Draves, W.A., *How to Teach Adults*, Manhattan, KS: Learning Resources Network, (1984).

Drumheller, S.J., *Handbook of Curriculum Design for Individualized Instruction: A Systems Approach*, Engelwood Cliffs, NJ: Educational Technology Publications, (1971).

Dwyer, F.M., *Strategies for Improving Visual Learning: A Handbook for the Effective Selection and Use of Visual Materials*, State College, PA: Learning Services, (1978).

Flemming, M.L. and W.H. Levie, *Instructional Message Design: Principles from the Behavioral Sciences*, Englewood Cliffs, NJ: Educational Technology Publications, (1978).

Gagne, R.M., L.E. Briggs and W.W. Wagner, *Principles of Instructional Design, 3rd Ed.*, New York, NY: Holt Rinehart and Winston, Inc., (1988).

Greer, C.E., H.G. Miller and J.R. Verduin, *Adults Teaching Adults: Principles and Strategies*, Austin, TX: Learning Concepts, (1977).

Kemp, J.E., *The Instructional Design Process*, New York, NY, Harper & Row, (1985).

Knowles, M.S., *Self-Directed Learning a Guide for Learners and Teachers*, Chicago, IL: Foller, (1978).

Merril, M.D. and R.D. Tennyson, *Teaching Concepts: An Instructional Design Guide*, Englewood Cliffs, NJ: Educational Technology Publishing, (1977).

Miller, H.L., *Teaching and Learning in Adult Education*, New York, NY: Macmillan, (1964).

Raulerson, J.D. and M.R. Wong, *A Guide to Systematic Instructional Design*, Englewood Cliff, NJ: Educational Technology Publications, (1974).

Smith, R.M., *Learning How to Learn Applied Theory for Adults*, Chicago, IL: Follert, (1982).

Verduin, J.R., *Curriculum Building for Adult Learning*, Carbondale, IL: Southern Illinois University Press, (1980).

Appendix A

OSHA Standards

PERMIT-REQUIRED CONFINED SPACE STANDARD: 29 CFR 1910.146

(a) *Scope and application.* This section contains requirements for practices and procedures to protect employees in general industry from the hazards of entry into permit-required confined spaces. This section does not apply to agriculture, to construction, or to shipyard employment (Parts 1928, 1926 and 1915 of this chapter, respectively).

(b) *Definitions.* "Acceptable entry conditions" means the conditions that must exist in a permit space to allow entry and to ensure that employees involved with a permit-required confined space entry can safely enter into and work within the space.

"Attendant" means an individual stationed outside one or more permit spaces who monitors the authorized entrants and who performs all attendant's duties assigned in the employer's permit space program.

"Authorized entrant" means an employee who is authorized by the employer to enter a permit space.

"Blanking or blinding" means the absolute closure of a pipe, line or duct by the fastening of a solid plate (such as a spectacle blind or a skillet blind) that completely covers the bore and that is capable of withstanding the maximum pressure of the pipe, line or duct with no leakage beyond the plate.

"Confined space" means a space that:

(1) Is large enough and so configured that an employee can bodily enter and perform assigned work

(2) Has limited or restricted means for entry or exit (for example, tanks, vessels, silos, storage bins, hoppers, vaults and pits are spaces that may have limited means of entry)

(3) Is not designed for continuous employee occupancy

"Double block and bleed" means the closure of a line, duct or pipe by closing and locking or tagging two in-line valves and by opening and locking or tagging a drain or vent valve in the line between the two closed valves.

"Emergency" means any occurrence (including any failure of hazard control or monitoring equipment) or event internal or external to the permit space that could endanger entrants.

"Engulfment" means the surrounding and effective capture of a person by a liquid or finely divided (flowable) solid substance that can be aspirated to cause death by filling or plugging the respiratory system or that can exert enough force on the body to cause death by strangulation, constriction, or crushing.

"Entry" means the action by which a person passes through an opening into a permit-required confined space. Entry includes ensuing work activities in that space and is considered to have occurred as soon as any part of the entrant's body breaks the plane of an opening into the space.

"Entry permit (permit)" means the written or printed document that is provided by the employer to allow and control entry into a permit space and that contains the information specified in paragraph (f) of this section.

"Entry supervisor" means the person (such as the employer, foreman or crew chief) responsible for determining if acceptable entry conditions are present at a permit space where entry is planned, for authorizing entry and overseeing entry operations, and for terminating entry as required by this section.

NOTE: An entry supervisor also may serve as an attendant or as an authorized entrant, as long as that person is trained and equipped as required by this section for each role he or she fills. Also, the duties of entry supervisor may be passed from one individual to another during the course of an entry operation.

"Hazardous atmosphere" means an atmosphere that may expose employees to the risk of death, incapacitation, impairment of ability to self-rescue (that is, escape unaided from a permit space), injury, or acute illness from one or more of the following causes:

(1) Flammable gas, vapor or mist in excess of 10% of its lower flammable limit (LFL)

(2) Airborne combustible dust at a concentration that meets or exceeds its LFL

NOTE: This concentration may be approximated as a condition in which the dust obscures vision at a distance of 5 feet (1.52 m) or less.

(3) Atmospheric oxygen concentration below 19.5% or above 23.5%

(4) Atmospheric concentration of any substance for which a dose or a permissible exposure limit is published in Subpart G, Occupational Health and Environmental Control, or in Subpart Z, Toxic and Hazardous Substances, of this Part and which could result in employee exposure in excess of its dose or permissible exposure limit;

NOTE: An atmospheric concentration of any substance that is not capable of causing death, incapacitation, impairment of ability to self-rescue, injury or acute illness due to its health effects is not covered by this provision.

(5) Any other atmospheric condition that is immediately dangerous to life or health.

NOTE: For air contaminants for which OSHA has not determined a dose or permissible exposure limit, other sources of information, such as Material Safety Data Sheets that comply with the Hazard Communication Standard, Section 1910.1200 of this Part, published information, and internal documents can provide guidance in establishing acceptable atmospheric conditions.

"Hot work permit" means the employer's written authorization to perform operations (for example, riveting, welding, cutting, burning, and heating) capable of providing a source of ignition.

"Immediately dangerous to life or health (IDLH)" means any condition that poses an immediate or delayed threat to life or that would cause irreversible adverse health effects or that would interfere with an individual's ability to escape unaided from a permit space.

NOTE: Some materials—hydrogen fluoride gas and cadmium vapor, for example—may produce immediate transient effects that, even if severe, may pass without medical attention, but are followed by sudden, possibly fatal collapse 12-72 hours after exposure. The victim "feels normal" from recovery from transient effects until collapse. Such materials in hazardous quantities are considered to be "immediately" dangerous to life or health.

"Inerting" means the displacement of the atmosphere in a permit space by a noncombustible gas (such as nitrogen) to such an extent that the resulting atmosphere is noncombustible.

NOTE: This procedure produces an IDLH oxygen-deficient atmosphere.

"Isolation" means the process by which a permit space is removed from service and completely protected against the release of energy and material into the space by such means as: blanking or blinding; misaligning or removing sections of lines, pipes, or ducts; a double block and bleed system; lockout or tagout of all sources of energy; or blocking or disconnecting all mechanical linkages.

"Line breaking" means the intentional opening of a pipe, line or duct that is or has been carrying flammable, corrosive or toxic material, an inert gas, or any fluid at a volume, pressure, or temperature capable of causing injury.

"Non-permit confined space" means a confined space that does not contain or, with respect to atmospheric hazards, have the potential to contain any hazard capable of causing death or serious physical harm.

"Oxygen deficient atmosphere" means an atmosphere containing less than 19.5% oxygen by volume.

"Oxygen enriched atmosphere" means an atmosphere containing more than 23.5% oxygen by volume.

"Permit-required confined space (permit space)" means a confined space that has one or more of the following characteristics:

(1) Contains or has a potential to contain a hazardous atmosphere

(2) Contains a material that has the potential for engulfing an entrant

(3) Has an internal configuration such that an entrant could be trapped or asphyxiated by inwardly converging walls or by a floor which slopes downward and tapers to a smaller cross-section

(4) Contains any other recognized serious safety or health hazard

"Permit-required confined space program (permit space program)" means the employer's overall program for controlling and, where appropriate, for protecting employees from permit space hazards and for regulating employee entry into permit spaces.

"Permit system" means the employer's written procedure for preparing and issuing permits for entry and for returning the permit space to service following termination of entry.

"Prohibited condition" means any condition in a permit space that is not allowed by the permit during the period when entry is authorized.

"Rescue service" means the personnel designated to rescue employees from permit spaces.

"Retrieval system" means the equipment (including a retrieval line, chest or full-body harness, wristlets, if appropriate, and a lifting device or anchor) used for non-entry rescue of persons from permit spaces.

"Testing" means the process by which the hazards that may confront entrants of a permit space are identified and evaluated. Testing includes specifying the tests that are to be performed in the permit space.

NOTE: Testing enables employers both to devise and implement adequate control measures for the protection of authorized entrants and to determine if acceptable entry conditions are present immediately prior to, and during, entry.

(c) *General requirements.*

(1) The employer shall evaluate the workplace to determine if any spaces are permit-required confined spaces.

NOTE: Proper application of the decision flow chart in Appendix A to Section 1910.146 would facilitate compliance with this requirement.

(2) If the workplace contains permit spaces, the employer shall inform exposed employees by posting danger signs or by any other equally effective means of the existence and location of and the danger posed by the permit spaces.

NOTE: A sign reading DANGER—PERMIT-REQUIRED CONFINED SPACE, DO NOT ENTER or using other similar language would satisfy the requirement for a sign.

(3) If the employer decides that its employees will not enter permit spaces, the employer shall take effective measures to prevent its employees from entering the permit spaces and shall comply with paragraphs (c)(1), (c)(2), (c)(6) and (c)(8) of this section.

(4) If the employer decides that its employees will enter permit spaces, the employer shall develop and implement a written permit space entry program that complies with this section. The written program shall be available for inspection by employees and their authorized representatives.

(5) An employer may use the alternate procedures specified in paragraph (c)(5)(ii) of this section for entering a permit space under the conditions set forth in paragraph (c)(5)(i) of this section.

(i) An employer whose employees enter a permit space need not comply with paragraphs (d) through (f) and (h) through (k) of this section, provided that:

(A) The employer can demonstrate that the only hazard posed by the permit space is an actual or potential hazardous atmosphere

(B) The employer can demonstrate that continuous forced air ventilation alone is sufficient to maintain that permit space safe for entry

(C) The employer develops monitoring and inspection data that supports the demonstrations required by paragraphs (c)(5)(i)(A) and (c)(5)(i)(B) of this section

(D) If an initial entry of the permit space is necessary to obtain the data required by paragraph (c)(5)(i)(C) of this section, the entry is performed in compliance with paragraphs (d) through (k) of this section

(E) The determinations and supporting data required by paragraphs (c)(5)(i)(A), (c)(5)(i)(B) and (c)(5)(i)(C) of this section are documented by the employer and are made available to each employee who enters the permit space under the terms of paragraph (c)(5) of this section

(F) Entry into the permit space under the terms of paragraph (c)(5)(i) of this section is performed in accordance with the requirements of paragraph (c)(5)(ii) of this section

NOTE: See paragraph (c)(7) of this section for reclassification of a permit space after all hazards within the space have been eliminated.

(ii) The following requirements apply to entry into permit spaces that meet the conditions set forth in paragraph (c)(5)(i) of this section.

(A) Any conditions making it unsafe to remove an entrance cover shall be eliminated before the cover is removed.

(B) When entrance covers are removed, the opening shall be promptly guarded by a railing, temporary cover or other temporary barrier that will prevent an accidental fall through the opening and that will protect each employee working in the space from foreign objects entering the space.

(C) Before an employee enters the space, the internal atmosphere shall be tested with a calibrated direct-reading instrument for the following conditions in the order given:

(1) Oxygen content

(2) Flammable gases and vapors

(3) Potential toxic air contaminants

(D) There may be no hazardous atmosphere within the space whenever any employee is inside the space.

(E) Continuous forced air ventilation shall be used, as follows:

(1) An employee may not enter the space until the forced air ventilation has eliminated any hazardous atmosphere

(2) The forced air ventilation shall be so directed as to ventilate the immediate areas where an employee is or will be present within the space and shall continue until all employees have left the space

(3) The air supply for the forced air ventilation shall be from a clean source and may not increase the hazards in the space

(F) The atmosphere within the space shall be periodically tested as necessary to ensure that the continuous forced air ventilation is preventing the accumulation of a hazardous atmosphere.

(G) If a hazardous atmosphere is detected during entry:

(1) Each employee shall leave the space immediately.

(2) The space shall be evaluated to determine how the hazardous atmosphere developed.

(3) Measures shall be implemented to protect employees from the hazardous atmosphere before any subsequent entry takes place.

(H) The employer shall verify that the space is safe for entry and that the measures required by paragraph (c)(5)(ii) of this section have been taken, through a written certification that contains the date, the location of the space and the signature of the person providing the certification. The certification shall be made before entry and shall be made available to each employee entering the space.

(6) When there are changes in the use or configuration of a non-permit confined space that might increase the hazards to entrants, the employer shall reevaluate that space and, if necessary, reclassify it as a permit-required confined space.

(7) A space classified by the employer as a permit-required confined space may be reclassified as a non-permit confined space under the following procedures:

(i) If the permit space poses no actual or potential atmospheric hazards and if all hazards within the space are eliminated without entry into the space, the permit space may be reclassified as a non-permit confined space for as long as the non-atmospheric hazards remain eliminated.

(ii) If it is necessary to enter the permit space to eliminate hazards, such entry shall be performed under paragraphs (d) through (k) of this section. If testing and inspection during that entry demonstrate that the hazards within the permit space have been eliminated, the permit space may be reclassified as a non-permit confined space for as long as the hazards remain eliminated.

NOTE: Control of atmospheric hazards through forced air ventilation does not constitute elimination of the hazards. Paragraph (c)(5) covers permit space entry where the employer can demonstrate that forced air ventilation alone will control all hazards in the space.

(iii) The employer shall document the basis for determining that all hazards in a permit space have been eliminated, through a certification that contains the date, the location of the space and the signature of the person making the determination. The certification shall be made available to each employee entering the space.

(iv) If hazards arise within a permit space that has been declassified to a non-permit space under paragraph (c)(7) of this section, each employee in the space shall exit the space. The employer shall then reevaluate the space and determine whether it must be reclassified as a permit space, in accordance with other applicable provisions of this section.

(8) When an employer (host employer) arranges to have employees of another employer (contractor) perform work that involves permit space entry, the host employer shall:

(i) Inform the contractor that the workplace contains permit spaces and that permit space entry is allowed only through compliance with an permit space program meeting the requirements of this section

(ii) Apprise the contractor of the elements, including the hazards identified and the host employer's experience with the space, that make the space in question a permit space

(iii) Apprise the contractor of any precautions or procedures that the host employer has implemented for the protection of employees in or near permit spaces where contractor personnel will be working

(iv) Coordinate entry operations with the contractor, when both host employer personnel and contractor personnel will be working in or near permit spaces, as required by paragraph (d)(11) of this section

(v) Debrief the contractor at the conclusion of the entry operations regarding the permit space program followed and regarding any hazards confronted or created in permit spaces during entry operations

(9) In addition to complying with the permit space requirements that apply to all employers, each contractor who is retained to perform permit space entry operations shall:

(i) Obtain any available information regarding permit space hazards and entry operations from the host employer

(ii) Coordinate entry operations with the host employer, when both host employer personnel and contractor personnel will be working in or near permit spaces, as required by paragraph (d)(11) of this section

(iii) Inform the host employer of the permit space program that the contractor will follow and of any hazards confronted or created in permit spaces, either through a debriefing or during the entry operation

(d) *Permit-required confined space program.* Under the permit-required confined space program required by paragraph (c)(4) of this section, the employer shall:

(1) Implement the measures necessary to prevent unauthorized entry.

(2) Identify and evaluate the hazards of permit spaces before employees enter them.

(3) Develop and implement the means, procedures, and practices necessary for safe permit space entry operations, including, but not limited to, the following:

(i) Specifying acceptable entry conditions

(ii) Isolating the permit space

(iii) Purging, inerting, flushing or ventilating the permit space as necessary to eliminate or control atmospheric hazards

(iv) Providing pedestrian, vehicle or other barriers as necessary to protect entrants from external hazards

(v) Verifying that conditions in the permit space are acceptable for entry throughout the duration of an authorized entry

(4) Provide the following equipment, (specified in paragraphs (d)(4)(i) through (d)(4)(ix) of this section) at no cost to employees, maintain that equipment properly and ensure that employees use that equipment properly:

(i) Testing and monitoring equipment needed to comply with paragraph (d)(5) of this section

(ii) Ventilating equipment needed to obtain acceptable entry conditions

(iii) Communications equipment necessary for compliance with paragraphs (h)(3) and (i)(5) of this section

(iv) Personal protective equipment insofar as feasible engineering and work practice controls do not adequately protect employees

(v) Lighting equipment needed to enable employees to see well enough to work safely and to exit the space quickly in an emergency

(vi) Barriers and shields as required by paragraph (d)(3)(iv) of this section

(vii) Equipment, such as ladders, needed for safe ingress and egress by authorized entrants

(viii) Rescue and emergency equipment needed to comply with paragraph (d)(9) of this section, except to the extent that the equipment is provided by rescue services

(ix) Any other equipment necessary for safe entry into and rescue from permit spaces

(5) Evaluate permit space conditions as follows when entry operations are conducted:

(i) Test conditions in the permit space to determine if acceptable entry conditions exist before entry is authorized to begin, except that if isolation of the space is infeasible because the space is large or is part of a continuous system (such as a sewer), pre-entry testing shall be performed to the extent feasible before entry is authorized and, if entry is authorized, entry conditions shall be continuously monitored in the areas where authorized entrants are working.

(ii) Test or monitor the permit space as necessary to determine if acceptable entry conditions are being maintained during the course of entry operations.

(iii) When testing for atmospheric hazards, test first for oxygen, then for combustible gases and vapors, and then for toxic gases and vapors.

NOTE: Atmospheric testing conducted in accordance with Appendix B to Section 1910.146 would be considered as satisfying the requirements of this paragraph. For permit space operations in sewers, atmospheric testing conducted in accordance with Appendix B, as supplemented by Appendix E to Section 1910.146, would be considered as satisfying the requirements of this paragraph.

(6) Provide at least one attendant outside the permit space into which entry is authorized for the duration of entry operations.

NOTE: Attendants may be assigned to monitor more than one permit space provided the duties described in paragraph (i) of this section can be effectively performed for each permit space that is monitored. Likewise, attendants may be stationed at any location outside the permit space to be monitored as long as the duties described in paragraph (i) of this section can be effectively performed for each permit space that is monitored.

(7) If multiple spaces are to be monitored by a single attendant, include in the permit program the means and procedures to enable the attendant to respond to an emergency affecting one or more of the permit spaces being monitored without distraction from the attendant's responsibilities under paragraph (i) of this section.

(8) Designate the persons who are to have active roles (as, for example, authorized entrants, attendants, entry supervisors, or persons who test or monitor the atmosphere in a permit space) in entry operations, identify the duties of each such employee, and provide each such employee with the training required by paragraph (g) of this section.

(9) Develop and implement procedures for summoning rescue and emergency services, for rescuing entrants from permit spaces, for providing necessary emergency services to rescued employees and for preventing unauthorized personnel from attempting a rescue.

(10) Develop and implement a system for the preparation, issuance, use and cancellation of entry permits as required by this section.

(11) Develop and implement procedures to coordinate entry operations when employees of more than one employer are working simultaneously as authorized entrants in a permit space, so that employees of one employer do not endanger the employees of any other employer.

(12) Develop and implement procedures (such as closing off a permit space and canceling the permit) necessary for concluding the entry after entry operations have been completed.

(13) Review entry operations when the employer has reason to believe that the measures taken under the permit space program may not protect employees and revise the program to correct deficiencies found to exist before subsequent entries are authorized.

NOTE: Examples of circumstances requiring the review of the permit-required confined space program are any unauthorized entry of a permit space, the detection of a permit space hazard not covered by the permit, the detection of a condition prohibited by the permit, the occurrence of an injury or near-miss during entry, a change in the use or configuration of a permit space, and employee complaints about the effectiveness of the program.

(14) Finally, review the permit-required confined space program, using the canceled permits retained under paragraph (e)(6) of this section within 1 year after each entry and revise the program as necessary, to ensure that employees participating in entry operations are protected from permit space hazards.

NOTE: Employers may perform a single annual review covering all entries performed during a 12-month period. If no entry is performed during a 12-month period, no review is necessary. Appendix C to Section 1910.146 presents examples of permit entry programs that are considered to comply with the requirements of paragraph (d) of this section.

(e) *Permit system.*

(1) Before entry is authorized, the employer shall document the completion of measures required by paragraph (d)(3) of this section by preparing an entry permit.

NOTE: Appendix D to Section 1910.146 presents examples of permits whose elements are considered to comply with the requirements of this section.

(2) Before entry begins, the entry supervisor identified on the permit shall sign the entry permit to authorize entry.

(3) The completed permit shall be made available at the time of entry to all authorized entrants by posting it at the entry portal or by any other equally effective means, so that the entrants can confirm that pre-entry preparations have been completed.

(4) The duration of the permit may not exceed the time required to complete the assigned task or job identified on the permit in accordance with paragraph (f)(2) of this section.

(5) The entry supervisor shall terminate entry and cancel the entry permit when:

(i) The entry operations covered by the entry permit have been completed.

(ii) A condition that is not allowed under the entry permit arises in or near the permit space.

(6) The employer shall retain each canceled entry permit for at least 1 year to facilitate the review of the permit-required confined space program required by paragraph (d)(14) of this section. Any problems encountered during an entry operation shall be noted on the pertinent permit so that appropriate revisions to the permit space program can be made.

(f) *Entry permit.* The entry permit that documents compliance with this section and authorizes entry to a permit space shall identify:

(1) The permit space to be entered

(2) The purpose of the entry

(3) The date and the authorized duration of the entry permit

(4) The authorized entrants within the permit space, by name or by such other means (for example, through the use of rosters or tracking systems) as will enable the attendant to determine quickly and accurately, for the duration of the permit, which authorized entrants are inside the permit space

NOTE: This requirement may be met by inserting a reference on the entry permit as to the means used, such as a roster or tracking system, to keep track of the authorized entrants within the permit space.

(5) The personnel, by name, currently serving as attendants

(6) The individual, by name, currently serving as entry supervisor, with a space for the signature or initials of the entry supervisor who originally authorized entry

(7) The hazards of the permit space to be entered;

(8) The measures used to isolate the permit space and to eliminate or control permit space hazards before entry

NOTE: Those measures can include the lockout or tagging of equipment and procedures for purging, inerting, ventilating and flushing permit spaces.

(9) The acceptable entry conditions

(10) The results of initial and periodic tests performed under paragraph (d)(5) of this section, accompanied by the names or initials of the testers and by an indication of when the tests were performed

(11) The rescue and emergency services that can be summoned and the means (such as the equipment

to use and the numbers to call) for summoning those services

(12) The communication procedures used by authorized entrants and attendants to maintain contact during the entry

(13) Equipment, such as personal protective equipment, testing equipment, communications equipment, alarm systems and rescue equipment, to be provided for compliance with this section

(14) Any other information whose inclusion is necessary, given the circumstances of the particular confined space, in order to ensure employee safety

(15) Any additional permits, such as for hot work, that have been issued to authorize work in the permit space.

(g) Training.

(1) The employer shall provide training so that all employees whose work is regulated by this section acquire the understanding, knowledge and skills necessary for the safe performance of the duties assigned under this section.

(2) Training shall be provided to each affected employee:

(i) Before the employee is first assigned duties under this section

(ii) Before there is a change in assigned duties

(iii) Whenever there is a change in permit space operations that presents a hazard about which an employee has not previously been trained

(iv) Whenever the employer has reason to believe either that there are deviations from the permit space entry procedures required by paragraph (d)(3) of this section or that there are inadequacies in the employee's knowledge or use of these procedures

(3) The training shall establish employee proficiency in the duties required by this section and shall introduce new or revised procedures, as necessary, for compliance with this section.

(4) The employer shall certify that the training required by paragraphs (g)(1) through (g)(3) of this section has been accomplished. The certification shall contain each employee's name, the signatures or initials of the trainers, and the dates of training. The certification shall be available for inspection by employees and their authorized representatives.

(h) Duties of authorized entrants. The employer shall ensure that all authorized entrants:

(1) Know the hazards that may be faced during entry, including information on the mode, signs or symptoms, and consequences of the exposure

(2) Properly use equipment as required by paragraph (d)(4) of this section

(3) Communicate with the attendant as necessary to enable the attendant to monitor entrant status and to enable the attendant to alert entrants of the need to evacuate the space as required by paragraph (i)(6) of this section

(4) Alert the attendant whenever:

(i) The entrant recognizes any warning sign or symptom of exposure to a dangerous situation

(ii) The entrant detects a prohibited condition

(5) Exit from the permit space as quickly as possible whenever:

(i) An order to evacuate is given by the attendant or the entry supervisor

(ii) The entrant recognizes any warning sign or symptom of exposure to a dangerous situation

(iii) The entrant detects a prohibited condition

(iv) An evacuation alarm is activated

(i) Duties of attendants. The employer shall ensure that each attendant:

(1) Knows the hazards that may be faced during entry, including information on the mode, signs or symptoms, and consequences of the exposure

(2) Is aware of possible behavioral effects of hazard exposure in authorized entrants

(3) Continuously maintains an accurate count of authorized entrants in the permit space and ensures that the means used to identify authorized entrants

under paragraph (f)(4) of this section accurately identifies who is in the permit space

(4) Remains outside the permit space during entry operations until relieved by another attendant
NOTE: When the employer's permit entry program allows attendant entry for rescue, attendants may enter a permit space to attempt a rescue if they have been trained and equipped for rescue operations as required by paragraph (k)(1) of this section and if they have been relieved as required by paragraph (i)(4) of this section.

(5) Communicates with authorized entrants as necessary to monitor entrant status and to alert entrants of the need to evacuate the space under paragraph (i)(6) of this section

(6) Monitors activities inside and outside the space to determine if it is safe for entrants to remain in the space and orders the authorized entrants to evacuate the permit space immediately under any of the following conditions:

(i) If the attendant detects a prohibited condition

(ii) If the attendant detects the behavioral effects of hazard exposure in an authorized entrant

(iii) If the attendant detects a situation outside the space that could endanger the authorized entrants

(iv) If the attendant cannot effectively and safely perform all the duties required under paragraph (i) of this section;

(7) Summon rescue and other emergency services as soon as the attendant determines that authorized entrants may need assistance to escape from permit space hazards

(8) Takes the following actions when unauthorized persons approach or enter a permit space while entry is underway:

(i) Warn the unauthorized persons that they must stay away from the permit space

(ii) Advise the unauthorized persons that they must exit immediately if they have entered the permit space

(iii) Inform the authorized entrants and the entry supervisor if unauthorized persons have entered the permit space

(9) Performs non-entry rescues as specified by the employer's rescue procedure

(10) Performs no duties that might interfere with the attendant's primary duty to monitor and protect the authorized entrants

(j) *Duties of entry supervisors.* The employer shall ensure that each entry supervisor:

(1) Knows the hazards that may be faced during entry, including information on the mode, signs or symptoms, and consequences of the exposure

(2) Verifies, by checking that the appropriate entries have been made on the permit, that all tests specified by the permit have been conducted and that all procedures and equipment specified by the permit are in place before endorsing the permit and allowing entry to begin

(3) Terminates the entry and cancels the permit as required by paragraph (e)(5) of this section

(4) Verifies that rescue services are available and that the means for summoning them are operable

(5) Removes unauthorized individuals who enter or who attempt to enter the permit space during entry operations

(6) Determines, whenever responsibility for a permit space entry operation is transferred and at intervals dictated by the hazards and operations performed within the space, that entry operations remain consistent with terms of the entry permit and that acceptable entry conditions are maintained

(k) *Rescue and emergency services.*

(1) The following requirements apply to employers who have employees enter permit spaces to perform rescue services:

(i) The employer shall ensure that each member of the rescue service is provided with, and is trained to use properly, the personal protective equipment and rescue equipment necessary for making rescues from permit spaces.

(ii) Each member of the rescue service shall be trained to perform the assigned rescue duties. Each member of the rescue service shall also receive the training required of authorized entrants under paragraph (g) of this section.

(iii) Each member of the rescue service shall practice making permit space rescues at least once every 12 months, by means of simulated rescue operations in which they remove dummies, manikins or actual persons from the actual permit spaces or from representative permit spaces. Representative permit spaces shall, with respect to opening size, configuration and accessibility, simulate the types of permit spaces from which rescue is to be performed.

(iv) Each member of the rescue service shall be trained in basic first-aid and in cardiopulmonary resuscitation (CPR). At least one member of the rescue service holding current certification in first aid and in CPR shall be available.

(2) When an employer (host employer) arranges to have persons other than the host employer's employees perform permit space rescue, the host employer shall:

(i) Inform the rescue service of the hazards they may confront when called on to perform rescue at the host employer's facility

(ii) Provide the rescue service with access to all permit spaces from which rescue may be necessary so that the rescue service can develop appropriate rescue plans and practice rescue operations.

(3) To facilitate non-entry rescue, retrieval systems or methods shall be used whenever an authorized entrant enters a permit space, unless the retrieval equipment would increase the overall risk of entry or would not contribute to the rescue of the entrant. Retrieval systems shall meet the following requirements.

(i) Each authorized entrant shall use a chest or full body harness, with a retrieval line attached at the center of the entrant's back near shoulder level, or above the entrant's head. Wristlets may be used in lieu of the chest or full body harness if the employer can demonstrate that the use of a chest or full body harness is infeasible or creates a greater hazard and that the use of wristlets is the safest and most effective alternative.

(ii) The other end of the retrieval line shall be attached to a mechanical device or fixed point outside the permit space in such a manner that rescue can begin as soon as the rescuer becomes aware that rescue is necessary. A mechanical device shall be available to retrieve personnel from vertical type permit spaces more than 5 feet deep.

(4) If an injured entrant is exposed to a substance for which a Material Safety Data Sheet (MSDS) or other similar written information is required to be kept at the worksite, that MSDS or written information shall be made available to the medical facility treating the exposed entrant.

OSHA LOCKOUT/TAGOUT STANDARD 29 CFR 1910.147

(a) *Scope, application and purpose*

(1) Scope

(i) This standard covers the servicing and maintenance of machines and equipment in which the unexpected energization or start up of the machines or equipment, or release of stored energy could cause injury to employees. This standard establishes minimum performance requirements for the control of such hazardous energy.

(ii) This standard does not cover the following:

(A) Construction, agriculture and maritime employment

(B) Installations under the exclusive control of electric utilities for the purpose of power generation, transmission and distribution, including related equipment for communication or metering

(C) Exposure to electrical hazards from work on, near, or with conductors or equipment in electric utilization installations, which is covered by Subpart S of this part

(D) Oil and gas well drilling and servicing

(2) Application

(i) This standard applies to the control of energy during servicing and/or maintenance of machines and equipment.

(ii) Normal production operations are not covered by this standard (See Subpart O of this Part). Servicing and/or maintenance which takes place during normal production operations is covered by this standard only if:

(A) An employee is required to remove or bypass a guard or other safety device

(B) An employee is required to place any part of his or her body into an area on a machine or piece of equipment where work is actually performed upon the material being processed (point of operation) or where an associated danger zone exists during a machine operating cycle.

NOTE: Exception to paragraph (a)(2)(ii): Minor tool changes and adjustments, and other minor servicing activities, which take place during normal production operations, are not covered by this standard if they are routine, repetitive and integral to the use of the equipment for production, provided that the work is performed using alternative measures which provide effective protection (See Subpart O of this Part).

(iii) This standard does not apply to the following:

(A) Work on cord and plug connected electric equipment for which exposure to the hazards of unexpected energization or start-up of the equipment is controlled by the unplugging of the equipment from the energy source and by the plug being under the exclusive control of the employee performing the servicing or maintenance

(B) Hot tap operations involving transmission and distribution systems for substances such as gas, steam, water or petroleum products when they are performed on pressurized pipelines, provided that the employer demonstrates that:

(1) Continuity of service is essential

(2) Shutdown of the system is impractical

(3) Documented procedures are followed, and special equipment is used which will provide proven effective protection for employees

(3) Purpose

(i) This section requires employers to establish a program and utilize procedures for affixing appropriate lockout devices or tagout devices to energy isolating devices, and to otherwise disable machines or equipment to prevent unexpected energization, start-up or release of stored energy in order to prevent injury to employees.

(ii) When other standards in this part require the use of lockout or tagout, they shall be used and supplemented by the procedural and training requirements of this section.

(b) *Definitions applicable to this section*

Affected employee. An employee whose job requires him/her to operate or use a machine or equipment on which servicing or maintenance is being performed under lockout or tagout, or whose job requires him/her to work in an area in which such servicing or maintenance is being performed.

Authorized employee. A person who locks out or tags out machines or equipment in order to perform servicing or maintenance on that machine or equipment. An affected employee becomes an authorized employee when that employee's duties include performing servicing or maintenance covered under this section.

Capable of being locked out. An energy isolating device is capable of being locked out if it has a hasp or other means of attachment to which, or through which, a lock can be affixed, or it has a locking mechanism built into it.

Other energy isolating devices are capable of being locked out, if lockout can be achieved without the need to dismantle, rebuild or replace the energy isolating device or permanently alter its energy control capability.

Energized. Connected to an energy source or containing residual or stored energy.

Energy isolating device. A mechanical device that physically prevents the transmission or release or energy, including but not limited to the following:

A manually operated electrical circuit breaker; a disconnect switch; a manually operated switch by which the conductors of a circuit can be disconnected from all ungrounded supply conductors and, in addition, no pole can be operated independently; a line valve; a block; and any similar device used to block or isolate energy. Push buttons, selector switches and other control circuit type devices are not energy isolating devices.

Energy source. Any source of electrical, mechanical, hydraulic, pneumatic, chemical, thermal, or other energy.

Hot tap. A procedure used in the repair maintenance and services activities which involves welding on a piece of equipment (pipelines, vessels or tanks) under pressure, in order to install connections or appurtenances. It is commonly used to replace or add sections of pipeline without the interruption of service for air, gas, water, steam and petrochemical distribution systems.

Lockout. The placement of a lockout device on an energy isolating device, in accordance with an established procedure, ensuring that the energy isolating device and the equipment being controlled cannot be operated until the lockout device is removed.

Lockout device. A device that utilizes a positive means such as a lock, either key or combination type, to hold an energy isolating device in the safe position and prevent the energizing of a machine or equipment. Included are blank flanges and bolted slip blinds.

Normal production operations. The utilization of a machine or equipment to perform its intended production function.

Servicing and/or maintenance. Workplace activities such as constructing, installing, setting up, adjusting, inspecting, modifying and maintaining and/or servicing machines or equipment. These activities include lubrication, cleaning or unjamming of machines or equipment and making adjustments or tool changes, where the employee may be exposed to the unexpected energization or start-up of the equipment or release of hazardous energy.

Setting up. Any work performed to prepare a machine or equipment to perform its normal production operation.

Tagout. The placement of a tagout device on an energy isolating device, in accordance with an established procedure, to indicate that the energy isolating device and the equipment being controlled may not be operated until the tagout device is removed.

Tagout device. A prominent warning device, such as a tag and a means of attachment, which can be securely fastened to an energy isolating device in accordance with an established procedure, to indicate that the energy isolating device and the equipment being controlled may not be operated until the tagout device is removed.

(c) *General*

(1) Energy control program. The employer shall establish a program consisting of energy control procedures, employee training and to periodic inspections to ensure that before any employee performs any servicing or maintenance on a machine or equipment where the unexpected energizing, start-up or

release of stored energy could occur and cause injury, the machine or equipment shall be isolated from the energy source and rendered inoperative.

(2) Lockout/tagout.

(i) If an energy-isolating device is not capable of being locked out, the employer's energy control program under paragraph (c)(1) of this section shall utilize a tagout system.

(ii) If an energy-isolating device is capable of being locked out, the employer's energy control program under paragraph (c)(1) of this section shall utilize lockout, unless the employer can demonstrate that the utilization of a tagout system will provide full employee protection as set forth in paragraph (c)(3) of this section.

(iii) After January 2, 1990, whenever replacement or major repair, renovation or modification of a machine or equipment is performed, and whenever new machines or equipment are installed, energy-isolating devices for such machine or equipment shall be designed to accept a lockout device.

(3) Full employee protection.

(i) When a tagout device is used on an energy-isolating device which is capable of being locked out, the tagout device shall be attached at the same location that the lockout device would have been attached, and the employer shall demonstrate that the tagout program will provide a level of safety equivalent to that obtained by using a lockout program.

(ii) In demonstrating that a level of safety is achieved in the tagout program which is equivalent to the level of safety obtained by using a lockout program, the employer shall demonstrate full compliance with all tagout-related provisions of this standard together with such additional elements as are necessary to provide the equivalent safety available from the use of a lockout device. Additional means to be considered as part of the demonstration of full employee protection shall include the implementation of additional safety measures such as the removal of an isolating circuit element, blocking of a controlling switch, opening of an extra disconnecting device or the removal of a valve handle to reduce the likelihood of inadvertent energization.

(4) Energy control procedure.

(i) Procedures shall be developed, documented and utilized for the control of potentially hazardous energy when employees are engaged in the activities covered by this section.

NOTE: Exception: the employer need not document the required procedure for a particular machine or equipment when all of the following elements exist:

(1) The machine or equipment has no potential for stored or residual energy or reaccumulation of stored energy after shutdown which could endanger employees.

(2) The machine or equipment has a single energy source which can be readily identified and isolated.

(3) The isolation and locking out of that energy source will completely deenergize and deactivate the machine or equipment.

(4) The machine or equipment is isolated from that energy source and locked out during servicing or maintenance.

(5) A single lockout device will achieve a locked-out condition.

(6) The lockout device is under the exclusive control of the authorized employee performing the servicing or maintenance.

(7) The servicing or maintenance does not create hazards for other employees.

(8) The employer, in utilizing this exception, has had no accidents involving the unexpected activation or reenergization of the machine or equipment during servicing or maintenance.

(ii) The procedures shall clearly and specifically outline the scope, purpose, authorization, rules and techniques to be utilized for the control of hazardous energy, and the means to enforce compliance including, but not limited to, the following:

(A) A specific statement of the intended use of the procedure

(B) Specific procedural steps for shutting down, isolating, blocking and securing machines or equipment to control hazardous energy

(C) Specific procedural steps for the placement, removal and transfer of lockout devices or tagout devices and the responsibility for them

(D) Specific requirements for testing a machine or equipment to determine and verify the effectiveness of lockout devices, tagout devices and other energy control measures

(5) Protective materials and hardware.

(i) Locks, tags, chains, wedges, key blocks, adapter pins, self-locking fasteners or other hardware shall be provided by the employer for isolating, securing or blocking of machines or equipment from energy sources.

(ii) Lockout devices and tagout devices shall be singularly identified; shall be the only device(s) used for controlling energy; shall not be used for other purposes; and shall meet the following requirements:

(A) Durable.

(1) Lockout and tagout devices shall be capable of withstanding the environment to which they are exposed for the maximum period of time that exposure is expected.

(2) Tagout devices shall be constructed and printed so that exposure to weather conditions or wet and damp locations will not cause the tag to deteriorate or the message on the tag to become illegible.

(3) Tags shall not deteriorate when used in corrosive environments such as areas where acid and alkali chemicals are handled and stored.

(B) Standardized. Lockout and tagout devices shall be standardized within the facility in at least one of the following criteria:

Color; shape; or size; and additionally, in the case of tagout devices, print and format shall be standardized.

(C) Substantial.

(1) Lockout devices. Lockout devices shall be substantial enough to prevent removal without the use of excessive force or unusual techniques, such as with the use of bolt cutters or other metal cutting tools.

(2) Tagout devices. Tagout devices, including their means of attachment, shall be substantial enough to prevent inadvertent or accidental removal. Tagout device attachment means shall be of a non-reusable type, attachable by hand, self-locking, and non-releasable with a minimum unlocking strength of no less than 50 pounds and having the general design and basic characteristics of being at least equivalent to a one-piece, all environment-tolerant nylon cable tie.

(D) Identifiable. Lockout devices and tagout devices shall indicate the identify of the employee applying the device(s).

(iii) Tagout devices shall warn against hazardous conditions if the machine or equipment is energized and shall include a legend such as the following: Do Not Start. Do Not Open. Do Not Close. Do Not Energize. Do Not Operate.

(6) Periodic inspection.

(i) The employer shall conduct a periodic inspection of the energy control procedure at least annually to ensure that the procedure and the requirements of this standard are being followed.

(A) The periodic inspection shall be performed by an authorized employee other than the one(s) utilizing the energy control procedure being inspected.

(B) The periodic inspection shall be conducted to correct any deviations or inadequacies identified.

(C) Where lockout is used for energy control, the periodic inspection shall include a review between the inspector and each authorized employee of that employee's responsibilities under the energy control procedure being inspected.

(D) Where tagout is used for energy control, the periodic inspection shall include a review between the inspector and each authorized and affected employee of that employee's responsibilities under the energy control procedure being inspected and the elements set forth in paragraph (c)(7)(ii) of this section.

(ii) The employer shall certify that the periodic inspections have been performed. The certification shall identify the machine or equipment on which the energy control procedure was being utilized, the date of the inspection, the employees included in the inspection and the person performing the inspection.

(7) Training and communication

(i) The employer shall provide training to ensure that the purpose and function of the energy control program are understood by employees and that the knowledge and skills required for the safe application, usage and removal of the energy controls are acquired by employees.

The training shall include the following:

(A) Each authorized employee shall receive training in the recognition of applicable hazardous energy sources, the type and magnitude of the energy available in the workplace, and the methods and means necessary for energy isolation and control.

(B) Each affected employee shall be instructed in the purpose and use of the energy control procedure.

(C) All other employees whose work operations are or may be in an area where energy control procedures may be utilized shall be instructed about the procedure and about the prohibition relating to attempts to restart or reenergize machines or equipment which are locked out or tagged out.

(ii) When tagout systems are used, employees shall also be trained in the following limitations of tags:

(A) Tags are essentially warning devices affixed to energy isolating devices and do not provide the physical restraint on those devices that is provided by a lock.

(B) When a tag is attached to an energy-isolating means, it is not to be removed without authorization of the authorized person responsible for it, and it is never to be bypassed, ignored or otherwise defeated.

(C) Tags must be legible and understandable by all authorized employees, affected employees and all other employees whose work operations are or may be in the area, in order to be effective.

(D) Tags and their means of attachment must be made of materials which will withstand the environmental conditions encountered in the workplace.

(E) Tags may evoke a false sense of security, and their meaning needs to be understood as part of the overall energy control program.

(F) Tags must be securely attached to energy isolating devices so that they cannot be inadvertently or accidentally detached during use.

(iii) Employee retraining.

(A) Retraining shall be provided for all authorized and affected employees whenever there is a change in their job assignments, a change in machines, equipment or processes that present a new hazard, or when there is a change in the energy control procedures.

(B) Additional retraining shall also be conducted whenever a periodic inspection under paragraph (c)(6) of this section reveals, or whenever the employer has reason to believe that there are deviations from or inadequacies in the employee's knowledge or use of the energy control procedures.

(C) The retraining shall reestablish employee proficiency and introduce new or revised control methods and procedures, as necessary.

(iv) The employer shall certify that employee training has been accomplished and is being kept up to date. The certification shall contain each employee's name and dates of training.

(8) Energy isolation. Lockout or tagout shall be performed only by the authorized employees who are performing the servicing or maintenance.

(9) Notification of employees. Affected employees shall be notified by the employer or authorized employee of the application and removal of lockout devices or tagout devices. Notification shall be given before the controls are applied and after they are removed from the machine or equipment.

(d) *Application of control.* The established procedures for the application of energy control (the lockout or tagout procedures) shall cover the following elements and actions and shall be done in the following sequence:

(1) Preparation for shutdown. Before an authorized or affected employee turns off a machine or equipment, the authorized employee shall have knowledge of the type and magnitude of the energy, the hazards

of the energy to be controlled and the method or means to control the energy.

(2) Machine or equipment shutdown. The machine or equipment shall be turned off or shut down using the procedures established for the machine or equipment. An orderly shutdown must be utilized to avoid any additional or increased hazard(s) to employees as a result of the equipment stoppage.

(3) Machine or equipment isolation. All energy isolating devices that are needed to control the energy to the machine or equipment shall be physically located and operated in such a manner as to isolate the machine or equipment from the energy source(s).

(4) Lockout or tagout device application.

(i) Lockout or tagout devices shall be affixed to each energy-isolating device by authorized employees.

(ii) Lockout devices, where used, shall be affixed in a manner to that will hold the energy-isolating devices in a "safe" or "off" position.

(iii) Tagout devices, where used, shall be affixed in such a manner as will clearly indicate that the operation or movement of energy isolating devices from the "safe" or "off" position is prohibited.

(A) Where tagout devices are used with energy-isolating devices designed with the capability of being locked, the tag attachment shall be fastened at the same point at which the lock would have been attached.

(B) Where a tag cannot be affixed directly to the energy-isolating device, the tag shall be located as close as safely possible to the device, in a position that will be immediately obvious to anyone attempting to operate the device.

(5) Stored energy.

(i) Following the application of logout or tagout devices to energy-isolating devices, all potentially hazardous stored or residual energy shall be relieved, disconnected, restrained and otherwise rendered safe.

(ii) If there is a possibility of reaccumulation of stored energy to a hazardous level, verification of isolation shall be continued until the servicing or maintenance is completed, or until the possibility of such accumulation no longer exists.

(6) Verification of isolation. Prior to starting work on machines or equipment that have been locked out or tagged out, the authorized employee shall verify that isolation and deenergization of the machine or equipment have been accomplished.

(e) *Release from lockout or tagout.* Before lockout or tagout devices are removed and energy is restored to the machine or equipment, procedures shall be followed and actions taken by the authorized employee(s) to ensure the following:

(1) The machine or equipment. The work area shall be inspected to ensure that nonessential items have been removed and to ensure that machine or equipment components are operationally intact.

(2) Employees.

(i) The work area shall be checked to ensure that all employees have been safely positioned or removed.

(ii) Before lockout or tagout devices are removed and before machines or equipment are energized, affected employees shall be notified that the lockout or tagout devices have been removed.

(iii) After lockout or tagout devices have been removed and before a machine or equipment is started, affected employees shall be notified that the lockout or tagout device(s) have been removed.

(3) Lockout or tagout devices removal. Each lockout or tagout device shall be removed from each energy isolating device by the employee who applied the device. Exception to paragraph (e)(3). When the authorized employee who applied the lockout or tagout device is not available to remove it, that device may be removed under the direction of the employer, provided that specific procedures and training for such removal have been developed, documented and incorporated into the employer's energy control program. The employer shall demonstrate that the specific procedure shall include at least the following elements:

(i) Verification by the employer that the authorized employee who applied the device is not at the facility

(ii) Making all reasonable efforts to contact the authorized employee to inform him/her that his/her lockout or tagout device has been moved

(iii) Ensuring that the authorized employee has this knowledge before he/she resumes work at that facility

(f) *Additional requirements.*

(1) Testing or positioning of machines, equipment or components thereof. In situations in which lockout or tagout devices must be temporarily removed from the energy isolating device and the machine or equipment energized to test or position the machine, equipment or component thereof, the following sequence of actions shall be followed:

(i) Clear the machine or equipment of tools and materials in accordance with paragraph (e)(1) of this section.

(ii) Remove employees from the machine or equipment area in accordance with paragraph (e)(2) of this section.

(iii) Remove the lockout or tagout devices as specified in paragraph (e)(3) of this section.

(iv) Energize and proceed with testing or positioning.

(v) Deenergize all systems and reapply energy control measures in accordance with paragraph (d) of this section to continue the servicing and/or maintenance.

(2) Outside personnel (contractors, etc.).

(i) Whenever outside servicing personnel are to be engaged in activities covered by the scope and application of this standard, the on-site employer and the outside employer shall inform each other of their respective lockout or tagout procedures.

(ii) The on-site employer shall ensure that his/her employees understand and comply with the restrictions and prohibitions of the outside employer's energy control program.

(3) Group lockout or tagout.

(i) When servicing and/or maintenance is performed by a crew, craft, department or other group, they shall utilize a procedure which affords the employees a level of protection equivalent to that provided by the implementation of a personal lockout or tagout device.

(ii) Group lockout or tagout devices shall be used in accordance with the procedures required by paragraph (c)(4) of this section including, but not necessarily limited to, the following specific requirements.

(A) Primary responsibility is vested in an authorized employee for a set number of employees working under the protection of a group lockout or tagout device (such as an operations lock).

(B) Provision for the authorized employee to ascertain the exposure status of individual group members with regard to the lockout or tagout of the machine or equipment.

(C) When more than one crew, craft, department, etc. is involved, assignment of overall job-associated lockout or tagout control responsibility to an authorized employee designated to coordinate affected work forces and ensure continuity of protection.

(D) Each authorized employee shall affix a personal lockout or tagout device to the group lockout device, group lockbox, or comparable mechanism when he or she begins work, and shall remove those devices when he or she stops working on the machine or equipment being serviced or maintained.

(4) Shift or personnel changes. Specific procedures shall be utilized during shift or personnel changes to ensure the continuity of lockout or tagout protection, including provision for the orderly transfer of lockout or tagout device protection between off-going and on-coming employees, to minimize exposure to hazards from the unexpected energization or start-up of the machine or equipment, or the release of stored energy.

29 CFR 1910.134

(a) *Permissible practice.*

(1) In the control of those occupational diseases caused by breathing air contaminated with harmful dusts, fogs, fumes, mists, gases, smokes, sprays or vapors, the primary objective shall be to prevent atmospheric contamination. This shall be accomplished as far as feasible by accepted engineering control measures (for example, enclosure or confinement of the operation, general and local ventilation, and substitution of less toxic materials). When effective engineering controls are not feasible, or while they are being instituted, appropriate respirators shall be used pursuant to the following requirements.

(2) Respirators shall be provided by the employer when such equipment is necessary to protect the health of the employee. The employer shall provide the respirators which are applicable and suitable for the purpose intended. The employer shall be responsible for the establishment and maintenance of a respiratory protective program which shall include the requirements outlined in paragraph (b) of this section.

(3) The employee shall use the provided respiratory protection in accordance with instructions and training received.

(b) *Requirements for a minimal acceptable program.*

(1) Written standard operating procedures governing the selection and use of respirators shall be established.

(2) Respirators shall be selected on the basis of hazards to which the worker is exposed.

(3) The user shall be instructed and trained in the proper use of respirators and their limitations.

(4) [Reserved]

(5) Respirators shall be regularly cleaned and disinfected. Those used by more than one worker shall be thoroughly cleaned and disinfected after each use.

(6) Respirators shall be stored in a convenient, clean and sanitary location.

(7) Respirators used routinely shall be inspected during cleaning. Worn or deteriorated parts shall be replaced. Respirators for emergency use such as self-contained devices shall be thoroughly inspected at least once a month and after each use.

(8) Appropriate surveillance of work area conditions and degree of employee exposure or stress shall be maintained.

(9) There shall be regular inspection and evaluation to determine the continued effectiveness of the program.

(10) Persons should not be assigned to tasks requiring use of respirators unless it has been determined that they are physically able to perform the work and use the equipment. The local physician shall determine what health and physical conditions are pertinent. The respirator user's medical status should be reviewed periodically (for instance, annually).

(11) Approved or accepted respirators shall be used when they are available. The respirator furnished shall provide adequate respiratory protection against the particular hazard for which it is designed in accordance with standards established by competent authorities. The U.S. Department of Interior, Bureau of Mines and the U.S. Department of Agriculture are recognized as such authorities. Although respirators listed by the U.S. Department of Agriculture continue to be acceptable for protection against specified pesticides, the U.S. Department of the Interior, Bureau of Mines, is the agency now responsible for testing and approving pesticide respirators.

(c) Selection of respirators. Proper selection of respirators shall be made according to the guidance of American National Standard Practices for Respiratory Protection Z88.2-1969.

(d) *Air quality.*

(1) Compressed air, compressed oxygen, liquid air and liquid oxygen used for respiration shall be of high purity. Oxygen shall meet the requirements of the U.S. Pharmacopoeia for medical or breathing oxygen. Breathing air shall meet at least the requirements of the specification for Grade D breathing air as described in Compressed Gas Association Commodity Specification G-7.1-1966. Compressed oxygen shall not be used in supplied-air respirators or in open circuit self-contained breathing apparatus that

have previously used compressed air. Oxygen must never be used with air-line respirators.

(2) Breathing air may be supplied to respirators from cylinders or air compressors.

(i) Cylinders shall be tested and maintained as prescribed in the Shipping Container Specification Regulations of the Department of Transportation (49 CFR Part 178).

(ii) The compressor for supplying air shall be equipped with necessary safety and standby devices. A breathing air-type compressor shall be used. Compressors shall be constructed and situated so as to avoid entry of contaminated air into the system and suitable in-line air purifying sorbent beds and filters installed to further assure breathing air quality. A receiver of sufficient capacity to enable the respirator wearer to escape from a contaminated atmosphere in event of compressor failure, and alarms to indicate compressor failure and overheating shall be installed in the system. If an oil-lubricated compressor is used, it shall have a high-temperature or carbon monoxide alarm, or both. If only a high-temperature alarm is used, the air from the compressor shall be frequently tested for carbon monoxide to insure that it meets the specifications in paragraph (d)(1) of this section.

(3) Air line couplings shall be incompatible with outlets for other gas systems to prevent inadvertent servicing of air line respirators with nonrespirable gases or oxygen.

(4) Breathing gas containers shall be marked in accordance with American National Standard Method of Marking Portable Compressed Gas Containers to Identify the Material Contained, Z48.1-1954; Federal Specification BB-A-1034a, June 21, 1968, Air, Compressed for Breathing Purposes; or Interim Federal Specification GG-B-00675b, April 27, 1965, Breathing Apparatus, Self-Contained.

(e) *Use of respirators.*

(1) Standard procedures shall be developed for respirator use. These should include all information and guidance necessary for their proper selection, use, and care. Possible emergency and routine uses of respirators should be anticipated and planned for.

(2) The correct respirator shall be specified for each job. The respirator type is usually specified in the work procedures by a qualified individual supervising the respiratory protective program. The individual issuing them shall be adequately instructed to insure that the correct respirator is issued.

(3) Written procedures shall be prepared covering safe use of respirators in dangerous atmospheres that might be encountered in normal operations or in emergencies. Personnel shall be familiar with these procedures and the available respirators.

(i) In areas where the wearer, with failure of the respirator, could be overcome by a toxic or oxygen-deficient atmosphere, at least one additional man shall be present. Communications (visual, voice or signal line) shall be maintained between both or all individuals present. Planning shall be such that one individual will be unaffected by any likely incident and have the proper rescue equipment to be able to assist the other(s) in case of emergency.

(ii) When self-contained breathing apparatus or hose masks with blowers are used in atmospheres immediately dangerous to life or health, standby men must be present with suitable rescue equipment.

(iii) Persons using air-line respirators in atmospheres immediately hazardous to life or health shall be equipped with safety harnesses and safety lines for lifting or removing persons from hazardous atmospheres or other and equivalent provisions for the rescue of persons from hazardous atmospheres shall be used. A standby man or men with suitable self-contained breathing apparatus shall be at the nearest fresh air base for emergency rescue.

(4) Respiratory protection is no better than the respirator in use, even though it is worn conscientiously. Frequent random inspections shall be conducted by a qualified individual to assure that respirators are properly selected, used, cleaned and maintained.

(5) For safe use of any respirator, it is essential that the user be properly instructed in its selection, use and maintenance. Both supervisors and workers shall be so instructed by competent persons. Training shall provide the men an opportunity to handle the respirator, have it fitted properly, test its face-piece-to-face seal, wear it in normal air for a long familiarity period and, finally, to wear it in a test atmosphere.

(i) Every respirator wearer shall receive fitting instructions including demonstrations and practice in how the respirator should be worn, how to adjust it, and how to determine if it fits properly. Respirators shall not be worn when conditions prevent a good face seal. Such conditions may be a growth of beard, sideburns, a skull cap that projects under the facepiece or temple pieces on glasses. Also, the absence of one or both dentures can seriously affect the fit of a facepiece. The worker's diligence in observing these factors shall be evaluated by periodic check. To assure proper protection, the facepiece fit shall be checked by the wearer each time he puts on the respirator. This may be done by following the manufacturer's facepiece fitting instructions.

(ii) Providing respiratory protection for individuals wearing corrective glasses is a serious problem. A proper seal cannot be established if the temple bars of eye glasses extend through the sealing edge of the full facepiece. As a temporary measure, glasses with short temple bars or without temple bars may be taped to the wearer's head. Wearing of contact lenses in contaminated atmospheres with a respirator shall not be allowed. Systems have been developed for mounting corrective lenses inside full facepieces. When a workman must wear corrective lenses as part of the facepiece, the facepiece and lenses shall be fitted by qualified individuals to provide good vision, comfort and a gas-tight seal.

(iii) If corrective spectacles or goggles are required, they shall be worn so as not to affect the fit of the facepiece. Proper selection of equipment will minimize or avoid this problem.

(f) *Maintenance and care of respirators.*

(1) A program for maintenance and care of respirators shall be adjusted to the type of plant, working conditions, and hazards involved and shall include the following basic services:

(i) Inspection for defects (including a leak check)

(ii) Cleaning and disinfecting

(iii) Repair

(iv) Storage

Equipment shall be properly maintained to retain its original effectiveness.

(2) (i) All respirators shall be inspected routinely before and after each use. A respirator that is not routinely used but is kept ready for emergency use shall be inspected after each use and at least monthly to assure that it is in satisfactory working condition.

(ii) Self-contained breathing apparatus shall be inspected monthly. Air and oxygen cylinders shall be fully charged according to the manufacturer's instructions. It shall be determined that the regulator and warning devices function properly.

(iii) Respirator inspection shall include a check of the tightness of connections and the condition of the facepiece, headbands, valves, connecting tube and canisters. Rubber or elastomer parts shall be inspected for pliability and signs of deterioration. Stretching and manipulating rubber or elastomer parts with a massaging action will keep them pliable and flexible and prevent them from taking a set during storage.

(iv) A record shall be kept of inspection dates and findings for respirators maintained for emergency use.

(3) Routinely used respirators shall be collected, cleaned and disinfected as frequently as necessary to insure that proper protection is provided for the wearer. Respirators maintained for emergency use shall be cleaned and disinfected after each use.

(4) Replacement or repairs shall be done only by experienced persons with parts designed for the respirator. No attempt shall be made to replace components or to make adjustment or repairs beyond the manufacturer's recommendations. Reducing or admission valves or regulators shall be returned to the manufacturer or to a trained technician for adjustment or repair.

(5) (i) After inspection, cleaning and necessary repair, respirators shall be stored to protect against dust, sunlight, heat, extreme cold, excessive moisture or damaging chemicals. Respirators placed at stations and work areas for emergency use should be quickly accessible at all times and should be stored in compartments built for the purpose. The compartments should be clearly marked. Routinely used respirators, such as dust respirators, may be placed in plastic bags. Respirators should not be stored in such places as lockers or tool boxes unless they are in carrying cases or cartons.

(ii) Respirators should be packed or stored so that the facepiece and exhalation valve will rest in a normal position and function will not be impaired by the elastomer setting in an abnormal position.

(iii) Instructions for proper storage of emergency respirators, such as gas masks and self-contained breathing apparatus, are found in "use and care" instructions usually mounted inside the carrying case lid.

(g) Identification of gas mask canisters.

(1) The primary means of identifying a gas mask canister shall be by means of properly worded labels. The secondary means of identifying a gas mask canister shall be by a color code.

(2) All who issue or use gas masks falling within the scope of this section shall see that all gas mask canisters purchased or used by them are properly labeled and colored in accordance with these requirements before they are placed in service and that the labels and colors are properly maintained at all times thereafter until the canisters have completely served their purpose.

(3) On each canister shall appear in bold letters the following:

(i) Canister for______________________

(Name for atmospheric contaminant)

or

Type N Gas Mask Canister

(ii) In addition, essentially the following wording shall appear beneath the appropriate phrase on the canister label: "For respiratory protection in atmospheres containing not more than _______ percent by volume of _____________."

(Name of atmospheric contaminant)

(4) Canisters having a special high-efficiency filter for protection against radionuclides and other highly toxic particulates shall be labeled with a statement of the type and degree of protection afforded by the filter. The label shall be affixed to the neck end of or to the gray stripe which is around and near the top of the canister. The degree of protection shall be marked as the percent of penetration of the canister by a 0.3-micron-diameter dioctyl phthalate (DOP) smoke at a flow rate of 85 liters per minute.

(5) Each canister shall have a label warning that gas masks should be used only in atmospheres containing sufficient oxygen to support life (at least 16% by volume), since gas mask canisters are only designed to neutralize or remove contaminants from the air.

(6) Each gas mask canister shall be painted a distinctive color or combination of colors indicated in Table I-1. All colors used shall be such that they are clearly identifiable by the user and clearly distinguishable from one another. The color coating used shall offer a high degree of resistance to chipping, scaling, peeling, blistering, fading, and the effects of the ordinary atmospheres to which they may be exposed under normal conditions of storage and use. Appropriately colored pressure-sensitive tape may be used for the stripes.

Atmospheric Contaminants to be Protected Against	**Colors Assigned (1)**
Acid gases	White
Hydrocyanic acid gas	White with 1/2-inch green stripe completely around the canister near the bottom
Chlorine gas	White with 1/2-inch yellow stripe completely around the canister near the bottom
Organic vapors	Black
Ammonia gas	Green
Acid gases and ammonia gases	Green with 1/2-inch white stripe completely around the canister near the bottom
Carbon monoxide	Blue
Acid gases and organic vapors	Yellow
Hydrocyanic acid gas and chloropicrin vapor	Yellow with 1/2-inch blue stripe completely around the canister near the bottom
Acid gases, organic vapors and ammonia gases	Brown
Radioactive materials, excepting tritium and noble gases	Purple (magenta)
Particulates (dusts, fumes, mists, fogs or smokes) in combination with any of the above gases or vapors	Canister color for contaminant, as designated above, with 1/2-inch gray stripe completely around the canister near the top.
All of the above atmospheric contaminants	Red with 1/2-inch gray stripe completely around the canister near the top

(1) Gray shall not be assigned as a main color for a canister designed to remove acids or vapors.

NOTE: Orange shall be used as a complete body or stripe color to represent gases not included in this table. The user will need to refer to the canister label to determine the degree of protection the canister will afford.

Appendix B

Sources of Help

PROFESSIONAL ASSOCIATIONS

American Conference of Governmental Industrial Hygienists
6500 Glenway Ave., Bldg. D-5
Cincinnati, OH 45211

American Industrial Hygiene Association
2700 Prosperity Ave., Suite 250
Fairfax, VA 22031

American Society of Safety Engineers
1800 East Oakton St.
Des Plaines, IL 60016

American Society for Training and Development
1630 Duke St.
Alexandria, VA 22313

Illuminating Engineering Society of North America
345 E 47th Street
New York, NY 10017

Society of Toxicology
1133 I Street, NW Suite 800
Washington, DC 20005

System Safety Society
14252 Culver Drive, Suite A-261
Irvine, CA 92714

INDUSTRIAL AND TRADE ASSOCIATIONS

Air Movement Control Association
30 West University Dr.
Arlington Heights, IL 60004

American National Standards Institute
1430 Broadway
New York, NY 10018

American Paper Institute
260 Madison Ave.
New York, NY 10016

American Petroleum Institute
1220 L Street NW
Washington, DC 20036

American Society for Testing and Materials
1916 Race Street
Philadelphia, PA 19103

American Welding Society
550 NW LeJeune Road
Miami, FL 33126

Chemical Manufacturers Association
2501 M Street, NW
Washington, DC 20037

Compressed Gas Association
1235 Jefferson Davis Highway
Arlington, VA 22202

Factory Mutual System
1151 Boston-Providence Turnpike
Norwood, MA 02062

National Fire Protection Association
Batterymarch Park
Quincy, MA 02269

National Paint and Coating Association
1500 Rhode Island Avenue, NW
Washington, DC 20005

National Safety Council
1121 Spring Lake Drive
Itasca, IL 60143-3201

Rubber Manufacturers Association
1400 K Street
Washington, DC 20005

Safety Equipment Institute
1901 N Monroe Street
Arlington, VA 22209

Underwriters Laboratories
333 Pfingsten Road
Northbrook, IL 60062

GOVERNMENT AGENCIES

Consumer Products Safety Commission
111 Eighteenth Street, NW
Washington, DC 20270

National Council on Radiation
Protection and Measurements
7910 Woodmont Ave., Suite 1016
Bethesda, MD 20014

Environmental Protection Agency
401 M Street, SW
Washington, DC 20460

National Center for Toxicological Research
Food and Drug Administration
5600 Fishers Lane
Rockville, MD 20857

National Technical Information Service
Department of Commerce
5285 Port Royal Road
Springfield, VA 22161

Occupational Safety and Health Administration
National Office
US Department of Labor
2000 Constitution Ave.
Washington, DC 20210

Region I
CT, MA, ME, NH, RI, VT
133 Portland Street, 1st Floor
Boston, MA 02114
(617) 565-7164

Region II
NJ, NY, PR, VI
201 Varick Street, Room 670
New York, NY 10014
(212) 337-2378

Region III
DC, DE, MD, PA, VA, WV
Gateway Building, Suite 2100
3535 Market Street
Philadelphia, PA 19104
(215) 596-1201

Region IV
AL, FL, GA, KY, MS, NC, SC, TN
1375 Peachtree Street, NE, Suite 587
Atlanta, GA 30367
(404) 347-3573

Region V
IL, IN, MI, NM, OH, WI
230 South Dearborn Street, Room 3244
Chicago, IL 60604
(312) 353-2220

Region VI
AR, LA, NM, OK, TX
525 Griffin Street, Room 602
Dallas, TX 75202
(214) 767-4731

Region VII
IA, KS, MO, NE
911 Walnut Street, Room 406
Kansas City, MO 64106
(816) 426-5861

Region VIII
CO, MT, ND, SD, UT, WY
1961 South Street, Room 1576
Denver, CO 80294
(303) 844-3061

Region IX
AZ, CA, Hi, NV, American Samoa, Guam, and
Pacific Trust Territories
71 Stevenson Street
San Francisco, CA 94105
(415) 744-6670

Region X
AK, ID, OR, WA
1111 Third Ave., Suite 715
Seattle WA 98101-3212
(206) 553-5930

OSHA State-Plans

Alaska Department of Labor
P.O. Box 21149
Juneau, AK 99801
(907) 465-2700

Industrial Commission of Arizona
800 W. Washington
Phoenix, AZ 85007
(602) 542-5795

California Dept. of Industrial Relations
455 Golden Gate Avenue, 4th Floor
S. San Francisco, CA 94102
(415) 703-4590

Connecticut Department of Labor
200 Folly Brook Blvd.
Wetherfield, CT 06109
(203) 566-5123

Hawaii Department of Labor
830 Punchbowl Street
Honolulu, HI 96813
(808) 548-3150

Indiana Department of Labor
1013 State Office Building
100 North Senate Avenue
Indianapolis, IN 46204-2287
(317) 232-2665

Iowa Division of Labor Services
1000 E. Grand Avenue
Des Moines, IA 50319
(515) 281-3447

Kentucky Labor Cabinet
1049 US Highway, 127 South
Frankfort, KY 40601
(502) 564-3070

Maryland Division of Labor and Industry
501 St. Paul Street
Baltimore, MD 20202
(410) 333-4179

Michigan Department of Labor
Victor Office Center
201 N. Washington Street
P.O. Box 30015
Lansing, MI 48933
(517) 373-9600

Michigan Department of Public Health
3423 North Logan Street B
Lansing, MI 48909
(517) 335-8022

Minnesota Department of Labor and Industry
443 Lafayette Road
St. Paul, MN 55155
(612) 296-2342

Nevada Department of Industrial Relations
Capitol Complex
1370 S. Curry St.
Carson City, NV 89710
(702) 687-3032

New Mexico Environment Department and
Occupational Health and Safety Bureau
1190 St. Francis Dr., P.O. Box 26110
Sante Fe, MN 87502
(505) 827-2850

New York Department of Labor
State Office Building
Campus 12, Room 457
Albany, NY 12240
(518) 457-2741

North Carolina Department of Labor
4 West Edenton Street
Raleigh, NC 27601
(919) 733-7166

Oregon Department of Insurance and Finance
Labor and Industries Building, Room 160
Salem, OR 97310
(503) 378-3232

Puerto Rico Department of Labor
Prudenco Rivera
Martinez Building
505 Nunoz Rivera Avenue
Hato Rey, PR 00918
(809) 754-2119

South Carolina Department of Labor
3600 Forest Drive
P.O. Box 11329
Columbia SC 29211-1329
(803) 734-9594

Tennessee Department of Labor
510 Union Building
Suite A, 2nd Floor
Nashville, TN 37243-0655
(615) 741-2582

National Institute for Occupational
Safety and Health (NIOSH)
Headquarters CDC
1600 Clifton Rd. NE
Atlanta, GA 30333

Laboratory
4676 Columbia Parkway
Cincinnati, OH 45226

National Institute for Standards and Technology
Gaithersburg, MD 20877

Appendix C

Glossary of Confined Space Terms

Acceptable Environmental Conditions: the conditions which must exist for that employee to safely enter and perform work within a confined space.

Acute Effects: physiological changes which occur rapidly as a result of short-term exposure to a hazardous substance. Acute effects usually manifest with in 72 hours of exposure.

Air Mover: a device used to force, draw or exhaust gases through a specific assembly in order to move them from one place to another.

Air-Purifying Respirator: a device designed to protect the wearer from inhalation of harmful dusts, mist, fumes, vapors or gases by removing contaminants from the ambient air by way of a filter or chemical sorbent.

Air-Supplied Respirator: a respirator that supplies the wearer with clean, breathable air provided by a compressor, air mover or compressed gas cylinders.

Attendant: a specially trained individual stationed outside of a confined space who monitors the authorized entrants inside (see also standby person).

Authorized Person: a person approved or assigned by an employer to perform specific duties or to be at specific locations at a job site. (See also competent person and qualified person.)

Biological Hazards: infectious agents which present a risk to the well-being of people or animals, either directly through infection or indirectly through disruption of the environment.

Blanking or Blinding: the absolute closure of a pipe, line or duct by inserting a solid plate or cap which completely covering the bore and is capable of withstanding the maximum upstream pressure.

Body Belt: a device fastened around the waist used for work positioning at an elevation. Body belts are fall-prevention devices, not fall-protection devices.

Calibration: a laboratory or test-bench resetting of instrument zero, span and alarm points according to a manufacturer's specification. Calibration is performed by a factory-authorized service center or a trained technician.

Ceiling Concentration: the highest concentration of an airborne contaminant to which workers may be exposed.

Chest-Waist Harness: a body harness consisting of two sets of straps, one of which is secured around the rib cage, the other over the shoulders. The back of the harness is equipped with a D-ring to permit lifting or retrieval from a confined space.

Confined Space: a space that is large enough and so configured that an employee can bodily enter and perform assigned work, has limited or restricted means for entry or exit and is not designed for continuous employee occupancy.

Confined Space Entry: any action that results if any part of an entrant breaks the plane of any opening of a confined space.

Combustion: a complex sequence of chemical reaction between fuel and an oxidant accompanied by the evolution of light and heat.

Competent Person: a person who is capable of identifying existing and predictable hazards in the surroundings or working conditions which are hazardous or dangerous to employees, and who has the authority to take prompt corrective action. (See also authorized person and qualified person).

Corrosive: a chemical that causes visible destruction or irreversible alterations in living tissue by action at the site of contact.

Deflagation: Rapid combustion where the flame front propagates slower than the speed of sound (1,100 feet per second) without the generation of a shock wave.

Detonation: Extremely rapid combustion where the flame front travels greater than or equal to the speed of sound with generation of a shock wave in the combustible mixture.

Detector Tube: a device consisting of a glass tube filled with a solid material that reacts chemically with an air contaminant drawn through it, usually with a hand pump, resulting in a color change the length of which is proportional to the contaminant concentration.

Double Block and Bleed: a method used to isolate a confined space from line, duct or pipe by closing two in-line valves in a piping system and opening a valve between them which is vented to a safe location.

Direct Reading Instrument: an air sampling device used to obtain information on the concentration of an air contaminant in real time. Direct reading instruments may include colorimetric indicating tubes, combustible gas indicators and any of a variety of meters used to measure oxygen, carbon monoxide, hydrogen sulfide, etc.

Enclosed Space: a space other than a confined space which is enclosed by walls and ceiling such as a cargo hold, room, or machinery and boiler spaces where workers may find otherwise ordinary hazards aggravated or intensified.

Energy Isolating Device: a mechanical device that physically prevents the transmission or release of energy, including but not limited to, a manually operated circuit breaker, a blind, a disconnect switch, an in-line valve, or similar device used to block or isolate energy.

Engulfment: the surrounding and effective capture of a person by a liquid or finely divided solid that can be aspirated to cause death by filling or plugging the respiratory system, or that can exert enough force on the body to cause death by strangulation, constriction or crushing.

Entrant: an authorized person who enters a confined space.

Entry: see confined space entry.

Entry Permit: a written or printed document provided by an employer to allow or control entry into a confined space under defined conditions for a stated purpose during a specified time.

Entry Supervisor: a person such as a foreman or crew chief who is responsible for determining if acceptable entry conditions are present at a permit space where entry is planned, for authorizing entry and overseeing entry operations, and for terminating entry as required by this section.

Emergency: any unexpected internal or external occurrence or event which could endanger the confined space occupants.

Explosive Atmosphere: atmosphere which contain, or could contain, contaminant concentrations greater than 10% of the lower explosive level.

Explosive Range: a concentration of flammable gas or vapors which falls between the upper and lower explosive limits.

Field Check: a simple pass-fail test used in the field to determine if an instrument is functioning and responding properly.

Flame Propagation Rate: the velocity with which the combustion front travels through a body of gas, measured at the highest gas velocity at which stable combustion can be maintained.

Flammable Limits: the minimum and maximum concentration of fuel vapor or gas in a fuel vapor or gaseous oxidant mixture (usually expressed as percent by volume) over which propagation of a flame will occur on contact with an ignition source.

Flash Point: the minimum temperature of a liquid or solid at which it gives off vapor sufficient to form an ignitable mixture with a gaseous oxidant (i.e., oxygen) near the surface of the liquid or solid under specified environmental conditions.

Full-Body Harness: a harness that is secured around the wearer in a manner that distributes the load over the thighs, buttocks, chest and shoulders. A back-mounted D-ring is provided at the back for lifting.

Gas Freeing: expression used particularly in maritime work to describe the process of purging and ventilating a space to remove flammable gases and vapors prior to entry.

General Ventilation: a system of ventilation that introduces fresh air into a confined space and relies on its movement to mix with and dilute air contaminates.

Hazardous Atmosphere: an atmosphere presenting a potential for death, injury or illness due to the presence of flammable gases or vapors, oxygen deficiency or enrichment, or toxic substances.

Hot Work: work involving welding or oxy-acetylene cutting and other operations that generate heat, flames, arcs, sparks or other sources of ignition in a confined space.

Hot Work Permit: a written permit issued authorizing hot work in a confined space.

Ignition Energy, Minimum: the minimum energy required to ignite a flammable mixture—usually the minimum energy of an electric spark or arc expressed in joules. The minimum ignition energy is different for the different flammable mixtures and varies with the concentration, temperature and pressure, as well as the geometry and material of the sparking or arcing electrodes.

Ignition Temperature: the minimum temperature required to initiate or cause self-sustaining combustion independently of the heating or heated element. Ignition temperature is commonly reported as Auto Ignition Temperature (AIT) or Spontaneous Ignition Temperature (SIT).

Immediately Dangerous to Life or Health (IDLH): any condition which poses an immediate threat of loss of life or which may result in irreversible or immediate-severe health effects or may result in eye damage, irritation or other conditions which could impair escape from a confined space.

Immediate Severe Health Effects: acute clinical sign of a serious, exposure-related reaction that occurs within 72 hours of exposure.

Inerting: rendering the atmosphere in a confined space non-flammable, non-explosive or otherwise chemically non-reactive by displacing or diluting the original atmosphere with an inert gas such as argon or nitrogen.

Irritant: a chemical which is not corrosive, but which causes reversible inflammatory effect on living tissue at the site of contact.

Isolation: a process whereby a confined space is removed from service and completely protected from inadvertent release of material or start-up of any power source.

Lean Mixture: a fuel and oxidizer mixture having less than the stoichiometric concentration of fuel.

Line Breaking: the intentional opening of a pipe, line or duct that is or has been carrying flammable, corrosive or toxic materials, an inert gas, or any fluid at pressures or temperatures capable of causing injury.

Local Exhaust: a method of ventilation that captures air contaminants at their point of generation and exhausts them to a remote location.

Lockout: the placement of a lockout device on an energy-isolating device in accordance with an established procedure ensuring that the energy-isolating device and the equipment being controlled cannot operate until the lockout device is removed.

Lockout Device: a device that utilizes a positive means such as a key or combination lock to hold an energy-isolating device in the safe position preventing energizing of equipment or machinery.

Lower Explosive Limit: the lowest concentration of a flammable gas or vapor which will ignite and burn in the presence of an ignition source.

Marine Chemist: the holder of a valid certificate issued by the National Fire Protection Association in accordance with the "Rules for Certification of Marine Chemists," establishing him as a person qualified to determine whether construction, alteration, repair or shipbreaking of vessels, which may involve flammable gas or vapors hazards, may be undertaken.

Oxygen Deficient Atmosphere: an atmosphere with less than 19.5% oxygen by volume at normal atmospheric pressure; or atmosphere with a partial pressure of oxygen less that 132 mm Hg. Normal air at sea level contains about 21% oxygen at partial pressure of 160 mm Hg (21% of 760 mm Hg atmospheric pressure at sea level).

Oxygen Enriched Atmosphere: an atmosphere containing more than 23.5% oxygen by volume or any atmosphere with a partial pressure of oxygen greater than 178 mm Hg.

Permissible Exposure Limit: airborne concentration of contaminant established by the Occupational Safety and Health Administration.

Permit-Required Confined Space: a confined space that has one or more of the following characteristics:

(1) Contains or has the potential to contain hazardous atmospheres
(2) Contains a material that has the potential for engulfing an entrant
(3) Has an internal configuration such that an entrant could be trapped or asphyxiated by inwardly converging walls or by a floor which slopes downward and tapers to a smaller cross-sectional area
(4) Contains any other recognized serious safety or health hazards.

Permit System: written procedures for preparing and issuing permits for entry and for returning the permit space to service following termination of entry.

Pressure, Absolute: total pressure being measured. Absolute pressure equals gauge pressure plus atmospheric pressure.

Pressure, Gauge: pressure being measured with reference to atmospheric pressure. Gauge pressure equals absolute pressure minus atmospheric pressure.

Purging: a method by which gases, vapors or other air contaminants are displaced from a confined space.

Pyrophoric Material: a chemical that will ignite spontaneously in air at a temperature at or below 130°F.

Qualified Person: a person, who by possession of a recognized degree, certificate or professional standing, or who by extensive knowledge, training and experience, has successfully demonstrated his ability to solve problems related to the subject matter, work or project (see also authorized person and competent person).

Rescue Team: a designated group of employees specially trained and equipped to perform rescue work.

Retrieval Line: a line or rope secured to an anchor point or lifting device outside of a confined space and attached to a full-body harness, chest harness or wristlets worn by employees entering the space.

Retrieval System: the equipment (including a retrieval line, chest or full-body harness, wristlets, if appropriate, and a lifting device or anchor) used for non-entry rescue of persons from permit spaces.

Rich Mixture: a fuel and oxidizer mixture having more than stoichiometric concentration of fuel.

Self-Contained Breathing Apparatus: an atmosphere-supplied respirator in which the wearer carries a personal air supply usually in a compressed gas cylinder worn on his back.

Sensitizer: a chemical that causes a substantial portion of exposed people or animals to develop an allergic reaction in normal tissue after repeated exposure.

Short-Term Exposure Limit: the concentration of a material that a worker may be exposed to for a short period of time without suffering from irritation, chronic or irreversible tissue damage, or narcosis to a degree sufficient to increase the likelihood of accidental injury, impair self-rescue or materially reduce worker efficiency.

Standby Person: a person trained in emergency rescue procedures and outside a confined space who remains in communication with those inside for the purpose of rendering assistance or effecting rescue.

Stoichiometric Mixture: a balanced mixture of fuel and oxidized such that no excess of either remains after combustion.

Tagout: the placement of a tagout device on an energy isolating device in accordance with an established procedure to indicate that the energy isolating device and the equipment being controlled may not be operated until the tagout device is removed.

Tagout Device: a prominent warning device such as a tag and means of attachment which can be securely fastened to an energy-isolating device in accordance with an established procedure to indicate that the energy-isolating device and the equipment being controlled may not be operated until the tagout device is removed.

Threshold Limit Value (TLV): Registered trademark of the American Conference of Governmental Industrial Hygienists (ACGIH). Refers to the airborne concentration of substances to which it is believed that nearly all workers can be repeatedly exposed eight-hours a day, over a working lifetime without adverse affect.

Toxic Atmosphere: any atmosphere where the level of air contaminants exceed OSHA Permissible Exposure Limits (PELs), ACGIH Threshold Limit Values (TLVs) or NIOSH Recommended Exposure Levels.

Tritector: a direct reading instrument that monitors three atmospheric contaminants, usually oxygen, combustible gases and either carbon monoxide or hydrogen sulfide.

Upper Explosive Limit: the highest concentration of flammable gas or vapor that will burn in the presence of an ignition source.

Upper Flammable Limit: see upper explosive limit.

Zero Mechanical State: the mechanical potential energy in all elements of a machine or piece of equipment is dissipated to that the opening of any pipe, tube or hose or actuation of any lever, switch, button or control will not produce a movement that could cause injury.

Appendix D

Suppliers of Confined Space Related Products

Audio-Visual Materials

BNA Communications
9439 Key West Avenue
Rockville, MD 20850-3396
(800) 233-6067

Coastal Video Communications
3083 Brickhouse Court
Virginia Beach, VA 23452
(800) 767-7703

Industrial Training Corp.
13515 Dulles Technology Drive
Herndon, VA 22071
(800) 638-3757

ITS Corp
9 E. Stow Road
Marlton, NJ 08053
(609) 983-4311

Interactive Media Communications, Inc.
100 Fifth Avenue
Waltham, MA 02145
(612) 533-3200

Marcom Group Ltd.
4 Denny Road, Suite 22
Wilmington, DE 19809
(800) 654-2448

National Audio-Visual Center
8700 Edgeworth Drive
Capitol Heights, MD 20743
(301) 763-1850

National Safety Council
2221 Spring Lake
Itasca, IL 60143-3201
(708) 775-2175

NUS Training Corporation
910 Clopper Road
P.O. Box 6032
Gaithersburg, MD 20877-0962
(800) 338-1505

The Roco Corporation
8254 One Calais, Suite 240
Baton Rouge, LA 70809
(504) 769-8889

Summit Training Source, Inc.
620 Three Mile Road, NW
Grand Rapids, MI 49504
(616) 784-4500

Tel-A-Train, Inc.
P.O. Box 4752
Chattanooga, TN 37405
(615) 266-0113

Video Training Source, Inc.
6865 Cascade Road, SE
Grand Rapids, MI 49546
(616) 942-1400

Breathing Air Systems

Air Systems International
814-P Greenbrier Circle
Chesapeake, VA 23320
(800) 424-3967

Biosystems, Inc.
P.O. Box 158
Rockfall, CT 06481
(203) 344-1079

Deltec Engineering Limited
P.O. Box 667
New Castle, DE 19720
(302) 328-1345

Ingersoll-Rand Company
P.O. Box 458
Pleasant Garden, NC 27313
(800) 633-0306

Neoterik Health Technology
Neoterik Center
Woodsboro, MD 21798
(301) 845-2777

Zeks Air Drier Corporation
184 Pennsylvania Avenue
Malvern, PA
(800) 888-2323

Calibration Gas

Alphagaz, A Division of Liquid Air
2121 North California Boulevard
Walnut Creek, CA 94596
(415) 977-6506

Matheson Gas Products
30 Seaweed Drive
Secausus, NJ 07096
(215) 641-2700

Chemical Protective Clothing

Ansell Edmont Industrial, Inc.
1300 Walnut Street
Coshocton, OH 43812
(614) 622-4311

Beta Shoe Company, Inc.
Route 40
Belcamp, MD 21017
(301) 272-2000

Best Manufacturing Company
Edison Street
Menlo, GA 30731
(404) 862-2302

Broner Glove & Safety Co.
359 Robbins Drive
Troy, MI 48083
(313) 589-1919

Brunswick Corporation
302 Cornwell Avenue
Willard, OH 44890
(419) 933-2711

Chemical Fabrics Corporation
701 Daniel Webster Highway
P.O. Box 1137
Merrimack, NH 03054
(603) 424-9000

Chemron, Inc.
954 Corporate Woods Parkway
Vernon Hills, IL 60061
(312) 520-7300

Comasec, Inc.
8 Niblick Road
Enfield, CT 06082
(800) 333-0219

DuPont Company
1007 North Market Street
Wilmington, DE 19898
(302) 774-6652

Durafab, Inc.
1102 East Kilpatrick
Cleburne, TX 76031
(817) 769-0573

W.L. Gore and Associates, Inc.
3 Blue Ball Road
Elkton, MD 21921
(301) 392-3700

Kappler, Inc.
P.O. Box 218
Guntesville, AL 35976
(800) 633-2410

Kimberly-Clark Corporation
1400 Holcomb Bridge Road
Roswell, GA 30076
(800) 241-0220

Perfect Fit Glove Co., Inc.
1675 South Park Avenue
Buffalo, NY 14220
(800) 245-6837

Pioneer Industrial Products Co.
512 East Tiffin Street
Willard, OH 44890
(800) 537-2897

Playtex Family Products Corporation
700 Fairfield Avenue
Stamford, CT 06904
(203) 356-8000

Safe 4, Inc.
2920 Wolff Street
Racine, WI 53404
(414) 632-8133

Standard Safety Equipment Co.
P.O. Box 188
Palatine, IL 60078
(708) 359-1400

Communications Equipment

Con-Space Communications Ltd.
1300 Boblett Street
Blaine, WA 98230
(206) 332-2020

David Clark
P.O. Box 15054
Worcester, MA 01615-005
(508) 756-6216

Earmark, Inc.
1125 Dixwell Avenue
Hamden, CT 06514
(203) 777-2130

Motorola Land Mobile Products
1301 East Algonquin Road
Schaumburg, IL 60196
(708) 576-1000

Peltor, Inc.
63 Commercial Way
East Providence, RI 02914
(401) 438-4800

Telex Communications, Inc.
9600 Aldrich Avenue, South
Minneapolis, MN 55420
(612) 887-5550

Explosion-proof Lighting

Crouse-Hinds Lighting Products
P.O. Box 4999
Syracuse, NY 13221
(315) 477-8185

Stewart E. Browne Manufacturing Company
1165 Hightower Trail
Atlanta, GA 30356
(404) 993-9600

Hubble, Inc.
Lighting Division
2000 Electric Way
Christiansburg, VA 24073-2500
(703) 382-6111

Killark Electrical Manufacturing Company
P.O. Box 5325
St. Louis, MO 63115
(314) 531-0460

Gas Detection Instruments

A.I.M. USA
P.O. Box 720540
Houston, TX 77272-0540
(800) ASK-4AIM

Bacharach, Inc.
625 Alpha Drive
Pittsburgh, PA 15238
(412) 963-2000

CEA Instrument, Inc.
16 Chestnut Street
Emerson, NJ 07630
(201) 967-5660

ENMET Analytical Instruments
P.O. Box 979
Ann Arbor, MI 48106
(313) 761-1270

Foxboro Company
600 North Bedford Street
East Bridgewater, MA 02333
(508) 378-5556

G.C. Industries
8976 Oso Avenue, Unit C
Chatsworth, CA 91311
(818) 882-7852

GFG Gas Electronics, Inc.
6617 Clayton Road, Suite 209
St. Louis, MO 63117
(314) 725-9050

GMD Systems, Inc.
Old Route 519
Hendersonville, PA 15238
(412) 746-1359

GasTech, Inc.
8445 Central Avenue
Newark, CA 94560
(415) 745-8700

HNU Systems, Inc.
160 Charlemont Street
Newton Highlands, MA 02161
(617) 964-9555

Industrial Scientific Corporation
355 Steubenville Pike
Oakland, PA 15071
(800) 338-3287

Interscan Corporation
P.O. Box 2496
Chatsworth, CA 91913-2496
(800) 458-6253

Lumidor Safety Products/ESP, Inc.
5364 NW 167th Street
Miami, FL 33014
(305) 625-6511

MDA Scientific, Inc.
405 Barclay Boulevard
Lincolnshire, IL 60069
(708) 634-2800

MSA
P.O. Box 426
Pittsburgh, PA 15230
(800) MSA-2222

Metrosonics, Inc.
P.O. Box 23075
Rochester, NY 14692
(716) 334-7300

National Draeger, Inc.
101 Technology Drive
P.O. Box 120
Pittsburgh, PA 15230-0120
(412) 787-8383

Neotronics N.A., Inc.
1244 Hilton Drive
Gainesville, GA 30501

PPM Enterprises
11428 Kingston Pike
Knoxville, TN 37922
(615) 966-8796

Sensidyne, Inc.
16333 Bay Vista Drive
Clearwater, FL 34620
(813) 530-3602

Lockout/Tagout Devices

American Lock Co.
3400 West Exchange Road
Crete, IL 60417
(708) 534-20000

Brady USA, Inc.
727 West Glendale Avenue
Milwaukee, WI 53209
(800) 445-7557

Idesco Corporation
37 West 26th Street
New York, NY 10010
(800) 336-1383

Label Master
5724 North Pulaski Road
Chicago, IL 60646
(800) 621-5808

Master Lock
2600 North 32nd Street
Milwaukee, WI 53210
(414) 444-2800

National Marker
P.O. Box 1659
Pawtucket, RI 02862
(800) 453-2727

Norgren
5400 South Delaware
Littleton, CO 80120
(303) 795-2611

Rescue Equipment

DBI/SALA DB Industries
P.O. Box 46
Red Wing, MN 55066
(612) 388-8282

Miller Equipment Division
1355 15th Street
Franklin, PA 16323
(800) 873-5242

MSA
P.O. Box 426
Pittsburgh, PA 15230
(800) 672-2222

Rescue Systems, Inc.
Rt. 2, Box RSI
Hocking, OH 45742
(800) 552-1133

Research & Trading Corporation
3101 North Market Street
Wilmington, DE
(800) 441-7593

Rose Manufacturing Company
2250 South Tejon Street
Englewood, CO 80110
(303) 922-6246

Respiratory Protection

Biomarine
456 Creameray Way
Exton, PA
(215) 524-8800

E.D. Bullard Company
Route 7, Box 596
Cynthia, KY 41031
(800) 227-0423

Glendale Protective Technologies, Inc.
130 Crossways Park Drive
Woodburry, NY 11797
(516) 921-5800

International Safety Instruments
922 Hurricane Shoals Road
Lawrenceville, GA 30243
(800) 235-7677

Interspiro
31 Business Park Drive
Branford, CT 06405
(203) 483-1879

Moldex-Metric
4671 Leahy Street
Culver City, CA 90232
(213) 870-9121

MSA
P.O. Box 426
Pittsburgh, PA 15230
(800) 672-2222

National Draeger
101 Technology Drive
Pittsburgh, PA 15203
(412) 787-8383

Neoterik Health Technologies
401 Main Street
Woodsboro, MD 21798
(301) 845-2777

North
2000 Plainfield Pike
Cranston, RI 02921
(401) 943-4400

Pro-Tec Respirators, Inc.
107 East Alexander Street
Buchanan, MI 49107
(616) 659-9663

Racal Health and Safety, Inc.
7305 Executive Way
Frederick, MD 21701
(301) 695-8200

Respiratory Systems, Inc.
18102 Skypark South, Suite J
Irvine, CA 92714

Scott Aviation
225 Erie Street
Lancaster, NY 14086
(716) 683-5100

Survivair
Division of Comasec, Inc.
3001 South Susan Street
Santa Anna, CA 92704
(714) 545-0410

3M/Occupational Health and
Environmental Safety Division
3M Center, Building 220-3E-04
St. Paul, MN 55144-1000
(612) 733-5608

Quantitative Fit Testing Equipment

Air Techniques, Inc.
1716 Whitehead Road
Baltimore, MD 21207
(301) 944-6037

Dynatech Nevada, Inc.
2000 Arrowhead Drive
Carson City, NV 89706
(702) 833-3400

TSI
500 Cardigan Road
St. Paul, MN 55126
(612) 490-2888

Velometers

Alnor Instrument Co.
7555 North Linder Avenue
Skokie, IL 60077
(708) 677-3500

TSI Incorporated
500 Cardigan Road
St. Paul, MN 55126
(612) 483-0900

Kurtz Instruments
2411 Garden Road
Monterey, CA 93940
(800) 424-7356

Ventilation Equipment

Air Systems International
814-P Greenbrier Circle
Chesapeake, VA 23320
(800) 866-8100

Coppus Engineering
P.O. Box 15003
Worcester, MA 01615
(508) 756-8391

General Equipment
1500 East Main Street
Owatonna, MN 55060
(800) 533-0524

LTC Americas
101-G Executive Drive
Sterling, VA 22170
(800) 822-2332

Pelsue
2500 South Tejon Street
Englewood, CO 80110
(303) 936-7432

Index